T0329877

SPECTRA AND
PSEUDOSPECTRA

SPECTRA AND PSEUDOSPECTRA

The Behavior of Nonnormal Matrices and Operators

Lloyd N. Trefethen and Mark Embree

Princeton University Press

Princeton and Oxford

Published by Princeton University Press
41 William Street, Princeton, New Jersey 08540

In the United Kingdom: Princeton University Press
3 Market Place, Woodstock, Oxfordshire OX20 1SY

Library of Congress Cataloging-in-Publication Data

Trefethen, Lloyd N. (Lloyd Nicholas)
    Spectra and pseudospectra: the behavior of nonnormal matrices
        and operators / Lloyd N. Trefethen and Mark Embree
    p. cm.
    Includes bibliographical references and index.
    ISBN-13: 978-0-691-11946-5 (alk. paper)
    ISBN-10: 0-691-11946-5 (alk. paper)
    1. Spectral theory (Mathematics). 2. Spectrum analysis.
        3. Eigenvalues. 4. Differential operators.
        I. Embree, Mark, 1974–. II. Title.

QA320.T67 2005
530.15'57222—dc22   2005046573

British Library Cataloging-in-Publication Data is available.

The publisher would like to acknowledge the authors of this volume
for providing the camera-ready copy from which this book was printed.

Printed on acid-free paper. ∞

pup.princeton.edu

Printed in the United States of America.

10 9 8 7 6 5 4 3 2 1

Quiz

Here is a plot of the 2-norms $\|e^{t\mathbf{A}}\|$ for the two matrices

$$\mathbf{A}_1 = \begin{pmatrix} -1 & 1 \\ 0 & -1 \end{pmatrix}, \quad \mathbf{A}_2 = \begin{pmatrix} -1 & 5 \\ 0 & -2 \end{pmatrix}.$$

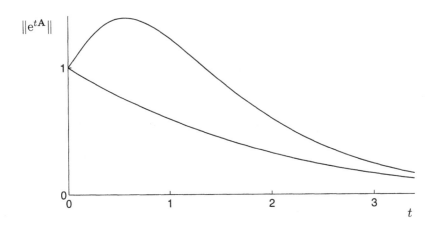

Which curve is which?

to Anne and Kristine

Note to the Reader

The style of this book is unusual. There are sixty short sections, and nobody will read them all. For easy study à la carte, each has been written as a self-contained essay.

But please, though you won't read all the words, look at all the figures! You will quickly feel the pleasure we have found in years of exploring spectra and pseudospectra.

Contents

Preface

This book sprang from computer plots of eigenvalues of matrices. In the late 1980s LNT was a junior faculty member at MIT in the habit of drawing plots in MATLAB on a workstation. Eigenvalues of nonsymmetric matrices were full of surprises. They might describe one pattern in the complex plane for a matrix **A**, then quite another if you added a small perturbation, or even if you just took the transpose! Yet mathematical scientists seemed to care about eigenvalues and reach conclusions based on them. Numerical analysts designed matrix iterations and judged stability of discretizations based on eigenvalues. Could this be right?

One thing was clear: Eigenvalues might be meaningful in theory, but they could not always be trusted on a computer. It took some time to realize that although rounding error effects are fascinating, the significance of fragile eigenvalues goes deeper. Indeed, the rounding errors are a sign that the eigenvalues are probably not so meaningful in theory after all. By 1990 LNT had converged to the view that is the theme of this book: what matters most about nonnormality is its effect on the behavior of the unperturbed matrix or operator. And to study nonnormality, a fruitful technique is to look beyond eigenvalues to pseudospectra, the sets in the complex plane bounded by level curves of the norm of the resolvent. Related work had been done by Varah, Landau, Godunov, Chatelin, Hinrichsen, and Pritchard. By 1990, pseudospectra were LNT's main research interest.

Nonnormal eigenvalue problems became a theme linking one field with another. Numerical iterations and discretizations were joined by Toeplitz matrices, random matrices, fluid mechanics, card shuffling, laser cavities, and food webs. In particular it was an extraordinary thing to learn from Brian Farrell and others that trouble with eigenvalues had been a pervasive theme through a century of fluid mechanics. With this new field in the picture, LNT, now at Cornell, became persuaded that tools for going beyond eigenvalues must take a permanent place in applied mathematics.

Meanwhile ME was finishing his degree at Virginia Tech and moving to Oxford for graduate school. He got hooked on pseudospectra after reading a blue A5-sized Oxford Numerical Analysis Group Report. A year later LNT moved to Oxford as the head of that group, and the two of us hit it off immediately. We found that we shared a love of linear algebra and a taste for careful numerical experiments, as well as a birthday in common with Carl Runge and Olga Taussky and Cameron Diaz. And new related topics kept appearing, including non-Hermitian quantum mechanics, explored by Hatano and Nelson and our UK colleague Brian Davies, and the Lewy–Hörmander theory of nonexistence of solutions to certain partial differential equations, whose connections with pseudospectra were pointed out by Zworski. The study of nonnormal effects in Toeplitz matrices expanded as ME pursued a number of such investigations with Albrecht Böttcher.

This book had been envisioned, even started, in 1990, but it was moving forward too slowly. Soon there were 50 publications on pseudospectra, then 100, and the book was not keeping up. By this time ME had completed his D.Phil. and become a post-doc working with LNT and had built the online Pseudospectra Gateway, which includes a bibliography of all papers we know of on pseudospectra (we will probably stop when the number reaches 1000). We decided to make the book a team effort. The collaboration continued electronically after ME moved to Rice University in 2002.

This would have been a very different book if it been been finished seven or eight years ago. The unifying treatment of Toeplitz and twisted Toeplitz matrices and of constant and variable coefficient differential operators, the subject of Parts II and III, had not been developed. Most of the lower bounds of Part IV that quantify the use of pseudospectra for analyzing transient behavior had not been derived. The preliminary triangularization idea of Shiu-Hong Lui had not yet speeded up calculations by a factor of 10 or more. Most important of all, Thomas Wright's extraordinary software system EigTool, whose influence can be seen in almost every section, had not been created. For any nonnormal matrix, whether of dimension 10 or 10,000, the best first step to examining its spectral properties is the same: check it out in EigTool!

Each of our sixty sections has been written as a self-contained essay. We have tried to make each of these documents the perfect introductory survey of its topic, complete with the right references and illustrations. We hardly mention any details of the computations that lie behind these illustrations, but we are proud of them. We will be glad if this volume serves as an example of a kind of mathematics that it is now possible to write, in which nontrivial computations help to communicate mathematical ideas without themselves being the subject.

At times we have felt oppressed by the volume of topics that we could not include, whether for reasons of space, time, or competence. This book says next to nothing about geophysical dynamos or plasma physics; about matrix preconditioners or biorthogonal iterations or differential-algebraic equations; about bases and diagonalization of linear operators in infinite-dimensional spaces; about multiplicity of pseudoeigenfunctions or their use in computation; about why perturbed eigenvalues tend to line up so beautifully along pseudospectra boundaries; about the use of adjoints in engineering design and optimization; about atmospheric science or chemical kinetics or pseudodifferential operators or C^*-algebras or functions of matrices or control theory. Logically, each of these topics should be the subject of a section of this book, or more. Luckily, it is not always best to push a book to its logical conclusion.

Acknowledgments

At the close of a project stretching back fifteen years, we cannot thank all those who helped, much as we would like to. Both of us are compulsive communicators and mad typists, and in those years we have exchanged thousands of email messages and computer files related to spectra and pseudospectra with our colleagues. These range from close friends and coauthors to people we have never met in fields and countries distant from our own, and they have helped us with articles and pointers to articles, suggestions of topics to include, proofs of theorems, corrections of errors, and many other things. It is one of the great pleasures of academic research in our age that one can take advantage so readily of this worldwide web of scholars.

A few people must be singled out for their sustained involvement in this work. One of these is Nick Higham of the University of Manchester, a leader in the field of numerical linear algebra as well as an expert on LATEX, MATLAB, and mathematical writing. Higham has taken an interest in this book since 1990, reading and rereading successive drafts and improving our mathematics and our presentation in hundreds of ways. Another generous critic from the beginning has been Michael Overton of New York University, who in recent years, jointly with Burke, Lewis, and Mengi, has connected the study of pseudospectra powerfully with problems and methods of optimization. Overton's influence can be felt especially in §§15, 42, and 49. A third invaluable colleague is that wizard of Toeplitz matrices and operators, Albrecht Böttcher of the Technische Universität Chemnitz. When Böttcher became interested in pseudospectra in the early 1990s, the field took a step towards greater mathematical depth, and in the years since then we have benefited from his wise and detailed comments on our various drafts. Böttcher has taught us much mathematics and treated with indulgence our 'linear algebra prestidigitator tricks'.

Another step forward for the theory of pseudospectra was the entry into the field of the eminent spectral theorist Brian Davies of King's College London. It was Davies' memorable investigation of the pseudospectra of complex harmonic oscillators that revealed the importance of wave packet pseudomodes of variable coefficient matrices and differential operators, one of the central themes in this book, and it has been an adventure to discuss and debate these and other matters with him. Probably no other colleague has contributed to so many areas of the theory of pseudospectra as Davies; his footprints can be tracked here in §§5, 10, 11, 12, 14, 15, and 43.

Finally, we didn't see it coming in advance, but the study of pseudospectra was transformed by the development of the EigTool system by Thomas Wright, now of Mirada Solutions, Ltd., when he was a graduate student at

Oxford. It is hard to imagine that we once planned to write a book without EigTool. During the happy years of 1999–2001, the three of us shared a hallway at Oxford and worked together on pseudospectra every day.

This book reaches across many subjects, and in each area we are grateful to a handful of colleagues whose contributions have been especially helpful. In numerical linear algebra, we thank Chris Beattie, Jim Demmel, Howard Elman, Anne Greenbaum, Rich Lehoucq, Lothar Reichel, Dan Sorensen, Pete Stewart, Kim-Chuan Toh, Henk van der Vorst, and Andy Wathen. In fluid mechanics we have relied on Jeff Baggett, Brian Farrell, Dan Henningson, Álvaro Meseguer, Satish Reddy, and Peter Schmid. For random matrices we are grateful to Alan Edelman, Ilya Goldsheid, and Boris Khoruzhenko, and for Toeplitz matrices and differential operators and their twisted variants we thank Jon Chapman, Alejandro Uribe, and Maciej Zworski. Concerning computation of pseudospectra and rounding error analysis on computers we thank Costas Bekas, Françoise Chaitin-Chatelin, Stratis Gallopoulos, and Shiu-Hong Lui. For discretization of differential operators and stability analysis of discretizations we are indebted to Desmond Higham, Satish Reddy, and Andre Weideman, and for fundamental theory of semigroups and stability for matrices and operators, to Hans Kraaijevanger, Rainer Nagel, Olavi Nevanlinna, and Marc Spijker. Tony Siegman taught us about lasers and Persi Diaconis about card shuffling, though in both cases, if we have not got it all right, it is our own fault.

There are many others we wish we could mention, but we shall content ourselves with thanking the following for good ideas and (in many cases) good times: Rafikul Alam, Rebecca Bolin, Carlo Cossu, Toby Driscoll, Michael Eiermann, Valérie Frayssé, Eduardo Gallestey, Didi Hinrichsen, Karel in 't Hout, Guðbjörn Jónsson, Viktor Kostin, Henry Landau, Marco Marletta, Nöel Nachtigal, Brynjulf Owren, Dimpy Pathria, Reinout Quispel, Kurt Riedel, Johannes Sjöstrand, Manfred Trummer, Alejandro Uribe, Divakar Viswanath, Harold Widom, Hongguo Xu, Akiva Yaglom, and David Zingg.

We wish to make special mention of three of our senior friends and colleagues whom we admire most and who have influenced us greatly: Gene Golub of Stanford, Cleve Moler of The MathWorks, Inc., and Gil Strang of MIT.

Our research has been supported by the U.S. National Science Foundation, the U.S. Department of Energy, and the U.K. Engineering and Physical Sciences Research Council. Much of LNT's work on the book has been made possible by two sabbaticals in Australia, one supported by Ian Sloan at the University of New South Wales and the other by Kevin Burrage at the University of Queensland with its superb library. At our home universities, we have been grateful for the support and friendship of Steve Cox, Dan Sorensen, Bill Symes, and Richard Tapia at Rice, Tom Coleman, Charlie Van Loan, and Steve Vavasis at Cornell, and Mike Giles, Endre

Süli, Andy Wathen, and the students and fellows of Balliol College at Oxford. And we must especially thank the woman who keeps the Numerical Analysis Group at Oxford running, the remarkable Shirley Day.

We approached Princeton University Press to publish this book because of its very special mathematics editor, Vickie Kearn. The book was brought into print by production editor Terri O'Prey, editorial assistant Alycia Somers, illustration specialist Dimitri Karetnikov, and the extraordinary copy editor LNT has relied upon twice before, Beth Gallagher.

Lastly, at home in Oxford and Houston, we depend always on Anne, Emma, and Jacob Trefethen and on Kristine, Cora, and Owen Embree.

I. Introduction

1 · Eigenvalues

Eigenvalues are among the most successful tools of applied mathematics. Here are some of the fields where they are important, with a representative citation from each.

acoustics [563],	chemistry [722]
control theory [443]	earthquake engineering [151]
ecology [130]	economics [739]
fluid mechanics [669]	functional analysis [630]
helioseismology [331]	magnetohydrodynamics [135]
Markov chains [582]	matrix iterations [338]
partial differential equations [178]	physics of music [279]
quantum mechanics [666]	spectroscopy [349]
structural analysis [154]	vibration analysis [376]
numerical solution of differential equations [639]	

Figures 1.1 and 1.2 present images of eigenvalues in two quite different applications.

In the simplest context of matrices, the definitions are as follows. Let $\mathbf{A}$ be an $N \times N$ matrix with real or complex coefficients; we write $\mathbf{A} \in \mathbb{C}^{N \times N}$. Let $\mathbf{v}$ be a nonzero real or complex column vector of length N, and let λ be a real or complex scalar; we write $\mathbf{v} \in \mathbb{C}^N$ and $\lambda \in \mathbb{C}$. Then $\mathbf{v}$ is an *eigenvector* of $\mathbf{A}$, and $\lambda \in \mathbb{C}$ is its corresponding *eigenvalue*, if

$$\mathbf{A}\mathbf{v} = \lambda\mathbf{v}. \tag{1.1}$$

(Even if $\mathbf{A}$ is real, its eigenvalues are in general complex unless $\mathbf{A}$ is self-adjoint.) The set of all the eigenvalues of $\mathbf{A}$ is the *spectrum* of $\mathbf{A}$, a nonempty subset of the complex plane $\mathbb{C}$ that we denote by $\sigma(\mathbf{A})$. The spectrum can also be defined as the set of points $z \in \mathbb{C}$ where the *resolvent* matrix,

$$(z - \mathbf{A})^{-1},$$

does not exist. Throughout this book, $z - \mathbf{A}$ is shorthand for $z\mathbf{I} - \mathbf{A}$, where $\mathbf{I}$ is the identity.

Unlike singular values [414, 776], eigenvalues conventionally make sense only for a matrix that is square. This reflects the fact that in applications, they are generally used where a matrix is to be compounded iteratively, for example, as a power $\mathbf{A}^k$ or an exponential $e^{t\mathbf{A}} = \mathbf{I} + t\mathbf{A} + \frac{1}{2}(t\mathbf{A})^2 + \cdots$.

For most matrices $\mathbf{A}$, there exists a *complete set of eigenvectors*, a set of N linearly independent vectors $\mathbf{v}_1, \ldots, \mathbf{v}_N$ with $\mathbf{A}\mathbf{v}_j = \lambda_j \mathbf{v}_j$. If $\mathbf{A}$ has N distinct eigenvalues, then it is guaranteed to have a complete set of

Figure 1.1: Spectroscopic image of light from the sun. The black 'Fraunhofer lines' correspond to various differences of eigenvalues of the Schrödinger operator for atoms such as H, Fe, Ca, Na, and Mg that are present in the solar atmosphere. Light at these frequencies resonates with frequencies of the transitions between energy states in these atoms and is absorbed. Spectroscopic measurements such as these are a crucial tool in chemical analysis, not only of astronomical bodies, and by making possible the measurement of redshifts of distant galaxies, they led to the discovery of the expanding universe. Original image courtesy of the Observatories of the Carnegie Institution of Washington.

eigenvectors, and they are unique up to normalization by scalar factors. For any matrix $\mathbf{A}$ with a complete set of eigenvectors $\{\mathbf{v}_j\}$, let $\mathbf{V}$ be the $N \times N$ matrix whose jth column is $\mathbf{v}_j$, a *matrix of eigenvectors*. Then we can write all N eigenvalue conditions at once by the matrix equation

$$\mathbf{A}\mathbf{V} = \mathbf{V}\mathbf{\Lambda}, \qquad (1.2)$$

where $\mathbf{\Lambda}$ is the diagonal $N \times N$ matrix whose jth diagonal entry is λ_j. Pictorially,

$$\mathbf{A} \begin{bmatrix} | & | & & | \\ \mathbf{v}_1 & \mathbf{v}_2 & \cdots & \mathbf{v}_N \\ | & | & & | \end{bmatrix} = \begin{bmatrix} | & | & & | \\ \mathbf{v}_1 & \mathbf{v}_2 & \cdots & \mathbf{v}_N \\ | & | & & | \end{bmatrix} \begin{pmatrix} \lambda_1 & & & \\ & \lambda_2 & & \\ & & \ddots & \\ & & & \lambda_N \end{pmatrix}.$$

Since the eigenvectors $\mathbf{v}_j$ are linearly independent, $\mathbf{V}$ is nonsingular, and thus we can multiply (1.2) on the right by $\mathbf{V}^{-1}$ to obtain the factorization

$$\mathbf{A} = \mathbf{V}\mathbf{\Lambda}\mathbf{V}^{-1}, \qquad (1.3)$$

known as an *eigenvalue decomposition* or a *diagonalization* of $\mathbf{A}$. In view of this formula, a matrix with a complete set of eigenvectors is said to be *diagonalizable*. An equivalent term is *nondefective*.

The eigenvalue decomposition expresses a change of basis to 'eigenvector coordinates', i.e., coefficients in an expansion in eigenvectors. If $\mathbf{A} = \mathbf{V}\mathbf{\Lambda}\mathbf{V}^{-1}$, for example, then we have

$$\mathbf{V}^{-1}(\mathbf{A}^k \mathbf{x}) = \mathbf{V}^{-1}(\mathbf{V}\mathbf{\Lambda}\mathbf{V}^{-1})^k \mathbf{x} = \mathbf{\Lambda}^k (\mathbf{V}^{-1}\mathbf{x}). \qquad (1.4)$$

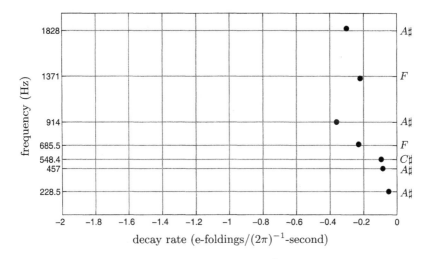

Figure 1.2: Measured eigenvalues in the complex plane of a minor third $A_4\sharp$ carillon bell (figure from [418] based on data from [696]). The grid lines show the positions of the frequencies corresponding to a minor third chord at 456.8 Hz, together with two octaves above the fundamental and one below. Immediately after the bell is struck, the ear hears all seven of the frequencies portrayed; a little later, the higher four have decayed and mostly the lowest three are heard; still later, the lowest mode, the 'hum', dominates. The simple rational relationships among these frequencies would not hold for arbitrarily shaped bells, but are the result of generations of evolution in bell shapes to achieve a pleasing effect.

Now the product $\mathbf{V}^{-1}(\mathbf{A}^k\mathbf{x})$ is equal to the vector $\mathbf{c}$ of coefficients in an expansion $\mathbf{A}^k\mathbf{x} = \mathbf{Vc} = \sum c_j\mathbf{v}_j$ of $\mathbf{A}^k\mathbf{x}$ as a linear combination of the eigenvectors $\{\mathbf{v}_j\}$, and similarly, $\mathbf{V}^{-1}\mathbf{x}$ is the vector of coefficients in an expansion of $\mathbf{x}$. Thus, (1.4) asserts that to compute $\mathbf{A}^k\mathbf{x}$, we can expand $\mathbf{x}$ in the basis of eigenvectors, apply the diagonal matrix $\mathbf{\Lambda}^k$, and interpret the result as the coefficients for another expansion in the basis of eigenvectors. In other words, the change of basis has rendered the problem diagonal and hence trivial. For $e^{t\mathbf{A}}\mathbf{x}$, similarly, we have

$$\mathbf{V}^{-1}(e^{t\mathbf{A}}\mathbf{x}) = \mathbf{V}^{-1}(\mathbf{V}e^{t\mathbf{\Lambda}}\mathbf{V}^{-1})\mathbf{x} = e^{t\mathbf{\Lambda}}(\mathbf{V}^{-1}\mathbf{x}), \qquad (1.5)$$

so diagonalization makes this problem trivial too, and likewise for other functions $f(\mathbf{A})$.

So far we have taken $\mathbf{A}$ to be a matrix, but eigenvalues are also important when $\mathbf{A}$ is a more general linear operator such as an infinite matrix, a differential operator, or an integral operator. Indeed, eigenvalue problems for matrices often come about through discretization of linear operators. The *spectrum* $\sigma(\mathbf{A})$ of a closed operator $\mathbf{A}$ defined in a Banach (or Hilbert) space is defined as the set of numbers $z \in \mathbb{C}$ for which the resolvent $(z - \mathbf{A})^{-1}$ does not exist as a bounded operator defined on the whole space (§4). It can be any closed set in the complex plane, including the empty set. Eigenvalues and eigenvectors (also called *eigenfunctions* or *eigenmodes*) are still defined by (1.1), but among the new features that arise in the operator case is the phenomenon that not every $z \in \sigma(\mathbf{A})$ is necessarily an eigenvalue. This book avoids fine points of spectral theory wherever possible, for the main issues to be investigated are orthogonal to the differences between matrices and operators. In particular, the distinction between spectra and pseudospectra has little to do with the distinction between point and continuous spectra. In certain contexts, of course, it will be necessary for us to be more precise.

This book is about the limitations of eigenvalues, and alternatives to them. In the remainder of this introductory section, let us accordingly consider the question, What are eigenvalues useful for? Why are eigenvalues and eigenfunctions—more generally, spectra and spectral theory—among the standard tools of applied mathematics? Various answers to these questions appear throughout this volume, but here, we shall make our best attempt to summarize them in a systematic way.

We begin with a one-paragraph history [96, 208, 721]. It is not too great an oversimplification to say that a major part of eigenvalue analysis originated early in the nineteenth century with Fourier's solution of the heat equation by series expansions. Fourier's ideas were extended by Poisson, and other highlights of the nineteenth century include Sturm and Liouville's treatment of more general second-order differential equations in the 1830s; Sylvester and Cayley's diagonalization of symmetric matrices in the 1850s (the origins of this idea go back to Cauchy, Jacobi, Lagrange, Euler, Fermat, and Descartes); Weber and Schwarz's treatment of a vibrating membrane in 1869 and 1885 (whose origins in vibrating strings go back to D. Bernoulli, Euler, d'Alembert, ..., Pythagoras); Lord Rayleigh's treatise *The Theory of Sound* in 1877 [618]; and further developments by Poincaré around 1890. By 1900, eigenvalues and eigenfunction expansions were well-known, especially in the context of differential equations. The new century brought the mathematical theory of linear operators due to Fredholm, Hilbert, Schmidt, von Neumann, and others; the terms 'eigenvalue' and 'spectral theory' appear to have been coined by Hilbert. The influential book by Courant and Hilbert, first published in 1924, surveyed a large amount of material concerning eigenvalues of differential equations and vibration problems [168]. Just two years later came the explosive ideas of quantum mechanics, which in a short time, in the hands of Heisenberg, Jordan, Schrödinger, Dirac, and others, moved matrices and operators to center stage of the scientific world. Quantum 'matrix mechanics' revealed that energy states of atoms and molecules could be viewed as eigenfunctions of a Schrödinger operator, thereby explaining Figure 1.1, the periodic table of the elements, and countless other scientific observations besides [384], and from that time on, every mathematical scientist has known the basics of matrices, operators, eigenvalues, and eigenfunctions.

What exactly do eigenvalues offer that makes them useful for so many problems? We believe there are three principal answers to this question, more than one of which may be important in a particular application.

1. *Diagonalization and separation of variables: use of the eigenfunctions as a basis.* One thing eigenvalues may accomplish is the decoupling, as in (1.3)–(1.5), of a problem involving vectors or functions into a collection of problems involving scalars, which may make subsequent computations easier. For example, in Fourier's problem of heat conduction in a solid bar with zero temperature at both ends, the eigenmodes are sine waves that decay independently as a function of time. If an arbitrary initial temperature distribution is expanded as a sum of these sine waves, then the solution at a later time can be calculated by summing the components of the expansion.

2. *Resonance: heightened response to selected inputs.* Diagonalization is an algorithmic idea; the other uses of eigenvalues are more physical. One is the analysis of the phenomenon of resonance, perhaps most familiar in the context of vibrating strings, drums, and mechanical structures. Any visitor to science museums has seen demonstrations showing that certain systems respond preferentially to vibrations at special frequencies. These frequencies are the eigenvalues of the linear or linearized operator that governs the system in question, and the form of the response is associated with the corresponding eigenfunctions. Examples of resonance are familiar: One thinks of soldiers breaking step as they cross bridges; of the less fortunate Tacoma Narrows Bridge in the 1940s, whose collapse was initiated by a wind-induced flow oscillation too close to a structural eigenfrequency; of buildings and their response to the vibrations of earthquakes—an application where eigenvalues are written into legal codes; of that old cartoon standby, the soprano whose high E shatters windows. In other examples resonance is desired rather than feared: examples include AM radio, where the signal from a far-off station is selected from a sea of background noise by a finely tuned resonant circuit, and the cochlea of the human ear, whose basilar membrane resonates preferentially in different locations according to the frequency of the sound input and thus in a sense tunes in all stations at once. These last two examples illustrate the wide range of complexity in applications of eigenvalue ideas, for the radio problem is straightforward and almost perfectly linear, whereas the ear is a complicated nonlinear system, not yet fully understood, for which eigenmodes are only a crude first step.

3. *Asymptotics and stability: dominant response to general inputs.* A related application of eigenvalues is to questions of the form, What will

happen as time elapses (or in the extreme, $t \to \infty$) to a system that has experienced some more or less random disturbance? Fourier's heat problem again affords an example: Whatever the shape of the initial temperature distribution, the higher sine waves decay faster than the lowest one, and therefore almost any initial distribution will eventually come to look like the half-wavelength sine with zeros just at the two ends of the interval. Similarly, what makes a church bell as in Figure 1.2 chime musically? As the clapper strikes, all frequencies are excited, but differential decay rates soon filter out all but a few dominant ones, and the result is a pleasing sound. Kettledrums operate on the same principle, as do Markov chains in probability theory. Sometimes the crucial issue is a question of stability: Are there modes that grow rather than decay with t? For example, in fluid mechanics a standard technique to determine whether small perturbations to a laminar flow will be amplified into large ones—which may then trigger the onset of turbulence—is to calculate whether the eigenvalues of the system all lie in the left half of the complex plane. (We shall see in §20 that this technique is not always successful.) Similar questions arise in control theory and in numerical analysis, where time is discrete and stability depends on eigenvalues being less than 1 in modulus. Problems of convergence of matrix iterations in numerical analysis are also related, the convergence rate being determined by how close certain eigenvalues are to zero.

Principles 1, 2, and 3 account for most applications of eigenvalues. (Sometimes the latter two are hard to distinguish, as, for example, in the operation of bowed or blown musical instruments. The significance of eigenvalues in quantum mechanics also may have special features, not well captured by 1–3.) In view of the ubiquity of vibrations, oscillations, and linear or approximately linear processes in the physical world, they amply justify the great attention that has been given to eigenvalues over the years.

And we think there is a fourth reason, too, for the success of eigenvalues.

4. *They give a matrix a personality.* We humans like images; our brains are specially adapted to interpret them. Eigenvalues enable us to take the abstraction of a matrix or linear operator, for whose analysis we possess no hardwired talent, and portray it as a picture.

This book is about a class of problems for which eigenvalue methods may fail: problems involving matrices or operators for which the matrix $\mathbf{V}^{-1}$ of (1.3)–(1.5), if it exists, contains very large entries:

$$\|\mathbf{V}^{-1}\| \gg 1. \tag{1.6}$$

('This often turns out to mean exponentially large with respect to a parameter.) This formulation of the matter assumes that the matrix $\mathbf{V}$ itself is in some sense reasonably scaled, with $\|\mathbf{V}\|$ roughly of order 1. If no assumptions are made about the scaling of $\|\mathbf{V}\|$, then (1.6) should be replaced by a statement about the *condition number* of $\mathbf{V}$ in the norm $\|\cdot\|$,

$$\|\mathbf{V}\|\|\mathbf{V}^{-1}\| \gg 1, \tag{1.7}$$

and to be still more precise we should require that (1.7) hold not just for some eigenvector matrix $\mathbf{V}$, whose eigenvector columns might be badly scaled relative to one another, but for any eigenvector matrix $\mathbf{V}$. For operators as opposed to matrices, a suitable generalization of (1.7) can be applied in some cases, but not all.

The conditions (1.6) and (1.7) depend upon the choice of norm $\|\cdot\|$. Though sometimes it is essential to consider other possibilities (see, e.g., §56 and §57), most of our examples will be based on the use of the 2-norm, defined by $\|\mathbf{x}\|_2 = (\sum |x_j|^2)^{1/2}$ for a vector $\mathbf{x}$ and then by

$$\|\mathbf{A}\|_2 = \max_{\mathbf{x}} \frac{\|\mathbf{A}\mathbf{x}\|_2}{\|\mathbf{x}\|_2} \tag{1.8}$$

for a matrix $\mathbf{A}$. This choice of norm corresponds mathematically to formulation in a Hilbert space and physically to consideration of energy defined by a sum of squares, and in this important special case, (1.7) amounts to the condition that the eigenvectors of $\mathbf{A}$, if they exist, are far from orthogonal. At the other extreme is a *normal* matrix, one that has a complete set of orthogonal eigenvectors; real symmetric and Hermitian matrices fall in this category. In this case, if each $\mathbf{v}_j$ is normalized by $\|\mathbf{v}_j\|_2 = 1$, then $\mathbf{V}$ is a *unitary* matrix (in the real case we say *orthogonal*), with $\mathbf{V}^{-1} = \mathbf{V}^*$ ($\mathbf{V}^*$ denotes the conjugate transpose) and $\|\mathbf{V}\|_2 = \|\mathbf{V}^{-1}\|_2 = 1$. Thus for $\|\cdot\| = \|\cdot\|_2$, (1.7) is a statement that $\mathbf{A}$ is in some sense far from normal. In this norm, it is the nonnormal matrices for which eigenvalue analysis may fail, and in this book, starting with the subtitle on the cover, we often speak of problems that are 'nonnormal' or 'far from normal' when a more careful statement would refer to a more general condition, such as (1.7).

The majority of the familiar applications of eigenvalue analysis involve matrices or operators that are normal or close to normal, having eigenfunctions orthogonal or nearly so. Among the examples mentioned so far, all of the physical ones are in this category except certain problems of fluid mechanics. The familiar mechanical oscillations are governed by normal operators, for example, and so are the oscillations of quantum mechanics, at least in their standard formulation. As a consequence, our intuition about eigenvalues has been formed by the normal case. Two centuries of successes have generated confidence that the eigenvalue idea is both powerful in practice and fundamental in concept. It has not always been noted that as most of these successes involve problems governed by normal or near-normal operators, our grounds for confidence in the nonnormal case are less solid.

With this in mind, we shall now briefly indicate what can go wrong with 1, 2, and 3 in certain applications.

First, consider 2. If a linear operator is normal, then the degree of resonant amplification that may occur in response to an input at frequency ω is equal to the inverse of the distance in the complex plane between ω and the nearest eigenvalue. (This formula can be found in first-year physics textbooks, usually without the word 'eigenvalue'.) For a nonnormal operator, however, the resonant amplification may be orders of magnitude greater. *The resonances of a nonnormal system are not determined by the eigenvalues alone.* This phenomenon is at the heart of the topic known as 'receptivity' in fluid mechanics (§23).

Next, consider 3. It is true that for a purely linear, constant-coefficient, homogeneous problem, eigenvalues govern the asymptotic behavior as $t \to \infty$. If the problem is normal, this statement is robust; the eigenvalues also have relevance to short-time or transient behavior, and moreover, their influence tends to persist if the problem is altered in small ways. If the problem is far from normal, however, conclusions based on eigenvalues are in general not robust. First, there may be a long transient that looks quite different from the asymptote and has no connection to the eigenvalues. Second, even the asymptote may change beyond recognition if the problem is modified slightly. *Eigenvalues do not always govern the transient behavior of a nonnormal system, nor the asymptotic behavior in the presence of nonlinear terms, variable coefficients, lower order terms, inhomogeneous forcing data, or other complications.* Few applied problems are free of all these effects. For those that are, it is rare that one is interested so purely in the limit $t \to \infty$ as one may at first imagine. These issues are at the heart of convergence and stability investigations in numerical analysis, and we discuss them, for example, in Parts VI and VII. For a high-level schema, see Figure 33.3.

This brings us to 1. Unlike 2 and 3, the algorithmic idea of diagonalization is not in general invalidated if $\|\mathbf{V}\| \, \|\mathbf{V}^{-1}\|$ is large (although in extreme cases there may be difficulties caused by rounding errors on a computer). On the other hand, there is a different difficulty that sometimes makes diagonalization less useful than one might expect, even for normal problems. In practice, for differential or other operators one works with truncated expansions; an infinite series is approximated by finite sum. The difficulty that arises sometimes is that the choice of the basis of eigenfunctions for such an expansion may necessitate taking an unacceptably large number of terms in the expansion to achieve the required accuracy. *Eigenfunction expansions may be exceedingly inefficient.* This fact was publicized by Orszag around 1970 in the context of spectral methods for the numerical solution of differential equations [588, 775]. Spectral methods, by contrast, are based on expansions in functions that have nothing to do with the eigenfunctions of the problem at hand, but which may converge geometrically, where an expansion in eigenfunctions converges only linearly. Thirty Chebyshev polynomials may resolve a problem as well as a thousand eigenfunctions. An example is considered in §59.

What about 4, a matrix or operator's personality? In the highly non-normal case, vivid though the image may be, the location of the eigenvalues may be as fragile an indicator of underlying character as the hair color of a Hollywood actor. We shall see that pseudospectra provide equally compelling images that may capture the spirit underneath more robustly.

In summary, eigenvalues and eigenfunctions have a distinguished history of application throughout the mathematical sciences; we could not get along without them. Their clearest successes, however, are associated with problems that involve well-behaved systems of eigenvectors, which in most contexts means matrices or operators that are normal or nearly so. This class of problems encompasses the majority of applications, but not all. For nonnormal problems, the record is less clear, and even the conceptual significance of eigenvalues is open to question.

2 · Pseudospectra of matrices

In certain applications, eigenvalue analysis proves to be misleading. Here are some examples, with a single citation in each case. (Listing a field like 'ecology', of course, does not mean that eigenvalue analysis is always misleading in that field, just that it sometimes is.) Most of these examples are discussed in detail later in the book, where many more references are provided.

atmospheric science [261]	control theory [401]
ecology [574]	hydrodynamic stability [780]
lasers [688]	magnetohydrodynamics [74]
Markov chains [435]	matrix iterations [342]
rounding error analysis [133]	operator theory [93]
non-Hermitian quantum mechanics [377]	
numerical solution of differential equations [796]	

This book is about phenomena that arise in such applications due to troublesome eigenvalues and about ways to understand them mathematically. Specifically, this book describes the mathematical tool known as pseudospectra. In this section we introduce pseudospectra of finite-dimensional matrices, and we generalize to linear operators in Banach space in §4. The history of these ideas is detailed in §6.

Let $\| \cdot \|$ denote a norm on $\mathbb{C}^N$, the space of complex N-vectors, and also the associated induced norm on $\mathbb{C}^{N \times N}$, the space of complex $N \times N$ matrices. (Vector and matrix norms are described, for example, in [414] and [776].) In most of the computed examples in this book, we take $\| \cdot \|$ to be the 2-norm $\| \cdot \|_2$ defined by the Euclidean inner product as in (1.8); norms induced by other inner products require only minor modifications. We shall generally let $\mathbf{A}$ denote a matrix in $\mathbb{C}^{N \times N}$ or a linear operator on an infinite-dimensional space.

One can motivate the idea of pseudospectra as follows. As observed throughout applied mathematics, the question 'Is $\mathbf{A}$ singular?' is not robust, for an arbitrarily small perturbation can change the answer from yes to no. For applied purposes, a better question is, 'Is $\|\mathbf{A}^{-1}\|$ large?' Now, the condition defining eigenvalues is a condition of matrix singularity. To ask, 'Is z an eigenvalue of $\mathbf{A}$?' is the same as to ask,

$$\text{Is } z - \mathbf{A} \text{ singular?}$$

Therefore, the property of being an eigenvalue of a matrix is also not robust. A better question may be,

$$\text{Is } \|(z - \mathbf{A})^{-1}\| \text{ large?}$$

This pattern of thinking leads naturally to our first definition of pseudo-spectra:[1]

First definition of pseudospectra
Let $\mathbf{A} \in \mathbb{C}^{N \times N}$ and $\varepsilon > 0$ be arbitrary. The ε-*pseudospectrum* $\sigma_\varepsilon(\mathbf{A})$ of $\mathbf{A}$ is the set of $z \in \mathbb{C}$ such that $$\|(z - \mathbf{A})^{-1}\| > \varepsilon^{-1}. \qquad (2.1)$$

The matrix $(z - \mathbf{A})^{-1}$ is known as the *resolvent* of $\mathbf{A}$ at z. In (2.1) and throughout this book we employ the convention that

$$\|(z - \mathbf{A})^{-1}\| = \infty \quad \text{for} \quad z \in \sigma(\mathbf{A}), \qquad (2.2)$$

where $\sigma(\mathbf{A})$ is the spectrum (set of eigenvalues) of $\mathbf{A}$, so that in particular, the spectrum is contained in the ε-pseudospectrum for every $\varepsilon > 0$. In words, *the ε-pseudospectrum is the open subset of the complex plane bounded by the ε^{-1} level curve of the norm of the resolvent.*[2]

It is perhaps not obvious at first whether the idea of pseudospectra serves much purpose. Is not $\|(z - \mathbf{A})^{-1}\|$ large precisely when z is close to an eigenvalue of $\mathbf{A}$? For a normal matrix, when $\| \cdot \| = \| \cdot \|_2$, as we shall see in a moment, this intuition is correct (see Figure 2.1). The importance of pseudospectra arises for matrices that are far from normal, for which $\|(z - \mathbf{A})^{-1}\|$ may be large even when z is far from the spectrum, or more generally for matrices satisfying the conditions (1.6) or (1.7) discussed in the last section (Figures 2.1 and 2.2).

Our second definition of pseudospectra is based on the connection between the resolvent norm and eigenvalue perturbation theory [448].

---

[1]In this book, reversing the pattern of our earlier papers on this subject, we define pseudospectra by strict rather than weak inequalities. This choice proves to be more convenient for infinite-dimensional operators, as has been pointed out by various authors, such as Davies [179], and emphasized particularly strongly by Chaitin-Chatelin and Harrabi [134].

[2]Thus, unlike the spectrum, the pseudospectra depend on the norm. At first sight this lack of norm-invariance may seem a defect in the idea of pseudospectra, and it has certainly contributed to the fact that the development of a theory of nonnormality has lagged far behind the development of standard spectral theory. (Pseudospectra are an idea of analysis; eigenvalues belong to algebra.) Yet what one really needs to know about an applied problem is usually norm-dependent. For example, many nonnormal operators can be made normal by a transformation to an exponentially weighted inner product and norm, but such a transformation will generally distort physical notions such as energy beyond recognition. We shall see examples in almost every section of this book.

(a) normal   (b) nonnormal

Figure 2.1: The geometry of pseudospectra: schematic view. In each plot, the contours represent the boundary of $\sigma_\varepsilon(\mathbf{A})$ for two values of ε.

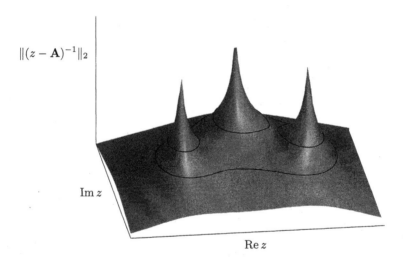

Figure 2.2: The resolvent norm as a function of $z \in \mathbb{C}$ for the matrix of Figure 2.1(b). The contours here—level sets of $\|(z - \mathbf{A})^{-1}\|_2$—match those shown in Figure 2.1(b). The spikes occur at eigenvalues of $\mathbf{A}$; in principle they extend to infinity.

Second definition of pseudospectra
$\sigma_\varepsilon(\mathbf{A})$ is the set of $z \in \mathbb{C}$ such that $$z \in \sigma(\mathbf{A} + \mathbf{E}) \qquad (2.3)$$ for some $\mathbf{E} \in \mathbb{C}^{N \times N}$ with $\|\mathbf{E}\| < \varepsilon$.

In words, *the ε-pseudospectrum is the set of numbers that are eigenvalues of some perturbed matrix $\mathbf{A} + \mathbf{E}$ with $\|\mathbf{E}\| < \varepsilon$.*

From either of these definitions, it follows that the pseudospectra associated with various ε are nested sets,

$$\sigma_{\varepsilon_1}(\mathbf{A}) \subseteq \sigma_{\varepsilon_2}(\mathbf{A}), \qquad 0 < \varepsilon_1 \leq \varepsilon_2, \tag{2.4}$$

and that the intersection of all the pseudospectra is the spectrum,

$$\bigcap_{\varepsilon > 0} \sigma_\varepsilon(\mathbf{A}) = \sigma(\mathbf{A}). \tag{2.5}$$

Figure 2.3 illustrates the equivalence of (2.1) and (2.6) for a highly non-normal 12×12 matrix arising in the field of spectral methods for partial differential equations [741, 782]; see §30. Contours of constant resolvent norm are plotted on the left, and eigenvalues of randomly perturbed matrices on the right. Evidently $(z - \mathbf{A})^{-1}$ has norm of order 10^5 or larger even when the distance from z to the spectrum is of order 1. Equivalently, the eigenvalues of $\mathbf{A}$ are highly sensitive to perturbations. This example also illustrates that there may be more geometric structure to a matrix or operator in the complex plane than is revealed by the spectrum alone. Here, a conspicuous geometric feature is the group of almost exactly straight sections of pseudospectral boundaries near the origin of the complex plane. This feature reflects the construction of this matrix as an approximation to a certain differential operator (see §§5 and 30).

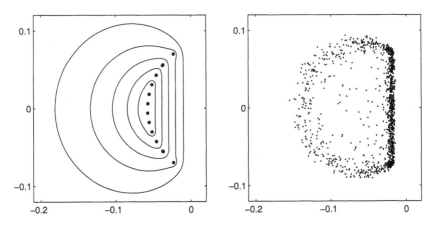

Figure 2.3: Pseudospectra of a 12×12 Legendre spectral differentiation matrix. The left plot shows the eigenvalues (solid dots) and the boundaries of the 2-norm ε-pseudospectra for $\varepsilon = 10^{-3}, 10^{-4}, \ldots, 10^{-7}$ (from outer to inner). The right side shows a superposition of the eigenvalues of 100 randomly perturbed matrices $\mathbf{A} + \mathbf{E}$, where each $\mathbf{E}$ is a matrix with independent normally distributed complex entries of mean 0 scaled so that $\|\mathbf{E}\| = 10^{-3}$. If all possible perturbations with $\|\mathbf{E}\| < 10^{-3}$ were considered, the dots would exactly fill the region bounded by the outermost curve on the left.

Here is a third characterization of the ε-pseudospectrum:

Third definition of pseudospectra

$\sigma_\varepsilon(\mathbf{A})$ is the set of $z \in \mathbb{C}$ such that

$$\|(z - \mathbf{A})\mathbf{v}\| < \varepsilon \qquad (2.6)$$

for some $\mathbf{v} \in \mathbb{C}^N$ with $\|\mathbf{v}\| = 1$.

The number z in (2.6) (or equivalently in any of our definitions) is an ε-*pseudoeigenvalue* of $\mathbf{A}$, and $\mathbf{v}$ is a corresponding ε-*pseudoeigenvector*. (Synonymous terms include *pseudoeigenfunction*, *pseudoeigenmode*, and *pseudomode*.) In words, *the ε-pseudospectrum is the set of ε-pseudoeigenvalues.*

We begin the substance of this book by establishing the equivalence of these three definitions.

Equivalence of the definitions of pseudospectra

Theorem 2.1 *For any matrix* $\mathbf{A} \in \mathbb{C}^{N \times N}$, *the three definitions above are equivalent.*

Proof. For $z \in \sigma(\mathbf{A})$ the equivalence is trivial, so assume $z \notin \sigma(\mathbf{A})$, implying the existence of $(z - \mathbf{A})^{-1}$. To prove (2.3)$\Rightarrow$(2.6), suppose that $(\mathbf{A} + \mathbf{E})\mathbf{v} = z\mathbf{v}$ for some $\mathbf{E} \in \mathbb{C}^{N \times N}$ with $\|\mathbf{E}\| < \varepsilon$ and some nonzero $\mathbf{v} \in \mathbb{C}^N$, which we may take to be normalized, $\|\mathbf{v}\| = 1$. Then $\|(z-\mathbf{A})\mathbf{v}\| = \|\mathbf{E}\mathbf{v}\| < \varepsilon$, as required. To prove (2.6)$\Rightarrow$(2.1), suppose $(z - \mathbf{A})\mathbf{v} = s\mathbf{u}$ for some $\mathbf{v}, \mathbf{u} \in \mathbb{C}^N$ with $\|\mathbf{v}\| = \|\mathbf{u}\| = 1$ and $s < \varepsilon$. Then $(z - \mathbf{A})^{-1}\mathbf{u} = s^{-1}\mathbf{v}$, so $\|(z - \mathbf{A})^{-1}\| \geq s^{-1} > \varepsilon^{-1}$. Finally, to prove (2.1)$\Rightarrow$(2.3), suppose $\|(z-\mathbf{A})^{-1}\| > \varepsilon^{-1}$. Then $(z - \mathbf{A})^{-1}\mathbf{u} = s^{-1}\mathbf{v}$ and consequently $z\mathbf{v} - \mathbf{A}\mathbf{v} = s\mathbf{u}$ for some $\mathbf{v}, \mathbf{u} \in \mathbb{C}^N$ with $\|\mathbf{v}\| = \|\mathbf{u}\| = 1$ and $s < \varepsilon$. To establish (2.3), it is enough to show that there exists a matrix $\mathbf{E} \in \mathbb{C}^{N \times N}$ with $\|\mathbf{E}\| = s$ and $\mathbf{E}\mathbf{v} = s\mathbf{u}$, for then $\mathbf{v}$ will be an eigenvector of $\mathbf{A} + \mathbf{E}$ with eigenvalue z. In fact, $\mathbf{E}$ can be taken to be a rank-1 matrix of the form $\mathbf{E} = s\mathbf{u}\mathbf{w}^*$ for some $\mathbf{w} \in \mathbb{C}^N$ with $\mathbf{w}^*\mathbf{v} = 1$. If $\| \cdot \|$ is the 2-norm, this is evident simply by taking $\mathbf{w} = \mathbf{v}$. In the case of an arbitrary norm $\| \cdot \|$, the existence of a vector $\mathbf{w}$ satisfying the required conditions can be interpreted as the existence of a linear functional $\mathbf{L}$ on $\mathbb{C}^N$ with $\|\mathbf{L}\mathbf{v}\| = 1$ and $\|\mathbf{L}\| = 1$, which is guaranteed by the Hahn–Banach theorem. A less highbrow version of the same proof can be carried out by the method of dual norms; see [414, Chap. 5] and [830].[3] ∎

So far we have taken $\| \cdot \|$ to be an arbitrary norm; when we come to

---

[3]The proof of (2.1)$\Rightarrow$(2.3) for arbitrary $\| \cdot \|$ first appeared in [830] using dual norms, and, using the Hahn–Banach theorem, in [796].

operators in §4, this will correspond to a setting of Banach spaces. Now, however, let us indicate some of the additional properties that emerge in the special case of Hilbert spaces. From now on in this section we shall often restrict attention to the case in which $\mathbb{C}^N$ is endowed with the standard inner product

$$(\mathbf{u}, \mathbf{v}) = \mathbf{v}^* \mathbf{u} \qquad (2.7)$$

and $\|\cdot\|$ is the corresponding 2-norm,

$$\|\mathbf{v}\| = \|\mathbf{v}\|_2 = \sqrt{\mathbf{v}^* \mathbf{v}}. \qquad (2.8)$$

With this choice of inner product and norm, the Hermitian conjugate (conjugate transpose) of a matrix is the same as its adjoint; we use the symbol $\mathbf{A}^*$. Applications where more general weighted inner products and norms are desired can be handled within the framework of the 2-norm by introducing a similarity transformation $\mathbf{A} \mapsto \mathbf{W}\mathbf{A}\mathbf{W}^{-1}$, where $\mathbf{W}$ is nonsingular; see §§45 and 51.

If $\|\cdot\| = \|\cdot\|_2$, the norm of a matrix is its largest singular value and the norm of the inverse is the inverse of the smallest singular value.[4] In particular,

$$\|(z - \mathbf{A})^{-1}\|_2 = [s_{\min}(z - \mathbf{A})]^{-1}, \qquad (2.9)$$

where $s_{\min}(z - \mathbf{A})$ denotes the smallest singular value of $z - \mathbf{A}$, suggesting a fourth definition.

Fourth definition of pseudospectra (2-norm)

For $\|\cdot\| = \|\cdot\|_2$, $\sigma_\varepsilon(\mathbf{A})$ is the set of $z \in \mathbb{C}$ such that

$$s_{\min}(z - \mathbf{A}) < \varepsilon. \qquad (2.10)$$

From (2.9) it is clear that (2.10) is equivalent to (2.1) and therefore also to our other characterizations of pseudospectra. In the proof of Theorem 2.1, the rank-1 matrix $\mathbf{E} = s\mathbf{u}\mathbf{v}^*$ can now be understood as follows: s, $\mathbf{u}$, and $\mathbf{v}$ are the smallest singular value and associated left and right singular vectors of $z - \mathbf{A}$.

We come now to the matter of nonnormality. First, note that if $\mathbf{U}$ is a unitary matrix (i.e., $\mathbf{U}^* = \mathbf{U}^{-1}$), then

$$(z - \mathbf{U}\mathbf{A}\mathbf{U}^*)^{-1} = [\mathbf{U}(z - \mathbf{A})\mathbf{U}^*]^{-1} = \mathbf{U}(z - \mathbf{A})^{-1}\mathbf{U}^*, \qquad (2.11)$$

and therefore

$$\|(z - \mathbf{U}\mathbf{A}\mathbf{U}^*)^{-1}\|_2 = \|(z - \mathbf{A})^{-1}\|_2 \qquad \forall z \in \mathbb{C}.$$

---

[4]We assume the reader is familiar with the singular value decomposition (SVD); see [415] or [776].

Thus the resolvent norm is invariant with respect to unitary similarity transformations, which implies that the same is true of the pseudospectra, still assuming $\| \cdot \| = \| \cdot \|_2$:

$$\sigma_\varepsilon(\mathbf{A}) = \sigma_\varepsilon(\mathbf{U A U}^*) \qquad \forall \varepsilon \geq 0. \tag{2.12}$$

A normal matrix is a matrix with the special property that there exists a unitary similarity transformation that makes it diagonal.

Normal matrix

A matrix $\mathbf{A} \in \mathbb{C}^{N \times N}$ is *normal*[5] if it has a complete set of orthogonal eigenvectors, that is, if it is unitarily diagonalizable:

$$\mathbf{A} = \mathbf{U \Lambda U}^*. \tag{2.13}$$

(Here $\mathbf{U}$ is unitary and $\mathbf{\Lambda}$ is a diagonal matrix of eigenvalues.) For a normal matrix, the ε-pseudospectrum is just the union of the open ε-balls about the points of the spectrum, as suggested in Figure 2.1. In other words, the eigenvalues all have condition number exactly 1 with respect to matrix perturbations (see §52); equivalently, the resolvent norm satisfies

$$\|(z - \mathbf{A})^{-1}\|_2 = \frac{1}{\text{dist}(z, \sigma(\mathbf{A}))}, \tag{2.14}$$

where $\text{dist}(z, \sigma(\mathbf{A}))$ denotes the usual distance of a point to a set in the complex plane. The next theorem expresses these facts with the aid of the following notation for an open ε-ball:

$$\Delta_\varepsilon = \{z \in \mathbb{C} : |z| < \varepsilon\}. \tag{2.15}$$

In this theorem a sum of sets has the usual meaning:

$$\sigma(\mathbf{A}) + \Delta_\varepsilon = \{z : z = z_1 + z_2,\ z_1 \in \sigma(\mathbf{A}),\ z_2 \in \Delta_\varepsilon\},$$

which is equal to $\{z : \text{dist}(z, \sigma(\mathbf{A})) < \varepsilon\}$.

---

[5]An equivalent characterization is that $\mathbf{A}$ is normal if it commutes with its adjoint: $\mathbf{A A}^* = \mathbf{A}^* \mathbf{A}$. This may seem a long way from (2.13), but the link is that both are equivalent to a third statement: $\mathbf{A}$ and $\mathbf{A}^*$ have the same eigenvectors; that is, they are *simultaneously diagonalizable*. For a catalogue of equivalent conditions for normality, see [242, 351].

Pseudospectra of a normal matrix

Theorem 2.2 *For any* $\mathbf{A} \in \mathbb{C}^{N \times N}$,

$$\sigma_\varepsilon(\mathbf{A}) \supseteq \sigma(\mathbf{A}) + \Delta_\varepsilon \qquad \forall \varepsilon > 0, \qquad (2.16)$$

and if $\mathbf{A}$ *is normal and* $\|\cdot\| = \|\cdot\|_2$, *then*

$$\sigma_\varepsilon(\mathbf{A}) = \sigma(\mathbf{A}) + \Delta_\varepsilon \qquad \forall \varepsilon > 0. \qquad (2.17)$$

Conversely, if $\|\cdot\| = \|\cdot\|_2$, *then* (2.17) *implies that* $\mathbf{A}$ *is normal.*

Proof. If z is an eigenvalue of $\mathbf{A}$, then $z + \delta$ is an eigenvalue of $\mathbf{A} + \delta$ for any $\delta \in \mathbb{C}$; since $\|\delta \mathbf{I}\| = |\delta|$, this establishes (2.16). For (2.17) we note that if $\mathbf{A}$ is normal, it can be assumed without loss of generality to be diagonal without any effect on norms if $\|\cdot\| = \|\cdot\|_2$, with diagonal elements a_{jj} equal to the eigenvalues λ_j. In this case the resolvent is also diagonal, which implies that it satisfies (2.14), and as noted above, (2.1) implies that this is equivalent to (2.17). Finally, for the converse, here is a sketch of a proof that can be made precise with the aid of results from §52. Equation (2.17) implies that each eigenvalue of $\mathbf{A}$ has condition number 1. By a standard formula for eigenvalue condition numbers (see §52), if $\|\cdot\| = \|\cdot\|_2$, it follows that each right eigenvector of $\mathbf{A}$ is also a left eigenvector, i.e., that $\mathbf{A}$ and $\mathbf{A}^*$ are simultaneously diagonalizable. Therefore $\mathbf{A}$ is normal. ∎

Now suppose $\mathbf{A}$ is diagonalizable but not necessarily normal, and let $\mathbf{V} \in \mathbb{C}^{N \times N}$ be a *matrix of eigenvectors* of $\mathbf{A}$ as in (1.2) and (1.3). With $\|\cdot\| = \|\cdot\|_2$, the condition number of this basis of eigenvectors, mentioned already in (1.7), is

$$\kappa(\mathbf{V}) \equiv \|\mathbf{V}\|_2 \|\mathbf{V}^{-1}\|_2 = \frac{s_{\max}(\mathbf{V})}{s_{\min}(\mathbf{V})}, \qquad (2.18)$$

where $s_{\max}(\mathbf{V})$ and $s_{\min}(\mathbf{V})$ are the largest and smallest singular values of $\mathbf{V}$.[6] In general, $\kappa(\mathbf{V})$ may be any number in the range $1 \leq \kappa(\mathbf{V}) < \infty$,[7] and the value $\kappa(\mathbf{V}) = 1$ is possible if and only if $\mathbf{A}$ is normal.

The condition number of $\mathbf{V}$ provides an upper bound for the condition numbers of the individual eigenvalues of $\mathbf{A}$. This fact is known as the Bauer–Fike theorem.

---

[6] Since $\mathbf{V}$ is not unique, $\kappa(\mathbf{V})$ is not uniquely defined for a given $\mathbf{A}$. If the eigenvalues of $\mathbf{A}$ are distinct, however, then $\kappa(\mathbf{V})$ becomes unique if the eigenvectors are normalized by $\|\mathbf{v}_j\| = 1$. Though this choice is not necessarily the one that minimizes $\kappa(\mathbf{V})$, it exceeds the optimal value by at most a factor of $\sqrt{N}$ [788]. For more details on scalar measures of nonnormality, see §48.

[7] We shall also write $\kappa(\mathbf{V}) = \infty$ as a convenient shorthand in the case of nondiagonalizable $\mathbf{A}$.

Bauer–Fike theorem

Theorem 2.3 *Suppose* $\mathbf{A} \in \mathbb{C}^{N \times N}$ *is diagonalizable,* $\mathbf{A} = \mathbf{V}\mathbf{\Lambda}\mathbf{V}^{-1}$. *Then for each* $\varepsilon > 0$, *with* $\|\cdot\| = \|\cdot\|_2$,

$$\sigma(\mathbf{A}) + \Delta_\varepsilon \subseteq \sigma_\varepsilon(\mathbf{A}) \subseteq \sigma(\mathbf{A}) + \Delta_{\varepsilon\kappa(\mathbf{V})}. \qquad (2.19)$$

Proof (cf. [32, 729, 827]). The first inclusion was established in (2.16). For the second we calculate

$$(z - \mathbf{A})^{-1} = (z - \mathbf{V}\mathbf{\Lambda}\mathbf{V}^{-1})^{-1} = [\mathbf{V}(z - \mathbf{\Lambda})\mathbf{V}^{-1}]^{-1} = \mathbf{V}(z - \mathbf{\Lambda})^{-1}\mathbf{V}^{-1},$$

which implies

$$\|(z - \mathbf{A})^{-1}\|_2 \le \kappa(\mathbf{V})\|(z - \mathbf{\Lambda})^{-1}\|_2 = \frac{\kappa(\mathbf{V})}{\mathrm{dist}(z, \sigma(\mathbf{A}))},$$

and the definition (2.1) completes the proof. ■

Theorem 2.3 holds for a more general class of norms than we have stated here, and in their famous paper, Bauer and Fike [32] prove stronger results along similar lines; see [729, p. 177].

The following theorem collects some basic properties of pseudospectra, whose proofs make good exercises in the basic definitions of this section.

Properties of pseudospectra

Theorem 2.4 *Let* $\mathbf{A} \in \mathbb{C}^{N \times N}$ *and* $\varepsilon > 0$ *be arbitrary.*

(i) $\sigma_\varepsilon(\mathbf{A})$ *is nonempty, open, and bounded, with at most N connected components, each containing one or more eigenvalues of $\mathbf{A}$.*

(ii) *If* $\|\cdot\| = \|\cdot\|_2$, *then* $\sigma_\varepsilon(\mathbf{A}^*) = \overline{\sigma_\varepsilon(\mathbf{A})}$.

(iii) *If* $\|\cdot\| = \|\cdot\|_2$, *then* $\sigma_\varepsilon(\mathbf{A}_1 \oplus \mathbf{A}_2) = \sigma_\varepsilon(\mathbf{A}_1) \cup \sigma_\varepsilon(\mathbf{A}_2)$.

(iv) *For any* $c \in \mathbb{C}$, $\sigma_\varepsilon(\mathbf{A} + c) = c + \sigma_\varepsilon(\mathbf{A})$.

(v) *For any nonzero* $c \in \mathbb{C}$, $\sigma_{|c|\varepsilon}(c\mathbf{A}) = c\sigma_\varepsilon(\mathbf{A})$.

In part (iii), $\mathbf{A}_1 \oplus \mathbf{A}_2$ denotes the direct sum of two square matrices $\mathbf{A}_1$ and $\mathbf{A}_2$, whose dimensions need not be equal; in other words it is the block diagonal matrix

$$\mathbf{A}_1 \oplus \mathbf{A}_2 = \begin{pmatrix} \mathbf{A}_1 & 0 \\ 0 & \mathbf{A}_2 \end{pmatrix}.$$

To prove the assertion about connected components in (i), one can use the fact that $\log \|(z - \mathbf{A})^{-1}\|$ is a subharmonic function and hence satisfies the maximum principle except at the eigenvalues of $\mathbf{A}$ (Theorem 4.2). Part (v) is also worth a comment, for although elementary, it is surprising at first: the ε-pseudospectrum of $2\mathbf{A}$ is not twice the ε-pseudospectrum of $\mathbf{A}$, but twice the $\varepsilon/2$-pseudospectrum of $\mathbf{A}$. Together parts (iv) and (v) describe the pseudospectra of linear functions of $\mathbf{A}$; Lui has established a mapping theorem for more general functions [521].

Where in the complex plane does a matrix $\mathbf{A}$ 'live'? If $\mathbf{A}$ is normal, then the spectrum $\sigma(\mathbf{A})$ is a satisfactory answer to that question for almost every purpose. The family of pseudospectra $\{\sigma_\varepsilon(\mathbf{A})\}$ is an attempt to provide a satisfactory answer in the case where $\mathbf{A}$ is not normal, perhaps far from normal. It is not a perfect answer; pseudospectra lack the simplicity of spectra, yet despite their complexity, they do not provide exact answers to the questions one would like to ask about the behavior of $\mathbf{A}$ (see §47). In this book we shall see that they do provide approximate answers, however, in the form of bounds that are often reasonably tight. (An overview of such bounds is given in §14.) Through the equivalence of definitions (2.1) and (2.3) they provide a reminder that eigenvalues that are sensitive to perturbations may be of limited significance in determining the behavior of $\mathbf{A}$. Finally, they provide an appealing geometric interpretation of non-normality. The fact is, nonnormal matrices and operators do *not* live in the complex plane, but one can get a good start in predicting their behavior if, in addition to the usual calculation of eigenvalues, one plots a few contour lines of the resolvent norm or the eigenvalues of a few randomly perturbed matrices. Beautiful plots can be obtained in seconds for matrices of dimensions in the hundreds with the MATLAB system EigTool [838].

Following the German *Eigenwert* and *Eigenvektor*, we find it convenient in informal work to use the abbreviations *ew* and *ev* for eigenvalue and eigenvector, ψew and ψev for pseudoeigenvalue and pseudoeigenvector. Those who find it worthy of remark that the word 'eigenvalue' is a blend of two languages may take pleasure in noting that 'pseudoeigenvalue', for better or worse, is a combination of three.

3 · A matrix example _____

Consider the tridiagonal Toeplitz matrix

$$
\mathbf{A} = \begin{pmatrix} 0 & 1 & & & \\ \frac{1}{4} & 0 & 1 & & \\ & \ddots & \ddots & \ddots & \\ & & \frac{1}{4} & 0 & 1 \\ & & & \frac{1}{4} & 0 \end{pmatrix} \in \mathbb{C}^{N \times N}. \tag{3.1}
$$

This matrix is nonsymmetric, but it can be symmetrized by the diagonal similarity transformation

$$
\mathbf{D} \mathbf{A} \mathbf{D}^{-1} = \mathbf{S} \tag{3.2}
$$

with $\mathbf{D} = \operatorname{diag}(2, 4, \ldots, 2^N)$ and

$$
\mathbf{S} = \begin{pmatrix} 0 & \frac{1}{2} & & & \\ \frac{1}{2} & 0 & \frac{1}{2} & & \\ & \ddots & \ddots & \ddots & \\ & & \frac{1}{2} & 0 & \frac{1}{2} \\ & & & \frac{1}{2} & 0 \end{pmatrix} \in \mathbb{C}^{N \times N}. \tag{3.3}
$$

It follows that the eigenvalues of $\mathbf{A}$ are the same as those of $\mathbf{S}$, namely

$$
\lambda_k(\mathbf{A}) = \lambda_k(\mathbf{S}) = \cos \frac{k\pi}{N+1}, \qquad 1 \leq k \leq N. \tag{3.4}
$$

Thus the spectrum of $\mathbf{A}$ consists of N distinct real numbers in the interval $(-1, 1)$.

The pseudospectra of $\mathbf{A}$, however, lie far from the real axis. For $N = 64$, Figure 3.1 plots the boundaries of $\sigma_\varepsilon(\mathbf{A})$ for $\varepsilon = 10^{-2}, 10^{-3}, \ldots, 10^{-8}$ with $\|\cdot\| = \|\cdot\|_2$, revealing wide oval-shaped regions in the complex plane. In fact, the ε-pseudospectrum of $\mathbf{A}$ is approximately equal to the region bounded by the ellipse that is the image of the circle $|z| = \varepsilon^{1/N}$ under the mapping

$$
f(z) = z^{-1} + \tfrac{1}{4} z, \tag{3.5}
$$

which is known as the *symbol* of $\mathbf{A}$. In §7 we make this statement precise by stating the following results, among others. For each z inside the ellipse $f(\mathbb{T})$, where $\mathbb{T} = \{z \in \mathbb{C} : |z| = 1\}$, the resolvent norm $\|(z - \mathbf{A})^{-1}\|$ grows exponentially as $N \to \infty$. On the other hand, for each z outside $f(\mathbb{T})$, $\|(z - \mathbf{A})^{-1}\|$ is bounded uniformly with respect to N.

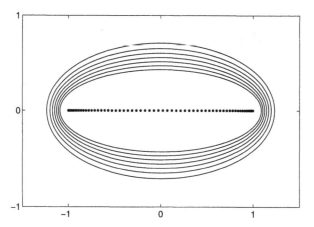

Figure 3.1: Boundaries of pseudospectra $\sigma_\varepsilon(\mathbf{A})$, $\varepsilon = 10^{-2}, 10^{-3}, \ldots, 10^{-8}$, for the matrix (3.1) of dimension $N = 64$. The eigenvalues are marked by solid dots.

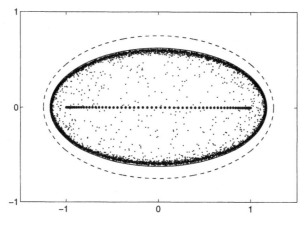

Figure 3.2: Superposition of eigenvalues of 100 matrices $\mathbf{A} + \mathbf{E}$, where $\mathbf{A}$ is the tridiagonal Toeplitz matrix (3.1) of dimension $N = 64$ and each $\mathbf{E}$ is a random matrix with $\|\mathbf{E}\| = 10^{-3}$. The eigenvalues of $\mathbf{A}$ are real (larger dots), but the perturbation introduced by $\mathbf{E}$ moves them far into the complex plane, close to the ellipse defined by (3.5) with $|z| = (.001)^{1/64}$ (solid curve). The dashed ellipse corresponds to $N \to \infty$ and $|z| = 1$.

Figure 3.2 illustrates the pseudospectra of $\mathbf{A}$ in another way by presenting eigenvalues of randomly perturbed matrices. The figure shows the eigenvalues (3.4) in the complex plane as solid dots. Superimposed as smaller dots on the same plot are 6400 10^{-3}-pseudoeigenvalues: the eigenvalues of 100 matrices $\mathbf{A} + \mathbf{E}$, where each $\mathbf{E}$ is a random matrix with $\|\mathbf{E}\| = 10^{-3}$. The connection with ellipses is again obvious.

These pictures change quantitatively but not qualitatively if N is varied. To illustrate this, Figure 3.3 shows nine sets of dots corresponding to the same experiment as in Figure 3.2, but for $N = 16$, 32, 64 and $\|\mathbf{E}\| = 10^{-2}$, 10^{-3}, 10^{-4}. Each tile of this figure depicts a superposition of eigenvalues of $640/N$ randomly perturbed matrices $\mathbf{A} + \mathbf{E}$—640 dots altogether. The sensitivity of the eigenvalues becomes more pronounced as N increases, but qualitatively, all nine pictures are much the same.

To appreciate how such small perturbations to $\mathbf{A}$ can move eigenvalues so dramatically, consider a modification to the $(N, 1)$ entry of $\mathbf{A}$:

$$\mathbf{A} + \mathbf{E} = \begin{pmatrix} 0 & 1 & & & \\ \frac{1}{4} & 0 & \ddots & & \\ & \ddots & \ddots & 1 & \\ \varepsilon & & \frac{1}{4} & 0 \end{pmatrix}.$$

Now apply the similarity transformation that symmetrized $\mathbf{A}$ to obtain

$$\mathbf{D}(\mathbf{A} + \mathbf{E})\mathbf{D}^{-1} = \begin{pmatrix} 0 & \frac{1}{2} & & & \\ \frac{1}{2} & 0 & \ddots & & \\ & \ddots & \ddots & \frac{1}{2} & \\ 2^{N-1}\varepsilon & & \frac{1}{2} & 0 \end{pmatrix}.$$

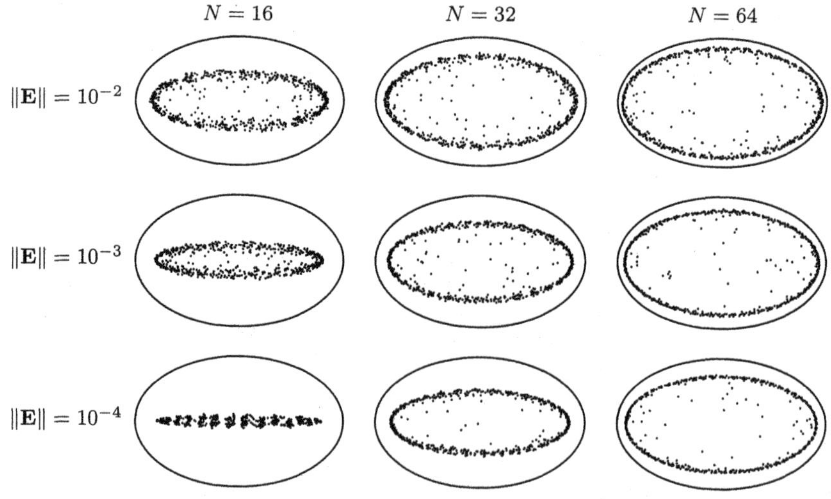

Figure 3.3: Nine plots as in Figure 3.2 corresponding to $N = 16$, 32, 64 and $\|\mathbf{E}\| = 10^{-2}, 10^{-3}, 10^{-4}$. Each plot shows eigenvalues of $640/N$ matrices $\mathbf{A} + \mathbf{E}$, i.e., 640 dots. The ellipse is the image of the unit circle under the symbol $f(z) = z^{-1} + \frac{1}{4}z$.

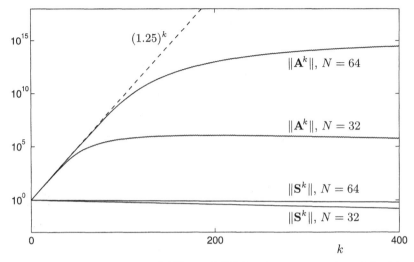

Figure 3.4: Norms of powers $\|\mathbf{A}^k\|$ and $\|\mathbf{S}^k\|$ for the matrices $\mathbf{A}$ and $\mathbf{S}$ of (3.1) and (3.3), dimensions $N = 32$ and 64. Since $\mathbf{A}$ and $\mathbf{S}$ are similar for each N, the curves approach zero as $k \to \infty$ with equal asymptotic slopes determined by the spectral radius. For finite k, however, $\|\mathbf{A}^k\|$ and $\|\mathbf{S}^k\|$ are very different, and in particular, the powers $\|\mathbf{A}^k\|$ are not bounded uniformly with respect to N. Note the logarithmic scale. Compare Figures 19.1 and 20.5.

Just as $\mathbf{A}$ and $\mathbf{S}$ have the same eigenvalues, so the spectrum of $\mathbf{A} + \mathbf{E}$ matches that of $\mathbf{D}(\mathbf{A} + \mathbf{E})\mathbf{D}^{-1}$. Thus the eigenvalues of $\mathbf{A} + \mathbf{E}$ correspond to those of an exponentially large perturbation of a symmetric matrix. In this light, it might seem remarkable that the eigenvalues of $\mathbf{A}$ move so little!

The theme of this book is that pseudospectra may reveal more than spectra about certain aspects of the behavior of matrices and operators. As an illustration of this principle for the present example, suppose we want to predict the norms of the powers $\mathbf{A}^k$ for various values of k. The traditional approach to this problem is to consider eigenvalues. By (3.4), the spectral radii of $\mathbf{A}$ and $\mathbf{S}$ are equal,

$$\rho(\mathbf{A}) = \rho(\mathbf{S}) = \cos \frac{\pi}{N+1}, \tag{3.6}$$

and since this quantity is less than 1, both $\mathbf{S}$ and $\mathbf{A}$ must be power-bounded:

$$\|\mathbf{S}^k\| \le C_S, \quad \|\mathbf{A}^k\| \le C_A \qquad \forall k \ge 0, \tag{3.7}$$

with $\|\mathbf{S}^k\| \to 0$ and $\|\mathbf{A}^k\| \to 0$ as $k \to \infty$. Figure 3.4, however, shows that although these statements are true, they are only a part of the truth. In actuality, $\|\mathbf{A}^k\|$ and $\|\mathbf{S}^k\|$ bear little resemblance to one another. Whereas the powers $\mathbf{S}^k$ decrease smoothly, so that (3.7) holds with $C_S = 1$, the powers $\mathbf{A}^k$ grow exponentially for $k < N$ and achieve huge norms. Obviously, although the matrices $\mathbf{A}$ are power-bounded for each dimension N, they are not uniformly power-bounded. Even for fixed N, if $\|\mathbf{A}^k\|$ becomes as great as 10^{15} for certain values of k, it is doubtful whether the power-boundedness of $\mathbf{A}$ has much practical meaning.

On the other hand, Figure 3.4 reveals that the rate of growth of $\|\mathbf{A}^k\|$ for $k < N$ is very close to $(1.25)^k$. This number 1.25 is the largest absolute value of the points on the dashed ellipse in Figure 3.2. Evidently in this example the slope of the $\|\mathbf{A}^k\|$ curve for modest values of k can be accurately predicted by considering a *pseudospectral radius*. We give details about such predictions in §§14 and 16; in particular, see Theorem 16.5. Figure 14.5 shows a similar example involving a more complicated Toeplitz matrix taken from [80].

4 · Pseudospectra of linear operators _____

All the ideas of the last two sections can be generalized to linear operators acting in infinite-dimensional spaces. The mathematics of this generalization is beautiful, but inevitably more technical than in the matrix case. In this section we set down the fundamentals of the theory of pseudospectra for linear operators in Banach space. Fortunately, the technical details do not matter for many applications, and in much of this book we are able to use a language closer to linear algebra than functional analysis.

Any researcher in the field of nonnormal operators will be aware of Kato's book *Perturbation Theory for Linear Operators*, whose second edition was published in 1976 [448]. This magnificent treatise covers almost all the aspects of functional analysis and spectral theory that we need here, and does so in a lucid style supported by hundreds of examples. For the present section, where many results are stated quickly, it may be helpful to the reader to have references to appropriate locations in Kato's book. In the next seven pages, accordingly, each page number listed in double square brackets is to be interpreted as a pointer to [448], where a proof or additional insight may be found. Other major general references include [221, 323, 395, 606, 630].

Let X be a complex Banach space, that is, a complete normed vector space over the complex field $\mathbb{C}$, with norm $\|\cdot\|$. We shall consider linear operators mapping X into itself. Such an operator $\mathbf{A}$ has a domain denoted by $\mathcal{D}(\mathbf{A}) \subseteq X$, which may or may not be all of X. We denote by $\mathcal{B}(X)$ the set of bounded operators on X; for $\mathbf{A} \in \mathcal{B}(X)$, we assume without loss of generality $\mathcal{D}(\mathbf{A}) = X$. We denote by $\mathcal{C}(X)$ the set of closed operators on X. (An operator $\mathbf{A}$ is *closed* provided that if $\{u_k\}$ is a sequence in $\mathcal{D}(\mathbf{A})$ converging to a limit $u \in X$ and if $\{\mathbf{A}u_k\}$ converges to a limit $v \in X$, then $u \in \mathcal{D}(\mathbf{A})$ and $\mathbf{A}u = v$.) An unbounded closed operator will necessarily have $\mathcal{D}(\mathbf{A}) \neq X$ [[p. 166]], although many such operators are *densely defined*, meaning that the closure of $\mathcal{D}(\mathbf{A})$ is X. Throughout this book we deal only with closed operators; bounded operators and matrices are special cases.

In the next section we consider an example in which $\mathbf{A}$ is a first derivative operator on an interval $[0, d]$. The Banach space is $L^2[0, d]$, and $\mathcal{D}(\mathbf{A})$ is the set of absolutely continuous functions on $[0, d]$ that satisfy the boundary condition $u(d) = 0$. (An absolutely continuous function is one that is the indefinite integral of a function that is locally Lebesgue measurable.) In this case $\mathcal{D}(\mathbf{A})$ is densely defined, because any function in $L^2[0, d]$ can be approximated in the L^2-norm by absolutely continuous functions. The same would be true in any space $L^p[0, d]$ with $1 \leq p < \infty$. In $L^\infty[0, d]$, however, it is not possible to approximate a step function, say, by absolutely continuous functions, and so this operator is not densely defined in $L^\infty[0,d]$ [[p. 145]].

For $\mathbf{A} \in \mathcal{C}(X)$ and $\mathbf{E} \in \mathcal{B}(X)$, $\mathbf{A} + \mathbf{E}$ is also in $\mathcal{C}(X)$, with domain $\mathcal{D}(\mathbf{A} + \mathbf{E}) = \mathcal{D}(\mathbf{A})$. A special case is the situation where $\mathbf{E}$ is a multiple of the identity. Thus perturbations of closed operators by bounded operators or shifts introduce no technical difficulties [[p. 164]].

Given $\mathbf{A} \in \mathcal{C}(X)$, a *bounded inverse* is an operator $\mathbf{A}^{-1} \in \mathcal{B}(X)$ such that $\mathbf{A}\mathbf{A}^{-1}$ is the identity on X and $\mathbf{A}^{-1}\mathbf{A}$ is the identity on $\mathcal{D}(\mathbf{A})$. This is the only kind of inverse we shall be concerned with, and when an expression like $\mathbf{A}^{-1}$ or $(z - \mathbf{A})^{-1}$ appears in this book, it should always be interpreted as a bounded inverse, defined on all of X.

The following theorem is the basis of all the developments of this section. It is the 'essence of the matter' of the theory of pseudospectra.

Invertibility and perturbation of closed operators

Theorem 4.1 *Suppose* $\mathbf{A} \in \mathcal{C}(X)$ *has a bounded inverse* $\mathbf{A}^{-1}$. *Then for any* $\mathbf{E} \in \mathcal{B}(X)$ *with* $\|\mathbf{E}\| < 1/\|\mathbf{A}^{-1}\|$, $\mathbf{A} + \mathbf{E}$ *has a bounded inverse* $(\mathbf{A} + \mathbf{E})^{-1}$ *satisfying*

$$\|(\mathbf{A} + \mathbf{E})^{-1}\| \leq \frac{\|\mathbf{A}^{-1}\|}{1 - \|\mathbf{E}\|\|\mathbf{A}^{-1}\|}. \tag{4.1}$$

Conversely, for any $\mu > 1/\|\mathbf{A}^{-1}\|$, *there exists* $\mathbf{E} \in \mathcal{B}(X)$ *with* $\|\mathbf{E}\| < \mu$ *such that* $(\mathbf{A} + \mathbf{E})u = 0$ *for some nonzero* $u \in X$.

Proof. Kato calls the first assertion the 'stability of bounded invertibility' [[p. 196]]. Given $\mathbf{E} \in \mathcal{B}(X)$ with $\|\mathbf{E}\| \leq 1/\|\mathbf{A}^{-1}\|$, we wish to establish the invertibility of $\mathbf{A} + \mathbf{E} = (\mathbf{I} + \mathbf{E}\mathbf{A}^{-1})\mathbf{A}$. Since $\|\mathbf{E}\mathbf{A}^{-1}\| \leq \|\mathbf{E}\|\|\mathbf{A}^{-1}\| < 1$, the Neumann series $\sum_{k=0}^{\infty}(-\mathbf{E}\mathbf{A}^{-1})^k$ converges and equals $(\mathbf{I} + \mathbf{E}\mathbf{A}^{-1})^{-1}$. Summing the norms of the terms of the Neumann series leads to the bound $\|(\mathbf{I} + \mathbf{E}\mathbf{A}^{-1})^{-1}\| \leq 1/(1 - \|\mathbf{E}\|\|\mathbf{A}^{-1}\|)$. It follows that $(\mathbf{A} + \mathbf{E})^{-1} = \mathbf{A}^{-1}(\mathbf{I} + \mathbf{E}\mathbf{A}^{-1})^{-1}$, which implies (4.1).

For the converse, following the same line of reasoning as in the proof of Theorem 2.1, we can argue as follows. By the definition of $\|\mathbf{A}^{-1}\|$, there exists $u \in X$ with $\|u\| = 1$ such that $v = \mathbf{A}u$ satisfies $\|v\| < \mu$. Now take $\mathbf{E} \in \mathcal{B}(X)$ to be an operator that maps u to $-v$ and has norm $\|v\|$; the existence of such an operator is ensured by the Hahn–Banach theorem. ∎

The theory of resolvents, spectra, and pseudospectra is derived by applying Theorem 4.1 to shifted operators $z - \mathbf{A}$, where z is a complex constant. Given $\mathbf{A} \in \mathcal{C}(X)$ and $z \in \mathbb{C}$, the *resolvent* of $\mathbf{A}$ at z is the operator $(z - \mathbf{A})^{-1} \in \mathcal{B}(X)$, if this exists. The *resolvent set* $\varrho(\mathbf{A})$ is the set of numbers $z \in \mathbb{C}$ for which $(z - \mathbf{A})^{-1}$ exists. Theorem 4.1 implies that $\varrho(\mathbf{A})$ is open. In fact, $(z - \mathbf{A})^{-1}$ is an analytic function of $z \in \varrho(\mathbf{A})$ [[p. 174]], which implies that $\|(z - \mathbf{A})^{-1}\|$ is an unbounded continuous subharmonic function of $z \in \varrho(\mathbf{A})$ and that it satisfies the maximum principle for $z \in \varrho(\mathbf{A})$ [395, Thm. 3.13.1].

The *spectrum* of $\mathbf{A} \in \mathcal{C}(X)$ is the complement of the resolvent set in the complex plane: $\sigma(\mathbf{A}) = \mathbb{C} \backslash \varrho(\mathbf{A})$. Since $\varrho(\mathbf{A})$ is open, $\sigma(\mathbf{A})$ is closed. For $\mathbf{A} \in \mathcal{B}(X)$, $\sigma(\mathbf{A})$ is bounded and nonempty. For $\mathbf{A} \in \mathcal{C}(X)$, $\sigma(\mathbf{A})$ may be unbounded or empty. For example, the operator $\mathbf{A} : u \mapsto u'$ in $L^2[0, 1]$ has an empty spectrum if $\mathcal{D}(\mathbf{A})$ is the set of absolutely continuous functions on $[0, 1]$ satisfying $u(1) = 0$. Without this boundary condition, the spectrum is the entire complex plane [[p. 174]].

If $\mathbf{A}u = \lambda u$ for some nonzero $u \in X$ and $\mathbf{A} \in \mathcal{C}(X)$, then u and λ are an *eigenvector* and *eigenvalue* of $\mathbf{A}$. The spectrum $\sigma(\mathbf{A})$ contains all the eigenvalues of $\mathbf{A}$, but it may be bigger than this. For example, let $\mathbf{A}$ be the left shift operator on ℓ^2, defined by

$$\mathbf{A}(u_1, u_2, \ldots) = (u_2, u_3, \ldots).$$

For any z in the open unit disk, i.e., with $|z| < 1$, the vector defined by $u_j = z^j$ is an eigenvector with eigenvalue z, and the spectrum of $\mathbf{A}$ is the closed unit disk. By contrast, the right shift operator in ℓ^2, defined by

$$\mathbf{A}(u_1, u_2, \ldots) = (0, u_1, u_2, \ldots),$$

also has spectrum equal to the closed unit disk, but it has no eigenvalues [[p. 176]].

For another example, consider the operator $\mathbf{A} : u \mapsto u'$ in $L^2[0, 1]$ with $\mathcal{D}(\mathbf{A})$ taken as the set of absolutely continuous functions on $[0, 1]$ with $u(0) = u(1) = 0$. In this case with 'too many boundary conditions', although $z - \mathbf{A}$ has an inverse, it is not densely defined. The reason is that as u ranges over $\mathcal{D}(\mathbf{A})$, the set of vectors $v = (z - \mathbf{A})u$ ranges only over a subset of functions in X satisfying the constraint $\int_0^1 e^{-zx} v(x) = 0$. To see this, given u and z, define $w \in \mathcal{D}(\mathbf{A})$ by $w(x) = -e^{-zx} u(x)$. Then $w' = e^{-zx}(zu - u') = e^{-zx} v$, and the integral of this function from 0 to 1 must be $w(1) - w(0) = 0$. We conclude that although $\mathbf{A}$ has no eigenvalues, its spectrum is all of $\mathbb{C}$ [[p. 174]].

Theorem 4.1 implies that for any $\mathbf{A} \in \mathcal{C}(X)$ and $z \in \varrho(\mathbf{A})$, we have $\|(z - \mathbf{A})^{-1}\| \geq 1/\text{dist}(z, \sigma(\mathbf{A}))$. (Proof: $z \in \sigma(\mathbf{A} + \mathbf{E})$ if $\mathbf{E}$ is the identity times the constant $e^{i\theta} \text{dist}(z, \sigma(\mathbf{A}))\mathbf{I}$ for some $\theta \in [0, 2\pi)$.) Thus $\|(z - \mathbf{A})^{-1}\|$ approaches ∞ as z approaches the spectrum. We now introduce a notational convention used throughout this book:

Convention: If $z \in \sigma(\mathbf{A})$, we write $\|(z - \mathbf{A})^{-1}\| = \infty$.

Thus for $z \in \sigma(\mathbf{A})$, we shall use the notation $\|(z - \mathbf{A})^{-1}\|$ even though $(z - \mathbf{A})^{-1}$ itself does not exist. If $\mathbf{A}$ is a matrix on a finite-dimensional space, this usage is very natural since any $z \in \sigma(\mathbf{A})$ must be an eigenvalue, so $(z - \mathbf{A})^{-1}$ is indeed 'infinite'. For operators, it is more artificial, since there may be points in the spectrum that are not eigenvalues.

This convention gives us the ability to derive a wide range of results about resolvents and perturbations from a single fact: $\|(z - \mathbf{A})^{-1}\|$ is a continuous function from the entire complex plane to $(0, \infty]$. The following results are implied by Theorem 4.1 and the other observations above.

Norm of the resolvent

Theorem 4.2 *Given* $\mathbf{A} \in \mathcal{C}(X)$, *and with* $\|(z - \mathbf{A})^{-1}\|$ *defined as* ∞ *for* $z \in \sigma(\mathbf{A})$, *the norm of resolvent* $\|(z - \mathbf{A})^{-1}\|$ *is a function from* $z \in \mathbb{C}$ *to* $(0, \infty]$ *with the following properties. It is continuous and unbounded and takes the value* ∞ *precisely on* $\sigma(\mathbf{A})$. *For* $z \notin \sigma(\mathbf{A})$ *it is subharmonic and satisfies the maximum principle as well as the bound*

$$\|(z - \mathbf{A})^{-1}\| \geq \frac{1}{\mathrm{dist}(z, \sigma(\mathbf{A}))}. \tag{4.2}$$

If $z \notin \sigma(\mathbf{A})$, *then* $z \notin \sigma(\mathbf{A} + \mathbf{E})$ *for any* $\mathbf{E} \in \mathcal{B}(X)$ *that satisfies* $\|\mathbf{E}\| \leq 1/\|(z - \mathbf{A})^{-1}\|$; *conversely, for any* $\mu > \|(z - \mathbf{A})^{-1}\|^{-1}$, *there exists* $\mathbf{E} \in \mathcal{B}(X)$ *with* $\|\mathbf{E}\| < \mu$ *such that* $(\mathbf{A} + \mathbf{E})u = zu$ *for some nonzero* $u \in X$.

According to Theorem 4.2, every z that is not in $\sigma(\mathbf{A})$ is also not in $\sigma(\mathbf{A}+\mathbf{E})$ for sufficiently small $\|\mathbf{E}\|$. Thus, loosely speaking, an infinitesimal perturbation of $\mathbf{A}$ can enlarge $\sigma(\mathbf{A})$ only infinitesimally. This principle is known as 'upper-semicontinuity of the spectrum' [[p. 208]]. By contrast, the spectrum may be lower-semidiscontinuous in the sense that an infinitesimal perturbation can shrink $\sigma(\mathbf{A})$ finitely. For example, let $\mathbf{A}$ be the doubly infinite matrix acting in $\ell^2(\mathbb{Z})$ with $a_{j,j+1} = 1$ for each j and other entries equal to zero, except that $a_{0,1}$ is equal to zero too. This operator is a kind of 'infinite Jordan block', and its spectrum is the closed unit disk. Now suppose $a_{0,1}$ is changed to any nonzero number, however small. The spectrum shrinks to the unit circle [[p. 210]].

In §2 we defined the ε-pseudospectrum of a matrix $\mathbf{A}$ in three equivalent ways. The same three definitions apply for linear operators in Banach space. Statements of aspects of this equivalence in various contexts can be found in a number of publications, including [78, 93, 134, 298, 375, 575, 773, 796, 835]. The closest to the formulation we give here appears in the unpublished technical report of Chaitin-Chatelin and Harrabi [134].

As always, we shall follow the convention that $\|(z - \mathbf{A})^{-1}\| = \infty$ for $z \in \sigma(\mathbf{A})$.

Three equivalent definitions of pseudospectra

Let $\mathbf{A} \in \mathcal{C}(X)$ and $\varepsilon > 0$ be arbitrary. The ε-*pseudospectrum* $\sigma_\varepsilon(\mathbf{A})$ of $\mathbf{A}$ is the set of $z \in \mathbb{C}$ defined equivalently by any of the conditions

$$\|(z - \mathbf{A})^{-1}\| > \varepsilon^{-1}, \tag{4.3}$$

$$z \in \sigma(\mathbf{A} + \mathbf{E}) \text{ for some } \mathbf{E} \in \mathcal{B}(X) \text{ with } \|\mathbf{E}\| < \varepsilon, \tag{4.4}$$

$$z \in \sigma(\mathbf{A}) \text{ } or \text{ } \|(z - \mathbf{A})u\| < \varepsilon \text{ for some } u \in \mathcal{D}(\mathbf{A}) \text{ with } \|u\| = 1. \tag{4.5}$$

If $\|(z - \mathbf{A})u\| < \varepsilon$ as in (4.5), then z is an ε-*pseudoeigenvalue* of $\mathbf{A}$ and u is a corresponding ε-*pseudoeigenvector* (or *pseudoeigenfunction* or *pseudomode*).

From the material above we obtain the following collection of facts about pseudospectra in Banach space. Figure 4.1 gives a schematic view.

Properties of pseudospectra

Theorem 4.3 *Given* $\mathbf{A} \in \mathcal{C}(X)$, *the pseudospectra* $\{\sigma_\varepsilon(\mathbf{A})\}_{\varepsilon>0}$ *have the following properties. They can be defined equivalently by any of the conditions* (4.3)–(4.5). *Each* $\sigma_\varepsilon(\mathbf{A})$ *is a nonempty open subset of* $\mathbb{C}$, *and any bounded connected component of* $\sigma_\varepsilon(\mathbf{A})$ *has a nonempty intersection with* $\sigma(\mathbf{A})$. *The pseudospectra are strictly nested supersets of the spectrum:* $\cap_{\varepsilon>0}\sigma_\varepsilon(\mathbf{A}) = \sigma(\mathbf{A})$, *and conversely, for any* $\delta > 0$, $\sigma_{\varepsilon+\delta}(\mathbf{A}) \supseteq \sigma_\varepsilon(\mathbf{A}) + \Delta_\delta$, *where* Δ_δ *is the open disk of radius* δ.

Proof. The equivalence of the three definitions follows from Theorem 4.2. The condition $\sup_{z\in\varrho(\mathbf{A})} \|(z-\mathbf{A})^{-1}\| = \infty$ implies that $\sigma_\varepsilon(\mathbf{A})$ is nonempty, and the maximum principle for $\|(z - \mathbf{A})^{-1}\|$ implies that any bounded component of $\sigma_\varepsilon(\mathbf{A})$ intersects the spectrum. (Thus if $\sigma(\mathbf{A})$ is empty, $\sigma_\varepsilon(\mathbf{A})$ is unbounded for *all* $\varepsilon > 0$.) The statement about $\cap \sigma_\varepsilon(\mathbf{A})$ follows from the upper-semicontinuity of the spectrum. ∎

We have developed the fundamental properties of pseudospectra in Banach spaces without any need to use adjoint operators. In applications, however, adjoints are of considerable interest, and we now record some of the main facts.

Given a Banach space X, the *adjoint space* or *dual space* of X is the set X^* of all bounded conjugate-linear functionals on X, i.e., linear functions $f : X \to \mathbb{C}$ with the property $f(\alpha u) = \overline{\alpha} f(u)$ for any $u \in X$ and $\alpha \in \mathbb{C}$. Like X, X^* is a Banach space ⟦p. 134⟧. We usually write $f(u)$ in the inner product notation (f, u). For any $u \in X$, the norm of u satisfies the identity $\|u\| = \sup_{f\in X^*, \|f\|=1} |(f, u)|$, and there exists an $f \in X^*$ for which this supremum is attained ⟦p. 135⟧.

For example, for $1 \leq p < \infty$ and with q defined by $p^{-1} + q^{-1} = 1$, the

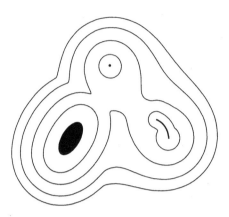

Figure 4.1: Schematic view of the geometry of pseudospectra for an operator in Banach space. The black region and the thick arc indicate components of the spectrum that are not just points. For z in the interior of an open region of spectrum, though z is in the ε-pseudospectrum for any $\varepsilon > 0$, eigenvectors and ε-pseudoeigenvectors may or may not exist. This sketch corresponds to a bounded operator. For an unbounded operator, though the spectrum may still be bounded, all the pseudospectra will be unbounded.

adjoint space of ℓ^p is (or, more precisely, can be identified with) ℓ^q, and the adjoint space of $L^p(E)$, where E is a compact subset of $\mathbb{C}$, is $L^q(E)$. Thus, for example, any linear functional on ℓ^1 can be interpreted as an inner product with a vector in ℓ^∞.

Given $\mathbf{A} \in \mathcal{C}(X)$, an *adjoint operator* of $\mathbf{A}$ is an operator $\mathbf{A}^* \in \mathcal{C}(X^*)$ with the property that $(f, \mathbf{A}u) = (\mathbf{A}^*f, u)$ for each $f \in \mathcal{D}(\mathbf{A}^*)$, $u \in \mathcal{D}(\mathbf{A})$. Every $\mathbf{A}$ has an adjoint, and if $\mathbf{A}$ is densely defined, then $\mathbf{A}^*$ is unique (assuming its domain is taken to be as large as possible) and closed [[p. 167]]. The spectrum of $\mathbf{A}^*$ is the complex conjugate of the spectrum of $\mathbf{A}$ [[p. 184]].

For any $\mathbf{A} \in \mathcal{B}(X)$, $\|\mathbf{A}\| = \|\mathbf{A}^*\|$ [[p. 154]]. It can also be shown that if $\mathbf{A} \in \mathcal{C}(X)$ has a bounded inverse $\mathbf{A}^{-1}$, then the same is true of $\mathbf{A}^*$, with $(\mathbf{A}^*)^{-1} = (\mathbf{A}^{-1})^*$ [[p. 169]]; it is usual to call this operator $\mathbf{A}^{-*}$. Together these facts imply that for any $\mathbf{A} \in \mathcal{C}(X)$ with a bounded inverse, $\|\mathbf{A}^{-1}\| = \|\mathbf{A}^{-*}\|$.

These apparently modest results have substantial consequences for pseudospectra of adjoints of closed operators.

Pseudospectra of the adjoint

Theorem 4.4 *For any* $\mathbf{A} \in \mathcal{C}(X)$, $z \in \mathbb{C}$, *and* $\varepsilon > 0$, *we have* $\|(\overline{z} - \mathbf{A}^*)^{-1}\| = \|(z - \mathbf{A})^{-1}\|$, $\sigma(\mathbf{A}^*) = \overline{\sigma(\mathbf{A})}$, *and* $\sigma_\varepsilon(\mathbf{A}^*) = \overline{\sigma_\varepsilon(\mathbf{A})}$. *Moreover, suppose* $\mathbf{A}$ *has an* ε-*pseudoeigenvector* $u \in \mathcal{D}(\mathbf{A})$ *corresponding to the* ε-*pseudoeigenvalue* z. *If* $z \notin \sigma(\mathbf{A})$, *then* $\mathbf{A}^*$ *has an* ε-*pseudoeigenvector* $f \in \mathcal{D}(\mathbf{A}^*)$ *corresponding to the* ε-*pseudoeigenvalue* $\overline{z}$.

In applications involving differential operators, the definition of the adjoint requires attention to boundary conditions. For example, consider the operator $\mathbf{A} : u \mapsto u''$ in $L^2[0,1]$ with $\mathcal{D}(\mathbf{A})$ equal to the set of functions in $L^2[0,1]$ with an absolutely continuous derivative [[p. 148]]. The adjoint is a second derivative operator with four boundary conditions: $\mathcal{D}(\mathbf{A}^*)$ is the set of functions in $L^2[0,1]$ with an absolutely continuous derivative and $u(0) = u'(0) = u(1) = u'(1) = 0$.

We must issue a warning about pseudoeigenvectors: The assumption $z \notin \sigma(\mathbf{A})$ at the end of Theorem 4.4 cannot be dispensed with. If z is in the ε-pseudospectrum of $\mathbf{A}$ but not the spectrum, then $\mathbf{A}$ and $\mathbf{A}^*$ must have ε-pseudoeigenvectors for z and $\bar{z}$, respectively. For $z \in \sigma(\mathbf{A})$, however, this need not be so; see Theorem 11.3.

One of the important applications of adjoints is to the definition of the *numerical range* of an operator $\mathbf{A} \in \mathcal{C}(X)$. The largest real part of the numerical range, known as the *numerical abscissa*, determines the initial growth rate of an evolution process (semigroup) $e^{t\mathbf{A}}$. This is a central topic in the study of time-dependent dynamical systems, and details are given in §§14 and 17.

In this section we have considered operators in a Banach space, not the more specialized case of a Hilbert space, i.e., a Banach space where the norm is derived from an inner product. The reason is that many of the fundamentals of pseudospectra are the same in both cases. However, some results are restricted to Hilbert space. One example is the Gearhart–Prüss theorem concerning growth bounds for semigroups, discussed in §19. Another may be the fact that in Hilbert space, $\|(z - \mathbf{A})^{-1}\|$ can never take a constant finite value on an open set. This theorem, conjectured by Böttcher and proved by Daniluk, is reported as Proposition 6.1 in [79] and has been generalized to L^p spaces ($1 < p < \infty$) [90, Thm. 5.1] and to Banach spaces of finite dimension [375], but it is not known if it is valid in infinite-dimensional Banach spaces. If it is not, then it follows that there exist examples of operators whose pseudospectra 'jump' for particular values of ε. This concern, discussed in [86] and [134], is one of the reasons we chose to formulate the definitions (4.3)–(4.4) in terms of strict inequalities. If pseudospectra can jump, the analogous definitions based on weak inequalities would not be equivalent.

In §2, having defined pseudospectra for general norms, we moved on to questions of bases, diagonalizability, condition numbers, and singular values. All of these matters have analogues for operators, and some of them are taken up in §§51 and 52.

5 · An operator example

Figure 5.1 shows the pseudospectra of a highly nonnormal differential operator, a Schrödinger operator for a harmonic potential acting in $L^2(\mathbb{R})$. Unlike the usual Schrödinger operator, this one has a potential that is complex rather than real. The operator is defined by

$$\mathbf{A}u = -\frac{\mathrm{d}^2 u}{\mathrm{d}x^2} + ix^2 u, \qquad x \in \mathbb{R}, \tag{5.1}$$

acting in $L^2(\mathbb{R})$ with domain $\mathcal{D}(\mathbf{A})$ equal to the set of $L^2(\mathbb{R})$ functions with an absolutely continuous derivative. The figure reveals that, like the matrix of the last section, this operator deviates strongly from normality, a discovery due to E. B. Davies [179, 180]. In fact, the norm of the resolvent grows exponentially as one moves out into the complex plane along any ray at angle θ from the real axis with $0 < \theta < \pi/2$. In §§11 and 13 we shall discuss examples of this kind and their connections with fundamental issues of the theory of partial differential equations.

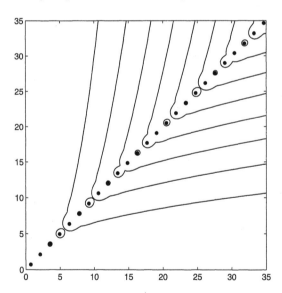

Figure 5.1: Spectrum and ε-pseudospectra of Davies' complex harmonic oscillator (5.1). From outside in, the curves correspond to $\varepsilon = 10^{-1}, 10^{-2}, \ldots, 10^{-8}$. The resolvent norm grows exponentially as $z \to \infty$ along rays in the complex plane satisfying $0 < \theta < \pi/2$. Consequently, this operator's higher eigenvalues would be of limited physical significance in applications. (This figure is based on a spectral discretization resulting in a 200×200 matrix.)

Rather than pursue this relatively complicated example here, we shall study a simpler example, already mentioned in the last section, that we can explain more easily.[1] Consider the first derivative operator

$$\mathbf{A}u = u' = \frac{\mathrm{d}u}{\mathrm{d}x} \tag{5.2}$$

in the space $L^2(0, d)$, subject to the boundary condition

$$u(d) = 0. \tag{5.3}$$

To be precise, $\mathbf{A}$ is the differentiation operator (5.2) with domain equal to the set of absolutely continuous functions $u \in L^2(0, d)$ that satisfy (5.3) (cf. Example III.2.7 of [448]).

The spectrum of $\mathbf{A}$ is empty: $\sigma(\mathbf{A}) = \emptyset$. Intuitively one sees this by noting that an eigenfunction would have to be of the form e^{zx} for some $z \in \mathbb{C}$, but since no such functions satisfy the boundary condition, there are no eigenfunctions. A proof can be obtained by showing that the resolvent $(z - \mathbf{A})^{-1}$ exists as a bounded operator for any $z \in \mathbb{C}$; it is given by

$$(z - \mathbf{A})^{-1}v(x) = \int_x^d \mathrm{e}^{z(x-s)}v(s)\,\mathrm{d}s. \tag{5.4}$$

This formula can be derived by the method of variation of parameters applied to the ordinary differential equation $zu - u' = v$. It can also be interpreted as the integral of $v(x)$ times the Green's function for the solution to $zu - u' = \delta(x)$, where $\delta(x)$ denotes the Dirac delta function.

The pseudospectra of $\mathbf{A}$, however, are another matter. It follows from (5.4) that although the resolvent norm $\|(z-\mathbf{A})^{-1}\|$ is finite for every z, it is enormous when z is well inside the left half-plane, growing exponentially as a function of $\exp(-d\,\mathrm{Re}\,z)$. It can also be seen from (5.4) that $\|(z-\mathbf{A})^{-1}\|$ depends only on $\mathrm{Re}\,z$, not $\mathrm{Im}\,z$. (Proof: For any $z \in \mathbb{C}$, $v(s)$, and $\alpha \in \mathbb{R}$, the pairs z, $v(s)$ and $z + \mathrm{i}\alpha$, $\mathrm{e}^{\mathrm{i}\alpha s}v(s)$ lead to the same norm of the integral in (5.4).) Therefore for each ε, $\sigma_\varepsilon(\mathbf{A})$ is equal to the half-plane lying to the left of some line $\mathrm{Re}\,z = c_\varepsilon$ in the complex plane. This situation is illustrated in Figure 5.2, where the striking thing to note is the rapid decrease of ε as one moves into the left half-plane.

Why does $\mathbf{A}$ have a huge resolvent norm in the left half-plane? One explanation is suggested by Figure 5.3. The function

$$u(x) = \mathrm{e}^{zx}, \qquad z \in \mathbb{C}$$

does not satisfy (5.3), but for $\mathrm{Re}\,z \ll 0$, it almost does. Thus u is not an eigenfunction of $\mathbf{A}$, nor near to any eigenfunction, but it is 'nearly an

---

[1] For discussions of the same example from other points of view, see [395, p. 537], [448, p. 174], and [606, p. 44].

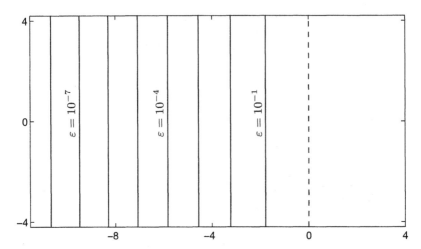

Figure 5.2: Pseudospectra of the differentiation operator $\mathbf{A}$ of (5.2)–(5.3) for an interval of length $d = 2$. The solid lines are the right-hand boundaries of $\sigma_\varepsilon(\mathbf{A})$ for $\varepsilon = 10^{-1}, 10^{-2}, \ldots, 10^{-8}$ (from right to left). The dashed line, the imaginary axis, is the right-hand boundary of the numerical range. If d were increased, the ε levels would decrease exponentially.

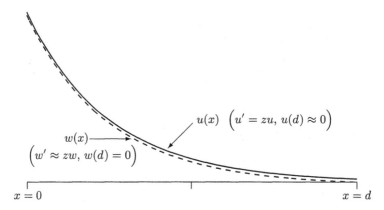

Figure 5.3: For $d\mathrm{Re}\,z \ll 0$, the functions $u(x) = \mathrm{e}^{zx}$ and $w(x) = \mathrm{e}^{zx} - \mathrm{e}^{d\mathrm{Re}\,z + ix\mathrm{Im}\,z}$ are 'nearly eigenfunctions' of $\mathbf{A}$, though neither is near any eigenfunction. Note that u satisfies the eigenvalue equation $u'(x) = zx$, but not the boundary condition; w satisfies the boundary condition, but not the eigenvalue equation. Here $d = 2$, $z = -2$.

eigenfunction' in the sense that it is an eigenfunction of a slightly perturbed problem. It is tempting to call it a pseudoeigenfunction, or pseudomode, but in keeping with the definition on page 31, we reserve this usage for functions that belong to the domain of the operator in question, which u

does not because it violates the boundary condition. However, $u(x)$ can be modified so as to become a true pseudomode by subtraction of a small term such as $c^{d\mathrm{Re}z+ix\mathrm{Im}z}$, and by this means a lower bound for $\|(z-\mathbf{A})^{-1}\|$ can be derived.

Instead of pursuing this idea, we shall obtain sharper estimates by working with (5.4) directly.[2]

Pseudospectra of the differentiation operator

Theorem 5.1 *The spectrum of the operator* $\mathbf{A}$ *is the empty set. The resolvent norm* $\|(z-\mathbf{A})^{-1}\|$ *depends on* $\mathrm{Re}z$ *but not* $\mathrm{Im}z$ *and satisfies*

$$\|(z-\mathbf{A})^{-1}\| \le \frac{1}{\mathrm{Re}z} \tag{5.5}$$

for $\mathrm{Re}z > 0$ *and*

$$\|(z-\mathbf{A})^{-1}\| = \frac{e^{d|\mathrm{Re}z|}}{2|\mathrm{Re}z|} + \mathcal{O}\left(\frac{1}{|\mathrm{Re}z|}\right) \tag{5.6}$$

for $\mathrm{Re}z < 0$, *where the constant in the '$\mathcal{O}$' is independent of* z *and* d. *The pseudospectra of* $\mathbf{A}$ *are half-planes of the form*

$$\sigma_\varepsilon(\mathbf{A}) = \{z \in \mathbb{C} : \mathrm{Re}z < c_\varepsilon\} \tag{5.7}$$

with

$$c_\varepsilon \sim \begin{cases} (\log \varepsilon)/d & as \ \varepsilon \to 0, \\ \varepsilon & as \ \varepsilon \to \infty. \end{cases} \tag{5.8}$$

Proof. If $u = (z - \mathbf{A})^{-1}v$ is given by (5.4), then $u(x)$ is the restriction to $(0, d)$ of the convolution $v * g$, where v and g are both regarded as functions in $L^2(-\infty, \infty)$ and $g(x) = e^{zx}$ for $x \in [-d, 0]$, 0 otherwise. Therefore by the Fourier transform, with $\|\cdot\|$ temporarily denoting the norm in $L^2(-\infty, \infty)$,

$$\|u\| \le \|v * g\| = \|\widehat{v * g}\| = \|\widehat{v}\widehat{g}\| \le \|\widehat{v}\| \sup_{\omega \in \mathbb{R}} |\widehat{g}(\omega)| = \|v\| \sup_{\omega \in \mathbb{R}} |\widehat{g}(\omega)|.$$

An elementary calculation gives $\widehat{g}(\omega) = (e^{d(i\omega - z)} - 1)/(i\omega - z)$, and this expression reaches a maximum at $\omega = \mathrm{Im}z$:

$$\sup_{\omega \in \mathbb{R}} |\widehat{g}(\omega)| = \frac{e^{-d\mathrm{Re}z} - 1}{|\mathrm{Re}z|}.$$

Thus we have

$$\|(z-\mathbf{A})^{-1}\| \le \frac{1 - e^{-d\mathrm{Re}z}}{\mathrm{Re}z} \qquad (\mathrm{Re}z > 0), \tag{5.9}$$

---

[2]It is also possible to determine $\|(z-\mathbf{A})^{-1}\|$ exactly by calculus of variations, though the result is not a closed formula. We are indebted to Satish Reddy for this observation.

which establishes (5.5).

On the other hand, assuming $\mathrm{Re}\,z < 0$, break (5.4) into two pieces

$$(z-\mathbf{A})^{-1}v(x) = \mathbf{R}_1 v(x) - \mathbf{R}_2 v(x) \equiv \int_0^d e^{z(x-s)}v(s)\,\mathrm{d}s - \int_0^x e^{z(x-s)}v(s)\,\mathrm{d}s.$$

Then we have

$$\|\mathbf{R}_1\| - \|\mathbf{R}_2\| \le \|(z-\mathbf{A})^{-1}\| \le \|\mathbf{R}_1\| + \|\mathbf{R}_2\|, \qquad (5.10)$$

and by an argument like the one just used for (5.9)

$$\|\mathbf{R}_2\| \le -\frac{1}{\mathrm{Re}\,z} \qquad (\mathrm{Re}\,z < 0). \qquad (5.11)$$

We can evaluate the norm of $\mathbf{R}_1$ exactly. Since $\mathbf{R}_1 v(x) = e^{zx}\int_0^d e^{-zs}v(s)\,\mathrm{d}s$, the dependence of $\mathbf{R}_1 v(x)$ on x is independent of the choice of v. Thus if we find a function $v(x)$ that maximizes $|\mathbf{R}_1 v(0)|/\|v\|$, this choice will also maximize $\|\mathbf{R}_1 v\|/\|v\|$. By the Cauchy–Schwarz inequality, an appropriate choice is $v(s) = \exp(-\bar{z}s)$ or $v(s) = e^{-\bar{z}s}$ or $v(s) = e^{-\bar{z}s}$, with which we calculate

$$\|\mathbf{R}_1\| = \frac{\|\mathbf{R}_1 v\|}{\|v\|} = \frac{|\mathbf{R}_1 v(0)|}{|v(d)|} = \frac{\int_0^d e^{-2s\mathrm{Re}z}\mathrm{d}s}{e^{-d\mathrm{Re}z}} = \frac{e^{-2d\,\mathrm{Re}z} - 1}{-2\mathrm{Re}z\,e^{-d\mathrm{Re}z}}. \qquad (5.12)$$

Combining this result with (5.10) and (5.11) establishes (5.6).

A proof of (5.7) was sketched in the text. Finally, the upper half of (5.8) follows from (5.6), and the lower half follows from (5.5) (upper bound) and further estimates based on (5.7) (lower bound), which we omit. ∎

In §3 we saw that the pseudospectra of the nonnormal Toeplitz matrix $\mathbf{A}$ provided insight into the behavior of $\mathbf{A}$ as measured by the norms of powers $\|\mathbf{A}^k\|$. For the present example, the analogous problem concerns the norms of the exponentials $\|e^{t\mathbf{A}}\|$, discussed in §15. If $\mathbf{A}$ were a bounded operator, then $e^{t\mathbf{A}}$ could be defined by the usual power series. Since $\mathbf{A}$ is unbounded, $e^{t\mathbf{A}}$ must be defined in a more general manner as the solution operator for the continuous evolution problem $\mathrm{d}u/\mathrm{d}t = \mathbf{A}u$, that is, the first-order partial differential equation $u_t = u_x$ on $(0,d)$ with boundary condition $u(d) = 0$. The solution to this problem is the leftward translation

$$e^{t\mathbf{A}}u(x) = \begin{cases} u(x+t) & \text{if } x+t < d, \\ 0 & \text{if } x+t \ge d. \end{cases} \qquad (5.13)$$

Although $\mathbf{A}$ is unbounded, $e^{t\mathbf{A}}$ is a bounded operator on $L^2(0,d)$ for all $t \ge 0$. In the theory of semigroups, $\{e^{t\mathbf{A}}\}$ is a familiar example of a translation semigroup and $\mathbf{A}$ is known as its infinitesimal generator; see §§15 and 19.

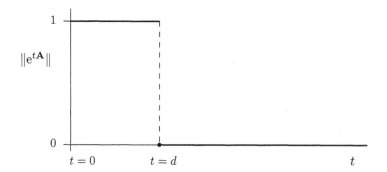

Figure 5.4: Norm of the evolution operator $e^{t\mathbf{A}}$ as a function of t. The fact that $\|e^{t\mathbf{A}}\| \leq 1$ for all $t \geq 0$ ($\mathbf{A}$ is dissipative) can be inferred from (5.5). The fact that $e^{t\mathbf{A}} = 0$ for $t \geq d$ ($\mathbf{A}$ is nilpotent) can be inferred from (5.6).

We consider two aspects of the behavior of $\mathbf{A}$, both of which are illustrated in Figure 5.4. First, from (5.13) we see that $\mathbf{A}$ is dissipative in the sense that the associated evolution process is a contraction: $\|e^{t\mathbf{A}}\| \leq 1$ for $t \geq 0$. This property can be inferred from the behavior of the ε-pseudospectra of $\mathbf{A}$ in the limit $\varepsilon \to \infty$. Specifically, by the Hille–Yosida theorem of semigroup theory, any closed operator that satisfies (5.5) must be dissipative. As discussed in §§14 and 17, (5.5) is also equivalent to the statement that the numerical range of $\mathbf{A}$ is contained in the left half-plane.

Second and perhaps more interesting, $\mathbf{A}$ is nilpotent in the sense that for $t \geq d$, $e^{t\mathbf{A}} = \mathbf{0}$. This property can be inferred from the behavior of the pseudospectra of $\mathbf{A}$ in the limit $\varepsilon \to 0$. In fact, by a Laplace transform argument analogous to the Paley–Wiener theorem, it follows that any closed operator with resolvent norm $\mathcal{O}(e^{-d\,\mathrm{Re}\,z})$ as $\mathrm{Re}\,z \to -\infty$, as holds for $\mathbf{A}$ by (5.6), must satisfy $e^{t\mathbf{A}} = \mathbf{0}$ for $t \geq d$; see Theorem 15.6.

The operator $\mathbf{A}$ has appeared in this book already: We saw a glimpse of it in Figure 2.3! The matrix considered in that figure is a spectral discretization of $\mathbf{A}/144$, and this explains why the pseudospectra portrayed there contain nearly straight boundary segments near the origin. These segments represent high-accuracy approximations to the exactly straight lines of Figure 5.2. Indeed, the data for Figure 5.2 were calculated numerically via matrix approximations of just this kind. For further details, see §§30 and 43.

The two examples considered in this section, illustrated in Figures 5.1 and 5.2, represent two major classes of nonnormal differential operators with links to many other subjects in mathematics and the sciences. In the first case, the nonnormality is introduced by variable coefficients and the pseudomodes have the form of wave packets. In the second case, with constant coefficients, the nonnormality is introduced by boundary conditions and the pseudomodes have the form of evanescent waves pinned at the boundaries. General theorems for such problems are presented with many more examples in §§11 and 10, respectively. For matrices, as opposed to differential operators, one finds analogous classes. Constant-coefficient Toeplitz matrices have pseudomodes pinned at the boundaries, treated in §7, such as our example with approximately elliptical pseudospectra presented in the last section. Variable coefficient 'twisted Toeplitz matrices' have pseudomodes in the form of wave packets, treated in §8. These four fundamental classes of matrices and differential operators lie at the heart of many of the instances of nonnormality that one encounters in applications. Examples appear throughout this book.

6 · History of pseudospectra _____

The importance of nonnormality has been recognized by certain researchers for many years. In this section we attempt to survey the history of the narrower subject of pseudospectra. The following discussion includes all authors we are aware of who made use of pseudospectra of nonnormal matrices or operators before the year 1992—at which point the idea took off, and it becomes difficult to be comprehensive. Very probably there are others of whom we are unaware.

In the context of Hermitian and near-Hermitian systems, mathematical physicists have for years spoken of *quasimodes*, which are the same as what we would call *pseudomodes* or *pseudoeigenvectors*. An early reference is a 1972 article, 'Modes and Quasimodes', by Arnol'd [11], and an even earlier one by Vishik and Lyusternik dates to 1957 [811].[1] The phenomena of interest in this literature differ from the main concerns that arise for nonnormal problems, however, and we shall not discuss quasimodes further.

In the nonnormal context, the earliest definition of pseudospectra we have encountered is given in J. M. Varah's 1967 thesis at Stanford University, *The Computation of Bounds for the Invariant Subspaces of a General Matrix Operator* [801]. Varah introduced the notion of an ε-pseudoeigenvalue under the name *r-approximate eigenvalue*, as shown in Figure 6.1. Motivated by his analysis of the accuracy of eigenpairs produced by computer implementations of the inverse iteration algorithm, Varah incorporated a parameter, η_1, in his definition to describe floating-point precision.

DEFINITION:

$\{^\lambda_y\}$ is an <u>r-approximate</u> $\{^{\text{eigenvalue}}_{\text{eigenvector}}\}$ of A if there exists a

matrix E with $\|E\|_2 = r \cdot \eta_1$ such that $\{^\lambda_y\}$ is an exact $\{^{\text{eigenvalue}}_{\text{eigenvector}}\}$

of A + E .

Figure 6.1: First definition of pseudospectra? From Varah's unpublished 1967 thesis [801, p. 47].

Varah returned to similar ideas in a 1979 paper, 'On the Separation of Two Matrices', whose starting point was an investigation of the conditioning of the Sylvester equation $\mathbf{AX} - \mathbf{XB} = \mathbf{C}$ [802]. In this context, he questioned the circumstances under which the spectra of two matrices

---

[1]In fact, Kato considers related problems in his treatise on the perturbation of matrices and operators [448], and even introduces the terms *pseudo-eigenvalue* and *pseudo-eigenvector*, but in a late section devoted to the Hermitian case.

A and **B** could be said to be well separated. Varah defined the 2-norm ε-pseudospectrum in terms of the minimal singular value $s_{min}(\mathbf{A} - \lambda)$, giving it the name ε-*spectrum* and the notation $S_\varepsilon(\mathbf{A})$. He noted that there is an equivalent definition in terms of matrix perturbations and emphasized that for a nonnormal matrix, the pseudospectra may be very different from the spectrum.

In 1975, between these two works of Varah, H. J. Landau of AT&T Bell Laboratories published a paper, 'On Szegő's Eigenvalue Distribution Theorem and Non-Hermitian Kernels', that independently introduced ε-pseudoeigenvalues under the name ε-*approximate eigenvalues* [478]; see Figure 6.2. Landau applied this concept to the theory of Toeplitz matrices and associated integral operators. Two papers quickly followed on loss in unstable resonators [479] and mode selection in lasers [480] (see §60). Besides appearing early in the history of pseudospectra, these papers were notable in that, unlike much of the numerical analysis literature on pseudospectra that followed in the decade and a half afterward, they recognized that the limitations of eigenvalues go deeper than rounding errors on computers. Landau wrote [480, p. 167]:

> When we remember that, for ε sufficiently small, we cannot distinguish operationally between true and ε-approximate, the possibility arises that in certain non-Hermitian contexts it is the second notion that should replace the first at the center of the stage, even for purposes of theory.

Definition. λ is an ε-approximate eigenvalue of A, if there exists $\varphi \in \mathscr{D}(rQ)$, with $\| \varphi \| = 1$, such that $\| A_r \varphi - \lambda \varphi \| \leqq \varepsilon$. We call φ an ε-approximate eigenfunction corresponding to λ.

Figure 6.2: First published definition of pseudospectra? From Landau, 1975 [478].

In Novosibirsk, S. K. Godunov and his colleagues conducted research related to pseudospectra throughout the 1980s. Those involved included A. G. Antonov, A. Y. Bulgakov (later Haydar Bulgak), O. P. Kirilyuk, V. I. Kostin, A. N. Malyshev, and S. I. Razzakov. Godunov, together with Ryabenkii and others, had made significant contributions in the 1960s to the study of how nonnormality affects the numerical stability of discretized differential equations. In fact, in their 1962 monograph, Godunov and Ryabenkii introduce the *spectrum of the family of operators* $\{R_h\}$, indexed by a mesh parameter h [318, p. 188]. A point $z \in \mathbb{C}$ is in this set if, for every $\varepsilon > 0$, z is an ε-pseudoeigenvalue of R_h for all sufficiently small h. (While less general than pseudospectra, this notion is sufficient to reconcile the discontinuity between the spectra of the finite and infinite banded Toeplitz matrices that arise in finite difference methods; see §§7 and 31.)

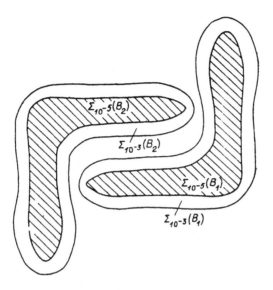

Figure 6.3: First published sketch of pseudospectra? From Kostin and Razzakov, 1985 [461].

Subsequent work in the 1980s explicitly involved pseudospectra and was oriented toward achieving 'guaranteed accuracy' in numerical linear algebra computations [316]. According to Malyshev and Kostin (personal communications, 1991), this work began around 1982. The Novosibirsk group defined the ε-*spectrum* by relative rather than absolute perturbations: $\sigma_\varepsilon(\mathbf{A}) = \{z \in \mathbb{C} : \|(z - \mathbf{A})^{-1}\| \geq (\varepsilon\|\mathbf{A}\|)^{-1}\}$.[2] In [317] and [460], computed plots of pseudospectra were presented and called 'spectral portraits of matrices' containing various 'patches of spectrum', and computations based on both contour plotting and curve tracing were discussed (see §41). A sketch of two pseudospectra, shown in Figure 6.3, had appeared earlier in a 1985 paper by Kostin and Razzakov [461]. Some of the further work by this group appeared in a book in Russian by Godunov in 1997, with a color picture of pseudospectra on the cover [315].

One of of the last papers of the eminent numerical analyst J. H. Wilkinson, 'Sensitivity of Eigenvalues II' (1986), defined the ε-pseudospectrum for an arbitrary matrix norm $\| \cdot \|$ induced by a vector norm [830]. The set was denoted by $D(\eta)$ and given no name other than 'the domain $D(\eta)$'. Wilkinson described in his usual lucid fashion how $D(\eta)$ could be inter-

---

[2] This definition is natural in those applications of backward error analysis where a numerical algorithm introduces perturbations on the scale of machine epsilon times $\|\mathbf{A}\|$; see §53. For other applications, its suitability is less clear, and it has the peculiarity that according to this definition, the question of whether a point $z \in \mathbb{C}$ belongs to a particular ε-pseudospectrum $\sigma_\varepsilon(\mathbf{A})$ depends, via the norm, on the behavior of $\mathbf{A}$ at distant points in $\mathbb{C}$.

preted equivalently in terms of matrix perturbations or the norm of the resolvent. He discussed various applications and examples of small dimension, and mentioned at the end the extension to generalized eigenvalue problems. The remarkable thing is that it took Wilkinson thirty years to come to the idea of pseudospectra, for, given his lifelong dual interests in eigenvalue problems and backward error analysis, the idea would seem to have been hard to avoid. We suggest four partial explanations. First, computer graphics was not the effortless tool in Wilkinson's day that it later became. Second, the matrices he could handle were small, making nonnormal effects less conspicuous. Third, perhaps Wilkinson was aware of the notion of pseudospectra for many years, but considered the idea a heuristic interpretation rather than something solid enough to be published. Fourth, a glance at any of his writings reveals that Wilkinson's habits of thought were resolutely algebraic, not visual. In the 662 pages of his magnum opus *The Algebraic Eigenvalue Problem* [827], there are only four figures, yet floating-point numbers seem to appear on every page.[3]

Pseudospectra were investigated in several papers by J. W. Demmel in the mid-1980s, appearing with the labels $S(\mathbf{A}, \varepsilon)$ in [196] and $\sigma(\varepsilon, \mathbf{A})$ in [198]. The former paper contains the first published computer plot that we know of, reproduced in Figure 6.4. Demmel's starting point was the problem discussed in his 1983 thesis [195]: to devise an analogue of the Jordan canonical form that is robust enough to have meaning in the presence of rounding errors and other perturbations. For example, under what circumstances does it make sense to view several eigenvalues of a matrix as belonging to a cluster, which may itself perhaps be viewed as a perturbation of a more highly defective set of Jordan blocks? As in Varah's paper [802], the question arose here of the separation of the spectra of two (sub)matrices. Demmel related pseudospectra to improvements of the Bauer–Fike theorem and to other results in matrix perturbation theory and discussed applications in control theory, numerical computation of eigenvalues, and other areas; see §49.

Beginning in the mid-1980s, D. Hinrichsen and A. J. Pritchard wrote a number of papers on the *stability radius* of a matrix, i.e., the distance to the set of unstable matrices (see §49). In a 1992 paper [400] they introduced the term *spectral value set* and notation $\sigma(\mathbf{A}, \rho)$ to denote the real structured ε-pseudospectrum of a nonnormal matrix. Another paper by Hinrichsen and Kelb in 1993 extended these ideas to complex perturbations and also to more general structured perturbations (§50) [397]. These articles presented examples of spectral value sets of several matrices illustrated by plots of superpositions of eigenvalues of random perturbations.

---

[3]In fact one of those figures, on p. 454 of [827], can be interpreted as a pseudospectrum—an example of what we call a *structured pseudospectrum*, defined by real perturbations of a real matrix; see §50. Wilkinson introduces this as a 'domain of indeterminacy' in his analysis of the method of bisection, but makes little use of the idea.

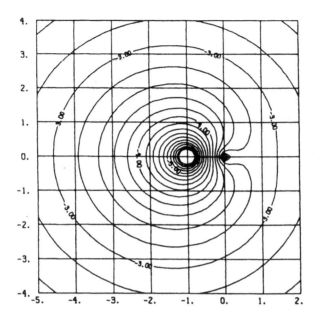

Figure 6.4: First published computer plot of pseudospectra? From Demmel [196], © 1987 IEEE. For an elaboration of this example, see §49.

Another body of early work related to pseudospectra was developed by F. Chatelin, later Chaitin–Chatelin, and her colleagues in France. Beginning in the 1980s, this group investigated questions of conditioning, stability, and floating-point arithmetic with the aid of random perturbations [133, 140]. By perturbing a problem at random, they pointed out, one can acquire knowledge about the properties of the unperturbed problem. One of the many applications they considered was to matrix eigenvalue problems, where random perturbations may reveal, for example, if the original matrix has a nontrivial Jordan block [142, 141]. Though pseudospectra were not explicitly defined in these writings, at least in the early years, the idea was implicit, and computed plots of eigenvalues of randomly perturbed matrices appeared in [140].

Trefethen's first publications that mentioned pseudospectra were [771] and [625] in 1990 (the first uses the term ε-*approximate eigenvalues*). These were an outgrowth of a 1987 paper by Trefethen and Trummer, who found eigenvalues that were extraordinarily sensitive to perturbations but failed fully to appreciate their significance [782]. Subsequent early papers by Trefethen and colleagues pertaining to pseudospectra included [569, 570, 626, 632]. A crucial collaborator in this work was Trefethen's student Satish Reddy, who began to work with pseudospectra in 1988 and made many contributions after that date. Reddy's early work on these topics was summarized in his 1991 thesis at MIT [620].

In 1992 Trefethen's paper 'Pseudospectra of Matrices' appeared, which presented the idea of pseudospectra and exhibited thirteen examples [772]. It was after this point that the idea began to be widely known. A later companion paper, 'Pseudospectra of Linear Operators', presented ten examples involving operators on infinite-dimensional spaces [773].

The papers by Demmel and Wilkinson mentioned above cite each other, and they both cite the paper of Varah [802]. Apart from these cases, none of the papers discussed here published before 1990 cite any of the others. These data suggest that pseudospectra have been invented at least five times:

J. M. Varah	1967	r-approximate eigenvalues
	1979	ε-spectrum
H. J. Landau	1975	ε-approximate eigenvalues
S. K. Godunov et al.	1982	spectral portrait
L. N. Trefethen	1990	ε-pseudospectrum
D. Hinrichsen and A. J. Pritchard	1992	spectral value set

One should not trust this table too much, however, as even recent history is notoriously hard to pin down. It is entirely possible that Godunov or Wilkinson thought about pseudospectra in the 1960s, and indeed, von Neumann may have thought about them in the 1930s. Nor were others such as Dunford and Schwartz, Gohberg, Halmos, Kato, Keldysch, or Kreiss far away.

II. Toeplitz Matrices

7 · Toeplitz matrices and boundary pseudomodes ___

We now embark upon two parts of this book—seven sections—that describe some of the best understood families of nonnormal matrices and linear operators: non-Hermitian Toeplitz matrices and non-self-adjoint differential operators. These two classes are closely related in that the starting point in each case is a translation-invariant action on a one-dimensional domain; the difference is in whether the domain is discrete or continuous. The non-normality may be introduced by boundary conditions, in which case we will find that there are exponentially good pseudoeigenvectors in the form of waves localized at the boundaries (as in Figure 5.2), or by variable coefficients, in which case there are exponentially good pseudoeigenvectors in the form of wave packets in the interior (as in Figure 5.1).

The parallel between the discrete and continuous cases is pervasive. In both situations the analysis depends on a *symbol*, $f(\theta)$ or $f(x,\theta)$, for constant or variable coefficients, respectively; if $z = e^{i\theta}$, this becomes $f(z)$ or $f(x,z)$. In both cases there are important generalizations from scalars to vectors (e.g., block Toeplitz matrices), from one to several space dimensions (e.g., Kronecker products), and from classical difference and differential operators to pseudodifference and pseudodifferential operators. In the interest of readability, however, we shall not attempt to present the two subjects in parallel but concentrate separately on Toeplitz and related matrices in this part of the book, then on differential and related operators in the next.

An $N \times N$ *Toeplitz matrix* is a matrix whose entries are constant along diagonals:

$$
\mathbf{A} = \begin{pmatrix}
a_0 & a_{-1} & & \cdots & a_{1-N} \\
a_1 & a_0 & & & \vdots \\
& & \ddots & \ddots & \ddots & \\
\vdots & & & a_0 & a_{-1} \\
a_{N-1} & \cdots & & a_1 & a_0
\end{pmatrix}.
\tag{7.1}
$$

A semi-infinite matrix of the same form is known as a *Toeplitz operator*, and a doubly infinite matrix of this kind is a *Laurent operator*.[1] A *circulant* matrix, which is the finite-dimensional analogue of a Laurent operator, is a special case of a Toeplitz matrix in which the entries wrap around periodically: $a_j = a_{j-N}$ for $1 \le j \le N - 1$. The standard references on these subjects are the books of Böttcher and Silbermann and their collaborators

---

[1] Following the usual convention in this book, we make no distinction between a matrix (finite or infinite) and the associated operator.

[85, 89, 91, 92, 93]; see especially [89, 93]. Valuable classical references include [213, 322, 367, 821], and in particular, Widom's article [821] remains a very appealing introduction to this subject.

The *symbol* of a Toeplitz matrix or Toeplitz operator or Laurent operator is the function

$$f(z) = \sum_k a_k z^k \, ; \qquad (7.2)$$

this is a finite sum or an infinite series depending on the context. As a running example in this section, we shall consider the family of banded Toeplitz matrices which take the following form for $N = 6$:

$$\mathbf{A} = \begin{pmatrix} 0 & 2i & -1 & 2 & & \\ 0 & 0 & 2i & -1 & 2 & \\ -4 & 0 & 0 & 2i & -1 & 2 \\ -2i & -4 & 0 & 0 & 2i & -1 \\ & -2i & -4 & 0 & 0 & 2i \\ & & -2i & -4 & 0 & 0 \end{pmatrix} ; \qquad (7.3)$$

the entries not shown are zero. With this matrix is associated the symbol

$$f(z) = 2z^{-3} - z^{-2} + 2iz^{-1} - 4z^2 - 2iz^3. \qquad (7.4)$$

Because the matrices are banded, $f(z)$ is just a finite linear combination of positive and negative powers of z, known (confusingly) as a *Laurent polynomial*. In such a case f is a rational function, so it is defined not only on the unit circle but throughout the complex plane, where it is analytic everywhere except at $\leq k$ poles, where k is the upper bandwidth (here, $k = 3$).

More generally, we shall assume that the vector $\mathbf{a} = (a_j)$ defining our Toeplitz or Laurent operator is in $\ell^2(\mathbb{Z})$, which ensures that the sum in (7.2) converges to a function $f \in L^2(\mathbb{T})$, where $\mathbb{T}$ denotes the unit circle, $\mathbb{T} = \{z \in \mathbb{C} : |z| = 1\}$. Furthermore, we shall assume that the symbol f defined in (7.2) is continuous on $\mathbb{T}$, which is sufficient for many applications but rules out some interesting cases discussed in the Toeplitz literature; see, e.g., [93, § 1.8]. The $\mathbf{A}$ associated with the symbol f is an operator on ℓ^2; the assumption that f is continuous is sufficient to ensure that $\mathbf{A}$ is bounded.[2] Often we shall fix f and consider the family of Toeplitz matrices $\{\mathbf{A}_N\}$ of various dimensions obtained as $N \times N$ finite sections of the infinite matrix $\mathbf{A}$ associated with this fixed choice.

---

[2]By ℓ^2 we denote the space of infinite-dimensional vectors $\mathbf{u}$ for which $\sum |u_j|^2$ is finite, leaving it to the context to make it clear whether the index ranges over positive integers (Toeplitz operators) or all integers (Laurent operators).

We can analyze Laurent operators in a few words by noting that the Laurent operator $\mathbf{A}$ defined by a vector $\mathbf{a} = (a_j)$ is equivalent to a convolution:

$$\mathbf{Au} = \mathbf{a} * \mathbf{u}.$$

This implies that Fourier transformation converts $\mathbf{A}$ to a pointwise multiplication:

$$\widehat{\mathbf{Au}}(\theta) = \widehat{\mathbf{a}}(\theta)\widehat{\mathbf{u}}(\theta), \quad \theta \in [0, 2\pi].$$

Roughly speaking, this amounts to the observation that if

$$\mathbf{u} = (\ldots, z^2, z, \underline{1}, z^{-1}, z^{-2}, \ldots)^{\mathrm{T}} \tag{7.5}$$

for some $z = \mathrm{e}^{\mathrm{i}\theta} \in \mathbb{T}$ (the underlined entry marks the central term of the infinite vector), then $\mathbf{Au}$ is given by

$$\mathbf{Au} = f(z)\mathbf{u}. \tag{7.6}$$

(More precisely, this calculation is valid provided the numbers a_j decay sufficiently rapidly, but for arbitrary $\mathbf{a} \in \ell^2(\mathbb{Z})$, we cannot convolve with a vector like (7.5) that is only in ℓ^∞.) The Laurent operator $\mathbf{A}$ is normal, and its spectrum consists of all the numbers $f(z)$ with $|z| = 1$. That is to say, $\sigma(\mathbf{A}) = f(\mathbb{T})$, a closed curve in the complex plane. We shall make extensive use of this *symbol curve*, sketched in Figure 7.1 for (7.4). If $\mathbf{A}$ is Hermitian, $f(\mathbb{T})$ is real and the curve degenerates to a subset of the real axis, but for non-Hermitian matrices, the behavior may be complicated. Given a point $\lambda \in \mathbb{C} \backslash f(\mathbb{T})$, we define the *winding number* $I(f, \lambda)$ to be the winding number of $f(\mathbb{T})$ about λ in the usual positive (counterclockwise) sense. If $\lambda \in f(\mathbb{T})$, $I(f, \lambda)$ is undefined.

The same arguments for Laurent operators also apply if $\mathbf{A}$ is an N-dimensional circulant matrix, except now the Fourier analysis is fully discrete rather than semidiscrete. This means that in (7.5), only values of z are appropriate for which $z^N = 1$, so that the vector is periodic. In other words, we now have $\sigma(\mathbf{A}) = f(\mathbb{T}_N)$, where $\mathbb{T}_N \subset \mathbb{T}$ denotes the set of Nth roots of unity.

We have proved most of parts (i) and (ii) of the following beautiful theorem.

Spectra of Toeplitz and Laurent operators

Theorem 7.1 *Let $\mathbf{A}$ be a circulant matrix or Laurent or Toeplitz operator with continuous symbol f.*

(i) *If $\mathbf{A}$ is a circulant matrix, then $\sigma(\mathbf{A}) = f(\mathbb{T}_N)$.*

(ii) *If $\mathbf{A}$ is a Laurent operator, then $\sigma(\mathbf{A}) = f(\mathbb{T})$.*

(iii) *If $\mathbf{A}$ is a Toeplitz operator, then $\sigma(\mathbf{A})$ is equal to $f(\mathbb{T})$ together with all the points enclosed by this curve with nonzero winding number.*

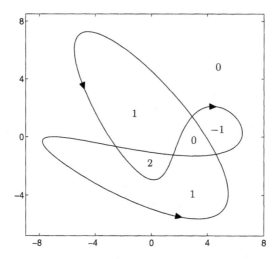

Figure 7.1: Symbol curve $f(\mathbb{T})$ in the complex plane for the symbol f of (7.4). The numbers indicate the winding numbers associated with various regions.

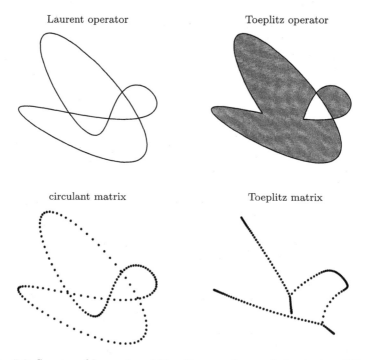

Figure 7.2: Spectra of Laurent and Toeplitz operators and circulant and Toeplitz matrices ($N = 150$) associated with the symbol (7.4).

Part (ii) of this theorem is due originally to Otto Toeplitz in 1911 (under stronger assumptions on the smoothness of f) [756]. Part (iii) was first obtained by Wintner in 1929 [834] for the case in which $\mathbf{A}$ is triangular, then for more general $\mathbf{A}$ essentially by Gohberg in 1952 [321], and in a form perhaps closer to what is written here, independently around 1958 by Krein [464] and Calderón, Spitzer, and Widom [125]. A proof based on C^*-algebras was provided by Coburn in the late 1960s [155, 156]. For a complete proof of Theorem 7.1(iii), a discussion of its generalization to discontinuous symbols, and further historical details, see [93].

   Figure 7.2 illustrates Theorem 7.1 for the example (7.4). For the Laurent operator and the circulant matrix, we see that the spectrum is confined to the symbol curve. The more interesting and complicated cases are the Toeplitz operator and the Toeplitz matrix. For Toeplitz matrices, there is no simple characterization of the eigenvalues, though it is known that if $\mathbf{A}_N$ is banded, then as $N \to \infty$, the eigenvalues cluster along curves in the complex plane, as is evident in the figure [402, 672, 786]. Of greater interest for us in this book and for many applications are the spectra of Toeplitz operators, for as we shall see, these are approximately the same as the pseudospectra of the corresponding Toeplitz matrices $\mathbf{A}_N$ for large N.

   What is going on in Theorem 7.1(iii), illustrated in the upper right plot of Figure 7.2, is as follows. Let $\mathbf{A}$ be a Toeplitz operator with continuous symbol f, and let $\lambda \in \mathbb{C}$ be any number for which $I(f, \lambda) < 0$; i.e., the winding number of the symbol curve is negative. Then not only is λ in the spectrum of $\mathbf{A}$, but it is an eigenvalue of $\mathbf{A}$, with a corresponding eigenvector $\mathbf{u} = (u_j)$ whose amplitude decreases as $j \to \infty$. If f is sufficiently smooth (e.g., a rational function), it decreases exponentially. We call this a *boundary eigenvector* or *boundary eigenmode*, since it is localized near the boundary $j = 1$. For example, if $\mathbf{A}$ is the Toeplitz operator given by (7.4), Figure 7.3 shows the eigenvector associated with the eigenvalue

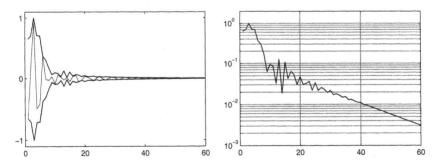

Figure 7.3: Eigenvector $\mathbf{v}$ (first 60 components) of the Toeplitz operator (7.3) associated with eigenvalue $\lambda = 5 + \frac{1}{2}\mathrm{i}$. The left side shows $|\mathbf{v}|$, $-|\mathbf{v}|$, and $\mathrm{Re}\,\mathbf{v}$ on a linear scale, and the right side shows $|\mathbf{v}|$ on a log scale. This eigenvector is exponentially localized at the boundary.

$\lambda = 5 + \frac{1}{2}i$, which is approximately the point at which the label '-1' appears in Figure 7.1.

On the other hand, suppose $\lambda \in \mathbb{C}$ is a number for which $I(f, \lambda) > 0$. Now $\mathbf{A}$ does not have a boundary eigenvector, but $\mathbf{A}^{\mathrm{T}}$ does, and this implies that λ is again in the spectrum. ($\mathbf{A}^{\mathrm{T}}$ is also a Toeplitz operator with the continuous symbol $f(z^{-1})$, so all the winding numbers are the negatives of those of $\mathbf{A}$.)

To explain where these boundary eigenmodes come from, it is helpful to start with the upper triangular case, for which $a_j = 0$ when $j > 0$. For example, consider the 'infinite Jordan block' shift operator

$$\mathbf{A} = \begin{pmatrix} 0 & 1 & & & \\ & 0 & 1 & & \\ & & 0 & 1 & \\ & & & 0 & \ddots \\ & & & & \ddots \end{pmatrix}, \tag{7.7}$$

with symbol $f(z) = z^{-1}$. If we apply $\mathbf{A}$ to the vector

$$\mathbf{u} = (1, z^{-1}, z^{-2}, \ldots)^{\mathrm{T}} \tag{7.8}$$

for any z with $0 < |z| \leq \infty$, the result is $\mathbf{A}\mathbf{u} = z^{-1}\mathbf{u}$. If $|z| > 1$, then $\mathbf{u} \in \ell^2$, and thus $\mathbf{u}$ is an eigenvector of $\mathbf{A}$ with eigenvalue $\lambda = z^{-1}$, in keeping with Theorem 7.1(iii). For a more general upper triangular matrix $\mathbf{A}$, if $\sum_k |a_k| < \infty$ (in which case one says that f is in the *Wiener class*), then $f(z)$ is an analytic function in $\{z : 1 < |z| \leq \infty\}$ and continuous in $\{z : 1 \leq |z| \leq \infty\}$. If the winding number $I(f, \lambda)$ is negative for some point $\lambda \in \mathbb{C}$, it follows from the principle of the argument of complex analysis that f takes the value λ at exactly $-I(f, \lambda)$ points in $\{z : 1 < |z| \leq \infty\}$, counted with multiplicity. If $\mathbf{u}$ is the vector (7.8) constructed from any of these points, then it is readily verified that $\mathbf{A}\mathbf{u} = f(z)\mathbf{u} = \lambda\mathbf{u}$, just as we saw for a Laurent operator in (7.6). (Indeed, because $\mathbf{A}$ is triangular, its behavior is the same as that of the associated Laurent operator, just restricted to rows $j \geq 1$.) Thus $\mathbf{u}$ is an eigenvector of $\mathbf{A}$ with eigenvalue λ.

More generally, consider a Toeplitz operator $\mathbf{A}$ that is banded or at least *semibanded*, by which we shall mean that $a_j = 0$ for $j > k$ and $\sum |a_j| < \infty$. The same reasoning generalizes immediately to this situation. Assume without loss of generality that $a_k \neq 0$. Now f is an analytic function in $\{z : 1 < |z| < \infty\}$, continuous in $\{z : 1 \leq |z| < \infty\}$, with k poles at ∞. If the winding number $I(f, \lambda)$ is negative for some point $\lambda \in \mathbb{C}$, the principle of the argument now implies that f takes the value λ at exactly $k - I(f, \lambda) \geq k + 1$ points in $\{z : 1 < |z| < \infty\}$, counted with multiplicity. Let $z_1, \ldots, z_{k+1}$ be any $k+1$ of these points, which we assume for simplicity are distinct, and consider the $(k + 1)$-dimensional subspace of ℓ^2 of linear combinations of the vectors

$$(1, z_j^{-1}, z_j^{-2}, \ldots)^{\mathrm{T}}, \qquad 1 \leq j \leq k+1. \tag{7.9}$$

If $\mathbf{u} \neq \mathbf{0}$ is one of these linear combinations, then because $\mathbf{A}$ has only k nonzero diagonals below the main diagonal, $\mathbf{u}$ satisfies the eigenvalue equation $\mathbf{A}\mathbf{u} = \lambda\mathbf{u}$ in rows $j \geq k+1$. For $\mathbf{u}$ to be an eigenvector of $\mathbf{A}$ with eigenvalue λ, it is necessary and sufficient that in addition it satisfy the same equation in rows $1 \leq j \leq k$. Since this is a homogeneous system of k equations in $k+1$ unknowns, there must be a nonzero solution. Without much effort, this argument for banded and semibanded arguments can be expanded to general continuous symbols. The remainder of the proof of Theorem 7.1(iii), a demonstration that $I(f, \lambda) = 0$ implies $\lambda \notin \sigma(\mathbf{A})$, is a nontrivial exercise in operator theory; see [93, Thm. 1.17].

To gain insight into the discontinuity between the spectra of the Toeplitz operator $\mathbf{A}$ and the finite-dimensional Toeplitz matrix $\mathbf{A}_N$, perform a similarity transformation of the latter with the matrix

$$\mathbf{D}_N = \begin{pmatrix} r & & & \\ & r^2 & & \\ & & \ddots & \\ & & & r^N \end{pmatrix}$$

for any fixed $r > 0$. The resulting matrix $\mathbf{D}_N \mathbf{A}_N \mathbf{D}_N^{-1}$ is also Toeplitz with the same eigenvalues as $\mathbf{A}_N$, but now with the symbol $f_r(z) := f(rz)$. Hence the spectrum of the associated Toeplitz operator consists of $f(r\mathbb{T})$ and all points this curve encloses with nonzero winding number. In short, this similarity transformation changes the spectrum of the Toeplitz operator while leaving the eigenvalues of $\mathbf{A}_N$ unmoved [672].[3]

We have reached the seventh page of this section and not mentioned pseudospectra! But all the work has been done. We have seen that if $\mathbf{A}$ is a banded or semibanded Toeplitz operator whose symbol curve encloses a point $\lambda \in \mathbb{C}$ with nonzero winding number, then $\mathbf{A}$ or $\mathbf{A}^{\mathrm{T}}$ has an eigenvector exponentially localized at the boundary with eigenvalue λ. What if $\mathbf{A}_N$ is a Toeplitz matrix with the same symbol curve? Then it is immediate that the first N components of the same eigenvector constitute an ε-pseudoeigenvector of $\mathbf{A}_N$ or $\mathbf{A}_N^{\mathrm{T}}$ for the same λ for a value of ε that shrinks exponentially with N. Alternatively, instead of taking the transpose in the case of positive winding number, we can construct pseudomodes at the right boundary ($j = N$) instead of the left one ($j = 1$). In words,

---

[3]This explains the example from §3, in which case $\mathbf{A}$ was a Toeplitz matrix with symbol $f(z) = \frac{1}{4}z + z^{-1}$. The similarity transformation $\mathbf{D}_N$ takes this matrix to one with symbol $f_r(z) = \frac{1}{4}rz + (rz)^{-1}$, and the critical curve $f_r(\mathbb{T}) = f(r\mathbb{T})$ is generally an ellipse, but when $r = 2$ this ellipse is degenerate, $f(2\mathbb{T}) = [-1, 1]$, a real interval, and $\mathbf{D}_N \mathbf{A}_N \mathbf{D}_N^{-1}$ is Hermitian.

every $\lambda \in \mathbb{C}$ enclosed by the symbol curve with nonzero winding number is an exponentially good pseudoeigenvalue of $\mathbf{A}_N$. Apart from a few details, we have proved the following theorem.

Pseudospectra of Toeplitz matrices

Theorem 7.2 *Let $\{\mathbf{A}_N\}$ be a family of banded or semibanded Toeplitz matrices as defined above, and let λ be any complex number with $I(f, \lambda) \neq 0$. Then for some $M > 1$ and all sufficiently large N,*

$$\|(\lambda - \mathbf{A}_N)^{-1}\| \geq M^N, \qquad (7.10)$$

and there exist nonzero pseudoeigenvectors $\mathbf{v}^{(N)}$ satisfying

$$\frac{\|(\mathbf{A}_N - \lambda)\mathbf{v}^{(N)}\|}{\|\mathbf{v}^{(N)}\|} \leq M^{-N}$$

such that

$$\frac{|v_j^{(N)}|}{\max_j |v_j^{(N)}|} \leq \begin{cases} M^{-j} & \text{if } I(f, \lambda) < 0, \\ M^{j-N} & \text{if } I(f, \lambda) > 0, \end{cases} \qquad 1 \leq j \leq N. \qquad (7.11)$$

The constant M can be taken to be any number for which $f(z) \neq \lambda$ in the annulus $1 \leq |z| \leq M$ (if $I(f, \lambda) < 0$) or $M^{-1} \leq |z| \leq 1$ (if $I(f, \lambda) > 0$).

As a corollary we note that if $I(f, 0) \neq 0$, then by Theorem 2.3 the condition numbers $\kappa(\mathbf{A}_N) = \|\mathbf{A}_N\|\|\mathbf{A}_N^{-1}\|$ must grow at least at the same exponential rates. Explicit bounds on $\kappa(\mathbf{A}_N)$ are developed in [84]. Böttcher and Grudsky have also obtained further results on the nature of the boundary pseudoeigenvectors $\mathbf{v}^{(N)}$ [88].

Theorem 7.2 is due to Reichel and Trefethen in [632] (for banded matrices; the extension to semibanded matrices is trivial). Since [632] appeared, the theorem has been greatly generalized by Böttcher and his coauthors, who have obtained a detailed understanding of how the smoothness of f relates to the rate of growth of the resolvent. It was pointed out in [83] that if f is just piecewise continuous, then the growth rate (7.10) may fall from exponential to algebraic, and in [87] it was shown that the same occurs even for continuous symbols that are not smooth. Just as the semibandedness and $1 \leq |z| \leq M$ conditions of Theorem 7.2 have a certain asymmetry in them, it was further shown in [87] that for points λ with $I(f, \lambda) = -1$, the resolvent norm grows faster than any polynomial if and only if the analytic (lower triangular) part of the symbol f is C^∞; the smoothness of the co-analytic (upper triangular) part does not matter, whereas for $I(f, \lambda) = +1$ the pattern is reversed.

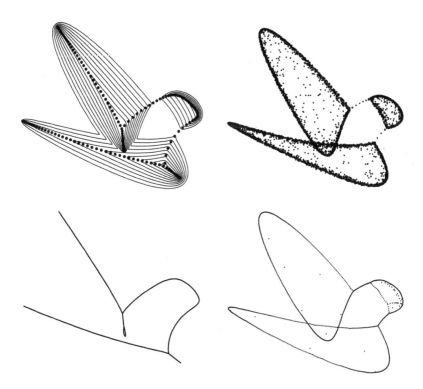

Figure 7.4: On the top left, ε-pseudospectra of the same Toeplitz matrix as in Figure 7.2 for $\varepsilon = 10^{-2}, 10^{-3}, \ldots, 10^{-10}$. On the top right, superimposed eigenvalues of fifty matrices $\mathbf{A} + \mathbf{E}$, where each $\mathbf{E}$ is a random complex dense matrix of norm 10^{-2}. The bottom left image shows the $N \to \infty$ limit of the finite-dimensional spectrum[4] $\sigma(\mathbf{A}_N)$, and the bottom right plot shows *computed* eigenvalues of $\mathbf{A}_{1000}$ as produced by MATLAB's `eig` command.

Figure 7.4 illustrates Theorem 7.2 for our example (7.4). The boundaries of the pseudospectra line up beautifully along curves determined by the symbol, and in the 'plot of dots' in the upper right, we see that dense random perturbations tend to trace out the pseudospectra strikingly. For numerically computed eigenvalues of matrices of large dimension, rounding errors tend to produce much the same effect, as appears in the bottom right plot; see §53.

Figures 7.5 and 7.6 show similar curves for six more examples. The

---

[4]The limiting spectrum can be computed using the following characterization of Schmidt and Spitzer [672]. Let f be the symbol for a banded Toeplitz matrix with upper bandwidth k. For any fixed $\lambda \in \mathbb{C}$, $z^k(f(z) - \lambda)$ is a polynomial; sort its roots by increasing modulus. If roots k and $k+1$ have the same modulus, then $\lambda \in \lim_{N \to \infty} \sigma(\mathbf{A}_N)$. For a numerical study, see Beam and Warming [37].

'limaçon matrix', from [632, 772], is a triangular Toeplitz matrix with symbol

$$f_{\text{limaçon}}(z) = z + z^2. \tag{7.12}$$

An interesting feature of this matrix is that it is mathematically similar to a Jordan block of the same dimension, but the similarity transformation has condition number exponentially large as a function of N; this illustrates how much can be hidden by the mathematical property of similarity. The 'bull's head matrix', from [632], has the symbol

$$f_{\text{bull's head}}(z) = 2\mathrm{i}z^{-1} + z^2 + \tfrac{7}{10}z^3; \tag{7.13}$$

it was designed just to give an interesting shape. The 'Grcar matrix' (pronounced 'Gur-chur'), with symbol

$$f_{\text{Grcar}}(z) = -z^{-1} + 1 + z + z^2 + z^3, \tag{7.14}$$

was devised by Grcar as a challenging example for matrix iterations [337]; its pseudospectra were first considered in [570, 772]. The 'triangle matrix',

$$f_{\text{triangle}}(z) = z^{-1} + \tfrac{1}{4}z^2, \tag{7.15}$$

comes from [632]. The 'whale matrix', with symbol

$$f_{\text{whale}}(z) = -z^{-4} - (3+2\mathrm{i})z^{-3} + \mathrm{i}z^{-2} + z^{-1} + 10z + (3+\mathrm{i})z^2 + 4z^3 + \mathrm{i}z^4, \tag{7.16}$$

appears on the cover of the book [93] by Böttcher and Silbermann and is discussed at length in §3.5 of that book; it originates in [79]. Finally, the 'butterfly matrix', with symbol

$$f_{\text{butterfly}}(z) = z^2 - \mathrm{i}z + \mathrm{i}z^{-1} - z^{-2}, \tag{7.17}$$

is from Böttcher and Grudsky [84].

One of the fruits of analysis of pseudospectra of Toeplitz matrices is a theorem that is more remarkable than it may look at first glance. For fifty years it was recognized that nonnormal Toeplitz matrices have a troublesome property, illustrated in Figures 7.2 and 7.4: As $N \to \infty$, the spectra of $\{\mathbf{A}_N\}$ do not converge to the spectrum of the Toeplitz operator $\mathbf{A}$. Theorem 7.3 asserts that by contrast, the ε-pseudospectrum has a well-behaved limit for any $\varepsilon > 0$. This result was asserted under various assumptions by Landau [478] and by Reichel and Trefethen [632]; the first full proof appeared in a 1994 paper of Böttcher [78].

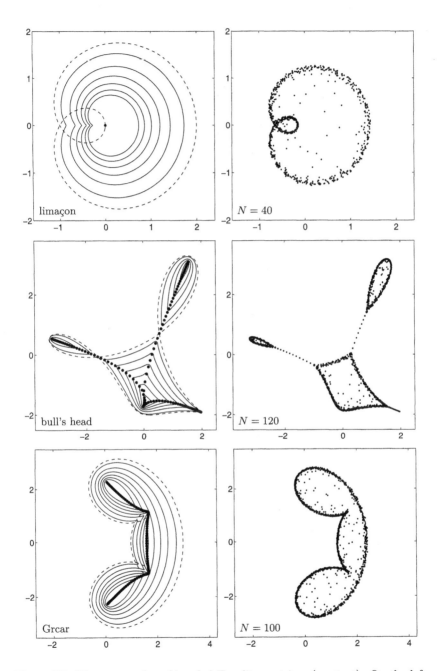

Figure 7.5: Three examples of banded Toeplitz matrices (see text). On the left, ε-pseudospectra for $\varepsilon = 10^{-2}, 10^{-4}, \ldots, 10^{-12}$, with the symbol curve marked by dashes. On the right, eigenvalues of 20 random perturbations of norm 10^{-3}.

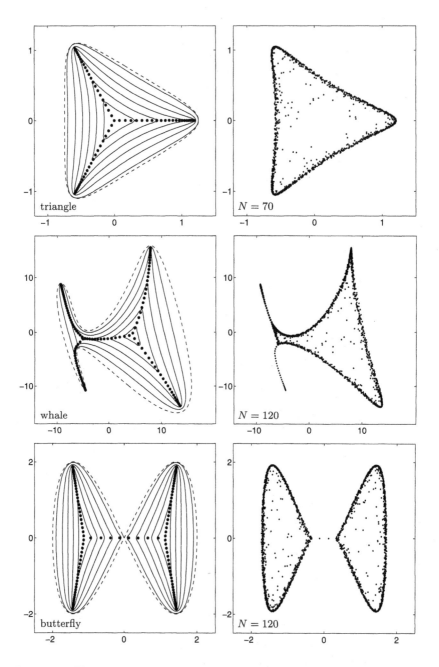

Figure 7.6: Three more examples of banded Toeplitz matrices, as in Figure 7.5.

Behavior of pseudospectra as $N \to \infty$

Theorem 7.3 *Let $\mathbf{A}$ be a Toeplitz operator with continuous symbol f and let $\{\mathbf{A}_N\}$ be the associated family of Toeplitz matrices. Then for any $\varepsilon > 0$,*

$$\lim_{N\to\infty} \sigma_\varepsilon(\mathbf{A}_N) = \sigma_\varepsilon(\mathbf{A}), \tag{7.18}$$

and thus

$$\lim_{\varepsilon\to 0} \lim_{N\to\infty} \sigma_\varepsilon(\mathbf{A}_N) = \sigma(\mathbf{A}). \tag{7.19}$$

The sets in Theorem 7.3 converge in the Hausdorff metric, which implies, for example,

$$\lim_{N\to\infty} S_N = \{z \in \mathbb{C} : z_N \to z \text{ for some sequence } \{z_N\} \text{ with } z_N \in S_N\}.$$

This theorem is part of a bigger story of how finite sections of Toeplitz operators behave as $N \to \infty$: The singular values converge to their infinite-dimensional counterparts, though the eigenvalues do not. The story is bigger in other ways, too, for we have not mentioned Wiener–Hopf factorization of symbols, LU factorization of infinite matrices, the Fredholm alternative and Fredholm indices, discontinuous symbols, or ℓ^p norms. For these subjects and more, see the works by Böttcher and Silbermann and their collaborators.

8 · Twisted Toeplitz matrices and wave packet pseudomodes _____

In the last section we considered Toeplitz matrices, which are constant along diagonals.[1] We found that a non-Hermitian Toeplitz matrix has exponentially large resolvent norms in the region of the complex plane enclosed by the symbol curve with nonzero winding number and that each point in this region is associated with exponentially decaying pseudoeigenvectors localized at the left or right boundary. Now, we extend this picture by considering 'twisted Toeplitz' matrices, in which the entries are permitted to vary continuously along each diagonal. The symbol $f(x, \theta)$ is now a function of two variables, with x-dependence describing variation along each diagonal, and θ-dependence, as in §7, giving variation across the diagonals. Again we shall find exponentially good pseudoeigenvectors, but now in the form of wave packets localized in the interior of the domain. The appearance of these wave packet pseudomodes is controlled by crossings of a symbol curve that depends on the position x as well as the wave number θ.

We begin with an example. For a positive integer N, define

$$x_j = \frac{2\pi j}{N}, \qquad 1 \le j \le N, \tag{8.1}$$

and consider the $N \times N$ bidiagonal matrix $\mathbf{A}$ defined by

$$a_{j,j} = x_j, \qquad a_{j,j+1} = \tfrac{1}{2}x_j. \tag{8.2}$$

The upper part of Figure 8.1 shows the eigenvalues and 2-norm pseudospectra of $\mathbf{A}$ for the case $N = 60$. The eigenvalues, simply the x_j values on the diagonal, are real numbers in $(0, 2\pi]$, and the pseudospectra expand in a wedge of angle 60° around this interval. The lower half of Figure 8.1 shows the corresponding optimal pseudomode $\mathbf{v}$ for $\lambda = 5 + 2\mathrm{i}$, satisfying

$$\frac{\|(\mathbf{A} - \lambda)\mathbf{v}\|}{\|\mathbf{v}\|} = \|(\lambda - \mathbf{A})^{-1}\|^{-1} \approx 0.0080.$$

We see a wave packet of classic shape approximately $e^{-C(x-x_*)^2 - \mathrm{i}\theta_* x}$ for some C; the central position and wave number (at least in the limit $N \to \infty$) are

$$x_* = \frac{58}{10 + \sqrt{13}} \approx 4.26, \quad \theta_* = -\cos^{-1}\left(\frac{\sqrt{325} - 8}{29}\right) \approx -1.22.$$

---

[1]This section is adapted from [777]. Early experiments in this direction were presented in the final section of [632].

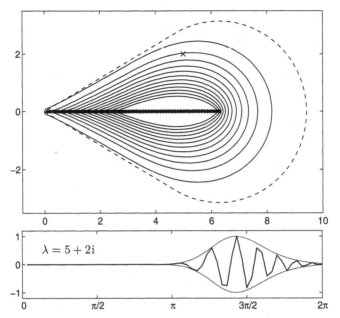

Figure 8.1: Top: Eigenvalues and ε-pseudospectra of the bidiagonal matrix (8.2) with $N = 60$ for $\varepsilon = 10^{-1}, 10^{-2}, \ldots, 10^{-12}$. Here and in the following figures, the dashed curve bounds the region in which $\|(\lambda - \mathbf{A}_N)^{-1}\|$ grows exponentially as $N \to \infty$. Bottom: Optimal pseudoeigenvector $\mathbf{v}$ of $\mathbf{A}$ corresponding to the pseudoeigenvalue $\lambda = 5 + 2i$ (marked by the cross), with $\|(\mathbf{A}_N - \lambda)\mathbf{v}\|/\|\mathbf{v}\| \approx 0.0080$. The real part, absolute value, and negative of the absolute value of $\mathbf{v}$ are shown. The data in question are discrete vectors of length N, but the dots are connected and thus appear as curves. The horizontal coordinate is x_j, ranging from $2\pi/N$ to 2π as j ranges from 1 to N.

This wave number corresponds to approximately $2\pi/1.22 \approx 5.2$ points per wavelength. We shall see that θ_*, x_*, and λ are related by the condition

$$f(x_*, \theta_*) = \lambda,$$

where $f(x, \theta) = x + \frac{1}{2}x e^{-i\theta}$.

In this section and the next we shall show that wave packet pseudomodes such as these appear universally with 'Toeplitz matrices' that have varying coefficients. Analogous effects for variable coefficient differential operators are discussed in §11.

First, a review of more or less standard definitions. A *Toeplitz matrix* is a matrix that is constant along diagonals: for some coefficients $\{c_j\}$,

$$a_{jk} = c_{j-k}, \qquad 1 \le j, k \le N.$$

A *circulant matrix* is a Toeplitz matrix that extends periodically around the boundaries:

$$a_{jk} = c_{(j-k)(\bmod N)}, \qquad 1 \le j, k \le N.$$

For integers m and n with $-m \le n$, an (m,n)-*banded matrix* is a matrix whose nonzero entries all lie within a band extending m entries below the main diagonal and n entries above:

$$a_{jk} \ne 0 \quad \text{only if} \quad -n \le j - k \le m.$$

An (m,n)-*periodic matrix* is the same, except that the nonzero entries wrap around periodically:

$$a_{jk} \ne 0 \quad \text{only if} \quad -n \le j - k \le m \ (\bmod N).$$

The *symbol* of an (m,n)-banded Toeplitz matrix or an (m,n)-periodic circulant matrix is the 2π-periodic trigonometric polynomial

$$f(\theta) = c_{-n} e^{-ni\theta} + \cdots + c_m e^{mi\theta}.$$

Now to the less standard definitions. For these, we move from individual matrices to families of matrices of dimensions $N \to \infty$. Given integers m and n with $-m \le n$, suppose we have $m+n+1$ real or complex 2π-periodic coefficient functions

$$c_j(x), \quad -n \le j \le m.$$

We make no assumptions about continuity or smoothness of c_j except as stated explicitly.

Twisted Toeplitz matrices

Let $c_{-n}, \dots, c_m$ be 2π-periodic coefficient functions. The associated family of *twisted Toeplitz matrices* is the set of (m,n)-periodic matrices $\{\mathbf{A}_N\}_{N \ge 1}$ with coefficients

$$a_{jk} = c_{(j-k)(\bmod N)}(x_j), \qquad (8.3)$$

where

$$x_j = 2\pi j / N, \quad 1 \le j \le N.$$

It might seem that such matrices should instead be called 'twisted circulant', but as we have not imposed any assumptions on the continuity of the c_j functions, the periodicity in the above definition is mainly just formal. In contrast to the conventional circulant and Toeplitz matrices discussed in §7, periodicity plays no role in the current setting, where essential effects are localized in the x variable. Twisted Toeplitz matrices are called 'Berezin–Toeplitz operators' by certain pure mathematicians. In particular, we mention an important paper of Borthwick and Uribe [76] after Theorem 8.1 and in the second half of the next section.

Given N, consider the vector

$$\mathbf{v} = (\mathrm{e}^{-\mathrm{i}\theta}, \mathrm{e}^{-2\mathrm{i}\theta}, \ldots, \mathrm{e}^{-N\mathrm{i}\theta})^{\mathrm{T}}.$$

For $m + 1 \leq j \leq N - n$, the jth entry of the matrix-vector product $\mathbf{A}_N \mathbf{v}$ can be written as

$$(\mathbf{A}_N \mathbf{v})_j = \left[c_{-n}(x_j)\mathrm{e}^{-ni\theta} + \cdots + c_m(x_j)\mathrm{e}^{mi\theta} \right] v_j.$$

In other words, we have

$$(\mathbf{A}_N \mathbf{v})_j = f(x_j, \theta) v_j, \qquad m + 1 \leq j \leq N - n, \tag{8.4}$$

where f is the x-dependent symbol defined as follows.

Symbol of twisted Toeplitz matrices

The *symbol* of the family of twisted Toeplitz matrices associated with $c_{-n}, \ldots, c_m$ is the function

$$f(x, \theta) = c_{-n}(x)\mathrm{e}^{-ni\theta} + \cdots + c_m(x)\mathrm{e}^{mi\theta}, \tag{8.5}$$

defined for $x \in \mathbb{R}$ and $\theta \in \mathbb{C}$.

By definition, $f(x, \theta)$ is 2π-periodic in both x and θ. The assumption of bandedness implies that for each fixed x, $f(x, \theta)$ depends smoothly on θ. Indeed, f is a trigonometric polynomial and thus an entire function of θ, i.e., analytic throughout the complex θ-plane. As for the dependence of f on x, in Theorem 8.1 we shall need nothing more than differentiability at a single point x_*, and in Theorem 9.1 of the next section we shall require even less.

Suppose that $\{\mathbf{A}_N\}$ is a family of (m, n)-periodic twisted Toeplitz matrices with symbol $f(x, \theta)$. Let x_* and θ_* be real numbers, and define $\lambda := f(x_*, \theta_*)$. If f were independent of x, then $\mathbf{A}_N$ would be a circulant matrix, and provided $N\theta_*/2\pi$ was an integer, the formula (8.4) would hold for all $1 \leq j \leq N$. Thus the vector

$$\mathbf{v} = (\mathrm{e}^{-\mathrm{i}\theta_*}, \mathrm{e}^{-2\mathrm{i}\theta_*}, \ldots, \mathrm{e}^{-N\mathrm{i}\theta_*})^{\mathrm{T}} \tag{8.6}$$

would be an eigenvector of $\mathbf{A}_N$ with eigenvalue λ, though $\mathbf{V}$ is, of course, not a wave packet but a global vector.

If f varies with x, however, (8.6) is no longer an eigenvector. As a first step to seeing how it must be modified to become an eigenvector, we may ask, How must the 'local wave number' θ perturb away from θ_* as x is perturbed away from x_*? In such a perturbation, λ must remain fixed. To leading order we accordingly have

$$0 = \frac{\mathrm{d}\lambda}{\mathrm{d}x} = \frac{\partial f}{\partial x} + \frac{\partial f}{\partial \theta}\frac{\mathrm{d}\theta}{\mathrm{d}x},$$

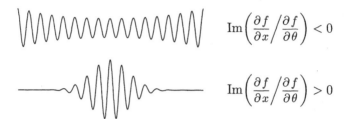

$$\mathrm{Im}\left(\frac{\partial f}{\partial x}\Big/\frac{\partial f}{\partial \theta}\right) < 0$$

$$\mathrm{Im}\left(\frac{\partial f}{\partial x}\Big/\frac{\partial f}{\partial \theta}\right) > 0$$

Figure 8.2: For a wave packet exponentially localized at $x \approx x_*$, $\theta \approx \theta_*$, the symbol must satisfy the twist condition (8.7).

which implies

$$\frac{\mathrm{d}\theta}{\mathrm{d}x} = -\frac{\partial f}{\partial x}\Big/\frac{\partial f}{\partial \theta}.$$

Two possibilities may now be distinguished, as illustrated in Figure 8.2. If $\mathrm{Im}\left(\frac{\partial f}{\partial x}\big/\frac{\partial f}{\partial \theta}\right) < 0$, then $\mathrm{Im}\,\theta$ becomes positive as x increases from x_*, and the amplitude of $\mathrm{e}^{-i\theta}$ grows exponentially. If $\mathrm{Im}\left(\frac{\partial f}{\partial x}\big/\frac{\partial f}{\partial \theta}\right) > 0$, then $\mathrm{Im}\,\theta$ becomes negative as x increases and the amplitude decays exponentially. The latter is the condition for a localized wave packet to be possible. (This argument can be systematically extended to higher orders by WKBJ analysis, but this requires an assumption of smooth dependence on x that is not needed for the theorem we are about to state.)

We formalize these conditions as follows.

Twist condition

Let $f(x, k)$ be a function of $x \in \mathbb{R}$ and $\theta \in \mathbb{C}$ that is 2π-periodic in both variables, and let x_* and θ_* be real numbers. We say that f satisfies the *twist condition* at $x = x_*$, $\theta = \theta_*$ if at this point it is differentiable with respect to x with $\partial f/\partial \theta \neq 0$ and

$$\mathrm{Im}\left(\frac{\partial f}{\partial x}\Big/\frac{\partial f}{\partial \theta}\right) > 0. \tag{8.7}$$

The function f satisfies the *antitwist condition* at $x = x_*$, $\theta = \theta_*$ if it has the same properties with (8.7) replaced by

$$\mathrm{Im}\left(\frac{\partial f}{\partial x}\Big/\frac{\partial f}{\partial \theta}\right) < 0. \tag{8.8}$$

Here is the basic theorem; it will be reinterpreted in Theorem 8.2 and generalized considerably in the next section. Here and throughout, the norm $\|\cdot\|$ can be any p-norm for $1 \le p \le \infty$, though our numerical illustrations always take $\|\cdot\| = \|\cdot\|_2$.

Wave packet pseudomodes of twisted Toeplitz matrices

Theorem 8.1 *Let $\{\mathbf{A}_N\}$ be a family of (m,n)-periodic twisted Toeplitz matrices with symbol $f(x,\theta)$. Let x_* and θ_* be real numbers, define $\lambda = f(x_*,\theta_*)$, and suppose that the twist condition (8.7) is satisfied at $x = x_*$, $\theta = \theta_*$. Suppose moreover that $f(x_*,\theta) \neq \lambda$ for all real $\theta \not\equiv \theta_*$ $(\mathrm{mod}\,2\pi)$ and that the extreme coefficients are nonzero in the following sense: if $n > 0$, $c_{-n}(x_*) \neq 0$; if $m > 0$, $c_m(x_*) \neq 0$; if $m = 0$ or $n = 0$, $c_0(x_*) \neq \lambda$; if $n < 0$ or $m < 0$, $\lambda \neq 0$. Then there exist constants C_1, $C_2 > 0$ and $M > 1$ such that for all sufficiently large N there exists a nonzero pseudomode $\mathbf{v}^{(N)}$ that is* exponentially good,

$$\frac{\|(\mathbf{A}_N - \lambda)\mathbf{v}^{(N)}\|}{\|\mathbf{v}^{(N)}\|} \leq M^{-N}, \tag{8.9}$$

and localized,

$$\frac{|v_j^{(N)}|}{\max_j |v_j^{(N)}|} \leq C_1 \exp(-C_2 N (x_j - x_*)^2) \quad (\mathrm{mod}\,2\pi). \tag{8.10}$$

A proof of Theorem 8.1 is given in [777]. The idea of the proof is to establish that, in a certain precise sense, there must be a linear space of solutions to the eigenvalue equation $\mathbf{A}_N \mathbf{v} = \lambda \mathbf{v}$ decaying to the right from $x = x_*$ and another space of solutions decaying to the left, and that the dimensions are such that these two spaces must have a nontrivial intersection of wave packet solutions decaying in both directions. The arguments resemble those that arise in proofs of the Stable Manifold Theorem and the Center Manifold Theorem in dynamical systems [682]. An alternative 'WKBJ' or 'microlocal' approach to such results has been pursued by Borthwick and Uribe [76], leading to a theorem that requires f to be smooth with respect to θ but does not require the condition $\theta \not\equiv \theta_*$ $(\mathrm{mod}\,2\pi)$. See the second half of §9 for examples illustrating the significance of this difference.

We shall now illustrate Theorem 8.1 with three more examples, and along the way, present an alternative and perhaps more memorable interpretation of the twist condition in terms of winding numbers of the symbol. We abbreviate the twist ratio by

$$\mathcal{T}(x,\theta) = \left(\frac{\partial f}{\partial x} \bigg/ \frac{\partial f}{\partial \theta}\right)(x,\theta) \tag{8.11}$$

and note that the twist condition (8.7) is the condition that $\mathcal{T}(x,\theta)$ is well defined at (x_*,θ_*) and has positive imaginary part.

First, let $\mathbf{A}_N$ be the $N \times N$ 'volcano matrix' that is zero everywhere except that the first superdiagonal contains the entries $1/N, 2/N, \ldots, (N-$

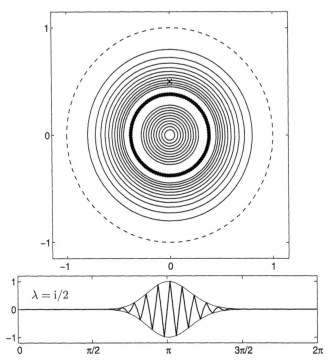

Figure 8.3: Spectrum and ε-pseudospectra of the 'volcano matrix' (8.12) with $N = 100$, $\varepsilon = 10^{-2}, 10^{-3}, \ldots, 10^{-12}$ (counting from outside in toward the circle of eigenvalues, or from the origin out toward the circle of eigenvalues). The lower curve shows the optimal ε-pseudoeigenvector for $\lambda = i/2$ (the cross in the top image), with $x_* = \pi$, $\theta_* = \pi/2$, and $\varepsilon \approx 3.53 \times 10^{-10}$.

1)$/N$, which wrap around periodically to $a_{N1} = 1$. These entries are samples of the function $x/2\pi$ on $(0, 2\pi]$; because they occur on the first superdiagonal, the symbol is

$$f_{\text{volcano}}(x, \theta) = \frac{x e^{-i\theta}}{2\pi}, \qquad (8.12)$$

with $m = -1$ and $n = 1$. As x and θ range over $(0, 2\pi]$, f ranges over the punctured unit disk. The twist ratio is

$$\mathcal{T}(x, \theta) = i/x, \qquad (8.13)$$

and since $\operatorname{Im} \mathcal{T}(x, \theta)$ is positive for all $x > 0$, we conclude from Theorem 8.1 that every point λ in the punctured disk is an exponentially good pseudo-eigenvalue. Figure 8.3 confirms this prediction. To relate the wave packet in the lower part of that figure quantitatively to Theorem 8.1, we note that from (8.12), $f(x, \theta) = \lambda = i/2$ will be achieved when $x = \pi$ and $\theta = -\pi/2$, and only with these values. Accordingly, the wave packet lies at the center of the interval with four points per wavelength.

Figure 8.4 confirms that, as predicted by (8.9), the resolvent norm at this point in the punctured disk grows exponentially as $N \to \infty$.

Theorem 8.1 does not guarantee that an optimal pseudoeigenvector has the form of a wave packet, merely that there exists an exponentially good pseudoeigenvector in that form. Figure 8.3 suggests, however, that in this case the optimal pseudoeigenvector does have the shape of a wave packet. We can see its shape more fully by looking at this vector on a logarithmic scale. The downward pointing curve of Figure 8.5, locally a parabola, is just the kind of structure described by (8.10).

Now, what about the winding number interpretation? Recall that in the theory of Toeplitz matrices and operators outlined in the last section, a key role is played by the winding number $I(f, \lambda)$ of the symbol with respect to a point $\lambda \in \mathbb{C}$. Geometrically, this is the number of times that the curve $f((0, 2\pi])$ winds around λ in the positive sense, where f is the symbol; the winding number is undefined if $f((0, 2\pi])$ passes through λ. Theorem 7.2 asserts that banded Toeplitz matrices have exponentially good

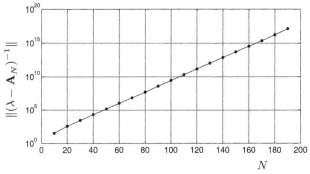

Figure 8.4: Resolvent norms for order-N volcano matrices at $\lambda = i/2$ as in Figure 8.3. The exponential growth confirms condition (8.9) of Theorem 8.1.

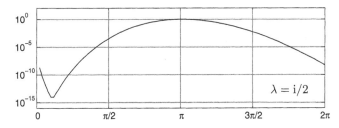

Figure 8.5: The absolute value of the pseudoeigenvector of Figure 8.3, plotted again on a logarithmic scale.

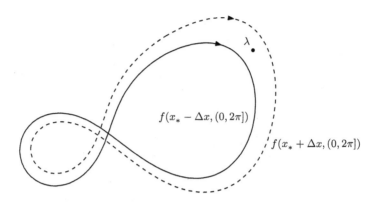

Figure 8.6: Winding number interpretation of Theorem 8.1. If the symbol curve $f(x, (0, 2\pi])$ crosses λ as x passes through x_* in such a way that the winding number about λ is decreased, then there is an exponentially good wave packet pseudomode centered at x_* with pseudoeigenvalue λ.

pseudomodes localized at the left boundary for any λ with $I(f, \lambda) < 0$ and at the right boundary for any λ with $I(f, \lambda) > 0$.

Theorem 8.1 can be interpreted in similar terms (see Figure 8.6). For a twisted Toeplitz matrix, whose symbol depends on x as well as θ, we consider the winding numbers of the symbol curves corresponding to coefficients frozen at each value of x. Suppose a number $\lambda \in \mathbb{C}$ satisfies $\lambda = f(x_*, \theta_*)$ for some $x_*, \theta_* \in (0, 2\pi]$. Then the curve $f(x_*, (0, 2\pi])$ passes through λ, and thus $I(f, \lambda, x)$ is not defined at $x = x_*$. Typically, however, it will be defined for all values of x sufficiently close to x_* to the left and right. The twist condition, together with the other conditions stated as assumptions in Theorem 8.1, amounts to the statement that the curve crosses λ just once and in such a way that as x increases through x_*, $I(f, \lambda, x)$ decreases by 1. This is essentially the following reformulation of Theorem 8.1; in the next section we shall generalize it further to discontinuous symbols.

Restatement of Theorem 8.1 in terms of winding numbers

Theorem 8.2 *Let $\{\mathbf{A}_N\}$ be a family of (m, n)-periodic twisted Toeplitz matrices with symbol $f(x, \theta)$ as in Theorem 8.1, with x_*, θ_*, and $\lambda = f(x_*, \theta_*)$ defined as in that theorem. Suppose that in place of the condition $f(x_*, \theta_*) \neq \lambda$ of that theorem, we require that the winding number $I(f, \lambda, x)$ is defined in a neighborhood of x_*, except at x_* itself, with $I(f, \lambda, x_*^+) = I(f, \lambda, x_*^-) - 1$. Then the conclusions of Theorem 8.1 hold.*

We can reinterpret Figures 8.1 and 8.3 in the light of winding numbers

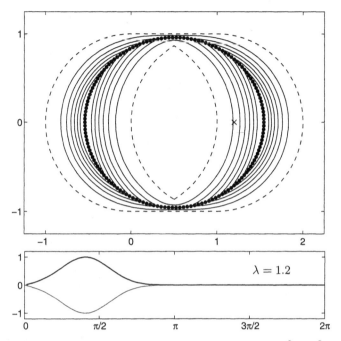

Figure 8.7: 'Wilkinson matrix' (8.14) with $N = 150$, $\varepsilon = 10^{-2}, 10^{-3}, \ldots, 10^{-7}$. The ε-pseudoeigenvector for $\lambda = 1.2$ (cross) has $x_* = 0.4\pi$, $\theta_* = 0$, and $\varepsilon \approx 0.0073$.

as follows. In Figure 8.1 we see a matrix whose symbol curve for each $x \in (0, 2\pi]$ is a positively oriented circle of radius $x/2$ centered at x. As x increases, these circles cross every point inside the ice cream cone–shaped region bounded by the dashes, reducing the corresponding winding numbers from 0 to -1. The configuration is shown schematically in the first panel of Figure 8.8. Similarly, for the volcano matrix of Figure 8.3, the symbol curve at $x \in (0, 2\pi]$ is the negatively oriented circle about 0 of radius $x/2\pi$. As x increases, these circles expand to cross each point in the unit disk, as suggested in the second panel of the figure.

Our third example, shown in Figure 8.7, is a 'Wilkinson matrix' of dimension $N = 150$, consisting of $1/N, \ldots, (N-1)/N$ on the main diagonal and 1 on the first superdiagonal and also in the $(N, 1)$ position. (Wilkinson proposed a multiple of the nonperiodic version of this matrix with $N = 20$ as an example of a matrix with ill-conditioned eigenvalues [827], and pseudospectra were considered in [772].) The symbol is

$$f_{\mathsf{Wilkinson}}(x, \theta) = \frac{x}{2\pi} + e^{-i\theta}, \qquad (8.14)$$

with $m = 0$ and $n = 1$, so the symbol curves are negatively oriented circles of radius 1 centered at $x \in (0, 2\pi]$. As x increases, the circles move right

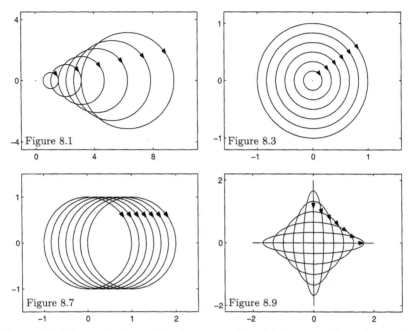

Figure 8.8: Schematic views of the four examples of this section. In each case the symbol curve is drawn for $x = 0, 2\pi/6, 2\pi/3, \ldots, 2\pi$. If the symbol curve crosses a point $\lambda \in \mathbb{C}$ as x increases in such a way that the winding number about λ decreases, then λ is an exponentially good pseudoeigenvalue with a corresponding wave packet pseudomode.

and cross each point in the crescent-shaped region bounded by C, $C + 1$, and the lines $\mathrm{Im}\,\lambda = \pm 1$, where C is the right half of the unit circle (third panel of Figure 8.8). By Theorem 8.2, each such point is accordingly an exponentially good pseudoeigenvalue. For the selected value $\lambda = 1.2$ we calculate $\theta_* = 0$ and $x_* = 0.4\pi \approx 1.26$, and this explains why the lower part of Figure 8.7 has a wave packet in the left of the interval with no oscillations inside the envelope: The wave packet is purely real.

Figure 8.7 demonstrates that the pseudospectra crescent in the half-plane $\mathrm{Re}\,\lambda \geq 1/2$ reflects to an identical pseudospectra crescent in the half-plane $\mathrm{Re}\,\lambda \leq 1/2$. Theorem 8.1 does not explain this, because the twist condition is not satisfied in this region, but as we shall see in the next section, it is enough for the antitwist condition to be satisfied instead.

Our final example, shown in Figure 8.9, is a 150×150 'target matrix', consisting of $-1 + x_j/\pi$ on the first subdiagonal and 1 on the first superdiagonal, with these patterns continued periodically to $a_{N1} = 1$ and $a_{1N} = x_1$. The symbol is

$$f_{\text{target}}(x, \theta) = (-1 + \frac{x}{\pi})e^{i\theta} + e^{-i\theta} = \frac{x}{\pi}\cos\theta - i\left(2 - \frac{x}{\pi}\right)\sin\theta. \quad (8.15)$$

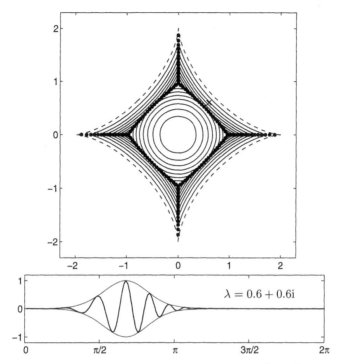

Figure 8.9: 'Target matrix' (8.15) with $N = 150$, $\varepsilon = 10^{-2}, 10^{-3}, \ldots, 10^{-8}$. The ε-pseudoeigenvector for $\lambda = 0.6 + 0.6\mathrm{i}$ (cross) has $x_* \approx 2.11$, $\theta_* \approx -0.47$, and $\varepsilon \approx 5.30 \times 10^{-4}$.

As x ranges over $(0, 2\pi]$, the symbol curves trace all ellipses centered at 0 with major and minor axes aligned with the real and imaginary axes and of lengths summing to 4 (fourth panel of Figure 8.8). Within the envelope of these ellipses, an astroid, the resolvent norm grows exponentially with N. For the choice $\lambda = 0.6 + 0.6\mathrm{i}$, the value of x that satisfies the twist condition is the one corresponding to an ellipse taller than it is wide. A little calculation shows that x_*/π is the root ≈ 0.67 of $25x^4 - 100x^3 + 82x^2 + 36x - 36$, leading to $x_* \approx 2.11317$ and $\theta_* \approx -0.46904$. This explains why the wave packet is located where it is and has about $2\pi/|\theta| \approx 13.4$ points per wavelength.

9 · Variations on twisted Toeplitz matrices _____

Theorems 8.1–8.2 may capture the essence of wave packet pseudomodes of Toeplitz-like matrices with variable coefficients, but when it comes to concrete examples, it is surprising how often these theorems fail to apply in cases where they 'ought to'. Fortunately, they can be extended in many ways. This section, like the last, is adapted from [777], where proofs of Theorems 9.1–9.4 can be found.

Section 8 was built around a precise definition (8.3) of twisted Toeplitz matrices. A simple example is the matrix

$$\mathbf{A}_N = \begin{pmatrix} 0 & c(x_1) & & \\ & 0 & \ddots & \\ & & \ddots & c(x_{N-1}) \\ c(x_N) & & & 0 \end{pmatrix},$$

where c is any continuous 2π-periodic function; here $a_{jk} = c(x_j)$ when $k \equiv j+1 \pmod{N}$. Though $\mathbf{B}_N = \mathbf{A}_N^{\mathrm{T}}$ must exhibit similar pseudospectral properties, it is *not* twisted Toeplitz according to (8.3). The function c is now sampled at x values indexed by the column, rather than the row:

$$b_{jk} = c(x_k), \qquad j \equiv k+1 \pmod{N}.$$

Our first generalization, related to work of Tilli [752], describes this and more exotic variations. We investigate matrices that are not exactly twisted Toeplitz as defined by (8.3) but close to that form locally near a point x_*. By 'near x_*', we mean throughout some neighborhood $x_* - \Delta x < x < x_* + \Delta x \pmod{2\pi}$, where Δx is independent of N.

Asymptotically twisted Toeplitz matrices

Let $\{\mathbf{A}_N\}$ be a family of matrices of degrees $N \to \infty$, let $x_* \in [0, 2\pi]$ be fixed, and let m and n be integers with $-m \le n$. $\{\mathbf{A}_N\}$ is *(m,n)-periodic near x_** if for all sufficiently large N, the rows of $\mathbf{A}_N$ corresponding to row indices j with x_j near x_* are zero outside the periodic (m,n)-band as defined in §8. The family $\{\mathbf{A}_N\}$ is *asymptotically (m,n)-twisted Toeplitz near x_** with symbol $f(x, \theta) = c_{-n}(x)e^{-ni\theta} + \cdots + c_m(x)e^{mi\theta}$ for some fixed functions $c_{-n}(x), \ldots, c_m(x)$ defined near x_* if it is (m,n)-periodic near x_* and if for all j with x_j near x_* and all k, the coefficients of $\mathbf{A}_N$ satisfy

$$a_{jk} = c_{(j-k)(\bmod N)}(x_j) + o(1) \tag{9.1}$$

uniformly as $N \to \infty$.

Since $x_j \to x_{j+1}$ uniformly as $N \to \infty$, our example $\mathbf{B} = \mathbf{A}_N^{\mathrm{T}}$ is asymptotically twisted Toeplitz near each $x \in [0, 2\pi]$.

The following theorem from [777] shows that the exponentially strong effects identified in Theorems 8.1–8.2 are *structurally stable*: They persist under small perturbations of the matrix entries.

Pseudomodes of asymptotically twisted Toeplitz matrices

Theorem 9.1 *Let $\{\mathbf{A}_N\}$ be a family of matrices that are asymptotically (m, n)-twisted Toeplitz near $x_* \in [0, 2\pi]$ with symbol $f(x, \theta)$ satisfying $\lambda = f(x_*, \theta_*)$ and the other conditions of Theorems 8.1 or 8.2 at $x = x_*$. Then the conclusions (8.9) and (8.10) of Theorem 8.1 hold.*

To illustrate Theorem 9.1, let us look ahead to the 'Ehrenfest matrices' discussed in §56. Set $N = n + 1 = 100$, and let $\mathbf{A}$ be the transpose of the $N \times N$ matrix $\mathbf{P}$ of (56.6). (The transpose is taken in order to undo the Markov chain convention followed in §56, in which matrices act on row vectors on the left rather than column vectors on the right.) This matrix is not twisted Toeplitz in the sense of (8.3), but it is asymptotically twisted Toeplitz near any $x_* \in (0, 2\pi)$, and the symbol is

$$f(x, \theta) = \mathrm{e}^{-\mathrm{i}\theta}(x/2\pi) + \mathrm{e}^{\mathrm{i}\theta}(1 - (x/2\pi)). \tag{9.2}$$

Note that the nonzero diagonal $a_{jj} = N^{-1}$ does not appear in the symbol at all, since it converges to zero as $N \to \infty$.

As in the example of Figure 8.9, the set of values attained by $f(x, \theta)$ for $x, \theta \in \mathbb{R}$ is a superposition of ellipses. We start at $x = 0$ with the positively oriented unit circle. As x increases to π, the circle flattens to the interval $[-1, 1]$, whereupon it begins to fatten again to ellipses, but now with negative orientation, reaching the unit circle once more at $x = 2\pi$. As a result, each point λ in the unit disk corresponds to *two* values x_* at which the twist condition is satisfied—first when the winding number jumps from 1 to 0, and then when it jumps from 0 to -1. This explains the appearance of two wave packets in the pseudomode of Figure 9.1. Note that the pseudospectra of this figure resemble those of Figure 56.4. These are defined in the 2-norm and those in the 1-norm, but since the effects on display are exponentially strong as $N \to \infty$, the distinction makes little difference.

Another application of Theorem 9.1 is that it makes possible the extension of our results to the antitwist condition.

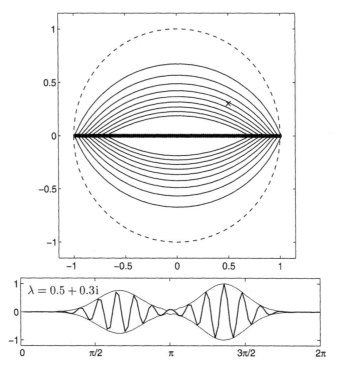

Figure 9.1: 'Ehrenfest matrix' (9.2) with $N = 100$, $\varepsilon = 10^{-2}$, 10^{-3}, ..., 10^{-10}. The ε-pseudomode for $\lambda = 0.5 + 0.3i$ (cross) has $\varepsilon \approx 5.81 \times 10^{-6}$. This pseudomode is a double wave packet because as x increases from 0 to 2π, the symbol curve crosses λ at two different values of x.

Pseudomodes and the antitwist condition

Theorem 9.2 *Let* $\{\mathbf{A}_N\}$ *be a family of matrices as in Theorems* 8.1 *or* 9.1 *but such that* $f(x,\theta)$ *satisfies the antitwist condition* (8.8) *at* (x_*, θ_*) *instead of the twist condition. Then* $\lambda = f(x_*, \theta_*)$ *is again an exponentially good pseudoeigenvalue, and the estimates* (8.9) *and* (8.10) *hold with* $\mathbf{A}_N$ *replaced by* $\mathbf{A}_N^T$.

Proof. This is a corollary of Theorems 8.1 and 9.1. As we have mentioned, if $\{\mathbf{A}_N\}$ is a twisted Toeplitz family, then $\{\mathbf{A}_N^T\}$ in general is not, because its diagonals are indexed by columns instead of rows. However, it is asymptotically twisted Toeplitz near any x_*, so Theorem 9.1 gives the desired conclusion. ∎

Theorem 9.2 explains the left half of Figure 8.7, where it is the antitwist rather than the twist condition that is satisfied. Of course, for such a simple example, one could also devise ad hoc explanations based on the symmetries of the matrix.

So far, we have considered symbols that depend continuously on x at $x = x_*$. For such problems, we have obtained wave packets of type $\exp(-N(x-x_*)^2)$ whenever the symbol curve crosses the point λ at $x = x_*$. However, the behavior at $x = x_*$ need not be continuous for wave packet pseudomodes to appear. One also gets exponentially good pseudomodes, now of the stronger type $\exp(-N|x - x_*|)$, if the symbol is discontinuous at x_* but well-behaved on both sides, provided that the winding number of the symbol curve is larger on the left side x_*^- than on the right x_*^+. The following theorem from [777] makes this precise. The winding number notation $I(f, \lambda, x)$ is the same as in the last section.

Twisted Toeplitz matrices with discontinuous symbols

Theorem 9.3 Let $\{\mathbf{A}_N\}$ be a family of matrices as in Theorems 8.1–8.2 or 9.1–9.2 whose symbol $f(x, \theta)$ is discontinuous at x_* but has left- and right-limits $f(x_*^-, \theta)$ and $f(x_*^+, \theta)$ with band widths (m^-, n^-) and (m^+, n^+), and suppose that the value $\lambda \in \mathbb{C}$ is not taken by $f(x_*^-, \mathbb{R})$ or $f(x_*^+, \mathbb{R})$, so that $I(f, \lambda, x_*^-)$ and $I(f, \lambda, x_*^+)$ are defined. Suppose also that the extreme coefficients of $f(x_*^-, \mathbb{R})$ and $f(x_*^+, \mathbb{R})$ are nonzero in the same sense as in Theorem 8.1. If $I(f, \lambda, x_*^-) > I(f, \lambda, x_*^+)$, then there exist constants C_1, $C_2 > 0$ and $M > 1$ such that for all sufficiently large N there exists a nonzero pseudomode $\mathbf{v}^{(N)}$ with

$$\frac{\|(\mathbf{A}_N - \lambda)\mathbf{v}^{(N)}\|}{\|\mathbf{v}^{(N)}\|} \leq M^{-N} \tag{9.3}$$

and

$$\frac{|v_j^{(N)}|}{\max_j |v_j^{(N)}|} \leq C_1 \exp(-C_2 N |x_j - x_*|) \pmod{2\pi}. \tag{9.4}$$

If $I(f, \lambda, x_*^-) < I(f, \lambda, x_*^+)$, then the same conclusions hold with $\mathbf{A}_N$ replaced by $\mathbf{A}_N^{\mathrm{T}}$.

Figure 9.2 shows an example of this theorem. Here $\mathbf{A}$ is a periodic tridiagonal matrix of dimension $N = 140$: the superdiagonal has constant value 2, and the subdiagonal is -1 in rows 2–71 and $+1$ in rows 72–140 and in the corner entry $a_{1,140}$. Thus the symbol has a jump at $x_* = \pi$:

$$f(x, \theta) = \begin{cases} -\exp(i\theta) + 2\exp(-i\theta) & \text{for } x < x_*, \\ +\exp(i\theta) + 2\exp(-i\theta) & \text{for } x > x_*; \end{cases} \tag{9.5}$$

as x and θ range over all real values, the symbol describes two ellipses but not the region interior to them. The selected value $\lambda = 1.6$ lies inside one ellipse but not the other, with a jump in winding number from 0 to -1

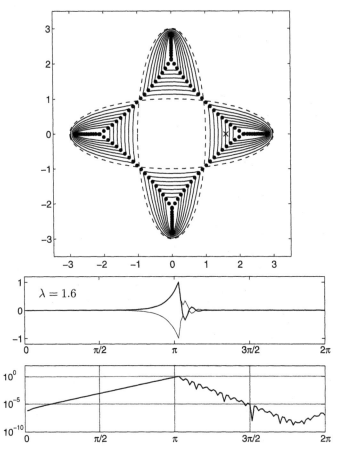

Figure 9.2: 'Two ellipses matrix' (9.5) with $N = 140$, $\varepsilon = 10^{-2}, 10^{-3}, \ldots, 10^{-10}$. The ε-pseudomode for $\lambda = 1.6$ (cross) has $x_* = \pi$ and $\varepsilon = 8.84 \times 10^{-7}$ (shown on linear and log scales). The left and right lobes of the pseudospectra correspond to wave packet pseudomodes of $\mathbf{A}_N$, and the top and bottom lobes to wave packet pseudomodes of $\mathbf{A}_N^{\mathrm{T}}$.

as x passes through x_*, and the figure shows the resulting localized wave packet. This matrix is a discontinuous analogue of the 'target matrix' of Figure 8.9.

Although Theorem 9.3 covers the example just presented, it would not apply to a similar kind of a matrix in which the discontinuity between $f(x_*^-, \theta)$ and $f(x_*^+, \theta)$ extended over several rows, corresponding, for example, to an 'antidiagonal' rather than 'horizontal' discontinuity in the matrix. In seeking a generalization in this direction one might expect that detailed attention to the nature of the discontinuity would be needed in order to ensure that the exponentially good wave packet pseudomodes persist. In

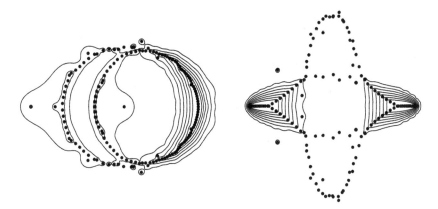

Figure 9.3: Repetition of Figures 8.7 and 9.2, but now with four rows in the middle of each matrix replaced by rows of independent random numbers from the standard normal distribution. As established by Theorem 9.4, the lobes of each figure corresponding to localized pseudomodes of $\mathbf{A}_N$ are hardly affected, while the other lobes, corresponding to wave packet pseudomodes of $\mathbf{A}_N^{\mathrm{T}}$, are entirely undone. These effects are reversed if columns rather than rows are altered.

fact, the opposite is true: They persist under arbitrary matrix alterations of any kind whatsoever, provided they are confined to a finite number of rows near the discontinuity. The following result from [777] sets forth this surprising state of affairs. This conclusion is consistent with the findings of [37] and [82] (see, e.g., Figure 4 of the latter) that whereas alterations of certain entries of large non-Hermitian banded Toeplitz matrices may move eigenvalues significantly, they have little effect on the overall nonnormality.

Twisted Toeplitz matrices with altered rows

Theorem 9.4 *Let $J \geq 0$ be a fixed integer. Let $\{\mathbf{A}_N\}$ be a family of matrices of any of the kinds described in Theorems 8.1 or 9.1–9.3, except that for each N, the J rows of $\mathbf{A}_N$ closest to x_* are modified arbitrarily, not only inside the band but potentially in any and all positions. (In the case of Theorem 9.2, replace 'rows' by 'columns'.) Then the conclusions of Theorems 8.1 and 9.3 still hold.*

Figure 9.3 illustrates this striking robustness of wave packet pseudomodes. As another application of this theorem, note that Theorem 7.2, the main result on pseudospectra of Toeplitz matrices first presented in [632], can be derived as a corollary [777].

Throughout the last section and this one, up to now, we have worked with problems in which the symbol curve changes winding number at a point $x = x_*$. For such problems we get wave packet pseudomodes that are robust with respect to perturbations. In the remainder of this section we shall consider a more delicate class of twisted Toeplitz problems, which we illustrate by one of our favorite matrices. For a positive integer N, define

$$s_j = 2 \sin x_j, \qquad 1 \le j \le N, \qquad (9.6)$$

with $x_j = 2\pi j/N$ as usual, and consider the $N \times N$ 'Scottish flag matrix' that in the case $N = 5$ takes the form

$$
\mathbf{A} = \begin{pmatrix}
s_1 & 1 & & & -1 \\
-1 & s_2 & 1 & & \\
& -1 & s_3 & 1 & \\
& & -1 & s_4 & 1 \\
1 & & & -1 & s_5
\end{pmatrix}. \qquad (9.7)
$$

The diagonal part of $\mathbf{A}$ is Hermitian, with real eigenvalues $\{s_j\}$. The off-diagonal part is skew-Hermitian, with imaginary eigenvalues $\{is_j\}$. However, $\mathbf{A}$ itself is strongly nonnormal. Figure 9.4 shows its eigenvalues and 2-norm pseudospectra for the case $N = 101$. Evidently the eigenvalues are neither real nor imaginary, lying instead on a cross at angle $\pi/4$ in the square $-2 \le \operatorname{Re}\lambda, \operatorname{Im}\lambda \le 2$. But whereas there are only N eigenvalues, the figure reveals that every number $-2 < \operatorname{Re}\lambda, \operatorname{Im}\lambda < 2$ is an ε-pseudoeigenvalue for a small value of ε.[1]

This is a twisted Toeplitz matrix, but the theorems presented so far do not apply to it. We can see the nature of the difficulty by examining the symbol, which is

$$f(x, \theta) = 2 \sin \theta - 2i \sin \theta. \qquad (9.8)$$

For each x, the symbol curve is an interval in the complex plane, enclosing no points at all with nonzero winding number. If the symbol curve crosses λ at $x = x_*$, the winding number is zero for $x < x_*$ and again zero for $x > x_*$. Thus the winding number hypothesis of our theorems is not satisfied; or algebraically, in Theorem 8.1, the condition $f(x_*, \theta) \ne \lambda$ for all $\theta \not\equiv \theta_*$ $(\operatorname{mod} 2\pi)$ is not satisfied.

The difficulty is not an artifact of our proofs but genuine. In general, if a family of twisted Toeplitz matrices has a symbol with a *double crossing*, by which we mean two or more values of θ_* for a single value of x_* associated with some number $\lambda = f(x_*, \theta_*)$, then exponentially good wave packet

---

[1] The spectrum and pseudospectra of $\mathbf{A}$ are exactly fourfold symmetric, as one can prove by showing that $\mathbf{A}$ is unitarily similar to $i\mathbf{A}$ via a Discrete Fourier Transform matrix: $i\mathbf{A} = \mathbf{F}\mathbf{A}\mathbf{F}^*$ with $f_{jk} = N^{-1/2}\omega^{(j-1)(k-1)}$, $1 \le j, k \le N$, where $\omega = \exp(-2\pi i/N)$. Thus for this matrix there is a perfect duality between $\mathbf{A}$ and its Fourier transform. More generally, one might consider that elegant family of matrices which have bandwidth m and whose Fourier transforms also have bandwidth m.

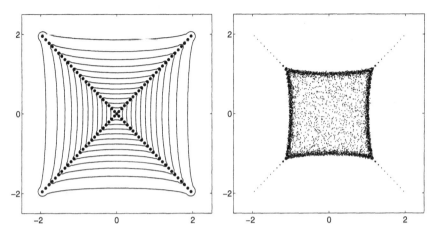

Figure 9.4: On the left, eigenvalues and ε-pseudospectra of the 'Scottish flag matrix' (9.7) for $N = 101$ and $\varepsilon = 10^{-1}, 10^{-2}, \ldots, 10^{-12}$. On the right, superposition of the eigenvalues of 100 matrices $\mathbf{A} + \mathbf{E}$, where each $\mathbf{E}$ is a random complex matrix of norm 10^{-4}.

pseudomodes need not exist for this λ. Figure 9.5 suggests this numerically by repeating Figures 8.9 and 9.4, but now with each entry of each matrix increased or decreased by 10%, at random. We see that the pseudospectra of Figure 9.4 are largely destroyed, while those of Figure 8.9 hardly change at all.

Why then does Figure 9.4 show such beautiful pseudospectra? The crucial fact is that the symbol (9.8) is *smooth* as a function of x. As a

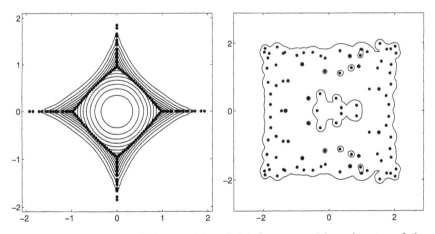

Figure 9.5: Repetition of Figures 8.9 and 9.4, but now with each entry of the matrix multiplied by 1.1 or 0.9 at random. The loss of smoothness has little effect in the first case, but destroys the pseudospectra in the second, where there is a double-crossing symbol.

result, good wave packets can be constructed entirely from energy at wave numbers close to θ_*, avoiding other wave numbers $\widehat{\theta}_*$ at which $f(x_*, \theta_*)$ takes the same value λ. The details can be worked out by methods of WKBJ asymptotics or microlocal analysis, and this is the subject of the paper by Borthwick and Uribe [76]. These authors establish a theorem in a more general context than ours that has the following special case.

Twisted Toeplitz matrices with smooth symbols

Theorem 9.5 *Under the same circumstances as in Theorem 8.1, instead of assuming that $f(x_*, \theta) \neq \lambda$ for all real $\theta \not\equiv \theta_*$ (mod 2π), assume that the dependence of f on x is C^∞. Then for any $M > 0$, there exist positive constants C_1, C_2, and C_3 such that for all sufficiently large N, there exists a nonzero pseudomode $\mathbf{v}^{(N)}$ satisfying*

$$\frac{\|(\mathbf{A}_N - \lambda)\mathbf{v}^{(N)}\|}{\|\mathbf{v}^{(N)}\|} \leq C_1 N^{-M}$$

and

$$\frac{|v_j^{(N)}|}{\max_j |v_j^{(N)}|} \leq C_2 \exp(-C_3 N (x_j - x_*)^2) \qquad (\text{mod } 2\pi).$$

This theorem explains the pseudospectra of Figure 9.4. Inside the unit square, it guarantees resolvent norm growth faster than any power of N, not actually exponential growth, though Borthwick and Uribe comment that similar techniques can probably be applied to prove exponential growth, too.

Figure 9.6 shows two further examples of twisted Toeplitz matrices with double crossing symbols. The 'dumbbell matrix' is tridiagonal with 0 on the main diagonal, $e^{ix_j/4}$ on the first superdiagonal, and the same entries reflected symmetrically to the first subdiagonal; the symbol is

$$f(x, \theta) = 2e^{ix/4} \cos\theta. \tag{9.9}$$

This example illustrates that a symmetric matrix can be exponentially non-normal and have interesting pseudospectra, provided it is not Hermitian. The 'scimitar matrix' is more complicated. If $\mathbf{J}$ is the matrix with $0.07x_j$ in the first superdiagonal and zero elsewhere, we set

$$\mathbf{A} = \text{diag}(e^{0.95ix_j})(\mathbf{J} + \mathbf{I} + \mathbf{J}^\mathrm{T});$$

the symbol is

$$f(x, \theta) = e^{0.95ix}(1 + 0.14x \cos\theta). \tag{9.10}$$

In our discussion of twisted Toeplitz matrices in this and the last section, we have made a distinction between symbols that satisfy a global winding

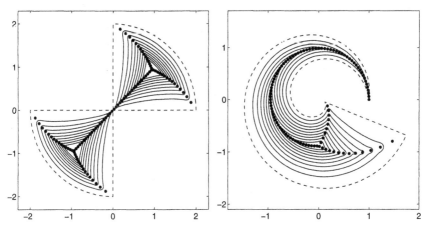

Figure 9.6: Spectra and ε-pseudospectra of the 'dumbbell' and 'scimitar' matrices of (9.9) and (9.10), both for $N = 100$ and $\varepsilon = 10^{-2}, 10^{-3}, \ldots, 10^{-10}$.

number condition (Theorem 8.1) and those that depend smoothly on x (Theorem 9.5). These two types of problems require different methods of analysis, and the genuineness of the difference between them has been highlighted in Figure 9.5. In §11 we shall see that the same distinction arises in the analysis of wave packet pseudomodes of variable coefficient linear differential operators, as illustrated in Figure 11.7.

III. Differential Operators

10 · Differential operators and boundary pseudomodes

In §7 we considered four types of matrices:

	no boundary	*boundary*
infinite	Laurent operator on $\{\ldots, -2, -1, 0, 1, 2, \ldots\}$	Toeplitz operator on $\{1, 2, 3, \ldots\}$
finite	circulant matrix on $\{1, 2, \ldots, N\}$ (periodic)	Toeplitz matrix on $\{1, 2, \ldots, N\}$ (nonperiodic)

The defining characteristics of these matrices are that they act on a domain that is *one-dimensional* and *discrete* and that the action is *translation-invariant* apart from boundary conditions (i.e., the matrices are constant along diagonals). Our analysis gave special attention to matrices that were banded, meaning that the operator also has a fourth property: It is *local*.

In this part of the book we turn to analogous problems where the domain is continuous. As before, we consider operators acting in a translation-invariant fashion in one dimension. The general term for such an object is a *convolution operator*, an integral operator defined by a kernel that depends on the difference $x - y$, i.e.,

$$(\mathbf{A}u)(x) = \int \kappa(x - y)u(y)\,\mathrm{d}y.$$

When the domain is a bounded or semibounded interval these are also called *Wiener–Hopf operators*. Virtually every result concerning the spectra or pseudospectra of Toeplitz and Laurent matrices and operators has a counterpart for convolution operators. Many of these were worked out originally in a 1993 paper of Reddy [621], which was extended by Böttcher in 1994 [78] and Davies in 2000 [181].

In the continuous case that fourth property, locality, assumes extra importance. At one extreme, one can consider a convolution operator for which $\kappa(x - y)$ is an arbitrary smooth function. This is analogous to a Toeplitz matrix or operator that is not banded. At the other extreme, corresponding to the banded case, one can consider a singular kernel $\kappa(x - y)$ consisting of nothing but a sum of powers of delta functions supported at $x - y = 0$. That is, $\mathbf{A}$ is a differential operator, and the property of translation-invariance becomes the condition that the operator has *constant coefficients*. Here we shall concentrate entirely on this case of constant-coefficient differential operators, bypassing the consideration of more general integral operators. Thus this section is devoted to the following four entities:

	no boundary	*boundary*
infinite	constant-coefficient differential operator on $(-\infty, \infty)$	constant-coefficient differential operator on $[0, \infty)$
finite	constant-coefficient differential operator on $[0, L]$ (periodic)	constant-coefficient differential operator on $[0, L]$ (nonperiodic)

Our principal interest is the bottom right case of differential operators on $[0, L]$, where we shall find resolvent norms that grow exponentially as $L \rightarrow \infty$, just as for Toeplitz matrices these norms grow exponentially as $N \rightarrow \infty$ (§7).

Let $a_0, \ldots, a_d$ $(a_d \neq 0)$ be a set of real or complex numbers, and let **A** denote the degree-d differential operator

$$a_0 + a_1 \frac{\mathrm{d}}{\mathrm{d}x} + \cdots + a_d \frac{\mathrm{d}^d}{\mathrm{d}x^d} \tag{10.1}$$

acting in L^2 on a domain and with boundary conditions to be specified.[1] The *symbol* of (10.1) is the function

$$f(k) = \sum_{j=0}^{d} a_j(-\mathrm{i}k)^j, \qquad k \in \mathbb{R}. \tag{10.2}$$

As a first example in this section, we shall consider the differential operator **A** defined by

$$\mathbf{A}u = \left(1 + \frac{\mathrm{d}}{\mathrm{d}x}\right)^3 u = u + 3u' + 3u'' + u''' \tag{10.3}$$

with symbol

$$f(k) = (1 - \mathrm{i}k)^3 = \mathrm{i}(k + \mathrm{i})^3. \tag{10.4}$$

For this, as for any constant-coefficient differential operator, $f(k)$ is a finite linear combination of powers of $\mathrm{i}k$. We regard **A** as an operator in L^2, defined, as described in §4, not on all of this Hilbert space but on a dense subdomain of sufficiently smooth functions.

---

[1]By L^2 we denote the space of square-integrable Lebesgue measurable functions on $[0, L]$, $[0, \infty)$, or $(-\infty, \infty)$, depending on the context. The domain $\mathcal{D}(\mathbf{A})$ in the operator sense of the term is in each case the subset of the appropriate L^2 space consisting of functions whose $(d-1)$st derivative is absolutely continuous and which satisfy the given boundary conditions, if any.

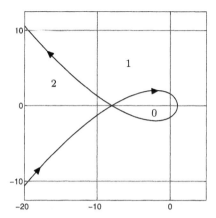

Figure 10.1: Symbol curve in the complex plane for the example (10.3)–(10.4). The numbers indicate regions associated with various winding numbers $I(f, \lambda)$. These are defined by completing the curve by a large circle near infinity that winds $3/2$ times around counterclockwise, since the order of the operator is 3.

To analyze constant-coefficient differential operators on $(-\infty, \infty)$, we note that in Fourier space, the operator is just multiplication by the symbol:

$$\widehat{\mathbf{A}u}(k) = f(k)\widehat{u}(k), \quad k \in [0, 2\pi].$$

Roughly speaking, subject to the same caveat mentioned on page 51 concerning ℓ^∞ versus ℓ^2, this amounts to the observation that if

$$u(x) = e^{-ikx}, \tag{10.5}$$

for some $k \in \mathbb{R}$, then $\mathbf{A}u$ is given by

$$\mathbf{A}u = f(k)u. \tag{10.6}$$

This operator $\mathbf{A}$ is normal, with spectrum consisting of all the numbers $f(k)$ with $k \in \mathbb{R}$: $\sigma(\mathbf{A}) = f(\mathbb{R})$. This is the *symbol curve*, shown in Figure 10.1 for (10.4). If $\mathbf{A}$ is formally self-adjoint (i.e., self-adjoint apart from the effects of boundary conditions), $f(\mathbb{R})$ is a subset of the real axis.

Given $\lambda \in \mathbb{C}\backslash f(\mathbb{R})$, we wish to define the winding number of f with respect to λ, but since $f(\mathbb{R})$ is not a closed curve, this requires some care. It is sufficient to replace $\mathbb{R}$ by closed contours Γ_R consisting of the interval $[-R, R]$ closed by a semicircle of radius R in the upper half of the complex plane. We consider the winding number associated with this closed contour traversed in the usual counterclockwise direction, and we define the *winding number* $I(f, \lambda)$ to be the limiting winding number of $f(\Gamma_R)$ about λ obtained for all sufficiently large R. If $\lambda \in f(\mathbb{R})$, $I(f, \lambda)$ is undefined.

The arguments used for operators on $(-\infty, \infty)$ carry over to the domain $[0, L]$ with periodic boundary conditions, if Fourier transforms are replaced by Fourier series. In (10.5), only values of k are now appropriate for which $e^{-ikL} = 1$, so that the function is periodic. Consequently we now have $\sigma(\mathbf{A}) = f(2\pi\mathbb{Z}/L)$, where $2\pi\mathbb{Z}/L$ denotes the integer multiples of $2\pi/L$.

We have proved most of parts (i) and (ii) of the following theorem. In this theorem, the statement that there are β homogeneous boundary conditions means that

$$u(0) = u'(0) = \cdots = u^{(\beta-1)}(0) = 0.$$

Part (iii) can be found, for example, in [229, Thm. 7.3] or [621, Thm. 7.1]. This result is related to the so-called Lopatinsky–Shapiro or complementing conditions that arise in the theory of elliptic boundary value problems for ordinary and partial differential equations [411]. For general spectral theory of non-self-adjoint two-point differential operators on finite intervals, see [512, 546].

Spectra of constant-coefficient differential operators

Theorem 10.1 *Let $\mathbf{A}$ be a degree-d constant-coefficient differential operator with symbol f: on $[0, L]$ with periodic boundary conditions, on $[0, \infty)$ with β homogeneous boundary conditions at $x = 0$ $(0 \le \beta \le d)$, or on $(-\infty, \infty)$.*

 (i) *On $[0, L]$, $\sigma(\mathbf{A}) = f(2\pi\mathbb{Z}/L)$.*

 (ii) *On $(-\infty, \infty)$, $\sigma(\mathbf{A}) = f(\mathbb{R})$.*

 (iii) *On $[0, \infty)$, $\sigma(\mathbf{A})$ is equal to $f(\mathbb{R})$ together with all the points enclosed by this curve with winding number that differs from $d - \beta$.*

The theorem is illustrated for the example (10.3)–(10.4) with $\beta = 2$ in Figure 10.2. For the unbounded and the periodic domains, the spectrum is a subset of the symbol curve. For the domains with boundaries, the behavior is not so simple. For differential operators with nonperiodic boundary conditions on $[0, L]$, there is no simple characterization of the spectrum. For $[0, \infty)$, the spectrum is generally a two-dimensional region.

Theorem 10.1(iii), illustrated in the upper right plot of Figure 10.2, can be explained in this way. Let $\mathbf{A}$ be a constant-coefficient differential operator with symbol f, and let $\lambda \in \mathbb{C}$ be any number for which $I(f, \lambda) < d - \beta$. Then not only is λ in the spectrum of $\mathbf{A}$, but it is an eigenvalue of $\mathbf{A}$, with a corresponding eigenfunction u whose amplitude decreases exponentially as $x \to \infty$. We call this a *boundary eigenfunction* or *boundary eigenmode*. For example, for the differential operator $\mathbf{A}$ given by (10.4), Figure 10.3 shows the eigenfunction associated with the eigenvalue $\lambda = 0$: $v(x) = x^2 e^{-x}$.

Now suppose $\lambda \in \mathbb{C}$ is such that $I(f, \lambda) > d - \beta$. In this case $\mathbf{A}$ has no boundary eigenfunction, but $\mathbf{A}^{\mathrm{T}}$ does, which implies that λ is again in the spectrum. (By $\mathbf{A}^{\mathrm{T}}$ we mean the complex conjugate of the adjoint of

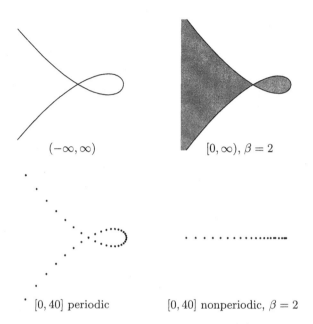

$(-\infty, \infty)$ $[0, \infty)$, $\beta = 2$

$[0, 40]$ periodic $[0, 40]$ nonperiodic, $\beta = 2$

Figure 10.2: Rightmost parts of the spectra of constant-coefficient differential operators of the four types associated with the symbol (10.4). In the final case there are two boundary conditions at the left and one at the right.

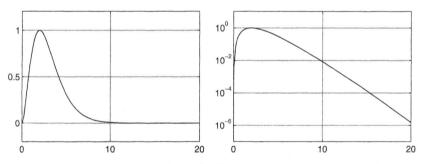

Figure 10.3: Eigenfunction $v(x) = x^2 e^{-x}$ of the differential operator (10.3) on $[0, \infty)$ with $\beta = 2$ associated with eigenvalue $\lambda = 0$ on a linear and a logarithmic scale. The eigenfunction is exponentially localized at the left boundary.

A, defined in §4.) It follows from (10.2) that $\mathbf{A}^{\mathrm{T}}$ is the degree-d constant-coefficient differential operator on $[0, \infty)$ with symbol $f(-k)$ and with $d - \beta$ homogeneous boundary conditions at $x = 0$. For this operator, the winding number is $\widehat{I} = d - I$ and $\widehat{\beta} = d - \beta$, and thus $\widehat{I} < d - \widehat{\beta}$ if $I > d - \beta$.

We can understand the origin of these boundary eigenmodes by considering the operator

$$\mathbf{A}u = u' \tag{10.7}$$

on $[0, \infty)$ with no boundary conditions, with the symbol $f(z) = -ik$. When $\mathbf{A}$ is applied to

$$u(x) = e^{-ikx} \tag{10.8}$$

for any $k \in \mathbb{C}$, the result is $\mathbf{A}u = -iku$. If $\mathrm{Im}\,k < 0$, then $u \in L^2$, and u is an eigenfunction of $\mathbf{A}$ with eigenvalue $\lambda = -ik$. This suggests that the spectrum of $\mathbf{A}$ is the left half-plane, and this is consistent with Theorem 10.1(iii), for with $f(k) = -ik$, the winding number is 0 in the left half-plane and 1 in the right. If we imposed a boundary condition at $x = 0$ and thus took $\beta = 1$, the spectrum would shift to the right half-plane.

For a general constant-coefficient differential operator $\mathbf{A}$, it follows from the principle of the argument of complex analysis that f takes the value λ at exactly $d - I(f, \lambda)$ points in the lower half-plane, counted with multiplicity. If u is the function (10.8) constructed from any of these points, then it is readily verified that $\mathbf{A}u = f(k)u = \lambda u$, just as we saw for an operator on $(-\infty, \infty)$ in (10.6). The question is, Does u satisfy the boundary conditions? In general, it does not. However, there are $d - I(f, \lambda)$ linearly independent possible choices of u, spanning a subspace of this dimension, with β boundary conditions to be satisfied. If $I(f, \lambda) > d - \beta$, there must exist a nonzero solution to this homogeneous system of equations, and this solution is an eigenfunction of $\mathbf{A}$ with eigenvalue λ on the interval $[0, \infty)$.

On the interval $[0, L]$, the same reasoning gives us good pseudoeigenfunctions. The following theorem, stated with slightly more general boundary conditions than in Theorem 10.1, is essentially due to Reddy [621].

Pseudospectra of constant-coefficient differential operators

Theorem 10.2 *Let $\{\mathbf{A}_L\}$ be a family of degree-d constant-coefficient differential operators on $[0, L]$ with β homogeneous boundary conditions at $x = 0$ and γ homogeneous boundary conditions at $x = L$, and let λ be any complex number with $I(f, \lambda) < d - \beta$ or $I(f, \lambda) > \gamma$. Then for some $M > 0$ and all sufficiently large L,*

$$\|(\lambda - \mathbf{A}_L)^{-1}\| \geq e^{LM}, \tag{10.9}$$

and there exist nonzero pseudomodes $v^{(L)}$ satisfying $\|(\mathbf{A}_L - \lambda)v^{(L)}\| / \|v^{(L)}\| \leq e^{-LM}$ such that for all $x \in [0, L]$,

$$\frac{|v^{(L)}(x)|}{\sup_x |v^{(L)}(x)|} \leq \begin{cases} e^{-Mx} & \text{if } I(f, \lambda) < d - \beta; \\ e^{-M(L-x)} & \text{if } I(f, \lambda) > \gamma. \end{cases} \tag{10.10}$$

The constant M can be taken to be any number for which $f(z) \neq \lambda$ in the strip $-M \leq \mathrm{Im}\,z \leq 0$ (if $I(f, \lambda) < d - \beta$) or $0 \leq \mathrm{Im}\,z \leq M$ (if $I(f, \lambda) > \gamma$).

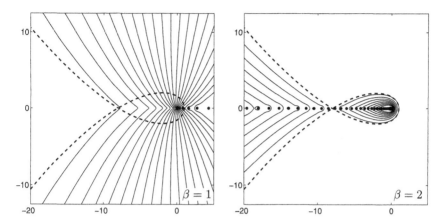

Figure 10.4: Spectrum and ε-pseudospectra of the constant-coefficient differential operator of Figure 10.2 on a finite interval $[0, L]$. On the left, one boundary condition at $x = 0$ and two at $x = L$ with $L = 16$ and $\varepsilon = 10^0$, $10^{-1/2}$, 10^{-1}, ..., 10^{-10}. On the right, two boundary conditions at $x = 0$ and one at $x = L$ with $L = 50$ and $\varepsilon = 10^{-1}$, 10^{-2}, ..., 10^{-8}. The dashed line is the symbol curve.

Extensions of this theorem for certain operators in multiple space dimensions or with variable coefficients (but still with pseudomodes localized at the boundary) are considered by Davies in [186].

We have already seen a simple example for Theorem 10.2: the operator (10.7) on $[0, L]$ with boundary condition $u(L) = 0$, considered in §5. There $d = 1$, $\beta = 0$, and $\gamma = 1$, and the theorem confirms that the resolvent norms must be exponentially large in the left half-plane.

Figure 10.4 illustrates Theorem 10.2 for the example (10.3)–(10.4). The boundaries of the pseudospectra line up beautifully along curves determined by applying the symbol to lines $\mathrm{Im}\, z = \text{constant}$ (not shown). These images, like Figures 10.7 and 10.8 below, were computed using Chebyshev spectral collocation methods on grids typically of about 100 points, as described in [775]; homogeneous boundary conditions of order higher than 1 at either end of the interval were handled by the method described at the beginning of Chapter 14 of that book. Techniques of this kind are reviewed in §43. For less carefully computed numerical eigenvalues of operators of large parameter L, truncation or rounding errors often produce much the same curves as are seen here for the pseudospectra.

Theorem 10.2 parallels Theorem 7.2 for Toeplitz matrices. To emphasize the connection between differential operators and Toeplitz matrices, suppose we consider finite difference discretizations of (10.3) on $[0, L]$. In the case $\beta = 1$ and $\gamma = 2$ (one boundary condition at $x = 0$ and two at $x = L$), $\mathbf{A}$ can be approximated by a Toeplitz matrix of dimension N with symbol

$$f_N(z) = \left(h^{-3} - \tfrac{1}{2}h^{-1}\right)z^{-2} + \left(-3h^{-3} + 3h^{-2} + 3h^{-1}\right)z^{-1}$$
$$+ \left(3h^{-3} - 6h^{-2} - \tfrac{3}{2}h^{-1} + 1\right) + \left(-h^{-3} + 3h^{-2} - h^{-1}\right)z,$$

where $h = L/(N+1)$; for $\beta = 2$ and $\gamma = 1$,

$$f_N(z) = \left(h^{-3} + 3h^{-2} + h^{-1}\right)z^{-1} - \left(3h^{-3} + 6h^{-2} - \tfrac{3}{2}h^{-1} - 1\right)$$
$$+ 3\left(h^{-3} + h^{-2} - h^{-1}\right)z - \left(h^{-3} - \tfrac{1}{2}h^{-1}\right)z^2.$$

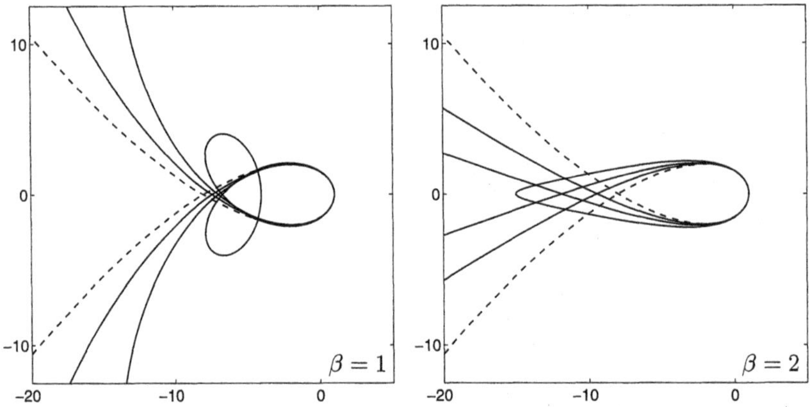

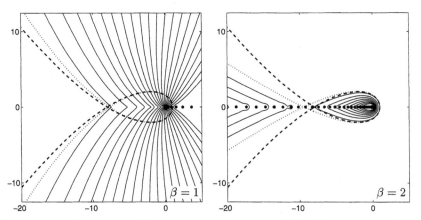

Figure 10.5: Symbol curves $f_N(\mathbb{T})$ for Toeplitz matrix approximations to the constant-coefficient differential operator (10.3) with β boundary conditions at $x = 0$ and $\gamma = d - \beta$ boundary conditions at $x = L$. The dashed curve marks the symbol curve for the differential operator; the solid curves show $f_N(\mathbb{T})$ for $N = 25, 50, 100$ (left) and $N = 50, 100, 200$ (right), read from the inside out.

Figure 10.6: Repetition of Figure 10.4, but now with Toeplitz matrices of dimension $N = 200$ approximating the constant-coefficient differential operators. The dashed lines denote the symbol curve for the differential operator; the dotted lines show the symbol curve for the finite-difference approximations. As N increases, the agreement with Figure 10.4 will improve.

These discretizations were computed by constructing cubic polynomial interpolants to the solution and then applying the operator (10.3) to the interpolants. Figure 10.5 shows the symbol curves $f_N(\mathbb{T})$ for both cases. As N increases, the symbol curves for the Toeplitz matrices become better approximations to those of the differential operator. The pseudospectra of the corresponding Toeplitz matrices, shown in Figure 10.6, approximate those of the differential operator seen in Figure 10.4. Note, however, that these finite difference discretizations would require far higher dimensions than Chebyshev spectral methods to achieve images correct to plotting accuracy.

We next turn to several other examples of constant-coefficient differential operators. Figure 10.7 shows pseudospectra for the advection-diffusion operator

$$\mathbf{A}u = u' + u'', \qquad f(k) = -\mathrm{i}k - k^2. \tag{10.11}$$

This operator (with $\beta = \gamma = 1$) is analyzed in detail in §12 and [627]; see also [185].

Figure 10.8 shows a more complicated example due to Davies (Example 2.4 of [181] with $c = 12$). Here we consider the sixth-order differential operator

$$\mathbf{A}u = -4u' + 6u'' - 15u''' - 12u^{(5)} - 2u^{(6)} \tag{10.12}$$

with symbol

$$f(k) = 4\mathrm{i}k - 6k^2 - 15\mathrm{i}k^3 + 12\mathrm{i}k^5 + 2k^6, \tag{10.13}$$

with $\beta = \gamma = 3$ homogeneous boundary conditions imposed at each endpoint of $[0, L]$. The coefficients are such that the symbol curve wraps around to leave a hole in the middle of the spectrum, as illustrated in the figure.

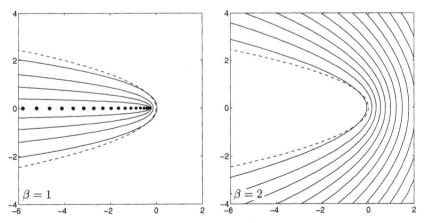

Figure 10.7: Spectrum and ε-pseudospectra of the advection-diffusion operator (10.11) on $[0, 24]$ for $\varepsilon = 10^{-1}, 10^{-2}, \ldots$ with one (left) and two (right) boundary conditions at $x = 0$.

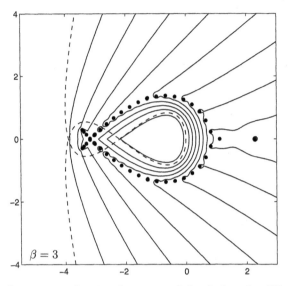

Figure 10.8: Spectrum and ε-pseudospectra of the sixth-order differential operator (10.12) on $[0, 120]$ with $\beta = \gamma = 3$ homogeneous boundary conditions at each
endpoint, for $\varepsilon = 10^{-1}, 10^{-2}, \ldots, 10^{-8}$.

Here is the analogue for constant-coefficient differential operators of
Theorem 7.3 for Toeplitz matrices. For $\beta = 0$ or $\beta = d$, this is Theorem 6.2
of [621]. For more general β, it is due to Davies [181] (who presents explicitly the case $\beta = d/2$ with d even).

Behavior of pseudospectra as $L \to \infty$

Theorem 10.3 *Let* $\mathbf{A}$ *be a degree-d constant-coefficient differential operator on* $[0, \infty)$ *with symbol f and β homogeneous boundary conditions
at $x = 0$ $(0 \leq \beta \leq d)$, and let $\{\mathbf{A}_L\}$ be the associated family of operators on $[0, L]$ with β homogeneous boundary conditions at $x = 0$ and
$\gamma = d - \beta$ homogeneous boundary conditions at $x = L$. Then for any
$\varepsilon > 0$,*

$$\lim_{L \to \infty} \sigma_\varepsilon(\mathbf{A}_L) = \sigma_\varepsilon(\mathbf{A}), \qquad (10.14)$$

and thus

$$\lim_{\varepsilon \to 0} \lim_{L \to \infty} \sigma_\varepsilon(\mathbf{A}_L) = \sigma(\mathbf{A}). \qquad (10.15)$$

One motivation for the study of spectral and pseudospectral properties of an operator $\mathbf{A}$ is to gain information about the behavior of the
associated time-dependent process (semigroup) $du/dt = \mathbf{A}u$. For example, if the spectrum of $\mathbf{A}$ extends infinitely far into the right half-plane,

then the time-dependent problem cannot be well-posed. The same is true of each ε-pseudospectrum: If $\mathbf{A}$ generates a C_0 semigroup, then each ε-pseudospectral abscissa $\alpha_\varepsilon(\mathbf{A})$ must be finite (Theorem 15.4). Conversely, if $\alpha_\varepsilon(\mathbf{A}) \leq \omega + \varepsilon$ for each $\varepsilon > 0$ for some ω, then $\mathbf{A}$ generates a C_0 semigroup with $\|e^{t\mathbf{A}}\| \leq e^{t\omega}$ for all $t \geq 0$ (Theorem 17.6). Such considerations provide interesting motivation for the problems discussed in this section. For example, it is clear from Figure 10.4 that the third-order differential operator (10.3) may generate a well-posed evolution process $du/dt = \mathbf{A}u$ in L^2 spaces on $[0, \infty)$ or $[0, L]$ if $\beta = 2$ boundary conditions are specified at $x = 0$, but that with just $\beta = 1$ boundary condition, the process will be ill-posed. This distinction is also noted by Fokas and Pelloni [281] for the same problem except with the lower order terms deleted (and with the conditions involving β interchanged since the third-order derivative has the opposite sign).

11 · Variable coefficients and wave packet pseudomodes

We now turn to the last of the four fundamental classes of matrices and operators whose discussion was initiated in §7.[1] Our concern here is variable coefficient non-self-adjoint linear differential operators, which under very general circumstances have exponentially good pseudoeigenfunctions in the form of localized wave packets. As we write, this material is far from well-known, but it seems likely that this situation may change, for this topic has connections with all kinds of other matters in mathematics and science, including hydrodynamic stability (§22), the theory of 'exponential dichotomy' in ordinary differential equations (ODEs) and their numerical discretizations [13, 46, 597], 'ghost solutions' of linear and nonlinear ODEs [211], Lewy's phenomenon of nonexistence of solutions to certain linear partial differential equations (PDEs) (§13), the 'non-Hermitian quantum mechanics' of Schrödinger operators with complex potentials [47, 48, 179], and just possibly, though these links have not been much investigated, any number of other problems involving non-Hermitian space-dependent systems, including atmospheric waves, wave propagation in stratified media, optical systems with an 'optical twist', 'quasi-modes' in variable waveguides, and the dynamics of the cochlea in the human ear. We comment on some of these applications at the end of the section.

The fact that variable coefficient non-self-adjoint differential operators have extended pseudospectra with wave packet pseudomodes was pointed out by Davies in 1999 [179, 180]. Shortly thereafter, Zworski [851] observed that Davies' discoveries could be related to long-established results in the theory of PDEs due to Hörmander, Duistermaat, Sjöstrand, and others [219, 409, 410]. Later related developments have appeared in [16, 95, 183, 188, 202, 612, 766, 851]; images in this book that correspond to problems of this kind include Figures 5.1 and 22.4–22.7.

We can see the essence of the matter in an elementary example. The differential equation

$$u' + xu = 0 \tag{11.1}$$

has the solution

$$u(x) = e^{-x^2/2}, \tag{11.2}$$

a Gaussian localized at $x = 0$. Suppose we are faced with the problem of solving (11.1) with one or both of the boundary conditions $u(-L) = u(L) = 0$ on the interval $[-L, L]$ for some large L. Then (11.2) is not a solution, but it comes very close. If we subtract the constant $\exp(-L^2/2)$ from it,

---

[1]This section is adapted from [766].

we get a function that satisfies the boundary conditions and also satisfies (11.1) for all x up to an error no greater than $L\exp(-L^2/2)$. This is what Domokos and Holmes call a *ghost solution* of the differential equation [211]. Equivalently, if we consider the linear differential operator

$$\mathbf{A}u = u' + xu \qquad (11.3)$$

acting on sufficiently smooth functions in $L^2(-L, L)$, the pseudoeigenfunction[2] (11.2) implies that 0 belongs to the ε-pseudospectrum of $\mathbf{A}$ for a value of ε that decreases exponentially as $L \to \infty$.

This consideration of intervals $[-L, L]$ with $L \to \infty$ follows the pattern of the last section, but we now switch to a different formulation that is more convenient for problems with variable coefficients. Instead of $[-L, L]$ we shall take the fixed interval $[-1, 1]$ and modify the differential operator so that it contains a small parameter h:

$$\mathbf{A}_h u = hu' + xu, \qquad u(-1) = u(1) = 0. \qquad (11.4)$$

This is known as a *semiclassical* formulation, and the letter h is used as an echo of Planck's constant [209, 449, 534]. The analogue of (11.2) becomes a Gaussian of width $\mathcal{O}(h^{1/2})$,

$$u(x) = e^{-x^2/2h}, \qquad (11.5)$$

and this pseudoeigenfunction shows that 0 is an ε-pseudoeigenvalue of $\mathbf{A}_h$ with $\varepsilon = \mathcal{O}(M^{-1/h})$ as $h \to 0$ for some $M > 1$. (Any value $1 < M < \sqrt{e}$ will do.) Moreover, the same is true for any number λ with $-1 < \operatorname{Re}\lambda < 1$, as is shown by the pseudoeigenfunction

$$u(x) = e^{-(x-\lambda)^2/2h} = Ce^{-(x-\operatorname{Re}\lambda)^2/2h}e^{ix\operatorname{Im}\lambda/h}. \qquad (11.6)$$

The situation is summarized in Figure 11.1 for $h = 1/50$. We see that the pseudospectra of $\mathbf{A}_h$ approximate the strip $-1 < \operatorname{Re}\lambda < 1$, and for the particular value $\lambda = 1/2 + i$, the optimal pseudoeigenfunction comes very close to the predicted form: a wave packet centered at $x = 1/2$ with wave number $1/h = 50$, i.e., wavelength $2\pi/50 \approx 0.13$.

This example can serve to illustrate the pattern of the general theory, which follows closely on that of the previous section. We associate the operator $\mathbf{A}_h$ of (11.4) with the x-dependent symbol

$$f(x, k) = -ik + x, \quad x \in (-1, 1), \qquad (11.7)$$

---

[2]More precisely, the pseudoeigenfunction is (11.2) minus the exponentially small constant $\exp(-L^2/2)$, with the correction necessary to satisfy the boundary conditions (see §4). The same qualification applies to many of the examples of this section, but for simplicity we shall use the term 'pseudoeigenfunction' without further comment.

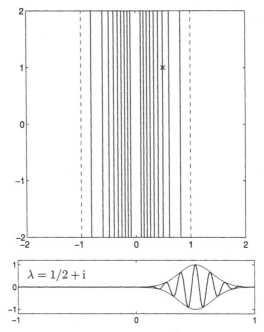

Figure 11.1: Above, ε-pseudospectra of the operator (11.4) with $h = 0.02$, $\varepsilon = 10^{-1}, \ldots, 10^{-9}$. (The spectrum is empty.) The resolvent norm $\|(\lambda - \mathbf{A}_h)^{-1}\|$ grows exponentially as $h \to 0$ for any λ lying in the strip $-1 < \mathrm{Re}\,\lambda < 1$, marked by the dashed lines. Below, an optimal pseudoeigenfunction for $\lambda = 1/2 + \mathrm{i}$ (marked with a cross in the top image) with $x_* = 1/2$ and $k_* = -1$. (Both real part and envelope are shown.)

which maps the real k-axis onto the negatively oriented vertical line $\mathrm{Re}\,f = x$ in the complex plane. We define the winding number of this symbol curve about any point $\lambda \in \mathbb{C}$ by completing it by a large semicircle traversed counterclockwise in the right half-plane. Thus for each x, the winding number is 0 if $\mathrm{Re}\,\lambda < x$, 1 if $\mathrm{Re}\,\lambda > x$, and undefined if $\mathrm{Re}\,\lambda = x$. As x increases from -1 to 1, each λ with $-1 < \mathrm{Re}\,\lambda < 1$ accordingly experiences a decrease in winding number when x passes through the value $\mathrm{Re}\,\lambda$. Just like Theorems 8.1 and 8.2 for the case of twisted Toeplitz matrices, Theorem 11.2 of this section will guarantee that each such value of λ is an ε-pseudoeigenvalue of $\mathbf{A}_h$ for a value of ε that shrinks exponentially as $h \to 0$.

We now formulate these ideas in general. Let an interval $[a, b]$ be given, $a < b$, and for a small parameter $h > 0$, let D_h be the scaled derivative operator[3]

$$D_h = h\frac{\mathrm{d}}{\mathrm{d}x}.$$

---

[3] Often a factor i is included in the definition of D_h.

For an integer $n \geq 0$, let continuous coefficient functions

$$a_j(x), \quad 0 \leq j \leq n,$$

be defined on (a, b), which may or may not be smooth. Following (10.1), we consider the family of linear operators $\{\mathbf{A}_h\}, h > 0$, defined by

$$(\mathbf{A}_h u)(x) = \sum_{j=0}^{n} a_j(x)(D_h^j u)(x), \quad a < x < b,$$

together with arbitrary homogeneous boundary conditions at $x = a$ and $x = b$ (the details of the boundary conditions will not matter), acting in a suitable dense domain in $L^p[a, b]$ for some p with $1 \leq p \leq \infty$.

Given $k \in \mathbb{C}$ and h, the action of $\mathbf{A}_h$ on the function

$$v(x) = \mathrm{e}^{-\mathrm{i}kx/h}$$

can be written as

$$(\mathbf{A}_h v)(x) = \sum_{j=0}^{n} a_j(x)(-\mathrm{i}k)^j v(x).$$

In other words, we have

$$(\mathbf{A}_h v)(x) = f(x, k)v(x),$$

where f is the *symbol* of $\{\mathbf{A}_h\}$, defined as in (10.2) as follows. For convenience of reference we also repeat here the definition of $\mathbf{A}_h$.

Differential operator, symbol, symbol curve, winding number

Let $\{\mathbf{A}_h\}$ be a family of variable coefficient differential operators as described above,

$$(\mathbf{A}_h u)(x) = \sum_{j=0}^{n} a_j(x)(D_h^j u)(x), \quad a < x < b. \tag{11.8}$$

The *symbol* of $\{\mathbf{A}_h\}$ is the function

$$f(x, k) = \sum_{j=0}^{n} a_j(x)(-\mathrm{i}k)^j, \tag{11.9}$$

defined for $x \in (a, b)$ and $k \in \mathbb{C}$. The *symbol curve* at $x \in (a, b)$ is the curve $f(x, \mathbb{R})$ interpreted as the oriented image in $\mathbb{C}$ of the positively oriented real k-axis under $f(x, k)$. For any $\lambda \in \mathbb{C} \backslash f(x, \mathbb{R})$, the *winding number* $I(f, x, \lambda)$ about λ is defined by completing the interval $[-R, R]$ of the k-axis by a semicircle of radius R in the upper half k-plane and taking the limit $R \to \infty$. If $\lambda \in f(x, \mathbb{R})$, $I(f, x, \lambda)$ is undefined.

This definition of the winding number is the same as in §10, apart from the x-dependence; for clarification see Figure 10.1.

Now let $x_* \in (a, b)$ and $k_* \in \mathbb{R}$ be given, and define $\lambda = f(x_*, k_*)$. If f were independent of x, the function $v(x) = e^{-ik_* x/h}$ would satisfy the eigenfunction equation for $\mathbf{A}_h$ with eigenvalue λ,

$$\mathbf{A}_h v = \lambda v. \tag{11.10}$$

The central idea of this section is that if f varies with x, then under suitable conditions, there will exist solutions to (11.10) near x_* in the form of wave packets, and if these decay exponentially as x deviates from x_*, they can be extended smoothly to zero so as to make exponentially good pseudoeigenfunctions, regardless of the boundary conditions.[4] The crucial condition is the following, the analogue for differential operators of the twist condition (8.7) for matrices.

Twist condition

The symbol $f = f(x, k)$ satisfies the *twist condition* at $x = x_* \in (a, b)$, $k = k_* \in \mathbb{R}$ if at this point it is differentiable with respect to x with $\partial f / \partial k \neq 0$ and

$$\mathrm{Im}\left(\frac{\partial f}{\partial x} \Big/ \frac{\partial f}{\partial k}\right) > 0. \tag{11.11}$$

Roughly speaking, if the twist condition is satisfied, then there are wave packet pseudomodes localized at $x = x_*$. Various theorems to this effect have been proved. Just as in the case of twisted Toeplitz matrices (§8), they follow two lines of reasoning.

The 'WKBJ' or 'microlocal' approach to these problems is the more standard one, with deep roots in the theoretical literature of partial differential equations and microlocal analysis and links also to physics and applied mathematics in connection with semiclassical mechanics and wave propagation.[5] The idea is to work with the equation (11.10) and use an asymptotic expansion to construct wave packet approximate solutions for $x \approx x_*$. This method requires f to depend smoothly on x. If the dependence is C^∞, one gets resolvent norms larger than any negative power of h

---

[4]To be precise, Theorem 11.2 is based upon solutions to (11.10) as just described, whereas Theorem 11.1 is based upon functions that satisfy (11.10) nearly but not exactly.

[5]There is an enormous literature on semiclassical analysis of various problems in wave propagation and quantum mechanics that sprang up with the work of Jeffreys, Wentzel, Kramer, and Brillouin in the 1920s and was carried on in later years by J. Keller and others [449, 450]. Most of the problems considered in this literature are self-adjoint, hence not exactly aligned with the phenomena discussed in this section, and yet the mathematical techniques involved are much the same. Even the self-adjoint case has relevance to the subject of this book; as mentioned in §6, an influential early publication was Arnol'd's 1972 article 'Modes and Quasimodes' [11].

as $h \to \infty$, and this is the standard assumption made in PDE theory going back to Hörmander and others [219, 410, 109]. If the dependence is analytic, the resolvent norms grow exponentially; a classic PDE reference for this case is a monograph-length paper of Sato, Kawai, and Kashiwara [662]. The process is purely local in k as well as x—this is the meaning of the word 'microlocal'—and thus yields a wave packet localized in wave number as well as frequency. This point of view was applied to pseudospectra by Davies [179, 180] and greatly generalized by Dencker, Sjöstrand, and Zworski [202]. The following is essentially a result of the latter authors, though expressed in a language quite different from theirs and restricted, unlike their theorems, to one space dimension and to differential rather than pseudodifferential operators. Our twist condition is a special case of the commutator condition in the general theory, due originally to Hörmander, which is expressed in terms of a Poisson bracket.

Wave packet pseudomodes (I)

Theorem 11.1 *Let $\{\mathbf{A}_h\}$ be a family of variable coefficient differential operators with symbol $f(x, k)$ as described above. Given $x_* \in (a, b)$ and $k_* \in \mathbb{R}$, define $\lambda = f(x_*, k_*)$, and suppose that the twist condition (11.11) is satisfied with $c_n(x_*) \neq 0$. Suppose also that the dependence of f on x is C^∞. Then for any $N > 0$, there exist constants C_1, C_2, and $C_3 > 0$ such that for all sufficiently small h, there exists a nonzero pseudomode $v^{(h)}$ satisfying*

$$\frac{\|(\mathbf{A}_h - \lambda)v^{(h)}\|}{\|v^{(h)}\|} \leq C_1 h^N \tag{11.12}$$

and

$$\frac{|v^{(h)}(x)|}{\max\limits_x |v^{(h)}(x)|} \leq C_2 \exp(-C_3(x - x_*)^2/h). \tag{11.13}$$

If f depends analytically on x in a neighborhood of x_*, then (11.12) can be improved to

$$\frac{\|(\mathbf{A}_h - \lambda)v^{(h)}\|}{\|v^{(h)}\|} \leq M^{-1/h} \tag{11.14}$$

for some $M > 1$.

The other approach to these problems could be called the 'winding number' or 'stable manifold' approach and appears in the paper [766], which was adapted from earlier results for twisted Toeplitz matrices [777]. Here, no smoothness of f with respect to x is assumed except for differentiability at a point. We can motivate this by noting that in the example (11.4), the existence of exponentially good localized pseudomodes certainly does

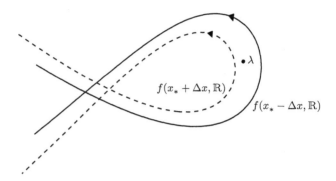

Figure 11.2: Winding number interpretation of Theorem 11.2. If the symbol curve $f(x, \mathbb{R})$ crosses λ as x increases through x_* in such a way that the winding number about λ decreases, then there is an exponentially good wave packet pseudomode localized at x_* with pseudoeigenvalue λ. Compare Figure 8.6.

not depend on smoothness with respect to x. The eigenvalue equation $\mathbf{A}_h u = \lambda u$ for that example is

$$\frac{u'}{u} = -\frac{1}{h}(x - \lambda),$$

and if the factor $x - \lambda$ is replaced by a function $\phi(x)$ that increases through 0 at $x = \lambda$, even if ϕ is not smooth, it is clear that the solution will still be an exponentially decaying wave packet.[6] In place of the smoothness assumption, such problems can be treated by a global assumption on k, which can be expressed algebraically or as a winding number condition. The proof now follows a very different tack involving the intersection of subspaces of exact solutions to (11.10) that decay to the right from the point x_* and those that decay to the left [766]. The process is not local with respect to k, and if f is not smooth, one must expect to find wave packets containing a wide spectrum of wave numbers. The result is summarized schematically in Figure 11.2.

It is interesting that winding numbers of symbol curves are used less often in the literature of differential operators than in that of Toeplitz matrices and operators (§§7–9), though they appear occasionally (see, e.g., Lemma 3.2 of [202]). Perhaps one reason is that winding numbers are particularly convenient for problems in one space dimension, which is the usual context for Toeplitz matrices but not for differential operators.

---

[6]Strictly speaking, the expression 'wave packet' implies localization in k as well as x, but in this book we use the term more loosely to refer to any function of amplitude $\mathcal{O}(\exp(-C(x - x_*)^2))$.

Wave packet pseudomodes (II)

Theorem 11.2 *Under the same circumstances as in Theorem 11.1, instead of assuming $f \in C^\infty$, suppose that $f(x_*, k) \neq \lambda$ for all real $k \neq k_*$. Equivalently, suppose that the symbol curve $f(x_*, \mathbb{R})$ passes just once through λ and that the winding number $I(f, x, \lambda)$ decreases by 1 as x increases through x_*. Then there exist constants C_1, $C_2 > 0$ and $M > 0$ such that for all sufficiently small h, there exists a nonzero pseudomode $v^{(h)}$ satisfying*

$$\frac{\|(\mathbf{A}_h - \lambda)v^{(h)}\|}{\|v^{(h)}\|} \leq M^{-1/h} \tag{11.15}$$

and

$$\frac{|v^{(h)}(x)|}{\max\limits_{x}|v^{(h)}(x)|} \leq C_1 \exp(-C_2(x - x_*)^2/h). \tag{11.16}$$

Similarly, Theorem 11.1 can be interpreted in terms of the symbol curve. The difference is that a global winding number is not defined; one just examines whether, as x increases through x_*, the portion of the symbol curve corresponding to $k \approx k_*$ sweeps across λ in the appropriate direction.

We have seen one example, and it is time for more. First, let us consider Davies' non-self-adjoint harmonic oscillator from [179, 180]. In (5.1) and Figure 5.1 we wrote this operator without the small parameter h. Including that parameter gives

$$\mathbf{A}_h u = -h^2 u'' + ix^2 u, \quad x \in (-\infty, \infty), \tag{11.17}$$

with symbol

$$f(x, k) = k^2 + ix^2. \tag{11.18}$$

For any fixed $x_* \in \mathbb{R}$, the symbol curve is the half-line $ix_*^2 + [0, \infty)$ in the complex plane traversed from ∞ to ix_*^2 and back again to ∞. For each λ along this half-line and corresponding choice of x_*, there are two values of k_*, one of which satisfies the twist condition (the one whose sign is the same as that of x_*). We can see this either by calculating the twist ratio $\frac{\partial f}{\partial x} / \frac{\partial f}{\partial k}$ of (11.11) as ix/k or by thinking of sections of the symbol curve. By Theorem 11.1 we conclude that every $\lambda \in \mathbb{C}$ with $\mathrm{Re}\,\lambda > 0$, $\mathrm{Im}\,\lambda > 0$ is an ε-pseudoeigenvalue of $\mathbf{A}_h$ for an exponentially small value of ε, as shown in Figure 11.3. There are two values of x_* for each λ, which explains why the optimal pseudomode in the figure consists of two wave packets rather than one. Because of the double crossing of the symbol curve, Theorem 11.2 does not apply to this example.

It is worth commenting further on the significance of a case like Figure 11.3 in which two distinct wave packets appear in a computed pseu-

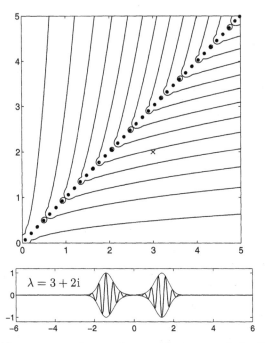

Figure 11.3: Above, eigenvalues and ε-pseudospectra of the Davies example (11.17) with $h = 1/10$, $\varepsilon = 10^{-1}, \ldots, 10^{-13}$. By Theorem 11.1, $\|(\lambda - \mathbf{A}_h)^{-1}\|$ grows exponentially as $h \to 0$ for any λ lying in the first quadrant of the complex plane. Below, an optimal pseudoeigenfunction for $\lambda = 3 + 2i$ (marked by the cross), with $x_* = \pm\sqrt{2}$ and $k_* = \pm\sqrt{3}$.

domode. The arguments on which Theorems 11.1 and 11.2 are based construct exponentially good pseudomodes in the form of single wave packets, not double ones. The present case is special because there are two values of x_* that are equally good for this construction: in some sense the multiplicity of the pseudoeigenvalue is 2, rather than the usual value of 1. Thus it would be equally valid to show a pseudomode with just one wave packet on the left, or just one on the right, except that the *optimal* pseudomode, just 0.003% better than these, is the odd function with two bumps. (The second-best would be another function with two bumps, but even instead of odd, and little different to the eye.)

Davies and Kuijlaars have analyzed the operator (11.17) in detail [188] (with i replaced by an arbitrary complex constant), basing their arguments on the theory of polynomials orthogonal with respect to a complex weight function. Among other results, their Theorem 3 implies that the condition numbers $\kappa(\lambda_n)$ of the eigenvalues of (11.17) (see §52) grow exponentially at the rate

$$\lim_{n\to\infty} \kappa(\lambda_n)^{1/n} = 1 + \sqrt{2},$$

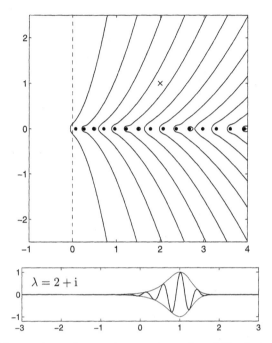

Figure 11.4: Above, eigenvalues and ε-pseudospectra of Bender's complex cubic oscillator (11.20) with $h = 1/10$, $\varepsilon = 10^{-1}, \ldots, 10^{-13}$. By Theorem 11.1, $\|(\lambda - \mathbf{A}_h)^{-1}\|$ grows exponentially as $h \to 0$ for any λ lying in the right half-plane. Below, an optimal pseudoeigenfunction for $\lambda = 2 + \mathrm{i}$ (cross), with $x_* = 1$ and $k_* = \sqrt{2}$.

with eigenvalues indexed with increasing distance from the origin. This precise estimate goes beyond the results presented here in (11.14) and (11.15), where the constant $M > 1$ is not specified.

The examples considered by Davies were not the first of this class whose pseudospectra were computed numerically. That distinction belongs to the Airy operator and related Orr–Sommerfeld operators considered by Reddy, Schmid, and Henningson in 1993 [624]. The Airy example is

$$\mathbf{A}_h u = h^2 u'' + \mathrm{i} x u, \quad x \in [-1, 1], \tag{11.19}$$

with boundary conditions $u(-1) = u(1) = 0$ and symbol $f(x, k) = -k^2 + \mathrm{i} x$. We shall not present a plot here, as an appropriate image appears as Figure 22.7 corresponding to the value $h = 1/50$. As mentioned in §22, the Airy operator has also been investigated by Stoller, Happer, and Dyson [731], Shkalikov [676, 677], and Redparth [628].

Our next example is closely related to that of Davies, the only difference being that the coefficient $\mathrm{i} x^2$ is replaced by $\mathrm{i} x^3$. This 'complex cubic oscillator' is a representative of a class of operators that have been discussed by

Bender and others, starting from unpublished work of D. Bessis in 1995, for applications in non-Hermitian quantum mechanics [47, 48, 194, 212, 370, 551]. The equation is

$$\mathbf{A}_h u = -h^2 u'' + ix^3 u, \quad x \in (-\infty, \infty), \tag{11.20}$$

with symbol $f(x, k) = k^2 + ix^3$. Mathematically, this is much the same as the Davies example, but the pseudospectra fill the right half-plane instead of the first quadrant since x^3 ranges over all of $\mathbb{R}$ rather than just $[0, \infty)$ (Figure 11.4). Most of this literature is concerned with establishing properties of the eigenvalues of (11.20) and related operators and does not question their physical significance. Again, Theorem 11.1 applies to this operator but Theorem 11.2 does not.

As a final example of this type, we mention the operator

$$\mathbf{A}_h u = h^2 u'' + (\alpha x^2 - \gamma x^4) u, \quad x \in (-\infty, \infty). \tag{11.21}$$

The pseudospectra of this operator, with parameter values $\alpha = 3 + 3i$ and $\gamma = 1/16$, are studied at length in the computational survey [774].

We now give two more examples with simply crossing symbol curves (the first was (11.4)), for which Theorem 11.2 is applicable as well as Theorem 11.1. Consider first the variable coefficient advection-diffusion equation

$$\mathbf{A}_h u = h^2 u'' + h(1 + \tfrac{2}{3} \sin x) u', \quad x \in [-\pi, \pi], \tag{11.22}$$

with periodic boundary conditions. The symbol is

$$f(x, k) = -k^2 - i(1 + \tfrac{2}{3} \sin x)k, \tag{11.23}$$

and for each x, the symbol curve is a parabola in the left half-plane traversed from the upper left to the origin and then down to the lower left. As x increases from $-\pi/2$ to $\pi/2$, the parabola widens from $-k^2 - \tfrac{1}{3}ik$ to $-k^2 - \tfrac{5}{3}ik$, causing a decrease in winding number from 1 to 0 for every point λ between these extremes. By Theorem 11.2, therefore, the region between the parabolas is one of exponentially large resolvent norm, as shown in Figure 11.5. We calculate that the pseudomode plotted has $x_* = \sin^{-1}(3/4) \approx 0.848$ and $k_* = -2$, i.e., wavelength $2\pi h/|k_*| = \pi/20 \approx 0.157$. The physical significance of the pseudospectra of (11.22) and related problems is discussed in §12.

Here is a higher order example. Consider the fourth-order differential operator

$$\mathbf{A}_h u = h^4 u^{(4)} - h \sin(x) u', \quad x \in (-\pi, \pi), \tag{11.24}$$

with periodic boundary conditions and symbol $f(x, k) = k^4 + i\sin(x)k$. For $x = -\pi/2$, the symbol curve is the quartic $k^4 - ik$, enclosing each point λ inside with winding number 3 (once by the quartic itself, twice

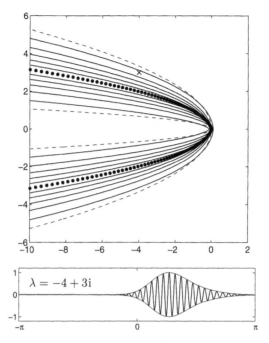

Figure 11.5: Above, eigenvalues and ε-pseudospectra of the advection-diffusion operator (11.22) with $h = 1/20$, $\varepsilon = 10^{-1}, \ldots, 10^{-5}$. By Theorems 11.1 or 11.2, $\|(\lambda - \mathbf{A}_h)^{-1}\|$ grows exponentially as $h \to 0$ for any λ lying between the two dashed parabolas. Below, an optimal pseudoeigenfunction of the same operator for $\lambda = -4 + 3i$ (cross), with $x_* = \sin^{-1}(3/4)$ and $k_* = -2$.

more by the fourth power of a large semicircle at ∞). As x increases, for any λ in this region, the winding number diminishes to 2 when the curve crosses once and then to 1 as it crosses a second time. Thus for each such λ we expect exponentially good pseudomodes consisting of a pair of wave packets (Figure 11.6). In the special case $\operatorname{Im}\lambda = 0$, both crossings occur at the same value of x_*. (Theorem 11.2 as written does not apply in this case, but that is just an accident of wording, for in fact its proof is valid in such cases of multiple crossings so long as there is a net decrease in winding number.) In this special case there will be pseudomodes consisting of two wave packets superimposed at the same x_* and with opposite values of k_*. This explains the lack of a smooth envelope in the figure.

We have discussed six examples, three governed by both Theorems 11.1 and 11.2 and three governed by Theorem 11.1 alone. The theorems lead us to expect that there should be a genuine difference between these cases: The first three examples should be robust with respect to nonsmooth per-turbations of the coefficients, while the others should be fragile. We found confirming this prediction numerically to be challenging, for the compu-

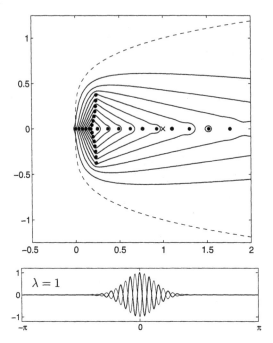

Figure 11.6: Above, eigenvalues and ε-pseudospectra of the fourth-order operator (11.24) with $h = 2/5$, $\varepsilon = 10^{-2},\ldots,10^{-10}$. By Theorems 11.1 or 11.2, $\|(\lambda - \mathbf{A}_h)^{-1}\|$ grows exponentially as $h \to 0$ for any λ lying in the quartic region marked by the dashed line (see text). Below, an optimal pseudomode for $\lambda = 1$ (cross), with $x_* = 0$ and k_* taking both values 1 and -1.

tations underlying Figures 11.1–11.6 are based on spectral methods (see §43), a technology that relies on smooth functions for its power, whereas if one reverts to simpler finite differences or finite elements, based on weaker smoothness assumptions, the accuracy may be too low to resolve ε-pseudospectra for small values of ε.[7] The compromise we eventually reached was to continue to use spectral methods but to choose perturbations that are somewhat smooth. We perturbed each ODE coefficient by multiplying it by a function having five but not six continuous derivatives, with amplitude varying between 0.9 and 1.1. The results appear in Figure 11.7. As predicted, three of the cases shown are robust and three are

---

[7]Indeed, an interesting subtlety arises here. Suppose one discretizes a smooth differential equation by a finite difference approximation on a uniform grid. The result is a twisted Toeplitz matrix of the kind considered in §§8 and 9. However, if the symbol curve for the differential equation has no crossings $f(x_*, k) = f(x_*, k_*)$ for $k \neq k_*$, so that Theorem 11.2 is applicable, this does not imply the same property for the twisted Toeplitz matrix approximation. Thus even when a differential equation has pseudospectra that are robust with respect to perturbations, those of its finite difference approximations will typically be fragile.

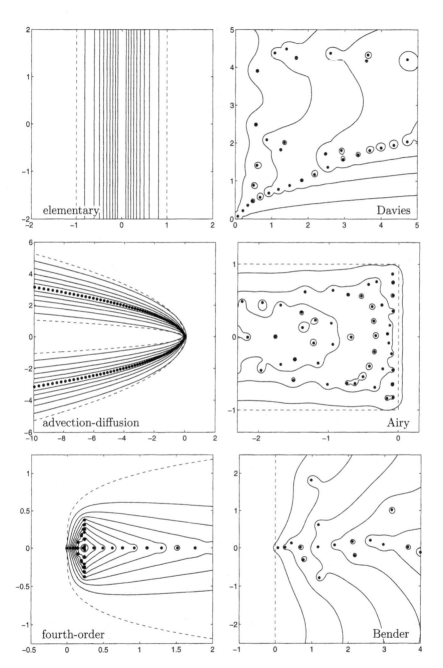

Figure 11.7: A repetition of six examples with each ODE coefficient modified by a C^5 multiplicative perturbation. The examples on the left have simply crossing symbol curves, and the perturbation has little effect on the pseudospectra (Theorem 11.2). For those on the right it changes them completely. Compare Figure 9.5 for twisted Toeplitz matrices.

fragile, with pseudospectra distorted almost beyond recognition by this C^5 perturbation. Theorem 11.2 ensures that the robust cases would in fact stand up to far rougher perturbations; it is just difficult to verify such cases numerically.

In our treatment of twisted Toeplitz matrices, Theorem 9.2 pointed out that exponentially good pseudoeigenvectors must exist not only if a problem satisfies the twist condition, but also if it satisfies the 'antitwist condition'. This conclusion is derived by considering transposed matrices, i.e., complex conjugates of the adjoints. The pseudoeigenvectors are concentrated not at the point where the antitwist condition is satisfied but at other internal or boundary points. A similar result, but with a crucial qualification, is true for differential operators.

Pseudomodes and the antitwist condition

Theorem 11.3 *Let* $\{\mathbf{A}_h\}$ *be a family of differential operators satisfying the conditions of Theorem 11.1 or 11.2, but with an antitwist condition instead of the twist condition (reversed inequality in (11.11)) or an increase in winding number instead of a decrease. Then provided* $\lambda \notin \sigma(\mathbf{A}_h)$, *there exist exponentially good ε-pseudomodes satisfying (11.12) and (11.14) or (11.15).*

The proof of this theorem consists of combining Theorem 4.4 with Theorem 11.1 or 11.2 applied to the family of transposed operators $A_h^{\mathrm{T}} = \overline{A_h^*}$, together with some further estimates [766]. The importance of the condition $\lambda \notin \sigma(\mathbf{A}_h)$ was made clear to us by Karel Pravda-Starov and can be motivated as follows. In a typical example satisfying the twist condition, the proof of Theorem 11.2 constructs a wave packet solution that is exponentially small at the boundaries and thus nearly satisfies the boundary conditions, making it an exponentially good pseudomode. The transposed operator will then have an equally good pseudomode concentrated (typically) at the boundaries. However, suppose the original problem has no boundary conditions, so that the wave packet solution is not just a pseudomode but an eigenmode. Then the transpose is an operator with 'too many boundary conditions', and may be unable to support modes pinned at the boundaries.

At the beginning of this section we mentioned some applications of the theory of wave packet pseudomodes of variable coefficient differential equations. Here at the end we briefly return to this list and make a few comments.

Lewy's phenomenon of nonexistence of solutions to certain linear PDEs is a fascinating story at the heart of the mathematical theory of PDEs, whose connection with pseudospectra was first pointed out by Zworski. This subject is discussed in §13. In brief, solutions to certain PDEs may lose uniqueness because of the possibility of ε-pseudosolutions for arbitrarily small ε, and this nonuniqueness for one problem carries over to nonexistence for the adjoint.

Hydrodynamic stability is the subject of several sections of this book. In §22 we shall see that the linear operators arising in the study of shear flows are non-self-adjoint with variable coefficients and have pseudomodes in the form of wave packets. The true eigenmodes also have this form, but often are of no greater physical significance than the pseudomodes. An extensive WKBJ analysis of modes and pseudomodes for some of these problems has been carried out by Chapman [138].

The phenomenon of 'ghost solutions' of certain differential equations is the subject of a recent article by Domokos and Holmes [211]. These are functions that one might say are 'nearly solutions, but not near solutions' to ODEs. Domokos and Holmes are concerned with nonlinear examples, their prototype being the looping of a long flexible rod, and they observe that although a ghost may be nowhere near a solution of the problem as posed, it may nevertheless be significant physically because it is a solution of a slightly perturbed problem. These are familiar themes for readers of this book. As this section has shown, the ghost phenomenon does not depend on nonlinearity.

Closely related to ghost solutions is the theory of dichotomy, a fundamental topic in the study of ODEs and their numerical approximations. In the mathematical literature this subject has roots going back to Perron and Lyapunov [46, 597], and its application to numerical computation was launched with the 1988 publication of the first edition of the leading textbook in this field, by Ascher, Mattheij, and Russell [13]. Ascher et al. showed that an ODE boundary-value problem is well-posed in a certain sense, rendering it solvable in practice, if and only if it exhibits an appropriate dichotomy between solutions decreasing from left to right and those decreasing from right to left. The ill-posed problems are those for which a solution may switch from increase to decrease at a midpoint, i.e., exactly our problems with wave packet pseudomodes. In fact, the first example that Ascher et al. give of a troublesome ODE, their Example 3.12, is precisely our equation (11.1).

We have not said much of the physical implications of the extended pseudospectra of variable coefficient non-self-adjoint differential operators, but these are considerable. For example, a non-self-adjoint system of this kind may exhibit resonant response to vibrations at pseudoeigenvalue input frequencies that is as strong in practice as if they were true eigenvalues. Unlike familiar eigenvalue phenomena, these responses involve continuous distributions of pseudoeigenvalues or pseudomodes, even for problems posed in bounded domains where the spectrum is in principle discrete.

We must emphasize that the theory and examples presented in this section involve one dimension, but the phenomena of wave packet pseudomodes extend readily to multiple dimensions, i.e., PDEs instead of ODEs, with the twist condition becoming an inequality involving a Poisson bracket. The generalization of Theorem 11.1 presented by Dencker, Sjöstrand, and Zworski is multidimensional from the start and holds for pseudodifferential, not just partial differential, operators [202]. Whether analogues of Theorem 11.2 exist in the multidimensional or pseudodifferential cases is not known.

12 · Advection-diffusion operators _____

A common source of nonnormality in applications is the blending of the phenomena of diffusion and advection, which occurs in fluid mechanics, financial mathematics, and many other fields. Mathematically this mix corresponds to the combination of a second derivative such as u_{xx} or Δu and a first derivative such as u_x or $\mathbf{a} \cdot \nabla u$. On an unbounded domain with constant coefficients, the result is a normal operator, but as soon as boundaries or variable coefficients are introduced, the operators become nonnormal, providing illustrations of the effects described in the last two sections. If the diffusion is weak relative to the advection, the singularly perturbed case, then the nonnormality is typically of magnitude $\mathcal{O}(C^{1/\eta})$ for some $C > 1$ as a function of the diffusion parameter η, whose inverse is known as the *Péclet number*.

Advection-diffusion equations are also called convection-diffusion, drift-diffusion, Fokker–Planck, and Ginzburg–Landau equations. The examples we shall consider are linear, though there are important nonlinear generalizations, notably the Navier–Stokes equations of fluid mechanics. Our focus will be on problems in one space dimension, which suffice to illustrate effects of nonnormality.

We begin with the time-dependent constant-coefficient partial differential equation

$$u_t = \eta u_{xx} + u_x, \qquad x \in (0,1), \tag{12.1}$$

where $\eta > 0$ is a constant, with boundary and initial data

$$u(0) = u(1) = 0, \qquad u(x,0) = u_0(x). \tag{12.2}$$

Equivalently, we write $u_t = \mathbf{L}u$, where $\mathbf{L}$ is the operator

$$\mathbf{L}u = \eta u_{xx} + u_x \tag{12.3}$$

acting on twice-differentiable functions satisfying the boundary conditions in a suitable function space. This problem is well-posed in $L^2(0,1)$ and $L^1(0,1)$, both of which may be of interest in applications. If u_0 is nonnegative and not identically zero, then for each $t > 0$, $u(x,t)$ is an analytic function of x with $u(x,t) > 0$ for all $x \in (0,1)$. Applying integration by parts to (u, u), where $(\cdot, \cdot)$ denotes the usual inner product on $(0,1)$, gives a decay estimate in L^2,

$$\frac{d}{dt}\|u\|_2^2 = -2\eta\|u_x\|_2^2 < 0, \tag{12.4}$$

and under the above assumptions on u_0, a similar consideration of $(u, 1)$ gives decay in L^1,

$$\frac{\mathrm{d}}{\mathrm{d}t}\|u\|_1 = -\eta(u_x(0) - u_x(1)) < 0. \tag{12.5}$$

The differences between (12.4) and (12.5) reflect fundamental physics. The first estimate reminds us that diffusion causes attenuation of mean-square energy throughout the interior of the domain. The second contains no interior term, since diffusion just redistributes the total volume of substance such as heat or a chemical; in this case attenuation occurs only at the boundary. From here on we shall take $\| \cdot \| = \| \cdot \|_2$, but the phenomena we focus on are essentially the same in both norms.

The behavior of (12.1)–(12.2) for small η is simple and easily understood. A signal well-separated from the boundary propagates leftward at speed 1, experiencing little alteration in shape except in the small wavelengths (of order $\mathcal{O}(\eta^{1/2})$). When it hits the boundary, the physics changes. The signal stops moving and shrinks to a boundary layer of thickness $\mathcal{O}(\eta)$, which proceeds to decay exponentially on a time scale of order η. Figure 12.1 shows this two-phase process for a problem with $\eta = 0.015$. Note how the initial Gaussian translates to the left, begins to change shape as it encounters the boundary, and then settles into its fixed asymptotic form. Figure 12.2 plots the operator norms for this and two other values of η.

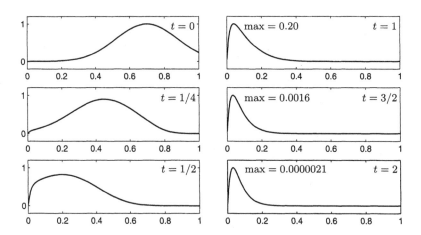

Figure 12.1: Behavior of (12.1)–(12.2) with diffusion constant $\eta = 0.015$: first advection unrelated to eigenmodes, then absorption at the boundary controlled by the dominant eigenmode and eigenvalue $\lambda \approx -16.8$. Note that the first three images, on the left, are plotted on a fixed scale, but the vertical scale varies in the next three images.

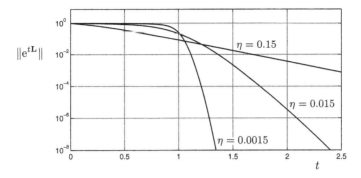

Figure 12.2: Operator 2-norms for (12.3). In the limit $\eta = 0$ the curve steepens to a cliff at $t = 1$ (compare Figure 5.4).

Again we see the two phases, first with slow decay, then much more rapid.[1]

The eigenvalues and eigenfunctions of $\mathbf{L}$ play a part in this story: They describe the second phase, the behavior for $t \gg 1$. In the images on the right in Figure 12.1, one can see that the curve is beginning to settle down to an asymptotically fixed shape, that of the dominant eigenfunction, with exponentially diminishing amplitude. The eigenvalues and eigenfunctions are

$$\lambda_n = -\tfrac{1}{4}\eta^{-1} - \eta\, n^2\pi^2, \quad u_n = \mathrm{e}^{-x/2\eta}\sin(n\pi x) \qquad (12.6)$$

for $n = 1, 2, 3, \ldots$; see Figure 12.3. The factor $\mathrm{e}^{-x/2\eta}$ defines the boundary layer that appears as $t \to \infty$. Notice that as η decreases toward 0, the eigenvalues move further from the origin, but they also get closer together. Davies has emphasized that, as a consequence, although the solution looks approximately like the dominant eigenmode for $t \gg 1$, the components in the second and higher eigenmodes remain relatively significant until $t \gg 1/\eta$, at which point the signal has largely died away [185].

The behavior of (12.1)–(12.2) for $t \ll 1$, on the other hand, has nothing to do with the eigenvalues and eigenfunctions. Mathematically, one could certainly describe the behavior in these terms. The eigenfunctions form a complete set in $L^2(0, 1)$ with a bounded condition number, a *Riesz basis* (see p. 472),[2] and therefore any initial function $u_0(x)$ is equivalent

---

[1]The initial part of the the $\|\mathrm{e}^{t\mathbf{L}}\|$ curve of Figure 12.2 is only algebraically flat: its slope is $-\eta\pi^2$, the numerical abscissa of $\mathbf{L}$. The corresponding curve for $L^1(0,1)$, by contrast, would be exponentially flat, with initial slope zero, since a narrow initial pulse experiences exponentially little attenuation for small times. Elsewhere in this book we shall see analogous weak initial decay for upwind Gauss–Seidel iterations (Figure 25.4), card shuffling (Figure 57.2), and laser cavities (Figures 60.4 and 60.8).

[2]This observation can be found, for example, in Example III.6.11 of the book by Kato [448]. Other information about advection-diffusion operators appears in that book in §III.2.3 and in Examples III.5.32, III.6.20, and VIII.1.19.

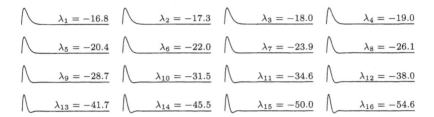

$$\lambda_1 = -16.8 \quad \lambda_2 = -17.3 \quad \lambda_3 = -18.0 \quad \lambda_4 = -19.0$$

$$\lambda_5 = -20.4 \quad \lambda_6 = -22.0 \quad \lambda_7 = -23.9 \quad \lambda_8 = -26.1$$

$$\lambda_9 = -28.7 \quad \lambda_{10} = -31.5 \quad \lambda_{11} = -34.6 \quad \lambda_{12} = -38.0$$

$$\lambda_{13} = -41.7 \quad \lambda_{14} = -45.5 \quad \lambda_{15} = -50.0 \quad \lambda_{16} = -54.6$$

Figure 12.3: First sixteen eigenfunctions u_n for (12.3) with $\eta = 0.015$, normalized so that $\|u_n\|_\infty = $ constant. The eigenvalues $\lambda_n = -\frac{1}{4}\eta^{-1} - \eta n^2 \pi^2$ are rounded to three digits.

to a superposition of them.[3] However, (12.6) shows that this is a highly unnatural basis in which to expand a general function f, as is confirmed in Figure 12.3. All the eigenfunctions have the same $e^{-x/2\eta}$ concentration near $x = 0$. It follows that if f is a function of amplitude of order 1 localized at the end of the interval near $x = 1$, then its representation in this basis will involve a sum of terms with huge coefficients, of order $e^{1/2\eta}$, that cancel almost perfectly. As time elapses and the eigencomponents decrease at their various rates, the cancellation will continue to be almost perfect, but the details will change in such a way as to generate leftward advection. We know this has to be so, but one would never figure it out by examining the eigenmodes.

If eigenvalues fail to explain the initial phase of an advection-diffusion process, can pseudospectra do better? The answer is half yes, for they explain the operator norms, and half no, for they do not reveal the leftward advection. Concerning the latter observation we note that despite the discussion above, the failure of eigenvalue analysis to capture the advection is not really caused by the nonnormality. The same failure occurs for a problem involving a normal operator, such as the dispersive wave equation $u_t = iu_{xx}$ on a domain $[-L, L]$ with periodic boundary conditions. The eigenfunctions of this problem are global complex exponentials e^{ikx}, which give no indication that a smooth wave packet will propagate steadily at the group velocity (see §54). That wave packet propagation is again an epiphenomenon of shifting patterns of interference among su-

---

[3]In fact, $\mathbf{L}$ is symmetrizable by a diagonal similarity transformation, which physicists call a gauge transformation. To see this we may define $u(x) = e^{-x/2\eta}v(x)$, which implies $u' = e^{-x/2\eta}(-v/2\eta + v')$, $u'' = e^{-x/2\eta}(v/4\eta^2 - v'/\eta + v'')$, and therefore $\mathbf{L}u = e^{-x/2\eta}(\eta v'' - v/4\eta)$. Thus if operators $\mathbf{K}$ and $\mathbf{M}$ are defined by $\mathbf{K}v = \eta v'' - v/4\eta$ and $\mathbf{M}v = e^{-x/2\eta}v$, then we have $\mathbf{L} = \mathbf{MKM}^{-1}$. As claimed, $\mathbf{K}$ is self-adjoint. However, since the norm of a multiplication operator is the supremum of its multiplier function, $\mathbf{M}$ has an exponentially large condition number: $\|\mathbf{M}\| \|\mathbf{M}^{-1}\| = e^{1/2\eta}$. Thus, although the similarity transformation is trouble-free in a theoretical sense, it distorts the physics of the advection-diffusion problem beyond recognition if η is small.

perposed eigenfunctions, though now without the huge coefficients. What distinguishes the advection phase of an advection-diffusion problem from simpler wave propagation phenomena, from the point of view of eigenvalue analysis, is a human factor: The former has a 'dominant' eigenmode, and this is distracting! Many have been misled into assuming that this eigenmode must have some significance for all t.

The pseudospectra of $\mathbf{L}$, however, do explain the initial flatness of the curve of $\|e^{t\mathbf{L}}\|$ against t. Figure 12.4 shows that whereas the eigenvalues lie deep inside the left half-plane, the pseudospectra crowd up toward the imaginary axis, filling the region of the λ-plane bounded by the parabola $\mathrm{Re}\,\lambda = -\eta(\mathrm{Im}\,\lambda)^2$. This can be explained as an application of the general theory presented in §10. To fit the framework of that section we note that if $t = \eta\tau$ and $x = \eta s$, then (12.1)–(12.2) can be rewritten as $u_\tau = u_{ss} + u_s$ for $x \in (0, \eta^{-1})$ with boundary conditions $u(0) = u(\eta^{-1}) = 0$ and initial condition $u(s,0) = u_0(\eta s)$. This differential operator in the variable s has symbol $f(k) = -k^2 - \mathrm{i}k$, given earlier in (10.11), which maps the real k-axis onto the parabola $\mathrm{Re}\,\lambda = -(\mathrm{Im}\,\lambda)^2$. It follows from Theorem 10.2

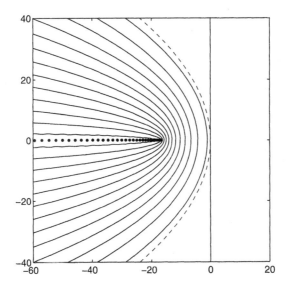

Figure 12.4: Eigenvalues and ε-pseudospectra of (12.3) for $\eta = 0.015$, $\varepsilon = 10^0, 10^{-1}, \ldots, 10^{-12}$; the condition numbers of the eigenvalues are all bounded by the condition number $e^{1/2\eta} \approx 3.4 \times 10^{14}$ of the system of eigenfunctions. The dashed line is the symbol curve $\mathrm{Re}\,\lambda = -\eta(\mathrm{Im}\,\lambda)^2$, and the vertical line is the imaginary axis. As the diffusion constant η is decreased, the eigenvalues move out of the frame to the left and the pseudospectra straighten up toward half-planes as in Figure 5.2, widening further the gap between initial and asymptotic behavior.

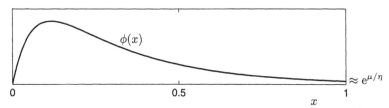

Figure 12.5: Explanation of the parabola $\mathrm{Re}\,z = -\eta(\mathrm{Im}\,z)^2$ of Figure 12.4, inside of which the resolvent norms are exponentially large. For any λ in the interior of the parabola, the function ϕ of (12.7) satisfies the eigenvalue equation $\mathbf{L}\phi = \lambda\phi$ and the boundary condition $\phi(0) = 0$, and it satisfies the boundary condition $\phi(1) = 0$ up to an error that decreases exponentially as $\eta \to 0$. Here $\eta = 0.05$ and $\lambda = -3$. Compare Figure 5.3.

that the norm of the associated resolvent is exponentially large inside this parabola, as shown in Figure 10.7, and this implies a similar assertion for the parabola $\mathrm{Re}\,\lambda = -\eta(\mathrm{Im}\,\lambda)^2$ in the x, t variables, as seen in Figure 12.4.

For an example of a quantitative link from pseudospectra to the $\|e^{t\mathbf{L}}\|$ curve, we may note in Figure 12.4 that the spectral abscissa is $\alpha(\mathbf{L}) \approx -16.8$, which implies $\|e^{0.1\mathbf{L}}\| > 0.186$, far below the true value of about 0.984. By contrast, the ε-pseudospectral abscissae for $\varepsilon = 1$ and 10 are $\alpha_\varepsilon(\mathbf{L}) \approx -1.046$ and 9.45, and if these are inserted in condition (15.12) of Theorem 15.4 with the value $M = 1$ in that theorem, we get the better estimates $\|e^{0.1\mathbf{L}}\| > 0.804$ and 0.908. (See Figure 15.2.)

We can explain these pseudospectra without appealing to the general theory as follows (Figure 12.5). For any $\lambda \in \mathbb{C}$, the function

$$\phi(x) = \frac{e^{\alpha_+ x/\eta} - e^{\alpha_- x/\eta}}{(\alpha_+ - \alpha_-)/\eta}, \quad a_\pm = -\tfrac{1}{2} \pm \tfrac{1}{2}\sqrt{1 + 4\eta\lambda} \qquad (12.7)$$

satisfies the eigenvalue equation $\mathbf{L}\phi = \lambda\phi$ and the boundary condition $\phi(0) = 0$. If λ is inside the parabola $\mathrm{Re}\,\lambda = -\eta(\mathrm{Im}\,\lambda)^2$, then α_+ and α_- have negative real parts, so ϕ decays exponentially with x. Thus although ϕ does not in general satisfy the right boundary condition exactly, it does so approximately within an error of exponentially small order $e^{\mu/\eta}$, where

$$\mu = \max\{\mathrm{Re}\,\alpha_+, \mathrm{Re}\,\alpha_-\} = -\tfrac{1}{2} + \left|\mathrm{Re}\,a_\pm + \tfrac{1}{2}\right|. \qquad (12.8)$$

Plots of the pseudospectra of advection-diffusion operators were first published in 1994 by Reddy and Trefethen [627], who established the pseudospectral estimates in the following theorem. Harrabi has also studied the computation of these pseudospectra under various discretizations [374]. The assertion below about the numerical range comes from a letter of Kato to Trefethen in March 1995, and further information about this example can be found in [185].

Advection-diffusion operator

Theorem 12.1 *The operator* **L** *of* (12.3) *has eigenvalues and eigenfunctions given by* (12.6), *and the condition number of the set of eigenfunctions normalized as in* (12.6) *is* $e^{1/2\eta}$. *The numerical range is the closed set of all points to the left of the parabola defined by* $\mathrm{Re}\,z = -\eta(\mathrm{Im}\,z)^2 - \eta\pi^2$. *Let* γ *be a fixed non-real number in the interior of the parabola* $\mathrm{Re}\,z = -(\mathrm{Im}\,z)^2$, *so that* $\lambda = \gamma/\eta$ *is a non-real number in the interior of the parabola* $\mathrm{Re}\,z = -\eta(\mathrm{Im}\,z)^2$, *and let* $a_\pm$ *and* ϕ *be defined by* (12.7), *with* $\alpha_+ - \alpha_- = \sqrt{1 + 4\gamma} = \sigma + \mathrm{i}\tau$. *Then*

$$\|(\lambda - \mathbf{L})^{-1}\| \sim \frac{2e^{-\mu/\eta}(\sigma^2 + \tau^2)^{1/2}}{(1 - \sigma^2)(1 + \tau^2)} \qquad (12.9)$$

as $\eta \to 0$, *where* $\mu = \max\{\mathrm{Re}\,\alpha_+, \mathrm{Re}\,\alpha_-\} < 0$ *as in* (12.8).

A proof of this theorem is carried out in [627] by means of Green's functions, the same technique used in Theorem 5.1 for a simpler operator without diffusion. Such a proof requires the estimation of the norm of the resolvent of **L**, which is the solution operator for the inhomogeneous problem $(\lambda - \mathbf{L})u = f$ with boundary conditions $u(0) = u(1) = 0$. For each fixed λ and y, the Green's function $G(x, y)$ for this problem is the solution $u(x)$ corresponding to the choice $f(x) = \delta(x - y)$, where δ is the Dirac delta function; in matrix language we may think of $G(\cdot, y)$ as being 'the yth column of the resolvent'. We can write $G(x, y)$ explicitly as

$$G(x, y) = a\phi(x) + \phi([x - y]_+), \qquad (12.10)$$

where a is a constant, with ϕ still defined by (12.7); the notation $[x - y]_+$ denotes 0 for $x - y \le 0$ and $x - y$ for $x - y \ge 0$. The term $\phi([x - y]_+)$ would be a solution to $(\lambda - \mathbf{L})u = f$ if there were no right-hand boundary condition; the term $a\phi(x)$ is a correction added to enforce that boundary condition. Figure 12.6 sketches $G(x, y)$ for $\eta = 0.05$, $y = 0.5$, and three values of λ, revealing that for $\lambda = -4$, this correction term is exponentially large in $0 < x < y$. It is by quantifying such effects that Theorem 12.1 can be proved.

We have concentrated on the time-dependent process generated by an advection-diffusion operator **L**, but of course, there are other aspects of such an operator, and pseudospectra are significant for a number of these. For example, whereas the eigenvalues of **L** imply that the inhomogeneous equation $(\lambda - \mathbf{L})u = f$ lacks a unique solution when λ is equal to certain negative real numbers, the pseudospectra show that the equation may be unsolvable in practice or have a solution of limited significance for applications, when λ is any number in a wide region of the complex plane. Physically one can interpret this as an effect of 'pseudoresonance' [780].

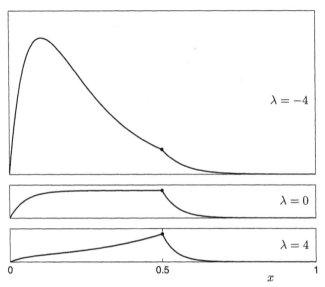

Figure 12.6: Green's functions $G(x, y)$ for (12.1)–(12.2) for three values of λ ($\eta = 0.05$, $y = 1/2$), normalized by $G(y, y) = 1$. The value $\lambda = -4$ lies in the interior of the parabola $\mathrm{Re}\, z = -\eta(\mathrm{Im}\, z)^2$, and this explains the exponentially large lobe of $G(x, y)$ in the first plot, giving a graphic view of the large pseudo-resonant response of the advection-diffusion system to certain stimuli.

The pseudospectra of these operators also have implications for numerical methods. One of these is that matrix iterations for such problems converge more slowly than the eigenvalues alone would suggest (§25). Another is that numerical time-stepping procedures have tighter time step restrictions than the eigenvalues would suggest, as pointed out by Morton in 1980 [564] (§§31 and 32). For these and other reasons, advection-diffusion equations have provided a challenge of lasting interest in numerical analysis, with three books on the subject published in 1996 alone [552, 565, 643]; see also [269, 813]. Davies has proposed the use of pseudoeigenfunctions for the numerical treatment of such equations [185]. Many other methods have also been put forward, including graded meshes, upwinding, exponential fitting, streamline diffusion, and Petrov–Galerkin finite elements.

Up to this point we have considered a constant-coefficient advection-diffusion operator with boundary pseudomodes, as in §10. Now let us look at some variable coefficient problems with interior wave packet pseudo-modes, as in §11. Following §11, we employ a semiclassical formulation with a small parameter h.

One example of this kind was presented in equation (11.22) and Figure 11.5. The operator was $\mathbf{L}_h u = h^2 u_{xx} + h(1 + \frac{2}{3}\sin x)u_x$, with periodic boundary conditions on $[-\pi, \pi]$ and symbol $f(x, k) = -k^2 - \mathrm{i}(1 + \frac{2}{3}\sin x)k$. For each x, the symbol curve is a parabola in the left half-plane, and

Figure 11.5 showed that the pseudospectra fill a region bounded by the two parabolas corresponding to the extremes $\sin x = \pm 1$. Pseudospectra bounded by parabolas are typical for all kinds of variable-coefficient advection-diffusion processes. Physically, one expects that any advection-diffusion system with variable coefficients is likely to respond to inputs at a continuum of frequencies, with a typical response taking the shape of a localized wave packet.

A simpler example of a variable coefficient advection-diffusion operator, examined in a paper of Cossu and Chomaz [164], is

$$\mathbf{L}_h u = h^2 u_{xx} + h u_x + (\tfrac{1}{4} - x^2)u, \tag{12.11}$$

acting on the whole real line.[4] Like Davies' example (11.17), this is a nonnormal variation on the theme of a Schrödinger operator for a harmonic oscillator. The eigenvalues and eigenfunctions are

$$\lambda_n = -(2n+1)h, \quad u_n = e^{-(x+x^2)/2h} H_n(x/\sqrt{h}) \tag{12.12}$$

for $n = 1, 2, 3, \ldots$, where H_n is the nth Hermite polynomial [150]. Pseudospectra for the case $h = 0.02$ are shown in Figure 12.7. Applying the theory of the last section, we note that the symbol is

$$f(x, k) = -k^2 - ik + (\tfrac{1}{4} - x^2) \tag{12.13}$$

and the symbol curves are parabolas adjusted by the variable horizontal offset $1/4 - x^2$. The winding number about a value λ decreases from 1 to 0 as this curve crosses λ from left to right, which occurs for a negative value of x, and this explains why the pseudoeigenfunction in the figure sits in the left half of the domain.

This is a very interesting example, for it illustrates fundamental behavior of wide physical importance and is simple enough for an elementary explanation. We have here a standard advection-diffusion process coupled with a factor that causes exponential amplification for $|x| < 1/2$ and exponential attenuation for $|x| > 1/2$. The associated time-dependent process $u_t = \mathbf{L}u$ must be susceptible to transient growth of order $\mathcal{O}(C^{1/h})$ on a time scale $\mathcal{O}(h^{-1})$ for some $C > 1$, for a pulse will grow exponentially during the time of order $\mathcal{O}(h^{-1})$ that it spends passing through the amplification region. Cossu and Chomaz relate this behavior to the notion of *local convective instability* used in plasma physics and fluid mechanics [421], which is relevant to a number of flows in unbounded domains, including wakes, jets, and boundary layers. The effect has nothing to do with eigenmodes, and indeed, if we set the diffusion to zero, the transient amplification and the

---

[4]Cossu and Chomaz consider more general coefficients, including a complex coefficient for the u_{xx} term, making the equation dispersive as well as diffusive and the pseudospectra asymmetric with respect to the real axis.

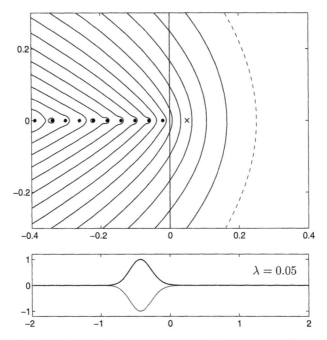

Figure 12.7: Eigenvalues and ε-pseudospectra of the operator (12.14) of Cossu and Chomaz [164] with $h = 0.02$, $\varepsilon = 10^{-2}, \ldots, 10^{-15}$. Below, an optimal pseudoeigenfunction for $\lambda = 0.05$ (marked by a cross in the top image) with central position and wave number $x_* = -1/\sqrt{5} \approx -0.45$ and $k_* = 0$.

pseudoeigenfunctions persist but the eigenvalues and eigenmodes vanish, since every signal attenuates at an accelerating rate as it travels toward $x = -\infty$. Diffusion creates eigenvalues by recycling a little energy back upstream, enabling a pulse to settle into a fixed form that attenuates at a fixed rate. In Figure 12.7 the eigenvalues are in the left half-plane, but by changing the constant $1/4$ to $1/2$, say, we could shift some of them into the right half-plane, in which case the amplification would exceed the diffusion on balance and there would be exponential growth. In fluid mechanics, this is a *global instability*.

A more exotic variable coefficient advection-diffusion operator has been investigated by Benilov, O'Brien, and Sazonov [49, 50]. These authors consider the instability of a thin viscous liquid film on the inner surface of a rotating cylinder in an approximation in which gravitational effects are included but inertial and capillary effects are ignored. They reduce their problem to the operator[5]

$$\mathbf{L}_h u = h^2 \sin(x) u_{xx} + h u_x, \qquad (12.14)$$

---

[5]More precisely, Benilov et al. consider $\mathbf{L}_h u = h \sin(x) u_{xx} + (1 + h \cos(x)) u_x$. The behavior is essentially the same as that of (12.14).

again with periodic boundary conditions on $[-\pi, \pi]$, with symbol

$$f(x, k) = -\sin(x)k^2 - \mathrm{i}k. \tag{12.15}$$

An unusual feature here is that for two values of x, the coefficient of $-\sin(x)$ passes through zero. For each x, the symbol curve is the parabola $\mathrm{Re}\,z = -\sin(x)(\mathrm{Im}\,z)^2$ described in the direction of decreasing imaginary part. Completing this curve by a semicircle at infinity, we see that its winding number is 1 about points to its right and 0 about points to its left. For any λ in the domain bounded by the two parabolas $\mathrm{Re}\,z = \pm(\mathrm{Im}\,z)^2$, the winding number accordingly decreases by 1 at some value of x in the interval $(-\pi, -\pi/2)$ or $(\pi/2, \pi)$, giving an exponentially good wave packet pseudoeigenfunction. Figure 12.8 confirms this geometry for the case $h = 1/10$.

From a dynamical point of view, a distinctive feature of (12.14) is that although the pseudospectra fill large expanses of the right half-plane, all the eigenvalues lie on the neutrally stable imaginary axis. Many operators with this property can be extracted from the examples in this book, including that of Figure 12.7, if multiplied by the factor i, but it is unusual to find such behavior arising in a fluid dynamics problem. Benilov et al. speak of a phenomenon of 'explosive instability'.

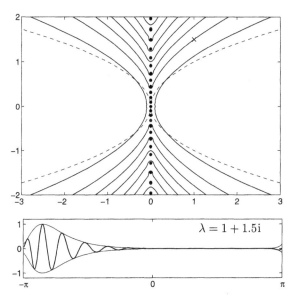

Figure 12.8: Above, eigenvalues and ε-pseudospectra of the operator (12.14) of Benilov et al. [49, 50] with $h = 1/10$ and $\varepsilon = 10^{-1}, \ldots, 10^{-7}$. The resolvent norm $\|(\lambda - \mathbf{L})^{-1}\|$ grows exponentially as $h \to 0$ for any λ lying in the hourglass shape bounded by the two dashed parabolas. Below, an optimal pseudoeigenfunction for $\lambda = 1 + 1.5\mathrm{i}$ (marked by a cross in the top image) with central position and wave number $x_* = \sin^{-1}(4/9) - \pi \approx -2.68$ and $k_* = -1.5$.

13 · Lewy–Hörmander nonexistence of solutions _____

One might expect that any linear partial differential equation with smooth coefficients should have some solutions, at least locally near a point and in the absence of boundary conditions. Many mathematicians were startled in 1957 when Hans Lewy published a four-page article that presented a counterexample [500]. From here, in the hands of Hörmander and others, sprang an extensive theory of existence and nonexistence of solutions to linear PDEs with smooth coefficients [36, 409, 412, 413, 497, 579, 580, 783].[1] In 2001, Zworski pointed out that Hörmander's construction could be interpreted in terms of pseudospectra [202, 850, 851], and the aim of this brief section is to explain this connection.

Lewy's example involved three independent variables x, y, and t, but for nonsolvability at a single point, only two are needed. Consider the 'Mizohata equation' $u_x + ixu_y = f(x,y)$, posed in a neighborhood of $(x,y) = (0,0)$ [553]. This is a linear first-order PDE with analytic coefficients. If the right-hand side f is also analytic, then the Cauchy–Kowalewski theorem ensures that there is a solution in a neighborhood of $(0,0)$.[2] If the assumption of analyticity of f is weakened, however, everything changes.

Example of Lewy nonexistence

Theorem 13.1 *There exists a C^∞ function $f(x,y)$ such that the partial differential equation*

$$\mathbf{L}u = u_x + ixu_y = f(x,y) \tag{13.1}$$

has no solution in any neighborhood of $(0,0)$.

By a 'solution' we mean a C^1 complex function of x and y, i.e., continuously differentiable, that satisfies the equation. More generally it can be shown that there are no solutions in the sense of distribution theory. The example (13.1) is analyzed in [300, 553, 578, 783], and [412, Section 26.3].

---

[1]Hörmander won the Fields Medal in 1962 for this and related work. Also among the authors of papers in this list, Charles Fefferman won the Fields Medal in 1978 and Louis Nirenberg won the Crafoord Prize in 1982 and the U.S. National Medal of Science in 1995.

[2]The Cauchy–Kowalewski theorem is formulated for a boundary value problem with boundary data on a noncharacteristic surface. Here a suitable surface is the line $x = 0$, and any analytic boundary condition can be specified there, such as $u(0,y) = 0$. The solution takes the form of a convergent Taylor series.

The key idea is that a Fourier transform in y reduces the adjoint of this equation to a family of equations

$$\widehat{u}_x + kx\widehat{u} = \widehat{f}$$

parametrized by the dual variable k. This is exactly the equation (11.4) that we examined as the simplest example of a variable coefficient problem with wave packet pseudomodes; the small parameter is $h = k^{-1}$.

Proof. Before looking at neighborhoods of $(0,0)$, let us consider the simpler problem of exhibiting a function f for which (13.1) has no bounded y-periodic solution in the strip $-\pi \le y \le \pi$ around the x-axis. Consider the adjoint $\mathbf{L}^*$ of $\mathbf{L}$, defined by

$$\mathbf{L}^*v = -v_x + \mathrm{i}xv_y, \tag{13.2}$$

and the particular function

$$v_k(x,y) = \mathrm{e}^{k(\mathrm{i}y - x^2/2)}, \tag{13.3}$$

for any integer $k > 0$. This function is $2\pi/k$-periodic (hence also 2π-periodic) with respect to y, and it satisfies $\mathbf{L}^*v_k = 0$. Define an inner product by

$$(u,v) = \int_{-\infty}^{\infty} \int_{-\pi}^{\pi} u(x,y)\,\overline{v(x,y)}\,\mathrm{d}y\,\mathrm{d}x;$$

for our purposes it is enough to assume that u and v are C^1 functions in the strip that are 2π-periodic with respect to y and that u is bounded and v is integrable over the strip. Now fix k and suppose u is a bounded C^1 function, 2π-periodic in y, that satisfies the equation $\mathbf{L}u = v_k$. Then using familiar manipulations of adjoints based on integration by parts, we calculate

$$0 = (u, \mathbf{L}^*v_k) = (\mathbf{L}u, v_k) = (v_k, v_k) \ne 0. \tag{13.4}$$

This contradiction implies that no such u can exist. Notice that (13.4) amounts to the Fredholm alternative: Lack of uniqueness for the problem $\mathbf{L}^*v = g$ implies lack of existence for the problem $\mathbf{L}u = f$.

To extend the argument to neighborhoods of $(0,0)$ we consider, What goes wrong with the contradiction (13.4) if $\mathbf{L}u = v_k$ holds only in such a neighborhood? It is the third equality that fails, $(\mathbf{L}u, v_k) = (v_k, v_k)$. However, the function v_k is localized with respect to x in a region of size $\mathcal{O}(k^{-1/2})$ around $(0,0)$, and by superposing a collection of such functions for various parameters k, we can construct a solution to $\mathbf{L}^*v = 0$ that is localized around $y = 0$ too. Here is one way to do it. First, we drop the assumption of periodicity and redefine the inner product to extend over the whole x-y plane,

$$(u,v) = \int_{-\infty}^{\infty} \int_{-\infty}^{\infty} u(x,y)\,\overline{v(x,y)}\,\mathrm{d}y\,\mathrm{d}x,$$

assuming now that u and v are C^1 and square-integrable over the plane. Next we redefine v_k so that instead of containing a single wave number k, it is a superposition of functions with wave numbers in a range about k (Figure 13.1):

$$v_k(x,y) = c_k \int_{-\infty}^{\infty} \phi(\kappa k^{-1/2}) e^{(k-\kappa)(iy - x^2/2)} \, d\kappa, \qquad (13.5)$$

where ϕ is a fixed nonzero C^∞ function with support in $[-1,1]$ and c_k is chosen so that $(v_k, v_k) = 1$. This integral is a convolution in the κ variable with a function of support in $[-k^{1/2}, k^{1/2}]$, and according to familiar properties of Fourier transforms, this is equivalent to a multiplication in the y variable by $\widehat{\phi}(yk^{1/2})$, where $\widehat{\phi}$ is an analytic function with $\widehat{\phi}(s) = \mathcal{O}(|s|^{-N})$ as $|s| \to \infty$ for any N. Thus v_k is a wave packet localized with respect to both x and y near $(0,0)$, and as before, it satisfies $\mathbf{L}^* v_k = 0$.

Now define $f(x,y)$ to be a superposition of the functions v_k for square-integer values $k \to \infty$:

$$f(x,y) = \sum_{j=4,9,16,\ldots} \alpha_j \, v_j(x,y).$$

If we choose coefficients with $\alpha_j = \mathcal{O}(j^{-N})$ as $j \to \infty$ for all N, then $f(x,y)$ is C^∞. The quadratic spacing of the indices j has the effect that the functions v_k in the sum are orthonormal, which implies

$$(f, v_k) = \alpha_k, \quad k = 4, 9, 16, \ldots. \qquad (13.6)$$

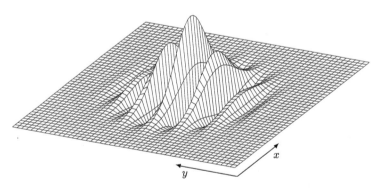

Figure 13.1: An example of the kind of multidimensional wave packet eigenfunction or pseudoeigenfunction that explains the phenomenon of Lewy–Hörmander nonexistence as in Theorem 13.1. The plot shows the real part of a typical function (13.5) for $k = 40$ on the square $[-0.7, 0.7]^2$. Because the adjoint equation $\mathbf{L}^* v = 0$ has solutions or approximate solutions like this, the primal equation (13.1) cannot be solved for all f.

Now suppose u is a bounded C^1 function that satisfies $\mathbf{L}u = f$ in a neighborhood of $(0,0)$. Then in view of the decay of v_k away from $(0,0)$ we have

$$0 = (u, \mathbf{L}^* v_k) = (\mathbf{L}u, v_k) = (f, v_k) + \beta_k \qquad (13.7)$$

with $\beta_k = \mathcal{O}(k^{-N})$ for all N as $k \to \infty$. Together, (13.6) and (13.7) imply $\alpha_k = -\beta_k$ for all k. However, the decay rate of the numbers β_k as $k \to \infty$ is determined by that of $\widehat{\phi}(s)$ as $|s| \to \infty$, whereas we are free to choose the coefficients α_k to decay more slowly. Thus we have a contradiction, and there can be no such function u. ∎

The sign of the imaginary coefficient in (13.1) does not matter. If the coefficient were negated, we would consider $k \to -\infty$ instead of $k \to \infty$ to obtain wave packet eigenfunctions.

Note that the above proof uses eigenfunctions rather than pseudoeigenfunctions. This is possible because solutions (13.3) to the adjoint equation happen to be available that are valid throughout the x-y plane. For a PDE with less simple coefficients, perhaps just C^∞ instead of analytic, this would no longer be the case. One would still make use of wave packets, but now they would satisfy $\mathbf{L}^* v = 0$ only approximately in a neighborhood of $(0,0)$.

In the theoretical PDE literature, the study of local solvability of linear equations has advanced far. First, unlike our example, this literature treats problems in an arbitrary number of space dimensions, where the wave packet pseudomode becomes higher dimensional. Hörmander's original condition that determines whether such pseudomodes may exist is known as the *commutator condition*, expressed in terms of a Poisson bracket. Hörmander showed that this condition is necessary for local existence, but did not settle the question of sufficiency. The theory was soon generalized to pseudodifferential operators, which means operators defined by a symbol that need not be a polynomial, and the commutator condition was generalized to the so-called Ψ condition, conjectured by Nirenberg and Treves in 1970 to be equivalent to local solvability for both partial differential and pseudodifferential operators of principal type [579, 580]. In 1973 Beals and Fefferman confirmed the Nirenberg–Treves conjecture for the special case of partial differential operators [36], and for pseudodifferential operators, after much work by many people, the conjecture has been proved recently by Dencker [201]. Surveys of developments in this field over half a century can be found in [413] and [497].

We shall not attempt to give any details of this extensive and rather technical subject, but instead offer an observation about how the wave packet argument presented here fits into the larger pattern of the theory of pseudospectra. Throughout the last eighty pages, we have emphasized that localized pseudomodes may appear in the interior of domains when there are variable coefficients, or at boundaries when the coefficients are constant. Theorem 13.1 is based on the former case, variable coefficients and wave packets. One might ask, does it have an analogue for a constant-coefficient problem with boundary pseudomodes? Here is an answer. This theorem is probably not new, but we do not know where it has appeared before.

Nonexistence near a boundary

Theorem 13.2 *There exists a C^∞ function $f(x, y)$ such that the PDE*

$$u_x + iu_y = f(x, y), \qquad (13.8)$$

together with the boundary condition

$$u(0, y) = 0, \qquad (13.9)$$

has no solution in any one-sided neighborhood of $(0, 0)$.

Notice that the coefficients in (13.8) are now constant. By a one-sided neighborhood, we mean the intersection of the half-plane $x \geq 0$ with a neighborhood of $(0, 0)$, and by a solution we mean a function that is C^1 in this set, including up to the boundary $x = 0$, and satisfies the equation. To prove this theorem we can follow the same pattern of argument as for Theorem 13.1, but instead of interior pseudomodes of the form $e^{-kx^2/2}$, we now use boundary pseudomodes of the form e^{-kx} (Figure 13.2). The adjoint operator now has no boundary condition at all and is essentially the first-derivative operator considered in §5.

In fact, Theorem 13.2 is more elementary than this, and (13.8) is well-known as the *inhomogeneous Cauchy–Riemann* or $\overline{\partial}$ *equation*. Harold Boas

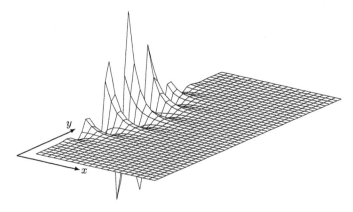

Figure 13.2: Similarly, boundary eigenmodes or pseudoeigenmodes can explain lack of solvability for boundary value problems like that of Theorem 13.2.

has pointed out to us that the theorem can be proved as follows. Let G be a real-valued C^∞ function defined in a neighborhood of 0. Define a function g in a one-sided neighborhood of $(0,0)$ by $g(x,y) = G(y)$, and define $f(x,y) = iG'(y)$. Then g satisfies

$$g_x + ig_y = f.$$

Now suppose that u is a solution to (13.8)–(13.9) in a one-sided neighborhood Ω of $(0,0)$. Then the difference $w = u - g$ satisfies the homogeneous Cauchy–Riemann equation $w_x + iw_y = 0$ in Ω, and thus w is an analytic function in Ω satisfying $w(0,y) = G(y)$. By the Schwarz reflection principle, it can be reflected to an analytic function in a neighborhood of $(0,0)$. But this implies that G is an analytic function of y and thus that $f(0,y)$ is an analytic function of y, not just C^∞. Thus (13.8)–(13.9) is not solvable for arbitrary C^∞ data f.

The use of analytic function theory to prove nonsolvability of certain differential equations is not new; it was the route followed by Lewy in his original paper [500] and by Garabedian for the Mizohata equation (13.1) in [300]. Ideas related to the nonsolvability of boundary-value problems like (13.8)–(13.9) actually go back much further, at least as far as Hadamard's analysis of ill-posedness of the Cauchy problem for the Laplace equation in the early twentieth century [363]. To derive solvability theorems for general partial differential or pseudodifferential operators, however, methods of analytic functions are not enough.

IV. Transient Effects
and Nonnormal Dynamics

14 · Overview of transients and pseudospectra _____

The quiz at the front of this book set the tone. Nonnormal matrices and operators have many aspects, but there is one that grabs our interest first and keeps appearing in applications: transient effects in time-dependent dynamical systems. We two authors know this from years of experience in lecturing and talking with colleagues. Everyone has seen a time-dependent problem where eigenvalues are misleading, and such effects are always intriguing.[1]

Sometimes when transient effects are conspicuous, the essence of the matter may be nonlinearity or time-varying coefficients. A schema is presented in Figure 33.3 that explains in a general way the relationship of nonlinearity, variable coefficients, and nonnormality for a time-dependent dynamical system. For this part of the book, however, we take it for granted that the problem at hand has been reduced to a linear dynamical system governed by a fixed matrix or operator. We are interested in cases in which the transient behavior of this system differs from the behavior at large times, for reasons of nonnormality.[2]

If eigenvalues fail to capture the transients, can pseudospectra do better? The answer is certainly yes: Though pseudospectra rarely give an exact answer, they detect and quantify transients that eigenvalues miss. But this subject is complex, for there are many ways in which such questions can be framed and many estimates that may be obtained; there are technicalities that arise for operators but not matrices; and most of the estimates appear once for continuous time and again for discrete time systems. Thus this part of the book, which is devoted to presenting some of these results, is inevitably somewhat complicated. This opening section gives an overview. Most of our estimates make use of the quantities

$\alpha(\mathbf{A})$, $\alpha_\varepsilon(\mathbf{A})$, $\omega(\mathbf{A})$: spectral, ε-pseudospectral, and
numerical abscissa of $\mathbf{A}$,

$\rho(\mathbf{A})$, $\rho_\varepsilon(\mathbf{A})$, $\mu(\mathbf{A})$: spectral, ε-pseudospectral, and
numerical radius of $\mathbf{A}$,

---

[1] For the sake of balance, perhaps it is worth recalling some aspects of nonnormality that are *not* just matters of transients. These include the pseudoresonant response of systems to external stimuli [780], the behavior of polynomial, rational, and other functions of matrices and operators (§§26 and 29), the effect of perturbations on eigenvalues (§52), conditioning and distortions introduced by eigenvector bases (§51), and backward error analysis of numerical algorithms (§53).

[2] As discussed in §2, strictly speaking we should not just refer to 'nonnormality', for this term is only correct in Hilbert space. More generally one is concerned with the conditioning of the set of eigenvectors in the norm of interest.

defined in each case as the supremum of the real part or absolute value of the spectrum, ε-pseudospectrum, or numerical range of $\mathbf{A}$. All six of these numbers are actually determined by the pseudospectra of $\mathbf{A}$: the spectrum by the limit $\varepsilon \to 0$ (at least for bounded operators), and the numerical range, as discussed in §17, by the limit $\varepsilon \to \infty$. For a matrix with $\|\cdot\| = \|\cdot\|_2$, one can compute them (except the numerical radius) with the EigTool system.

To emphasize the variety of applications in which linear nonnormal transient effects arise, we provide a list of examples that appear in this book.

Continuous time
   §5. 'Disappearing solutions' in a leftward drift
   §10. Well- or ill-posedness of time-dependent PDEs on intervals
   §12. The advection phase of an advection-diffusion process
   §12. Local convective instability and the Cossu–Chomaz operator
   §12. Explosive instability and the operator of Benilov et al.
   §19. Misleading eigenvalues as $t \to \infty$ for unbounded operators
   §20. How small vortices generate large streaks in shear flows
   §21. Basins of attraction and transition to turbulence
   §22. Orr–Sommerfeld eigenvalues, 2D transients, 3D transients
   §23. Linear transient effects throughout fluid dynamics
   §58. Food webs stable in theory, unstable in practice

Discrete time
   §3. Is a tridiagonal Toeplitz matrix power-bounded?
   §24. Convergence of Gauss–Seidel iterations
   §25. Why don't upwind SOR sweeps work as well as downwind?
   §27. Hybrid iterations cannot be based on eigenvalues alone
   §28. Convergence of the power method
   §31. Nonmodal instabilities of numerical discretizations
   §32. Convergence as the mesh is refined, if transients are bounded
   §33. ODEs may be stiff even when the eigenvalues are harmless
   §34. Boundary instabilities caused by linear 'algebraic' terms
   §56. The cutoff phenomenon for Markov chains
   §57. How many shuffles to randomize a deck of cards?
   §60. Bouncing photon packets and Petermann excess noise

We shall discuss continuous time first, our goal being to review the main questions of interest and point to sections of this book where relevant theorems may be found. Then we turn to the case of discrete time, always analogous yet never so close that one can dispense with a separate discussion. Our goal is to survey a broad range of results, most of which are proved in the following sections.

Figure 14.1 sets the framework. We imagine a matrix or linear operator $\mathbf{A}$ and are concerned with the growth and decay of solutions $\mathbf{u}(t)$ to the

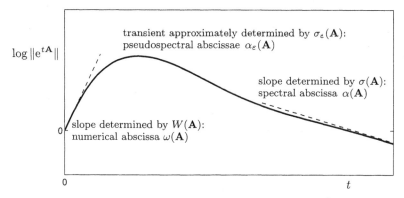

Figure 14.1: Initial, transient, and asymptotic behavior of $\|e^{t\mathbf{A}}\|$ for a nonnormal matrix or operator $\mathbf{A}$.

time-dependent equation $d\mathbf{u}/dt = \mathbf{A}\mathbf{u}$, that is, $\mathbf{u}(t) = e^{t\mathbf{A}}\mathbf{u}(0)$.[3] Specifically, we want to know something about the size of $\|e^{t\mathbf{A}}\|$ as a function of t.[4]

One familiar limit is $t \to \infty$. Here one ordinarily expects the eigenvalues, or more generally the spectrum, to be decisive, with the asymptotic growth rate of $\|e^{t\mathbf{A}}\|$ being determined by the spectral abscissa of $\mathbf{A}$ (called the *spectral bound* in semigroup theory). In fact, the equation

$$\lim_{t \to \infty} t^{-1} \log \|e^{t\mathbf{A}}\| = \alpha(\mathbf{A}) \qquad (14.1)$$

holds for any matrix or bounded operator $\mathbf{A}$ in a Banach space. Difficulties can arise for unbounded operators, where the left-hand side of (14.1), known as the *growth bound*, may be greater than the right-hand side. In such cases, in Hilbert space, it is enough to replace $\alpha(\mathbf{A})$ by $\lim_{\varepsilon \to 0} \alpha_\varepsilon(\mathbf{A})$, where $\alpha_\varepsilon(\mathbf{A})$ is the ε-pseudospectral abscissa of $\mathbf{A}$; this is the *Gearhart–Prüss theorem*, discussed in §19. (In the special case $e^{\tau\mathbf{A}} = \mathbf{0}$ for some finite τ, we take the limit on the left of (14.1) to be $-\infty$.)

The other interesting limit is $t \to 0$. Provided $\mathbf{A}$ is a matrix or linear operator that generates a C_0 semigroup in a Banach space, there is an equally sharp and simple result, discussed in §17. The initial growth rate of $\|e^{t\mathbf{A}}\|$ (we assume here a one-sided derivative based on $t \downarrow 0$) is

$$\frac{d}{dt}\|e^{t\mathbf{A}}\|\bigg|_{t=0} = \lim_{t \downarrow 0} t^{-1} \log \|e^{t\mathbf{A}}\| = \omega(\mathbf{A}), \qquad (14.2)$$

---

[3] We use the matrix notation $e^{t\mathbf{A}}$ throughout, though this is nonstandard if $\mathbf{A}$ is an unbounded operator (and hence not a matrix). In this case we assume that $\mathbf{A}$ is closed and densely defined (see §4) and generates a C_0 semigroup (see §15).

[4] The norm $\|e^{t\mathbf{A}}\|$ is defined via the supremum of $\|e^{t\mathbf{A}}\mathbf{u}\|/\|\mathbf{u}\|$ over nonzero vectors $\mathbf{u}$. By reversing time, all of our results can be transformed to results for the infimum of the same ratio, known as the *lower bound* of $e^{t\mathbf{A}}$.

where $\omega(\mathbf{A})$ is the numerical abscissa; in semigroup theory this is the circle of ideas related to the Hille–Yosida and Lumer–Phillips theorems [606]. In the Hilbert space case, the numerical abscissa is given by the explicit formula

$$\omega(\mathbf{A}) \;=\; \sup \sigma(\tfrac{1}{2}(\mathbf{A}+\mathbf{A}^*)).$$

(In the literature of numerical solutions of ordinary differential equations, an alternative terminology has arisen: one says that (14.2) holds with ω replaced by the *logarithmic norm* of $\mathbf{A}$, defined by $\lim_{\varepsilon \to 0} \varepsilon^{-1}(\|\mathbf{I}+\varepsilon\mathbf{A}\|-1)$; see [555] and [575, p. 43] and references therein.)

Our main interest is not $t \to \infty$ or $t \to 0$ but intermediate values of t. Here there are a variety of estimates that can be derived. Eigenvalues alone give a lower bound,

$$\|e^{t\mathbf{A}}\| \;\geq\; e^{t\alpha(\mathbf{A})} \qquad \forall t \geq 0, \tag{14.3}$$

valid in any norm for a matrix or linear operator. (In the case of a linear operator, $\alpha(\mathbf{A})$ is defined by the spectrum, not just the eigenvalues.) The companion upper bound comes from the numerical abscissa,

$$\|e^{t\mathbf{A}}\| \;\leq\; e^{t\omega(\mathbf{A})} \qquad \forall t \geq 0. \tag{14.4}$$

Eigenvalues, together with a finite condition number of a matrix of eigenvectors $\mathbf{V}$ (in the operator case, the 'columns of $\mathbf{V}$' are a complete set of eigenvectors of $\mathbf{A}$ that form a Riesz basis; see §51), give the alternative upper bound

$$\|e^{t\mathbf{A}}\| \;\leq\; \kappa(\mathbf{V})\,e^{t\alpha(\mathbf{A})} \qquad \forall t \geq 0. \tag{14.5}$$

This bound provides an interesting reference point, but in most applications with $\kappa(\mathbf{V}) \gg 1$, it is too loose to be very helpful. And of course if $\mathbf{A}$ is not diagonalizable, it is no bound at all.

For sharper information we turn to resolvent norms, i.e., pseudospectra. An extremely useful lower bound in practice is the simple inequality

$$\sup_{t\geq 0} \|e^{t\mathbf{A}}\| \;\geq\; \alpha_\varepsilon(\mathbf{A})/\varepsilon \qquad \forall \varepsilon > 0. \tag{14.6}$$

This estimate shows that if the pseudospectra of a matrix with $\alpha(\mathbf{A}) \leq 0$ protrude significantly into the right half-plane in the sense that $\alpha_\varepsilon(\mathbf{A}) > \varepsilon$ for some ε, there must always be transient growth. If $\mathbf{A}$ has $\|(z-\mathbf{A})^{-1}\| = 10^5$ for some z with $\mathrm{Re}\,z = 0.01$, for example, then $\alpha_\varepsilon(\mathbf{A}) \geq 0.01$ for $\varepsilon = 10^{-5}$, and thus there must be transient growth of magnitude at least 10^3. This estimate is sometimes known as the 'easy half of the Kreiss matrix theorem'. If we define the *Kreiss constant* of $\mathbf{A}$ with respect to the left half-plane by

$$\mathcal{K}(\mathbf{A}) \;\equiv\; \sup_{\varepsilon>0} \alpha_\varepsilon(\mathbf{A})/\varepsilon \;=\; \sup_{\mathrm{Re}\,z>0} (\mathrm{Re}\,z)\|(z-\mathbf{A})^{-1}\|, \tag{14.7}$$

then (14.6) implies

$$\sup_{t \geq 0} \|e^{t\mathbf{A}}\| \geq \mathcal{K}(\mathbf{A}). \tag{14.8}$$

To derive converses of such results, upper bounds on $\|e^{t\mathbf{A}}\|$, the natural tool is the definition of $e^{t\mathbf{A}}$ as a Cauchy integral of the resolvent. As the same technique will be useful for bounding norms of other functions of operators in later sections (§§16, 26, and 28), we shall begin with a general function f that is analytic in a neighborhood of the spectrum, $\sigma(\mathbf{A})$. Suppose $\mathbf{A}$ is a matrix or bounded operator, and let Γ denote a closed contour or union of closed contours enclosing $\sigma(\mathbf{A})$ once in the positive sense and contained in the region of analyticity of f. Then $f(\mathbf{A})$ can be defined by the operator analogue of the Cauchy integral formula, sometimes called a *Dunford–Taylor integral*:

$$f(\mathbf{A}) = \frac{1}{2\pi i} \int_\Gamma (z - \mathbf{A})^{-1} f(z) \, dz; \tag{14.9}$$

see, e.g., [139, 161, 221, 415, 448, 641]. This equation yields the same result that one gets by defining $f(\mathbf{A})$ by an eigenvalue decomposition (if $\mathbf{A}$ is a diagonalizable matrix) or the Jordan canonical form (if $\mathbf{A}$ is an arbitrary matrix) or direct calculation (if $\mathbf{A}$ is bounded and f is a polynomial) or a power series (if $\mathbf{A}$ is bounded and f is analytic in a sufficiently large disk).

To estimate $\|f(\mathbf{A})\|$, one can simply bound the norm of the Cauchy integral by the integral of $|f(z)| \|(z - \mathbf{A})^{-1}\|$. When Γ encloses $\sigma_\varepsilon(\mathbf{A})$, the resolvent norm is bounded by ε^{-1}, and hence

$$\|f(\mathbf{A})\| \leq \frac{L}{2\pi\varepsilon} \max_{z \in \Gamma} |f(z)|, \tag{14.10}$$

where L denotes the arc length of Γ.

To bound $\|e^{t\mathbf{A}}\|$, it would be natural to integrate along the contour $\text{Re}\, z = $ constant, but as this is infinitely long, no upper bound can come from a finite pseudospectral abscissa $\alpha_\varepsilon(\mathbf{A})$ in the absence of further information. On the other hand, if $\sigma_\varepsilon(\mathbf{A})$ has a boundary with finite arc length L_ε, we have

$$\|e^{t\mathbf{A}}\| \leq \frac{L_\varepsilon e^{t\alpha_\varepsilon(\mathbf{A})}}{2\pi\varepsilon} \qquad \forall \varepsilon > 0, \forall t \geq 0, \tag{14.11}$$

a bound that can readily be strengthened in most particular cases.

The Kreiss Matrix Theorem (for continuous time), also derived from a contour integral (§18), is a less elementary upper bound: if $\mathbf{A}$ is a matrix of dimension N, then

$$\|e^{t\mathbf{A}}\| \leq eN\mathcal{K}(\mathbf{A}) \qquad \forall t \geq 0. \tag{14.12}$$

This inequality implies that for a matrix or family of matrices of fixed dimension, any transient growth must be reflected in the pseudospectra, up to a constant factor eN. This factor cannot be dispensed with and is in certain senses sharp. In particular, there exist infinite-dimensional operators for which $\|e^{t\mathbf{A}}\|$ can exceed $\mathcal{K}(\mathbf{A})$ by an arbitrary factor.[5]

Missing from (14.6)–(14.12) is an indication of the time scale on which transient growth must occur. Our experience with eigenvalues is a good guide here. We know that if a matrix or operator has an eigenvalue z in the right half-plane, there must be exponential amplification on the time scale $1/\mathrm{Re}\,z$. The same is true, transiently, if z is an ε-pseudoeigenvalue for a sufficiently small value of ε. Thus for the example mentioned above with $\mathrm{Re}\,z = 0.01$, the transient growth of order 10^3 will unfold on a time scale of approximately 10^2, whereas if $\mathbf{A}$ has $\|(z - \mathbf{A})^{-1}\| = 10^{-1}$ for some z with $\mathrm{Re}\,z = 100$, there will be transient growth of the same order on the time scale approximately 10^{-2}. The following estimate making these ideas precise was derived by Trefethen in 2002 (unpublished) and reported in Wright's D.Phil. thesis [837] and is also implemented in the EigTool system [838]. Given z with $\mathrm{Re}\,z = a > 0$, suppose $\|(z - \mathbf{A})^{-1}\| = K/a$ for some $K > 1$. Then for any $\tau > 0$,

$$\sup_{0 < t \leq \tau} \|e^{t\mathbf{A}}\| \geq e^{a\tau} \Big/ \left(1 + \frac{e^{a\tau} - 1}{K}\right). \tag{14.13}$$

This inequality may look complicated, but its implications become clear if one notes that the expression in parentheses is close to 1 when τ is small enough that $e^{a\tau} \ll K$. Thus (14.13) asserts that for any such $\tau > 0$, there exists some t in the interval $[0, \tau]$ for which $\|e^{t\mathbf{A}}\|$ is approximately as big as $e^{a\tau}$, or bigger. In other words, on this time scale, z behaves approximately like an eigenvalue. As τ increases toward ∞, the right-hand side increases monotonically, and in the limit $\tau = \infty$ we get (14.6). This behavior is depicted graphically in Figure 15.1.

A similar bound has also been derived by Davies [187]. Let $N(t)$ denote the upper log-concave envelope of $\|e^{t\mathbf{A}}\|$, i.e., the smallest function such that $\log N(t)$ is concave and greater than or equal to $\log \|e^{t\mathbf{A}}\|$ for all $t \geq 0$. Assume also that the growth bound of $\mathbf{A}$ is zero, i.e., $\lim_{t\to\infty} t^{-1} \log \|e^{t\mathbf{A}}\| = 0$, which for a matrix or Hilbert space operator is equivalent to the condition $\alpha(\mathbf{A}) = 0$. Then

$$N(\tau) \geq \min\{e^{a\tau(1-1/K)}, K\} \qquad \forall \tau \geq 0. \tag{14.14}$$

On the whole (14.13) and (14.14) are rather similar, the former being perhaps easier to interpret and the latter perhaps more elegant and slightly sharper. Some remarks comparing the two can be found in [187].

---

[5] This statement refers to general values of $\mathcal{K}(\mathbf{A})$. The special value $\mathcal{K}(\mathbf{A}) = 1$ implies that the numerical range $W(\mathbf{A})$ is contained in the closed left half-plane, in which case $\omega(\mathbf{A}) \leq 0$ and the semigroup is a contraction: $\|e^{t\mathbf{A}}\| \leq 1$ for all $t \geq 0$, regardless of N.

The bounds (14.13) and (14.14) have the unfortunate feature that they apply not to a particular time t but to a supremum over a time interval $[0, \tau]$ (explicitly in the first case, implicitly in the second). By incorporating an upper bound in the derivation, we can get a pointwise bound. If $\|e^{t\mathbf{A}}\| \leq M$ for all $t \geq 0$, for example, following the same notation as above, but now with $K/M \leq 1$, we find for all $t \geq 0$

$$\|e^{t\mathbf{A}}\| \geq e^{at} - \frac{e^{at} - 1}{K/M} = 1 - \frac{(e^{at} - 1)(1 - K/M)}{K/M}. \qquad (14.15)$$

Suppose, for example, that $M = 1$ (and hence $e^{t\mathbf{A}}$ is a contraction) and that $K = 0.99$ for some z with $a = \mathrm{Re}\, z > 0$. Then (14.15) implies that $\|e^{t\mathbf{A}}\|$ cannot fall below 0.99 for any $t < 0.688/a$. For a fuller picture, see Figure 15.2.

Finally we note a theorem concerning a special class of operators. A matrix or operator that drives a nilpotent time-dependent process can be recognized from the pseudospectra:

$$e^{\tau\mathbf{A}} = \mathbf{0} \iff \sigma(\mathbf{A}) = \emptyset \text{ and } \|(z - \mathbf{A})^{-1}\| = \mathcal{O}(e^{-\tau\mathrm{Re}\,z}), \qquad (14.16)$$

with the limit implicit in the '$\mathcal{O}$' being $\mathrm{Re}\,z \to -\infty$. This result is given by Driscoll and Trefethen [216] and as Theorem 3 of [773]. It implies that $e^{\tau\mathbf{A}} = \mathbf{0}$ is possible only if $\mathbf{A}$ is an unbounded operator, since no bounded operator can have an empty spectrum.

We illustrate these results by an example, the 2×2 matrix

$$\mathbf{A} = \begin{pmatrix} 0 & 1 & 2 \\ -0.01 & 0 & 3 \\ 0 & 0 & 0 \end{pmatrix}. \qquad (14.17)$$

The left half of Figure 14.2 shows the pseudospectra of $\mathbf{A}$, and the right half shows the 2-norms $\|e^{t\mathbf{A}}\|$. Now for a 3×3 matrix like this it would be possible to analyze the behavior exactly, but let us imagine that we knew only spectral and pseudospectral information and wished to draw inferences about $\|e^{t\mathbf{A}}\|$. The eigenvalues are 0 and $\pm i/10$, so equations (14.1) and (14.3) tell us that $\|e^{t\mathbf{A}}\|$ is everywhere bigger than 1, approaching this value in a logarithmic sense as $t \to \infty$. Equations (14.2) and (14.4) tell us that $\|e^{t\mathbf{A}}\|$ is everywhere smaller than about $e^{2.054t}$, approaching this value as $t \to 0$. Equation (14.5) tells us that $\|e^{t\mathbf{A}}\|$ is less than about 635.8, which for this example is an impressively sharp estimate, since the actual maximum is about 600.7 at $t \approx 30.7$. In applying equation (14.6), we have a choice of a value of ε. The choice $\varepsilon = 10^{-4}$ is a good one, since the boundary of the 10^{-4}-pseudospectrum extends a distance close to 0.03 into the right half-plane. The actual distance is about 0.02831, so (14.6) implies that $\|e^{t\mathbf{A}}\| \geq 283$ for some t; this is about a factor of 2.14 below the actual

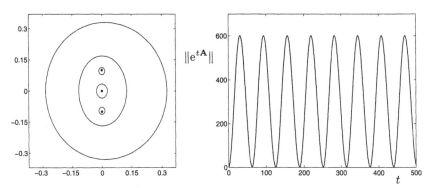

Figure 14.2: On the left, ε-pseudospectra of the matrix (14.17) for $\varepsilon = 10^{-2}$, 10^{-3}, 10^{-4}. On the right, norms of $e^{t\mathbf{A}}$ for the same matrix.

maximum value. Optimizing over ε gives the Kreiss constant $\mathcal{K}(\mathbf{A}) \approx 300.3$ (achieved with $\varepsilon \approx .00332$), so from (14.8) we get the slightly better lower bound of 300.3, and (14.12) gives the upper bound 2449. Finally, we can get information about time scales from (14.13) or (14.14). For example, with $\varepsilon = 10^{-4}$ and $\alpha_\varepsilon(\mathbf{A}) \approx 0.02831$ as before, (14.13) gives the lower bound 16.0 for the supremum of $\|e^{t\mathbf{A}}\|$ in $[0, 100]$ and the lower bound 267 in $[0, 300]$.

A more complicated and interesting example of norms of matrix exponentials is presented in §15, involving matrices derived from a model related to the Boeing 767 aircraft. The Orr–Sommerfeld differential operator also exhibits similar behavior; see Figure 22.3. Such examples highlight the fact that the pseudospectra of $\mathbf{A}$ may contain components lying at very different distances from the imaginary axis, corresponding to behavior of $\|e^{t\mathbf{A}}\|$ on very different time scales.

Now we turn to discrete time, that is, the behavior of norms of powers $\mathbf{A}^k$ of a matrix $\mathbf{A}$. We present lower and upper bounds for $\|\mathbf{A}^k\|$ following the same sequence as in (14.1)–(14.16), with a few inevitable differences. Figure 14.3 summarizes the situation.

For the limit $k \to \infty$, the spectrum is again decisive. The asymptotic growth rate of $\|\mathbf{A}^k\|$ is determined by $\rho(\mathbf{A})$, the spectral radius of $\mathbf{A}$. The equation

$$\lim_{k \to \infty} \|\mathbf{A}^k\|^{1/k} = \rho(\mathbf{A}) \tag{14.18}$$

holds for any matrix or bounded operator $\mathbf{A}$ in a Banach space; see §16. (In the special case $\mathbf{A}^\kappa = 0$ for some κ, we take the left-hand side to be 0.) The other limit $k \to 0$ is vacuous: there is nothing more to say than $\|\mathbf{A}^1\| = \|\mathbf{A}\|$.

Again our main interest is intermediate values of k. Eigenvalues (more generally, the spectrum) give the lower bound based on the spectral radius

$$\|\mathbf{A}^k\| \geq (\rho(\mathbf{A}))^k \qquad \forall k \geq 0, \tag{14.19}$$

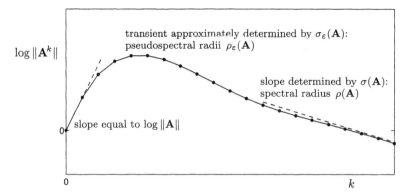

Figure 14.3: Analogue of Figure 14.1 for discrete time.

valid in any norm for a matrix or linear operator. One upper bound analogous to (14.4) is

$$\|\mathbf{A}^k\| \leq \|\mathbf{A}\|^k \qquad \forall k \geq 0, \tag{14.20}$$

and another one, related to the numerical range, is

$$\|\mathbf{A}^k\| \leq e(\mu(\mathbf{A}))^k \qquad \forall k \geq 0. \tag{14.21}$$

(In Hilbert space, the constant e can be improved to 2.) Eigenvalues, together with a finite condition number of a matrix of eigenvectors (or its infinite-dimensional generalization), give the upper bound

$$\|\mathbf{A}^k\| \leq \kappa(\mathbf{V})\,(\rho(\mathbf{A}))^k \qquad \forall k \geq 0. \tag{14.22}$$

For sharper information we again turn to pseudospectra. In analogy to (14.6) we have the simple but very useful bound

$$\sup_{k \geq 0} \|\mathbf{A}^k\| \geq (\rho_\varepsilon(\mathbf{A}) - 1)/\varepsilon \qquad \forall \varepsilon > 0, \tag{14.23}$$

which can be improved to the slightly stronger but less memorable

$$\sup_{k \geq 0} \|\mathbf{A}^k\| \geq (\rho_\varepsilon(\mathbf{A}) - 1)(\rho_\varepsilon(\mathbf{A})/\varepsilon - 1) \qquad \forall \varepsilon > 0. \tag{14.24}$$

These estimates show that if the pseudospectra protrude significantly outside the unit disk in the sense that $\rho_\varepsilon(\mathbf{A}) > 1 + \varepsilon$ for some ε, there must be transient growth. For a matrix with $\|(z - \mathbf{A})^{-1}\| = 10^5$ for some z with $|z| = 1.01$, for example, there must be growth of magnitude at least 10^3. If we define the *Kreiss constant* of $\mathbf{A}$ with respect to the unit disk by

$$\mathcal{K}(\mathbf{A}) \equiv \sup_{\varepsilon > 0} (\rho_\varepsilon(\mathbf{A}) - 1)/\varepsilon = \sup_{|z| > 1} (|z| - 1)\|(z - \mathbf{A})^{-1}\|,$$

then (14.23) implies

$$\sup_{k \geq 0} \|\mathbf{A}^k\| \geq \mathcal{K}(\mathbf{A}). \tag{14.25}$$

The simplest converse of these results comes from a contour integral around a circle of radius $\rho_\varepsilon(\mathbf{A})$:

$$\|\mathbf{A}^k\| \leq \rho_\varepsilon(\mathbf{A})^{k+1}/\varepsilon \qquad \forall \varepsilon > 0, \forall k \geq 0. \tag{14.26}$$

The particular choice of radius $1 + n^{-1}$ leads to

$$\|\mathbf{A}^k\| \leq e(k+1)\mathcal{K}(\mathbf{A}) \qquad \forall k \geq 0, \tag{14.27}$$

and by a further integration by parts and other estimates described in §18, one can derive the Kreiss Matrix Theorem for discrete time:

$$\|\mathbf{A}^k\| \leq eN\mathcal{K}(\mathbf{A}) \qquad \forall k \geq 0, \tag{14.28}$$

if $\mathbf{A}$ is a matrix of dimension N.

Again, the next step is to consider the time scale on which transient growth must occur. If a matrix or operator has an eigenvalue z outside the unit circle, there must be exponential amplification on a time scale $1/(|z| - 1)$. The same is true, transiently, if z is an ε-pseudoeigenvalue for a sufficiently small value of ε. The following estimate, like (14.13), was derived by Trefethen in 2002, reported in Wright's D.Phil. thesis [837], and incorporated in EigTool [838]. Suppose that for some z with $|z| = r > 1$, $\|(z - \mathbf{A})^{-1}\| = K/(r-1)$. Then for any $\nu > 0$,

$$\max_{0 < k \leq \kappa} \|\mathbf{A}^k\| \geq r^\kappa \Big/ \left(1 + \frac{r^\kappa - 1}{rK - r + 1}\right). \tag{14.29}$$

The expression in parentheses is close to 1 when κ is small enough that $r^\kappa \ll rR$. Thus the inequality asserts that for any κ, there exists some $k \in \{0, 1, \dots, \kappa\}$ for which $\|\mathbf{A}^k\|$ is approximately as big as r^κ, or bigger. In other words, z behaves approximately like an eigenvalue on this time scale. In the limit $\kappa = \infty$ we get (14.24). For more details, see Figure 16.1.

An analogue for discrete time of the Davies estimate (14.14) does not appear to have been derived. An analogue of the pointwise bound (14.15) is

$$\|\mathbf{A}^k\| \geq r^k - \frac{r^k - 1}{K/M} = 1 - \frac{(r^k - 1)(1 - K/M)}{K/M}; \tag{14.30}$$

see Figure 16.2. Another weaker bound is given as (16.25).

As before, we note a theorem concerning nilpotency, which can occur in this context for matrices as well as operators. If $\mathbf{A}$ is a matrix or bounded linear operator one can show

$$\mathbf{A}^\kappa = \mathbf{0} \iff \sigma(\mathbf{A}) = \{0\} \text{ and } \|(z - \mathbf{A})^{-1}\| = \mathcal{O}(|z|^{-\kappa}) \tag{14.31}$$

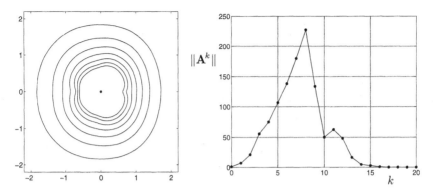

Figure 14.4: On the left, ε-pseudospectra of a 20×20 strictly upper random triangular matrix for $\varepsilon = 10^{-1}, 10^{-1.5}, 10^{-2}, \ldots, 10^{-4}$. On the right, norms of powers of $\mathbf{A}$. See §38.

as $|z| \to 0$. In the matrix case, the estimate on the resolvent norm follows from the perturbation theory described in §52.

For a numerical example, let us take $\mathbf{A}$ to be a 20×20 strictly upper triangular matrix whose nonzero entries are independent samples from the standard $N(0, 1)$ normal distribution (see §38). Since $\mathbf{A}$ is strictly triangular, it is nilpotent, with all eigenvalues equal to zero, but the pseudospectra protrude significantly outside the unit disk, and there will be transient effects in the matrix powers. The left half of Figure 14.4 shows the pseudospectra of $\mathbf{A}$, and the right half shows the norms $\|\mathbf{A}^k\|$.

Again let us imagine that we know only the pseudospectra of $\mathbf{A}$ and wish to draw inferences about $\|\mathbf{A}^k\|$. Equations (14.18) and (14.19) tell us that $\|\mathbf{A}^k\|$ approaches zero. Equation (14.20) tells us it is everywhere smaller than about $(7.13)^k$, and the sharpened form of equation (14.21) tells us it is smaller than about $2(4.30)^k$. Equation (14.22) tells us nothing since $\mathbf{A}$ is not diagonalizable. For (14.23) we have a choice of ε. The value $\varepsilon = 0.01$ gives $\rho_\varepsilon(\mathbf{A}) \approx 1.269$, so (14.23) implies that $\|\mathbf{A}^k\|$ is at least as great as 26.9 for some k; this is nearly a factor of 10 below the actual maximum value. Optimizing over ε gives the Kreiss constant $\mathcal{K}(\mathbf{A}) \approx 31.3$ (achieved with $\varepsilon \approx 0.0041$), so from (14.25) we get the slightly better lower bound of 31.3, and (14.28) gives the upper bound 1702. To get some information about time scales from (14.29), we have a choice of κ. For example, with $\varepsilon = 0.01$ and $\rho_\varepsilon(\mathbf{A}) \approx 1.269$ as before, (14.29) gives the lower bound 8.39 for the supremum of $\|\mathbf{A}^k\|$ for $0 \le k \le 10$ and the lower bound 26.5 for $0 \le k \le 20$.

Böttcher has pointed out that interesting results can be obtained by combining the estimates of this section with the theorems of §7 about Toeplitz matrices [80]. One of the examples he considers is the family of

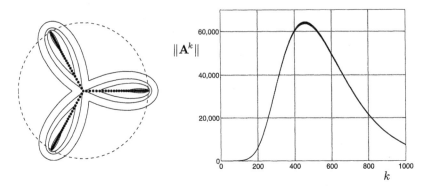

Figure 14.5: On the left, ε-pseudospectra of the 100×100 Toeplitz matrix of the form (14.32) for $\varepsilon = 10^{-1}, 10^{-2}, 10^{-4}, 10^{-8}$; the dashed curve is the unit circle. On the right, norms of powers of $\mathbf{A}$. The apparent thickening of the curve is caused by oscillations of up to about 1.4% over every three successive data points.

Toeplitz matrices $\mathbf{A}$ defined for each dimension N by

$$a_{j+1,j} = 10/19, \quad a_{j,j+2} = 10/19, \tag{14.32}$$

with all other entries equal to zero. The spectral radius of such a matrix is less than 1, approaching $15\sqrt[3]{2}/19 \approx 0.9947$ as $N \to \infty$. However, the symbol curve of the matrix, the image of the unit circle under the map $\frac{10}{19}(z + z^{-2})$, has three petal-like components that extend outside the unit circle with a maximal radius 20/19. According to Theorem 7.2, $\|(z-\mathbf{A})^{-1}\|$ will grow exponentially as $N \to \infty$ for any point enclosed by this curve. It follows from (14.29) that for large dimensions N, the curve of $\|\mathbf{A}^k\|$ against k must have an exponential initial transient growing approximately at the rate $(20/19)^k$, reaching values exponentially large with respect to N before eventually decaying to zero. This effect is confirmed in Figure 14.5. Analogous exponentially strong transients can be observed with continuous time and for any of the families of matrices and operators considered in sections §§7–12.

We have presented 26 bounds, most of which will be discussed at greater length in one of the next several sections:

§15: (14.1), (14.3), (14.6)–(14.11), (14.13), (14.15)–(14.16)
§16: (14.18)–(14.20), (14.23)–(14.27), (14.29)–(14.31)
§17: (14.2), (14.4), (14.21)
§18: (14.12), (14.28)

It remains to make some observations about generalizations, implications, and related matters.

One important generalization concerns left-right shifts (for continuous time) and rescalings of $\mathbf{A}$ (for discrete time). A bound on $\|e^{t\mathbf{A}}\|$ such as (14.6), for example, essentially compares $\|e^{t\mathbf{A}}\|$ to a constant. What if we wish to compare it instead to an exponential $e^{\omega t}$ for some $\omega \in \mathbb{R}$? We can do this by shifting $\mathbf{A}$ by a multiple of the identity; in fact, we have

$$\sup_{t\geq 0} \|e^{-\omega t}e^{t\mathbf{A}}\| \geq (\alpha_\varepsilon(\mathbf{A}) + \omega)/\varepsilon \qquad \forall \varepsilon > 0 \qquad (14.33)$$

for any $\omega \in \mathbb{R}$. Similarly, in discrete time, we may compare $\|\mathbf{A}^k\|$ to γ^k for some $\gamma > 0$ by noting that (14.23) generalizes to

$$\sup_{k\geq 0} \|\gamma^{-k}\mathbf{A}^k\| \geq (\rho_\varepsilon(\gamma\mathbf{A}) - 1)/\varepsilon \qquad \forall \varepsilon > 0. \qquad (14.34)$$

Among the bounds we have given for continuous time, (14.6), (14.8), and (14.12)–(14.15) can be generalized in this fashion, whereas (14.1)–(14.5), (14.11), and (14.16) are already in a form that incorporates such translations. For discrete time, (14.23)–(14.25), (14.27)–(14.30), (14.33), and (14.34) can be generalized, whereas (14.18)–(14.22), (14.26), and (14.31) already incorporate rescaling. In §15, simple versions of lower bounds are given in Theorem 15.4 and generalized versions in Theorem 15.5; similarly, in §16, we have Theorems 16.4 and 16.5.

We have concentrated on estimates related to resolvent norms or pseudospectra, but this is not the only approach that can be taken to estimating $\|e^{t\mathbf{A}}\|$ or $\|\mathbf{A}^k\|$. Another major approach to evolution problems, hardly touched upon in this book, is by means of Lyapunov equations and related algebraic techniques. For example, a matrix $\mathbf{A}$ has $\alpha(\mathbf{A}) < 0$ if and only if the equation

$$\mathbf{AX} + \mathbf{XA}^* + \mathbf{I} = \mathbf{0} \qquad (14.35)$$

has a unique self-adjoint positive definite solution. The calculation of such a matrix $\mathbf{X}$ is related to the derivation of an alternative norm in which $\{e^{t\mathbf{A}}\}$ is a contraction; if such an $\mathbf{X}$ exists, we have [317]

$$\|e^{t\mathbf{A}}\| \leq (\kappa(\mathbf{X}))^{1/2}e^{-t/2\|\mathbf{X}\|}, \qquad (14.36)$$

with $\kappa(\mathbf{X}) = \|\mathbf{X}\|\|\mathbf{X}^{-1}\|$. The generalization of such bounds to semigroups has been investigated by Veselić [806, 807, 808]. For alternative bounds for $\|e^{t\mathbf{A}}\|$ and $\|\mathbf{A}^k\|$ based on scalar metrics for nonnormality and other algebraic quantities, see [203, 309, 380].

15 · Exponentials of matrices and operators _____

This section and the next three spell out the details of most of the results summarized in §14. In particular, here we provide proofs, or pointers to proofs, for the inequalities (14.1), (14.3), (14.6)–(14.11), (14.13), and (14.15)–(14.16) concerning $\|e^{t\mathbf{A}}\|$, where $\mathbf{A}$ is a matrix or a closed linear operator acting in a Banach space X with norm $\|\cdot\|$. Some further results that apply only in Hilbert space are presented in §19.

The first thing we must do is specify the meaning of $e^{t\mathbf{A}}$. If $\mathbf{A}$ is a matrix or bounded operator, this is straightforward: $e^{t\mathbf{A}}$ can be defined by a convergent power series, an idea going back to Peano in 1887. For an unbounded operator $\mathbf{A} \in \mathcal{C}(X)$ with domain $\mathcal{D}(\mathbf{A})$ (these notations were introduced in §4), the meaning of $e^{t\mathbf{A}}$ comes from the mathematical theory of semigroups. General references on this subject include [177, 248, 395, 606]; the following discussion most closely follows Pazy [606]. A C_0 *semigroup* is a family of bounded operators $\{\mathbf{T}(t)\}_{0 \le t < \infty}$ with the properties that $\mathbf{T}(0)$ is the identity, $\mathbf{T}(t+s) = \mathbf{T}(t)\mathbf{T}(s)$ for $s, t \ge 0$, and $\mathbf{T}(t)\mathbf{u}$ is a continuous function of t for each $\mathbf{u} \in X$. The *infinitesimal generator* of the semigroup is the operator $\mathbf{A}$ defined by the condition

$$\mathbf{Au} = \lim_{t \to 0} \frac{\mathbf{T}(t)\mathbf{u} - \mathbf{u}}{t},$$

with the domain $\mathcal{D}(\mathbf{A})$ taken to be the set of all vectors $\mathbf{u} \in X$ for which this limit exists. $\mathbf{A}$ is a densely defined closed operator, and it determines the semigroup uniquely. Throughout this book, whenever the notation $e^{t\mathbf{A}}$ appears and $\mathbf{A}$ is an unbounded operator, this is shorthand for the C_0 semigroup generated by $\mathbf{A}$.

It is known that if a densely defined operator $\mathbf{A} \in \mathcal{C}(X)$ generates a C_0 semigroup, then there are constants $\omega \in \mathbb{R}$ and $M \ge 1$ such that $\|e^{t\mathbf{A}}\| \le Me^{\omega t}$ for all $t \ge 0$. In the special case $M = 1$, the semigroup is $e^{\omega t}$ times a contraction, and this situation can be characterized by means of the numerical range or the pseudospectra of $\mathbf{A}$; the details are given in §17. If $\mathbf{A}$ is a matrix, one can always follow this route by taking ω large enough, though for processes with transient effects in the norm of applied interest, it would generally be more useful to take $M > 1$ and a smaller value of ω. For some semigroups generated by unbounded operators, M cannot be taken to be 1 no matter how large ω is. For example, in $L^2(\mathbb{R})$, we might consider the semigroup of leftward translations defined by the condition $(e^{t\mathbf{A}}\mathbf{u})(x) = \mathbf{u}(x+t)$; obviously $\|e^{t\mathbf{A}}\| - 1$ for all t. On the other hand, again in $L^2(\mathbb{R})$, suppose we consider the semigroup defined by leftward translation except that a function doubles as it passes through

$x = 0$ [248, § 1.5.7]:

$$(e^{t\mathbf{A}}\mathbf{u})(x) = \begin{cases} 2\mathbf{u}(x+t) & \text{if } x \in [-t, 0]; \\ \mathbf{u}(x+t) & \text{otherwise.} \end{cases}$$

Now $\|e^{t\mathbf{A}}\| = 2$ for all t. What are the generators for these examples? In the first case $\mathbf{A}$ is the differentiation operator $\mathbf{u} \mapsto \mathbf{u}'$ with $\mathcal{D}(\mathbf{A})$ equal to the set of absolutely continuous functions in $L^2(\mathbb{R})$. For the second, $\mathbf{A}$ is again defined by $\mathbf{u} \mapsto \mathbf{u}'$, but now with a different domain: the set of absolutely continuous functions $\mathbf{u}$ with $\mathbf{u}(0) = 0$ and such that $\mathbf{u}'$ has left and right limits at $x = 0$ satisfying $\mathbf{u}'(0^-) = 2\mathbf{u}'(0^+)$.

The following theorem concerning Laplace transforms sets out the fundamentals from which our bounds are derived. For proofs, see, e.g., [606].

Relationships between $e^{t\mathbf{A}}$ and $(z - \mathbf{A})^{-1}$

Theorem 15.1 *Let $\mathbf{A}$ be a matrix or a closed linear operator generating a C_0 semigroup. There exist $\omega \in \mathbb{R}$ and $M \geq 1$ such that*

$$\|e^{t\mathbf{A}}\| \leq Me^{\omega t} \qquad \forall t \geq 0. \tag{15.1}$$

Any $z \in \mathbb{C}$ with $\mathrm{Re}\, z > \omega$ is in the resolvent set of $\mathbf{A}$, with

$$(z - \mathbf{A})^{-1} = \int_0^\infty e^{-zt} e^{t\mathbf{A}}\,\mathrm{d}t. \tag{15.2}$$

If $\mathbf{A}$ is a matrix or bounded operator, then[1]

$$e^{t\mathbf{A}} = \frac{1}{2\pi\mathrm{i}} \int_\Gamma e^{zt}(z - \mathbf{A})^{-1}\,\mathrm{d}z, \tag{15.3}$$

where Γ is any closed contour enclosing $\sigma(\mathbf{A})$ in its interior.

Our first bound follows immediately from (15.3); we do not give the calculations. The remark about the convex hull takes advantage of the fact that on the boundary of the convex hull of $\sigma_\varepsilon(\mathbf{A})$, $\|(z - \mathbf{A})^{-1}\| \geq \varepsilon^{-1}$, and by replacing $\sigma_\varepsilon(\mathbf{A})$ by its convex hull, one may reduce the constant L_ε. If $\sigma_\varepsilon(\mathbf{A})$ has several components, one can take the convex hull of each.

---

[1] The identity (15.3) holds more generally if $\mathbf{A}$ is *sectorial*, which means that its spectrum is contained in a sector of angle $<\pi$ in the left half-plane with resolvent norms decreasing outside the sector inverse-linearly with distance from the apex. Equation (15.6) below also holds for sectorial operators, as does a suitably modified form of (15.4). The semigroup generated by a sectorial operator is *analytic*, meaning that the family $\{e^{t\mathbf{A}}\}$ depends analytically on t.

Upper bound on $\|e^{t\mathbf{A}}\|$

Theorem 15.2 *If $\mathbf{A}$ is a matrix or bounded linear operator and L_ε is the arc length of the boundary of $\sigma_\varepsilon(\mathbf{A})$ or of its convex hull for some $\varepsilon > 0$, then*

$$\|e^{t\mathbf{A}}\| \leq \frac{L_\varepsilon e^{t\alpha_\varepsilon(\mathbf{A})}}{2\pi\varepsilon} \qquad \forall t \geq 0. \tag{15.4}$$

Next come perhaps the most familiar of all results involving $\|e^{t\mathbf{A}}\|$. As usual, $\alpha(\mathbf{A})$ denotes the spectral abscissa.

$\|e^{t\mathbf{A}}\|$ and the spectrum

Theorem 15.3 *Let $\mathbf{A}$ be a matrix or a closed linear operator generating a C_0 semigroup. Then*

$$\|e^{t\mathbf{A}}\| \geq e^{t\alpha(\mathbf{A})} \qquad \forall t \geq 0, \tag{15.5}$$

and if $\mathbf{A}$ is a matrix or bounded operator,

$$\lim_{t\to\infty} t^{-1} \log \|e^{t\mathbf{A}}\| = \alpha(\mathbf{A}). \tag{15.6}$$

Proof. To establish (15.5), suppose to the contrary that for some $\tau > 0$, $\|e^{\tau\mathbf{A}}\| = \nu < e^{\tau\alpha(\mathbf{A})}$, and assume $\omega \leq 0$ in (15.1) for simplicity; the case $\omega > 0$ is similar. From (15.1), we find that $\|e^{t\mathbf{A}}\|$ is bounded by M for $0 \leq t < \tau$, by $M\nu$ for $\tau \leq t < 2\tau$, by $M\nu^2$ for $2\tau \leq t < 3\tau$, and so on. Thus for all $t \geq 0$, $\|e^{t\mathbf{A}}\|$ is bounded by a function $\widetilde{M}e^{t\widehat{\omega}}$ with $\widehat{\omega} < \alpha(\mathbf{A})$, contradicting the assertion of Theorem 15.1 concerning the resolvent set.

To prove (15.6), note that $\liminf_{t\to\infty} t^{-1} \log \|e^{t\mathbf{A}}\| \geq \alpha(\mathbf{A})$ by (15.5), and thus our task is to prove $\limsup_{t\to\infty} t^{-1} \log \|e^{t\mathbf{A}}\| \leq \alpha(\mathbf{A})$. For each $\varepsilon > 0$, (15.4) implies that this 'lim sup' is $\leq \alpha_\varepsilon(\mathbf{A})$, and taking $\varepsilon \to 0$ completes the proof. ∎

We now present an important theorem containing many parts. It could be argued that Theorem 15.4 is the centerpiece of the theory of pseudospectra as applied to problems in continuous time. When dealing with nonnormal matrices and operators, we always face the questions, What do the pseudospectra tell us about dynamics? Are there significant transient effects? These bounds provide some answers.

The first few inequalities in this theorem have been known for years and represent the most straightforward lower bounds on $\|e^{t\mathbf{A}}\|$ that one gets if the resolvent norm is 'surprisingly large' at one or more points in the right half-plane, that is, larger than the inverse of the distance to the left half-plane, which implies that the semigroup cannot be a contraction. The final three inequalities, though also elementary, are not well-known. The inequality (15.11) was derived by Trefethen in 2002 (see (14.14) for a similar bound due to Davies [187]); the inequality (15.12) is new. For applications, these latter bounds have the valuable property that they identify a time scale on which transient effects must occur.

Lower bounds on $\|e^{t\mathbf{A}}\|$

Theorem 15.4 *Let $\mathbf{A}$ be a matrix or closed linear operator generating a C_0 semigroup. If $\|(z - \mathbf{A})^{-1}\| = K/\mathrm{Re}\,z$ for some z with $\mathrm{Re}\,z > 0$ and $K > 1$, then*

$$\sup_{t \geq 0} \|e^{t\mathbf{A}}\| \geq K. \tag{15.7}$$

The ε-pseudospectral abscissa $\alpha_\varepsilon(\mathbf{A})$ is finite for each $\varepsilon > 0$. Taking the rightmost value of z in the complex plane with the same value of $\|(z - \mathbf{A})^{-1}\|$ gives

$$\sup_{t \geq 0} \|e^{t\mathbf{A}}\| \geq \alpha_\varepsilon(\mathbf{A})/\varepsilon \qquad \forall \varepsilon > 0, \tag{15.8}$$

and maximizing over ε gives

$$\sup_{t \geq 0} \|e^{t\mathbf{A}}\| \geq \mathcal{K}(\mathbf{A}), \tag{15.9}$$

where the Kreiss constant is defined by

$$\mathcal{K}(\mathbf{A}) \equiv \sup_{\varepsilon > 0} \alpha_\varepsilon(\mathbf{A})/\varepsilon = \sup_{\mathrm{Re}\,z > 0} (\mathrm{Re}\,z)\|(z - \mathbf{A})^{-1}\|. \tag{15.10}$$

If $a = \mathrm{Re}\,z$, then for any $\tau > 0$,

$$\sup_{0 < t \leq \tau} \|e^{t\mathbf{A}}\| \geq e^{a\tau} \Big/ \left(1 + \frac{e^{a\tau} - 1}{K}\right), \tag{15.11}$$

and if $\|e^{t\mathbf{A}}\| \leq M$ for all $t \geq 0$, then for any $\tau \geq 0$, with K defined as before but now with $a < 0$ permitted and $-\infty < K/M \leq 1$,

$$\|e^{\tau\mathbf{A}}\| \geq e^{a\tau} - \frac{e^{a\tau} - 1}{K/M} = 1 - \frac{(e^{a\tau} - 1)(1 - K/M)}{K/M}. \tag{15.12}$$

In the particular case $a = K = 0$, (15.12) reduces by l'Hôpital's rule to

$$\|e^{\tau\mathbf{A}}\| \geq 1 - \frac{\tau M}{\|(z - \mathbf{A})^{-1}\|}. \tag{15.13}$$

Proof. If $\sup_{t \geq 0} \|e^{t\mathbf{A}}\| = M$, then by (15.2), for any z with $\mathrm{Re}\,z > 0$,

$$\frac{K}{\mathrm{Re}\,z} = \|(z - \mathbf{A})^{-1}\| \leq M \int_0^\infty |e^{-zt}| \, \mathrm{d}t = \frac{M}{\mathrm{Re}\,z}.$$

This implies (15.7), from which (15.8) and (15.9) follow.

To prove (15.11), we define $M_\tau = \sup_{0 < t \leq \tau} \|e^{t\mathbf{A}}\|$, which gives $\|e^{t\mathbf{A}}\| \leq$

M_τ for $0 < t \leq \tau$, $\|e^{t\mathbf{A}}\| \leq M_\tau^2$ for $\tau < t \leq 2\tau$, and so on. By (15.2) this implies

$$\|(z - \mathbf{A})^{-1}\| \leq \sum_{j=0}^{\infty} \int_{j\tau}^{(j+1)\tau} e^{-at} M_\tau^{j+1}\, dt = \int_0^\tau e^{-at} dt \sum_{j=0}^{\infty} e^{-aj\tau} M_\tau^{j+1}.$$

If $M_\tau \geq e^{a\tau}$, (15.11) certainly holds, so assume $M_\tau < e^{a\tau}$. Then we may sum the series to get

$$\|(z - \mathbf{A})^{-1}\| \leq \left(\frac{1 - e^{-a\tau}}{a}\right)\left(\frac{M_\tau}{1 - e^{-a\tau} M_\tau}\right) = \frac{e^{a\tau} - 1}{a(e^{a\tau}/M_\tau - 1)}.$$

Inverting this formula gives

$$\frac{a}{K} = \|(z - \mathbf{A})^{-1}\|^{-1} \geq \frac{a(e^{a\tau}/M_\tau - 1)}{e^{a\tau} - 1},$$

that is,

$$e^{a\tau}/M_\tau - 1 \leq \frac{e^{a\tau} - 1}{K},$$

which implies (15.11). The assertion that $\alpha_\varepsilon(\mathbf{A})$ is finite for all $\varepsilon > 0$ follows as a corollary, for suppose to the contrary that some value of $\|(z - \mathbf{A})^{-1}\| = \varepsilon^{-1}$ is achieved or exceeded for points z with arbitrarily large values of $a = \mathrm{Re}\, z$. Taking $\tau = c/a$ in (15.11) for some $c > 0$ gives

$$\sup_{0 \leq t \leq c/a} \|e^{t\mathbf{A}}\| \geq \frac{e^c}{1 + (e^c - 1)/(a/\varepsilon)},$$

and taking $a \to \infty$ shows that $\|e^{t\mathbf{A}}\|$ must be arbitrarily large for arbitrarily small t, contradicting the semigroup property (15.1).

Finally, to prove (15.12), we no longer assume $a > 0$; the bound may be interesting for $a \leq 0$. Set $\|e^{\tau\mathbf{A}}\| = P$. By (15.1) we have, for $0 \leq t \leq \tau$,

$$\|e^{t\mathbf{A}}\| \leq M, \quad \|e^{(\tau+t)\mathbf{A}}\| \leq PM, \quad \|e^{(2\tau+t)\mathbf{A}}\| \leq P^2 M,$$

and so on. If $P \geq e^{\tau a}$, (15.12) is trivial, so assume $P < e^{\tau a}$. Then from (15.2) we get

$$K/M \leq \frac{1 - e^{-\tau a}}{1 - Pe^{-\tau a}}$$

or

$$Pe^{-\tau a} \geq 1 - \frac{1 - e^{-a\tau}}{K/M},$$

and hence (15.12). ∎

The bounds (15.11)–(15.13) are important, but their significance may be hard to judge directly from the formulas. Figures 15.1 and 15.2 give an idea of what is going on and may also serve as references for quantitative application to particular problems. The context of (15.11) and Figure 15.1 is a matrix or operator $\mathbf{A}$ for which the resolvent norm is 'surprisingly

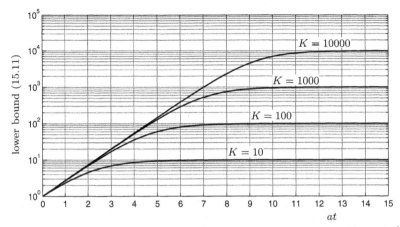

Figure 15.1: Illustration of the bound (15.11) with resolvent norm $\|(z-\mathbf{A})^{-1}\| = K/a$ for a single value z with $a = \mathrm{Re}\, z > 0$, for various values of K. The horizontal axis at reflects the fact that the time scale implied by the bound is inversely proportional to a. Each point on a curve indicates that somewhere for a value of at less than this value, $\mathrm{e}^{t\mathbf{A}}$ must be at least this large. The larger K is, the longer z must behave like an eigenvalue.

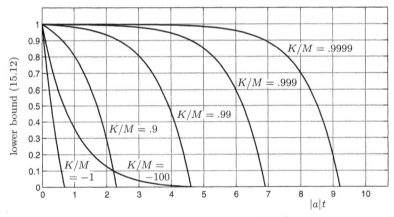

Figure 15.2: Illustration of the pointwise bound (15.12). Again the time scale is inversely proportional to a. Each point on a curve indicates that at this value of t, $\|\mathrm{e}^{t\mathbf{A}}\|$ must be at least this large. For $a > 0$ and $K > 0$, the closer K/M is to 1, the longer $\mathrm{e}^{t\mathbf{A}}$ must remain close to 1 or greater. For $a < 0$ and $K < 0$, the larger $|K/M|$ is, the longer $\mathrm{e}^{t\mathbf{A}}$ must remain close to e^{ta} or greater.

large' at some point z with $a = \mathrm{Re}\, z > 0$ as measured by a parameter $K \gg 1$. For large K, $\|\mathrm{e}^{t\mathbf{A}}\|$ must evolve at first at a rate close to e^{at}, and the larger K is, the longer this behavior must continue. In Figure 15.2 we assume $\|\mathrm{e}^{t\mathbf{A}}\| \le M$ for all $t \ge 0$. The context is now a situation in which the resolvent norm is 'surprisingly close to M/a' at a point z, as before. In such a situation $\|e^{t\mathbf{A}}\|$ must stay nearly as great as 1 or greater for all early values of t, falling below 1 only on a time scale proportional to $a^{-1}|\log(1 - K/M)|$.

Having presented one theorem with many parts, we shall now do it again. The next theorem is the same as Theorem 15.4, except generalized to account for left-right translations of $\mathbf{A}$ in the complex plane, as discussed on page 147. Mathematically, there is no new content here, but these results are so important in practice that it seems worthwhile to present them both in the simple form of Theorem 15.4, then again in this fuller generality.

Translated lower bounds on $\|e^{t\mathbf{A}}\|$

Theorem 15.5 *Let* $\mathbf{A}$ *be a matrix or closed linear operator generating a* C_0 *semigroup, and let* $\omega \in \mathbb{R}$ *denote a fixed shift parameter. If* $\|(z - \mathbf{A})^{-1}\| = K/(\mathrm{Re}\,z - \omega)$ *for some* z *with* $\mathrm{Re}\,z > \omega$ *and* $K > 1$*, then*

$$\sup_{t \geq 0} \|e^{-\omega t} e^{t\mathbf{A}}\| \geq K \qquad (15.14)$$

and

$$\sup_{t \geq 0} \|e^{-\omega t} e^{t\mathbf{A}}\| \geq (\alpha_\varepsilon(\mathbf{A}) - \omega)/\varepsilon \qquad \forall \varepsilon > 0 \qquad (15.15)$$

and

$$\sup_{t \geq 0} \|e^{-\omega t} e^{t\mathbf{A}}\| \geq \mathcal{K}(\mathbf{A}), \qquad (15.16)$$

where the Kreiss constant with respect to the half-plane $\mathrm{Re}\,z \leq \omega$ *is defined by*

$$\mathcal{K}(\mathbf{A}) \equiv \sup_{\varepsilon > 0} (\alpha_\varepsilon(\mathbf{A}) - \omega)/\varepsilon = \sup_{\mathrm{Re}\,z > \omega} (\mathrm{Re}\,z - \omega)\|(z - \mathbf{A})^{-1}\|. \quad (15.17)$$

If $a = \mathrm{Re}\,z$*, then for any* $\tau > 0$*,*

$$\sup_{0 < t \leq \tau} \|e^{-\omega t} e^{t\mathbf{A}}\| \geq e^{(a-\omega)\tau} \bigg/ \left(1 + \frac{e^{(a-\omega)\tau} - 1}{K} \right), \qquad (15.18)$$

and if $\|e^{t\mathbf{A}}\| \leq M e^{\omega t}$ *for all* $t \geq 0$ *as in* (15.1)*, then for any* $\tau \geq 0$*, with* K *defined as before but now with* $a < \omega$ *permitted and* $-\infty < K/M \leq 1$*,*

$$\|e^{\tau\mathbf{A}}\| \geq e^{a\tau} - \frac{e^{a\tau} - e^{\omega\tau}}{K/M} = e^{\omega\tau} - \frac{(e^{a\tau} - e^{\omega\tau})(1 - K/M)}{K/M}. \quad (15.19)$$

In particular, taking ω *to be the numerical abscissa,* $\omega(\mathbf{A}) = \max_{z \in W(\mathbf{A})} \mathrm{Re}\,z$*, so that* $\|e^{t\mathbf{A}}\| \leq e^{\omega t}$ *for all* $t \geq 0$ *and thus* $M = 1$ *in* (15.19) *(cf. §17),*

$$\|e^{\tau\mathbf{A}}\| \geq e^{a\tau} - \frac{e^{a\tau} - e^{\omega(\mathbf{A})\tau}}{K} = e^{\omega(\mathbf{A})\tau} - \frac{(e^{a\tau} - e^{\omega(\mathbf{A})\tau})(1 - K)}{K}.$$

$$(15.20)$$

Finally we present a theorem to the effect that nilpotent operators can be recognized from their pseudospectra; see Theorem 6.11 of [695], Example 4 on page 123 of [248], and [216].

Nilpotent operators

Theorem 15.6 *Let* $\mathbf{A}$ *be a closed linear operator generating a* C_0 *semi-group. For any* $\tau > 0$,

$$e^{\tau \mathbf{A}} = \mathbf{0} \iff \sigma(\mathbf{A}) = \emptyset \quad and \quad \|(z - \mathbf{A})^{-1}\| = \mathcal{O}(e^{-\tau \operatorname{Re} z}), \quad (15.21)$$

with the '$\mathcal{O}$' referring to the limit $\operatorname{Re} z \to -\infty$.

Proof. With M and ω as in (15.1), choose any $\mathbf{u} \in X$ and $\mathbf{v} \in X^*$ with $\|\mathbf{u}\| = \|\mathbf{v}\| = 1$ (see §4). Taking $z = \gamma + i\xi$ in (15.2) with $\gamma > \omega$ gives

$$(\mathbf{v}, ((\gamma + i\xi) - \mathbf{A})^{-1}\mathbf{u}) = \int_0^\infty e^{-it\xi}(\mathbf{v}, e^{-t\gamma}e^{t\mathbf{A}}\mathbf{u})\mathrm{d}t = \int_0^\infty e^{-it\xi} f(t)\mathrm{d}t$$

with $f(t) = (\mathbf{v}, e^{-t\gamma}e^{t\mathbf{A}}\mathbf{u})$. The notation $(\mathbf{v}, \mathbf{u})$ in this formula represents an inner product in Hilbert space and more generally, in Banach space, the action of the semilinear functional $\mathbf{v} \in X^*$ on $\mathbf{u} \in X$; see §4. The rapid decay of the term $e^{-t\gamma}e^{t\mathbf{A}}$ justifies the rearrangement of terms and implies that f is in L^2. This calculation shows that $F(\xi) = (\mathbf{v}, ((\gamma + i\xi) - \mathbf{A})^{-1}\mathbf{u})$ is the Fourier transform of f, and the Paley–Wiener theorem asserts that $f(t) = 0$ for all $t > \tau$ if and only if F is entire and $|F(\xi)| \leq Ce^{|\xi|\tau}$ for all complex ξ [222]. By one of the Phragmén–Lindelöf theorems, this latter bound is equivalent to $|F(\xi)| \leq Ce^{\tau \operatorname{Im}\xi}$ for $\operatorname{Im}\xi > 0$, which in turn is equivalent to $|(\mathbf{v}, (z - \mathbf{A})^{-1}\mathbf{u})| \leq Ce^{-\tau(\operatorname{Re} z - \gamma)}$ for $\operatorname{Re} z < \gamma$. The conclusion $f(t) = 0$ for $t > \tau$ can be extended to $t = \tau$ by the continuity property of C_0 semigroups. Since $\mathbf{u}$ and $\mathbf{v}$ are arbitrary unit elements, the proof is complete. ∎

This book is filled with examples of matrices $\mathbf{A}$ and their associated dynamical systems $\{e^{t\mathbf{A}}\}$ and pseudospectra $\{\sigma_\varepsilon(\mathbf{A})\}$, and in a number of cases, we draw connections between the two via some of the bounds (15.4)–(15.21). Here in this section we shall give just one example involving two matrices. It is an exceptionally interesting example because of its complexity and behavior unfolding on many time scales.

The example comes from a 2003 paper in the field of control theory by Burke, Lewis, and Overton [116], and it is also implemented as the 'Boeing' demonstration in EigTool. The problem originates in the analysis of a model of a Boeing 767 aircraft at a flutter condition; see [190, Problem 90-06] and [698]. The problem addressed in [116] leads to a family of real sparse matrices of dimension 55 of the form $\mathbf{A}_u + \mathbf{BKC}$, where $\mathbf{B}$ and $\mathbf{C}$ are fixed matrices of dimensions 55×2 and 2×55, respectively, and $\mathbf{K}$ is a

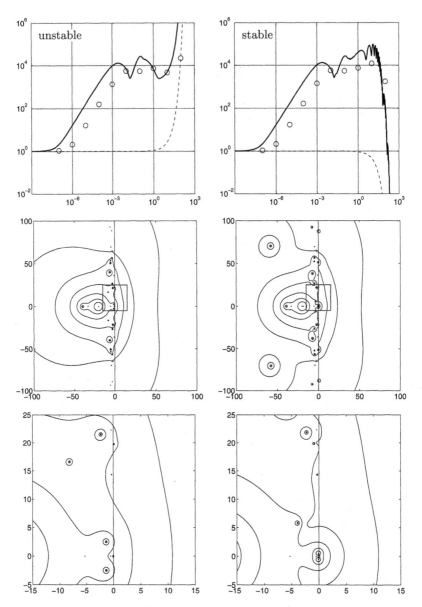

Figure 15.3: At the top, $\|e^{t\mathbf{A}}\|$ against t for the unstable and stable Boeing 767 flutter matrices $\mathbf{A}_u$ and $\mathbf{A}_s$, shown on log-log axes to highlight behavior on multiple time scales. The dashed lines show the eigenvalue lower bound (15.5), and the circles represent lower bounds (15.11) based on pseudospectra with $\tau = 10/a$, as discussed in the text. Below, ε-pseudospectra of each matrix with closeups as indicated by the boxes; the levels are $\varepsilon = 10^{-2}, 10^{-2.5}, \ldots$ in the first pair of plots and $\varepsilon = 10^{-3}, 10^{-3.5}, \ldots$ in the closeups. Although the second matrix is stable in theory, the huge transients would likely make it unstable in practice.

2×2 matrix containing four adjustable parameters. Each nonzero choice of parameters corresponds to a different feedback control on the physical system, and one may ask about the dynamical properties of the associated matrix. In particular, one may attempt to choose parameters so that the evolution process $\{e^{t\mathbf{A}}\}$ is as well-behaved as possible.

The uncontrolled matrix $\mathbf{A}_u$ from this class is unstable, with a pair of eigenvalues in the right half-plane and spectral abscissa $\alpha(\mathbf{A}_u) \approx 0.1015$. By applying techniques of nonsmooth optimization, Burke et al. found a set of control parameters so that $\mathbf{A}_s = \mathbf{A}_u + \mathbf{BKC}$ is stable, with $\alpha(\mathbf{A}_s) \approx -0.0788$. Classically, one would conclude that in a physical system governed by this model, $\{e^{t\mathbf{A}_u}\}$ will diverge to infinity whereas $\{e^{t\mathbf{A}_s}\}$ will behave well.

The actual operator norms tell a different story. The top row of Figure 15.3 shows $\|e^{t\mathbf{A}_u}\|$ and $\|e^{t\mathbf{A}_s}\|$ against t on a log-log scale. We see that for $t \gg 10$, $\|e^{t\mathbf{A}_u}\|$ diverges as expected; this time scale is consistent with the spectral abscissa of approximately 0.1. For $t < 10$, however, $\|e^{t\mathbf{A}_u}\|$ is four orders of magnitude larger than the eigenvalues can explain. As for $\|e^{t\mathbf{A}_s}\|$, for $t \gg 10$ it decays as expected, on a time scale consistent with the spectral abscissa of approximately -0.1. Again there is transient behavior, however, and this time it is even stronger, reaching a magnitude greater than 96,000 at $t \approx 12.5$. For $t \approx 1$, $\|e^{t\mathbf{A}_s}\|$ is about six times larger than $\|e^{t\mathbf{A}_u}\|$. In a physical system governed by these matrices, there is every possibility that $\mathbf{A}_s$ would behave as dangerously as $\mathbf{A}_u$.

These transient effects can be inferred from the pseudospectra, shown in the figure. The unstable eigenvalue for $\mathbf{A}_u$ has imaginary part about 19.8, corresponding to oscillatory behavior, but the closeup box shows that the resolvent norm is not especially large in this vicinity. Along the real axis, by contrast, the pseudospectra of $\mathbf{A}_s$ protrude into the right half-plane much more than those of $\mathbf{A}_u$. For example, the 10^{-2}-pseudospectral abscissa of $\mathbf{A}_s$ is about 54.15, implying that $\|e^{t\mathbf{A}_s}\|$ must exceed 5,000 on a time scale not much bigger than 0.02. (We compute $\alpha_\varepsilon(\mathbf{A})$ in EigTool using an algorithm described in §42, also due to Burke, Lewis, and Overton.) Similarly, the 10^{-4}-pseudospectral abscissa is about 2.11, implying that $\|e^{t\mathbf{A}_s}\|$ must exceed 20,000 on a time scale not much bigger than 0.5. To display such estimates graphically we considered the inequality (15.11), which provides lower bounds of the supremum over an interval $[0, \tau]$, for $z = 10^8, 10^7, \ldots, 10^{-1}$ (except with $z = 10^{-1}$ replaced by $z = 0.1016 + 19.77i$ for $\mathbf{A}_u$). In each case we took $\tau = 10a = 10\mathrm{Re}z$, so that $e^{a\tau} \approx 22,000$ in (15.11) and thus the bound is close to its large-τ asymptote. The results are plotted as circles in Figure 15.3. Each circle can be interpreted as follows: Somewhere for t smaller than this, the value of $\|e^{t\mathbf{A}}\|$ must be at least this high. The pseudospectral bounds track the transient convincingly.

16 · Powers of matrices and operators _____

This section, the analogue of the last one for powers instead of exponentials, provides proofs of the results (14.18)–(14.20), (14.23)–(14.27) and (14.29)–(14.31), as well as a pair of figures that summarize some of the most important lower bounds. Our concern now is $\|\mathbf{A}^k\|$, where $\mathbf{A}$ is a matrix or a bounded linear operator on a Banach space X with norm $\|\cdot\|$. For exponentials, we faced a technical challenge if $\mathbf{A}$ was unbounded. Here, since $\mathbf{A}$ is bounded, there are no technical difficulties.

The following standard results can be found, for example, in Sections 4.8 and 5.2 of Hille and Phillips [395].

Relationships between $\mathbf{A}^k$ and $(z - \mathbf{A})^{-1}$

Theorem 16.1 *Let $\mathbf{A}$ be a matrix or a bounded linear operator. There exist $\gamma > 0$ and $M \geq 1$ such that*

$$\|\mathbf{A}^k\| \leq M\gamma^k \qquad \forall k \geq 0. \tag{16.1}$$

Any $z \in \mathbb{C}$ with $|z| > \gamma$ is in the resolvent set of $\mathbf{A}$, and the resolvent for such z is given by the convergent series

$$(z - \mathbf{A})^{-1} = z^{-1}\left(\mathbf{I} + z^{-1}\mathbf{A} + (z^{-1}\mathbf{A})^2 + \cdots\right). \tag{16.2}$$

Conversely, for any $k \geq 0$,

$$\mathbf{A}^k = \frac{1}{2\pi\mathrm{i}} \int_\Gamma z^k (z - \mathbf{A})^{-1} \mathrm{d}z, \tag{16.3}$$

where Γ is any closed contour enclosing $\sigma(\mathbf{A})$ in its interior.

For our first set of bounds we take $\rho_\varepsilon(\mathbf{A})$ to be the ε-pseudospectral abscissa of $\mathbf{A}$, as usual, and we define the *Kreiss constant* of $\mathbf{A}$ (with respect to the unit disk) by

$$\mathcal{K}(\mathbf{A}) \equiv \sup_{\varepsilon > 0} \frac{\rho_\varepsilon(\mathbf{A}) - 1}{\varepsilon} = \sup_{|z| > 1} (|z| - 1)\|(z - \mathbf{A})^{-1}\|. \tag{16.4}$$

In dealing with powers of matrices we are coming to the original context of the Kreiss Matrix Theorem itself, and thus what we are now calling the Kreiss constant is the same quantity that Kreiss was concerned with in 1962 [465]; see §18.

Upper bounds on $\|\mathbf{A}^k\|$

Theorem 16.2 *If* $\mathbf{A}$ *is a matrix or bounded linear operator and* $k \geq 0$ *is arbitrary, then*

$$\|\mathbf{A}^k\| \leq \|\mathbf{A}\|^k \tag{16.5}$$

and, for any $\varepsilon > 0$,

$$\|\mathbf{A}^k\| \leq \frac{(\rho_\varepsilon(\mathbf{A}))^{k+1}}{\varepsilon}. \tag{16.6}$$

If L_ε *is the arc length of the boundary of* $\sigma_\varepsilon(\mathbf{A})$ *or of its convex hull for some* $\varepsilon > 0$, *then*

$$\|\mathbf{A}^k\| \leq \frac{L_\varepsilon(\rho_\varepsilon(\mathbf{A}))^k}{2\pi\varepsilon}, \tag{16.7}$$

and with $\mathcal{K}(\mathbf{A})$ *defined by* (16.4),

$$\|\mathbf{A}^k\| < \mathrm{e}(k+1)\mathcal{K}(\mathbf{A}). \tag{16.8}$$

Proof. The first estimate is trivial, and the next two follow from (16.3) by taking Γ to be the boundary of $\sigma_\varepsilon(\mathbf{A})$, its convex hull, or the circle about the origin of radius $\rho_\varepsilon(\mathbf{A})$. Taking $\rho_\varepsilon(\mathbf{A}) = 1 + k^{-1}$ in (16.6) and using the identity $(1 + k^{-1})^k < \mathrm{e}$ gives (16.8). ∎

From here we can derive familiar results relating $\|\mathbf{A}^k\|$ to the spectral radius $\rho(\mathbf{A})$.

$\|\mathbf{A}^k\|$ and the spectrum

Theorem 16.3 *If* $\mathbf{A}$ *is a matrix or bounded linear operator, then*

$$\|\mathbf{A}^k\| \geq \rho(\mathbf{A})^k \qquad \forall k \geq 0 \tag{16.9}$$

and

$$\lim_{k \to \infty} \|\mathbf{A}^k\|^{1/k} = \rho(\mathbf{A}). \tag{16.10}$$

Proof. To establish (16.9), suppose to the contrary that for some $\kappa > 0$, $\|\mathbf{A}^\kappa\| = \nu < \rho(\mathbf{A})^\kappa$, and assume $\gamma \leq 1$ in (16.1) for simplicity; the case $\gamma > 1$ is similar. From (16.1), we find that $\|\mathbf{A}^k\|$ is bounded by M for $0 \leq k < \kappa$, by $M\nu$ for $\kappa \leq k < 2\kappa$, by $M\nu^2$ for $2\kappa \leq k < 3\kappa$, and so on. Thus for all $k \geq 0$, $\|\mathbf{A}^k\|$ is bounded by a function $\hat{M}\hat{\gamma}^k$ with $\hat{\gamma} < \rho(\mathbf{A})$, contradicting the assertion of Theorem 16.1 concerning the resolvent set. To prove (16.10), we note that $\liminf_{k\to\infty} \|\mathbf{A}^k\|^{1/k} \geq \rho(\mathbf{A})$ by (16.9), and thus our task is to prove $\limsup_{k\to\infty} \|\mathbf{A}^k\|^{1/k} \leq \rho(\mathbf{A})$. This follows by taking $\varepsilon \to 0$ in (16.7). ∎

We now present the lengthy analogue for powers of Theorem 15.4 for exponentials. The first few inequalities represent the easiest lower bounds on $\|\mathbf{A}^k\|$ that one gets if the resolvent norm is large at one or more points outside the unit disk. The bound (16.15) is implemented computationally among the 'Transients' features in EigTool [838], and its application was illustrated in the discussion of the examples of Figures 14.4 and 14.5.

Lower bounds on $\|\mathbf{A}^k\|$

Theorem 16.4 *Let $\mathbf{A}$ be a matrix or bounded operator. If $\|(z-\mathbf{A})^{-1}\| = K/(|z|-1)$ for some z with $|z| = r > 1$ and $K > 1$, then*

$$\sup_{k \geq 0} \|\mathbf{A}^k\| \geq rK - r + 1 > K. \tag{16.11}$$

Taking the largest-modulus value of z in the complex plane with the same value of $\|(z-\mathbf{A})^{-1}\|$ gives

$$\sup_{k \geq 0} \|\mathbf{A}^k\| \geq (\rho_\varepsilon(\mathbf{A}) - 1)(\rho_\varepsilon(\mathbf{A})/\varepsilon - 1) \geq \frac{\rho_\varepsilon(\mathbf{A}) - 1}{\varepsilon} \tag{16.12}$$

for all $\varepsilon > 0$, and maximizing over ε gives

$$\sup_{k \geq 0} \|\mathbf{A}^k\| \geq \mathcal{K}(\mathbf{A}), \tag{16.13}$$

where the Kreiss constant *is defined by*

$$\mathcal{K}(\mathbf{A}) \equiv \sup_{\varepsilon > 0}(\rho_\varepsilon(\mathbf{A}) - 1)/\varepsilon = \sup_{|z| > 1}(|z| - 1)\|(z - \mathbf{A})^{-1}\|. \tag{16.14}$$

For any $\kappa > 0$,

$$\sup_{0 < k \leq \kappa} \|\mathbf{A}^k\| \geq r^\kappa \Big/ \left(1 + \frac{r^\kappa - 1}{rK - r + 1}\right), \tag{16.15}$$

and if $\|\mathbf{A}^k\| \leq M$ for all $k \geq 0$, then for any $\kappa \geq 0$, with K defined as before but now with $r < 1$ permitted and $-\infty < K/M \leq 1$,

$$\|\mathbf{A}^\kappa\| \geq r^\kappa - \frac{r^\kappa - 1}{K/M} = 1 - \frac{(r^\kappa - 1)(1 - K/M)}{K/M}. \tag{16.16}$$

When $r = 1$ and $K = 0$, (16.16) reduces by l'Hôpital's rule to

$$\|\mathbf{A}^\kappa\| \geq 1 - \frac{\kappa M}{\|(z - \mathbf{A})^{-1}\|}. \tag{16.17}$$

Proof. If $\sup_{k\geq 0}\|\mathbf{A}^k\| = M$, then by (16.2), for any z with $|z| = r > 1$,

$$\frac{rK}{r-1} = r\|(z-\mathbf{A})^{-1}\| \leq 1 + M\sum_{k=1}^{\infty} r^{-k} = 1 + \frac{M}{r-1},$$

which implies the middle bound of (16.11); this middle bound is in turn greater than K since $rK - r + 1 = K + (r-1)(K-1)$. Equations (16.12) and (16.13) follow from this weaker bound involving K.

To prove (16.15) we define $M_\kappa = \sup_{0<k\leq\kappa}\|\mathbf{A}^k\|$, which gives $\|\mathbf{A}^k\| \leq M_\kappa$ for $0 < k \leq \kappa$, $\|\mathbf{A}^k\| \leq M_\kappa^2$ for $\kappa < k \leq 2\kappa$, and so on. This implies

$$r\|(z-\mathbf{A})^{-1}\| \leq 1 + \sum_{j=0}^{\infty}\sum_{k=j\kappa+1}^{(j+1)\kappa} r^{-k}M_\kappa^{j+1} = 1 + \sum_{k=1}^{\kappa} r^{-k}\sum_{j=0}^{\infty} r^{-jk}M_\kappa^{j+1}.$$

If $M_\kappa \geq r^\kappa$, (16.15) holds, so assume $M_\kappa < r^\kappa$. Then we may sum these series to get

$$r\|(z-\mathbf{A})^{-1}\| \leq 1 + \left(\frac{1-r^{-\kappa}}{r-1}\right)\left(\frac{r^\kappa}{r^\kappa/M_\kappa - 1}\right) = \frac{r^\kappa - 1}{(r-1)(r^\kappa/M_\kappa - 1)}.$$

Subtracting 1 from both sides of the inequality yields $(rK - r + 1)/(r-1)$ on the left, so inverting gives

$$\frac{1}{rK-r+1} \geq \frac{r^\kappa/M_\kappa - 1}{r^\kappa - 1},$$

which implies

$$r^\kappa/M_\kappa - 1 \leq \frac{r^\kappa - 1}{rK - r + 1}$$

and hence (16.15).

To prove (16.16), we no longer assume $r > 1$; the bound is also interesting for $r < 1$. Set $\|\mathbf{A}^\kappa\| = P$. By (16.1) we have for $0 \leq k < \kappa$,

$$\|\mathbf{A}^k\| \leq M, \quad \|\mathbf{A}^{\kappa+k}\| \leq PM, \quad \|\mathbf{A}^{2\kappa+k}\| \leq P^2 M,$$

and so on. If $P \geq r^\kappa$, (16.16) is trivial, so assume $P < r^\kappa$. Then from (16.2) we get

$$K/M \leq \frac{1-r^{-\kappa}}{1-Pr^{-\kappa}}$$

or

$$Pr^{-\kappa} \geq 1 - \frac{1-r^{-\kappa}}{K/M},$$

which implies (16.16). ∎

Figures 16.1 and 16.2 illustrate the bounds (16.15) and (16.16), following the pattern of Figures 15.1 and 15.2 in the last section. This time, the

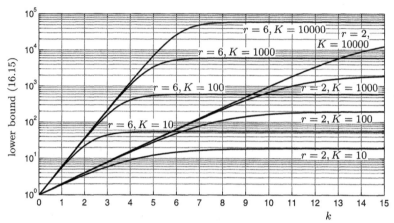

Figure 16.1: Illustration of the bound (16.15) with resolvent norm $\|(z - \mathbf{A})^{-1}\| = K/(r - 1)$ for a single value z with $r = |z| > 1$, for various values of K and r. Each point on a curve indicates that somewhere for a value of k less than or equal to this value, $\|\mathbf{A}^k\|$ must be at least this large. The larger K is, the longer z must behave like an eigenvalue.

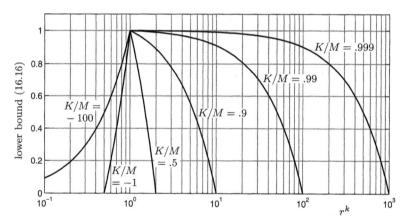

Figure 16.2: Illustration of the pointwise bound (16.16). Each point on a curve indicates that at this value of r^k, $\|\mathbf{A}^k\|$ must be at least this large. For $r > 1$ and $K > 0$, the closer K/M is to 1, the longer $\|\mathbf{A}^k\|$ must remain close to 1 or greater. For $r < 1$ and $K < 0$, the larger $|K/M|$ is, the longer $\|\mathbf{A}^k\|$ must remain close to r^k or greater.

bounds are slightly more complicated because there is a scale-dependence introduced by the discreteness of the time (i.e., k) axis. Though these figures are complex, the bounds they depict are of great importance in applications. It will be interesting to see whether alternative bounds are derived in the years ahead that reduce some of this complexity.

Like Theorem 15.4 in the last section, Theorem 16.4 has a built-in scaling: to the unit disk. As discussed at the end of §14, each of these results can be generalized by an arbitrary dilation so as to be scaled to the disk of radius γ. The next theorem records the details.

Dilated lower bounds on $\|\mathbf{A}^k\|$

Theorem 16.5 *Let $\mathbf{A}$ be a matrix or bounded operator, and let $\gamma > 0$ be a fixed dilation parameter. If $\|(z - \mathbf{A})^{-1}\| = K/(|z| - \gamma)$ for some z with $|z| = r > \gamma$ and $K > 1$, then*

$$\sup_{k \geq 0} \|\gamma^{-k}\mathbf{A}^k\| \geq rK/\gamma - r/\gamma + 1 > K \qquad (16.18)$$

and

$$\sup_{k \geq 0} \|\gamma^{-k}\mathbf{A}^k\| \geq (\rho_\varepsilon(\mathbf{A}) - \gamma)/\varepsilon \qquad \forall \varepsilon > 0 \qquad (16.19)$$

and

$$\sup_{k \geq 0} \|\gamma^{-k}\mathbf{A}^k\| \geq \mathcal{K}(\mathbf{A}), \qquad (16.20)$$

where the Kreiss constant *with respect to the disk of radius γ is defined by*

$$\mathcal{K}(\mathbf{A}) \equiv \sup_{\varepsilon > 0}(\rho_\varepsilon(\mathbf{A}) - \gamma)/\varepsilon = \sup_{|z| > \gamma}(|z| - \gamma)\|(z - \mathbf{A})^{-1}\|. \qquad (16.21)$$

For any $\kappa > 0$,

$$\sup_{0 < k \leq \kappa} \|\gamma^{-k}\mathbf{A}^k\| \geq r^\kappa\gamma^{-\kappa} \bigg/ \left(1 + \frac{r^\kappa\gamma^{-\kappa} - \gamma^\kappa}{rK/\gamma - r/\gamma + 1}\right), \qquad (16.22)$$

and if $\|\mathbf{A}^k\| \leq M\gamma^k$ for all $k \geq 0$ as in (16.1), then for any $\kappa \geq 0$, with K defined as before but now with $r < \gamma$ permitted and $-\infty < K/M \leq 1$,

$$\|\mathbf{A}^\kappa\| \geq r^\kappa - \frac{r^\kappa - \gamma^\kappa}{K/M} = \gamma^\kappa - \frac{(r^\kappa - \gamma^\kappa)(1 - K/M)}{K/M}. \qquad (16.23)$$

In particular, taking $M = 1$ and $\gamma = \|\mathbf{A}\|$ gives

$$\|\mathbf{A}^\kappa\| \geq r^\kappa - \frac{r^\kappa - \|A\|^\kappa}{K}. \qquad (16.24)$$

Another bound of the same flavor as (16.24), circulated by Trefethen in unpublished notes in the 1990s, is

$$\|\mathbf{A}\|^\kappa \geq (\rho_\varepsilon(\mathbf{A}))^\kappa - [(\|\mathbf{A}\| + \varepsilon)^\kappa - \|\mathbf{A}\|^\kappa] \qquad (16.25)$$

for any $\kappa \geq 0$ and $\varepsilon > 0$, which can be derived by noting that there must be a matrix or operator $\mathbf{B}$ with $\|\mathbf{B} - \mathbf{A}\| \leq \varepsilon$ whose spectral radius is $\rho_\varepsilon(\mathbf{A})$. It can be shown that (16.24) is a stronger estimate, implying (16.25) as a corollary.

Our final theorem is quite obvious if $\mathbf{A}$ is a matrix, thanks to the Jordan canonical form. The theorem reveals that no new behavior can arise with bounded operators.

Nilpotent matrices and operators

Theorem 16.6 *Let $\mathbf{A}$ be a matrix or bounded linear operator. For any $\kappa > 0$,*

$$\mathbf{A}^\kappa = \mathbf{0} \iff \sigma(\mathbf{A}) = \{0\} \ and \ \|(z - \mathbf{A})^{-1}\| = \mathcal{O}(|z|^{-\kappa}), \quad (16.26)$$

with the '$\mathcal{O}$' referring to the limit $z \to 0$.

Proof. If $\|(z - \mathbf{A})^{-1}\| = \mathcal{O}(|z|^{-\kappa})$, then $\mathbf{A}^\kappa = \mathbf{0}$ follows from (16.2). Conversely, suppose $\mathbf{A}^\kappa = \mathbf{0}$ for some κ. Define $M = \sup_{k \geq 0} \|\mathbf{A}^k\|$, and for any $z \neq 0$, define $r = |z|$ and $\gamma = r/2$. Then in the estimate (16.18), we have $K = (r/2)\|(z - \mathbf{A})^{-1}\|$, and, provided $|z| < 2$, that bound implies $K < M(r/2)^{1-\kappa}$. Hence as $z \to 0$, $\|(z - \mathbf{A})^{-1}\| = \mathcal{O}(|z|^{-\kappa})$. ∎

Illustrations of the results of this section appear throughout this book. Here we give just two examples, shown in Figure 16.3, based on the 8×8 Jordan matrix

$$\mathbf{A} = \begin{pmatrix} 0 & 2 & & & & & & \\ & 0 & 2 & & & & & \\ & & 0 & 2 & & & & \\ & & & 0 & 2 & & & \\ & & & & 0 & 2 & & \\ & & & & & 0 & 2 & \\ & & & & & & 0 & 2 \\ & & & & & & & 0 \end{pmatrix} \quad (16.27)$$

and the 5×5 Toeplitz matrix

$$\mathbf{A} = \begin{pmatrix} 0 & 10^2 & & & \\ 0 & 0 & 10^2 & & \\ 10^{-4} & 0 & 0 & 10^2 & \\ & 10^{-4} & 0 & 0 & 10^2 \\ & & 10^{-4} & 0 & 0 \end{pmatrix}. \quad (16.28)$$

The upper and lower shaded regions in each image correspond to the pseudospectral bounds (16.6) and (16.24). These bounds depend on parameters $\varepsilon > 0$ in the first case and $r > \|\mathbf{A}\|$ in the second. For each choice of parameters, one gets an upper bound in the form of a straight line on this log scale and a lower bound of a more complicated shape. The shaded region

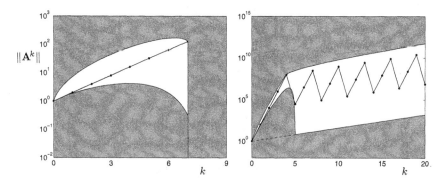

Figure 16.3: Norms of powers of the matrices (16.27) and (16.28), together with upper and lower bounds (16.6) and (16.24) based on pseudospectra, as discussed in the text. The dashed line is the eigenvalue lower bound (16.9).

has been obtained by taking the lower and upper envelopes of these curves, respectively, where ε and r range over all possible values. Similar figures for other matrices are presented in §§25 and 26.

Note that for the Jordan block (16.27), on the left of Figure 16.3, the upper bound has captured exactly the fact that $\|\mathbf{A}^k\| = 0$ for $k \geq 8$. This good fortune corresponds to the limits $z \to 0$ in (16.2) or $\varepsilon \to 0$ in (16.6), and matches Theorem 16.6.

Our discussion has concerned exact mathematics, not the results that might be obtained on a computer. However, as pointed out in [632] and examined more fully in [391, 392] and [728], rounding errors may have pronounced effects on the powers of nonnormal matrices. With rounding errors, a nilpotent matrix may cease to be nilpotent and a power-bounded matrix may have powers that grow exponentially. For example, the spectral differentiation matrix $\mathbf{D}_N$ of §30 is nilpotent, but on a computer, its powers grow exponentially for dimensions as low as $N = 10$. In many cases $\|\mathbf{A}^k\|$ grows in floating-point arithmetic approximately at the rate $(\rho_\varepsilon(\mathbf{A}))^k$, where ε is on the order of machine epsilon; see §53.

17 · Numerical range, abscissa, and radius _____

The numerical range is most familiar for matrices or linear operators in
Hilbert space, but the definition and the key properties (14.2) and (14.4)
generalize also to Banach space. Here, for simplicity, we first develop the
subject and the main theorems for a matrix in Hilbert space, i.e., with
$\| \cdot \| = \| \cdot \|_2$. In the final four pages of the section we indicate how these
ideas can be generalized.

The (2-norm) *numerical range* or *field of values* of a matrix $\mathbf{A} \in \mathbb{C}^{N \times N}$
is the set of all of its Rayleigh quotients:

$$W(\mathbf{A}) = \{\mathbf{x}^* \mathbf{A} \mathbf{x} \colon \mathbf{x} \in \mathbb{C}^N, \ \|\mathbf{x}\| = 1\}, \qquad (17.1)$$

where $\| \cdot \| = \| \cdot \|_2$. This is a closed, convex subset of $\mathbb{C}$ that contains
the convex hull of the spectrum $\sigma(\mathbf{A})$; if $\mathbf{A}$ is normal, the containment
is an identity.[1] The main aims of this section are, first, to discuss the
significance of the numerical range, which is narrower than is sometimes
supposed, and second, to show that just as the spectrum is determined by
the behavior of the pseudospectra $\sigma_\varepsilon(\mathbf{A})$ in the limit $\varepsilon \to 0$, the numerical
range is determined by the behavior of $\sigma_\varepsilon(\mathbf{A})$ in the limit $\varepsilon \to \infty$. If you
know the pseudospectra of a matrix, you know its numerical range.[2]

The principal application of the numerical range is to estimating the
behavior of $\|e^{t\mathbf{A}}\|$ as a function of t. We know that $\|e^{t\mathbf{A}}\|$ may have quite
different behaviors in the initial, transient, and asymptotic phases $t \to 0$, t
finite, and $t \to \infty$. According to (14.1) and (14.3) or Theorem 15.3, the behavior as $t \to \infty$ is determined by the spectral abscissa $\alpha(\mathbf{A})$. Analogously,
it was indicated in (14.2) that the behavior as $t \to 0$ is determined by the
numerical abscissa $\omega(\mathbf{A})$. This connection was illustrated schematically in
Figure 14.1, and later in this section, we shall examine the numerical range
of the matrix we used to generate that 'schematic' figure.

---

[1]Excellent sources of detailed information about the numerical range are [72, 73, 354,
367, 415]. The convexity property, due to Hausdorff in 1919, is not trivial: it can be
proved by projecting $\mathbf{A}$ to a two-dimensional subspace spanned by any two vectors $\mathbf{x}$
and $\mathbf{y}$, where the numerical range must be the closed set bounded by an ellipse or a line
segment. The symbol W originates in the German *Wertevorrat*, the term introduced by
Toeplitz and Hausdorff.

[2]Here is a fanciful way to describe a different relationship between the pseudospectra
and the numerical range: the ε-pseudospectrum is the spectrum as measured by a blunt
instrument, of precision ε, and the numerical range is the spectrum as measured by a
one-dimensional instrument. The meaning of the latter statement is that $W(\mathbf{A})$ is the
union of all the spectra obtained by projecting $\mathbf{A}$ onto one-dimensional subspaces in the
same sense as in §40. (Equivalently, $W(\mathbf{A})$ is the set of all Ritz values that might be
obtained after one step of Arnoldi iteration (§28).) The extension of this idea to higher
dimensional subspaces is immediate.

$\|e^{t\mathbf{A}}\|$ and the numerical range

Theorem 17.1 *For any* $\mathbf{A} \in \mathbb{C}^{N \times N}$ *with numerical abscissa* $\omega(\mathbf{A})$ *and* $\|\cdot\| = \|\cdot\|_2$,

$$\|e^{t\mathbf{A}}\| \le e^{t\omega(\mathbf{A})} \qquad \forall t \ge 0 \tag{17.2}$$

and

$$\|e^{t\mathbf{A}}\| = e^{t\omega(\mathbf{A})} + o(t) \quad as\ t \to 0. \tag{17.3}$$

In particular, $\|e^{t\mathbf{A}}\| \le 1$ *for all* $t \ge 0$ *(i.e.,* $\mathbf{A}$ *is contractive) if and only if* $\omega(\mathbf{A}) \le 0$.

Proof. If $\tau = kt$ for some positive integer k, then $\|e^{\tau\mathbf{A}}\| \le \|e^{t\mathbf{A}}\|^k$. By taking the limit $t \to 0$ we see that (17.3) implies (17.2) and the final assertion of the theorem. The proof of (17.3) is given after Theorem 17.4, below. ∎

Equation (17.3) shows that the numerical range answers a certain question about matrix behavior exactly. So far as we know, this is the only question about matrix behavior that it answers exactly. In other applications one finds that either the problem is equivalent to (17.3) or the information obtained from the numerical range is approximate. For example, the numerical range does not answer the question of whether there exists a constant C such that $\|e^{t\mathbf{A}}\| \le C$ for all $t \ge 0$ (i.e., $\mathbf{A}$ is *stable*; see §49). A sufficient condition for this is $\omega(\mathbf{A}) \le 0$, but one can see that there is no such necessary and sufficient condition by noting that the matrices

$$\mathbf{A} = \begin{pmatrix} -1 & 4 \\ 0 & -1 \end{pmatrix}, \qquad \mathbf{B} = \begin{pmatrix} \mathbf{A} & 0 \\ 0 & 1 \end{pmatrix}$$

have the same numerical ranges, namely the closed disk about -1 of radius 2, but $\mathbf{A}$ is stable whereas $\mathbf{B}$ is unstable.

Ideas related to the numerical range have been applied extensively in several fields since the days of Lyapunov, and the notion of contractivity (local) as a sufficient condition for stability (global) is usually at the root of these applications. The hypothesis of contractivity is so strong that it rarely leads to stability conditions that are necessary as well as sufficient, but by the same token, the conditions it does lead to are robust enough to extend to nonlinear problems (cf. Figure 33.3). One highly developed field of applications is the analysis of discrete numerical methods for ordinary differential equations, especially Runge–Kutta methods [193, 365, 796]. Another is fluid mechanics, where the theory of 'nonlinear stability' is again really a theory of contractivity [214, 436, 622]. A third is semigroup theory in mathematics, where contractive semigroups are characterized by the Hille–Yosida and Lumer–Phillips theorems, as we shall discuss in §19. The numerical range is also a familiar tool in numerical linear algebra [232, 249, 528, 719]. Terms related to contractivity and the numerical range that appear in various fields include *dissipativity, accretiveness, semi-boundedness*, the *one-sided Lipschitz constant*, and the *logarithmic norm*. The logarithmic norm, a real number that may be positive or zero or negative, is defined by [364, 575]

$$\beta(\mathbf{A}) = \lim_{\varepsilon \downarrow 0} \frac{\|\mathbf{I} + \varepsilon \mathbf{A}\| - 1}{\varepsilon}. \tag{17.4}$$

Another application of the numerical range is to the problem of contractivity for the powers $\|\mathbf{A}^k\|$. If $\|\mathbf{A}^k\| \leq 1$ for all $k \geq 0$, then $W(\mathbf{A})$ is contained in the closed unit disk (proof: take $k = 1$ and consider the definition (17.1)), but if $W(\mathbf{A})$ is contained in the closed unit disk, the best one can conclude is $\|\mathbf{A}^k\| \leq 2$. In general, the *numerical radius* of $\mathbf{A}$ is defined by

$$\mu(\mathbf{A}) = \sup_{z \in W(\mathbf{A})} |z|, \tag{17.5}$$

and the analogue of (17.2) is

$$\|\mathbf{A}^k\| \leq 2(\mu(\mathbf{A}))^k, \tag{17.6}$$

a result due to Berger [54, 367, 607]. Thus the numerical range does not give a complete answer to questions of norms of powers, but it gives partial information.

Now we turn to an example, the matrix that was used to generate the schematic illustration shown in Figure 14.1:

$$\mathbf{A} = \begin{pmatrix} -5 & 4 & 4 \\ 0 & -2 - 2i & 4 \\ 0 & 0 & -.3 + i \end{pmatrix}. \tag{17.7}$$

Figure 17.1 displays the spectrum, pseudospectra, and numerical range of this matrix. The spectral abscissa is negative, equal to -0.3, and this is the slope of the dashed line at the right in Figure 14.1. The numerical abscissa is positive, approximately 2.11, and this is the slope of the dashed line at the left. If $\mathbf{A}$ were a normal matrix with the same eigenvalues, such as (17.7) with the off-diagonal elements set to zero, the numerical range would be just a triangular region (the convex hull of the spectrum) and the spectral and numerical abscissas would both be equal to -0.3, as shown on the right side of Figure 17.1.

We come now to the relationship between the numerical range of a matrix and its pseudospectra. It is a corollary of Theorem 17.3, below, that the resolvent norm of any matrix satisfies

$$\|(z - \mathbf{A})^{-1}\| \leq \frac{1}{\text{dist}(z, W(\mathbf{A}))} \qquad \forall z \notin W(\mathbf{A}), \tag{17.8}$$

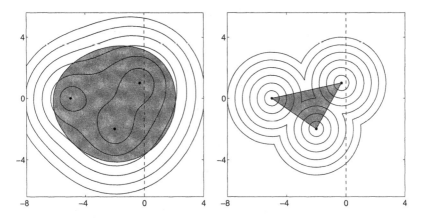

Figure 17.1: Spectrum, ε-pseudospectra ($\varepsilon = 3.0, 2.5, \ldots, 0.5$), and numerical range (shaded) of the 3×3 matrix (17.7) used to generate Figure 14.1 (left), and of the same matrix with the off-diagonal entries set to zero (right). The dashed line is the imaginary axis.

where dist denotes the usual distance from a point to a set (compare (4.2)). This resolvent bound, which has been known for years [732, Thm. 4.20], implies that pseudospectra cannot be much larger than the numerical range:

$$\sigma_\varepsilon(\mathbf{A}) \subseteq W(\mathbf{A}) + \Delta_\varepsilon, \tag{17.9}$$

where Δ_ε is the open disk of radius ε about the origin [624]. Conversely, the following theorem establishes that $W(\mathbf{A})$ is determined by (17.8).

The numerical range and the pseudospectra

Theorem 17.2 *For any $\mathbf{A} \in \mathbb{C}^{N \times N}$, with $\|\cdot\| = \|\cdot\|_2$, $W(\mathbf{A})$ is equal to the intersection of all the closed half-planes $H \subseteq \mathbb{C}$ satisfying*

$$\|(z - \mathbf{A})^{-1}\| \leq \frac{1}{\text{dist}(z, H)} \qquad \forall z \notin H, \tag{17.10}$$

or equivalently,

$$\text{dist}(\lambda, H) < \varepsilon \qquad \forall \varepsilon > 0, \quad \forall \lambda \in \sigma_\varepsilon(\mathbf{A}) \tag{17.11}$$

or

$$H + \Delta_\varepsilon \supseteq \sigma_\varepsilon(\mathbf{A}) \qquad \forall \varepsilon > 0. \tag{17.12}$$

Proof. The equivalence of (17.10)–(17.12) follows from the definition of $\sigma_\varepsilon(\mathbf{A})$. The connection of these formulas with $W(\mathbf{A})$ is proved after Theorem 17.4, below. ∎

The essential feature of (17.10) is that the numerator on the right-hand side is 1, not an arbitrary Kreiss constant $\mathcal{K} \geq 1$ as in §§14, 15, and 18. Equivalently, it is ε rather than $\mathcal{K}\varepsilon$ that appears on the right in (17.11) and (17.12). These are further reflections of the association of the numerical range with contractivity, not stability.

Theorems 17.1 and 17.2 relate the numerical range to the exponentials $\|e^{t\mathbf{A}}\|$ and the resolvent norms $\|(z - \mathbf{A})^{-1}\|$, or equivalently, the pseudospectra $\sigma_\varepsilon(\mathbf{A})$. We have deferred the proofs of these results because it is easiest to treat them together. For simplicity, the following theorem is formulated for the particular value $\omega(\mathbf{A}) = 0$, and nonzero values $\omega(\mathbf{A})$ are then treated as a corollary in Theorem 17.4. As always in this book, $\alpha_\varepsilon(\mathbf{A})$ denotes the ε-pseudospectral abscissa of $\mathbf{A}$, $\alpha_\varepsilon(\mathbf{A}) = \sup_{z \in \sigma_\varepsilon(\mathbf{A})} \mathrm{Re}\,z$.

Equivalent conditions for $\omega(\mathbf{A}) = 0$

Theorem 17.3 *Given $\mathbf{A} \in \mathbb{C}^{N \times N}$ with $\|\cdot\| = \|\cdot\|_2$, let z, ε, and t be positive real numbers. The following statements are equivalent:*

(i) $\omega(\mathbf{A}) = 0,$ $\qquad\qquad\qquad\qquad\qquad\qquad$ (17.13)

(ii) $\|(z - \mathbf{A})^{-1}\|^{-1} = z + o(1)$ *as* $z \to \infty,$ $\qquad$ (17.14)

(iii) $\alpha_\varepsilon(\mathbf{A}) = \varepsilon + o(1)$ *as* $\varepsilon \to \infty,$ $\qquad\qquad$ (17.15)

(iv) $\beta(\mathbf{A}) = 0,$ $\qquad\qquad\qquad\qquad\qquad\qquad$ (17.16)

(v) $(\frac{\mathrm{d}}{\mathrm{d}t}\|e^{t\mathbf{A}}\|)_{t=0} = 0.$ $\qquad\qquad\qquad\qquad$ (17.17)

Here ω, α_ε, and β denote the numerical abscissa, ε-pseudospectral abscissa, and logarithmic norm, respectively. The derivative in (17.17) is defined by the one-sided limit $t \downarrow 0$.

In semigroup theory, approximately speaking, the equivalence of (i) and (v) is the Lumer–Phillips theorem, and the equivalence of (ii) and (v) is the Hille–Yosida theorem.

Proof. Throughout the proof we use the convention $z = t^{-1}$ and omit the qualifiers 'as $t \to 0$' and 'as $z \to \infty$'. First, the following sequence establishes (iv)$\Longleftrightarrow$(ii):

$$\beta(\mathbf{A}) = 0 \iff \|\mathbf{I} + t\mathbf{A}\| = 1 + o(t)$$

$$\iff \|\mathbf{I} + t\mathbf{A} + t^2\mathbf{A}^2 + \cdots\| = 1 + o(t)$$

$$\iff \|(\mathbf{I} - t\mathbf{A})^{-1}\| = 1 + o(t)$$

$$\iff \|z(z - \mathbf{A})^{-1}\| = 1 + o(z^{-1})$$

$$\iff \|(z - \mathbf{A})^{-1}\|^{-1} = z + o(1).$$

For (iv) $\Longleftrightarrow$ (i) we argue as follows:

$$\beta(\mathbf{\Lambda}) = 0 \iff \|\mathbf{I} + t\mathbf{A}\| - 1 + o(t)$$

$$\iff \sup_{\|\mathbf{x}\|=1} \|\mathbf{x} + t\mathbf{A}\mathbf{x}\| = 1 + o(t)$$

$$\iff \sup_{\|\mathbf{x}\|=1} \mathbf{x}^*\mathbf{x} + t\mathbf{x}^*(\mathbf{A} + \mathbf{A}^*)\mathbf{x} = 1 + o(t)$$

$$\iff \sup_{\|\mathbf{x}\|=1} \mathbf{x}^*(\mathbf{A} + \mathbf{A}^*)\mathbf{x} = 0$$

$$\iff \sup_{\|\mathbf{x}\|=1} \mathrm{Re}(\mathbf{x}^*\mathbf{A}\mathbf{x}) = 0$$

$$\iff \omega(\mathbf{A}) = 0.$$

To prove (iv) $\Longleftrightarrow$ (v) it is enough to note that $\|e^{t\mathbf{A}}\|$ and $\|\mathbf{I} + t\mathbf{A}\|$ have equal one-sided derivatives with respect to t at $t = 0$. Finally, since $\|e^{t\mathbf{A}}\| = \|e^{t(\mathbf{A}+is)}\|$ for any real s, the equivalence of (v) and (ii) implies that (ii) can be strengthened to $\|(z + is - \mathbf{A})^{-1}\| = z + o(1)$ for any real s, and by the definition of $\alpha_\varepsilon(\mathbf{A})$, this is equivalent to (iii). ∎

By a translation in the complex plane, whose details we omit, the five equivalent conditions of Theorem 17.3 generalize to five equal expressions for $\omega(\mathbf{A})$, whether zero or nonzero.

Characterizations of the numerical abscissa

Theorem 17.4 *For any* $\mathbf{A} \in \mathbb{C}^{N \times N}$, *with* $\|\cdot\| = \|\cdot\|_2$,

$$\omega(\mathbf{A}) = \lim_{z \to +\infty} \left[z - \|(z - \mathbf{A})^{-1}\|^{-1} \right]$$

$$= \lim_{\varepsilon \to \infty} [\alpha_\varepsilon(\mathbf{A}) - \varepsilon] = \beta(\mathbf{A}) = \left(\tfrac{\mathrm{d}}{\mathrm{d}t} \|e^{t\mathbf{A}}\| \right)_{t=0},$$

where ω, α_ε, *and* β *denote the logarithmic norm, pseudospectral abscissa, and numerical abscissa, respectively, and the derivative* $\mathrm{d}/\mathrm{d}t$ *is again defined by a one-side limit.*

Completion of the proofs of Theorems 17.1 *and* 17.2. We can now complete the proofs of the first two theorems of this section. Equation (17.3) of Theorem 17.1 is equivalent to the equality of $\omega(\mathbf{A})$ and $\left(\tfrac{\mathrm{d}}{\mathrm{d}t} \|e^{t\mathbf{A}}\| \right)_{t=0}$ above. As for Theorem 17.2, the equality of $\omega(\mathbf{A})$ and $\lim_{\varepsilon \to \infty} [\alpha_\varepsilon(\mathbf{A}) - \varepsilon]$, which in turn is equal to $\sup_{\varepsilon > 0} [\alpha_\varepsilon(\mathbf{A}) - \varepsilon]$ since $\alpha_\varepsilon(\mathbf{A}) - \varepsilon$ is a nondecreasing function of ε, shows that the rightmost point of $W(\mathbf{A})$ in $\mathbb{C}$ just touches the boundary of the intersection of all half-planes $H = \{z : \mathrm{Re}\, z \leq \text{constant}\}$ that satisfy (17.11). The proof is completed by rotating this result, $\mathbf{A} \to e^{i\theta}\mathbf{A}$. ∎

The equality

$$\omega(\mathbf{A}) = \alpha(\tfrac{1}{2}(\mathbf{A} + \mathbf{A}^*)) \tag{17.18}$$

is implicit in our proof of Theorem 17.3 and forms the basis of the standard algorithm for computing the numerical range of a matrix [389, 433]. By computing the largest and smallest eigenvalues and associated eigenvectors of $\tfrac{1}{2}(\mathbf{A} + \mathbf{A}^*)$ one finds the vectors that achieve the maximum and minimum real part of $W(\mathbf{A})$;[3] rotating this calculation by various angles $\mathbf{A} \to e^{i\theta}\mathbf{A}$ traces out the boundary of $W(\mathbf{A})$. Plots of computed numerical ranges can be found, for example, in [354, 389], and are also available in EigTool.

The results presented so far have indicated that in two respects, the spectrum and the numerical range lie at complementary extremes. Schematically:

$$\|(z - \mathbf{A})^{-1}\| \to \infty \quad \longleftrightarrow \quad \sigma(\mathbf{A}) \quad \longleftrightarrow \quad \|e^{t\mathbf{A}}\| \text{ as } t \to \infty,$$

$$\|(z - \mathbf{A})^{-1}\| \to 0 \quad \longleftrightarrow \quad W(\mathbf{A}) \quad \longleftrightarrow \quad \|e^{t\mathbf{A}}\| \text{ as } t \to 0.$$

A third respect in which the spectrum and the numerical range are opposite is that whereas $\sigma(\mathbf{A})$ may be arbitrarily sensitive to perturbations, $W(\mathbf{A})$ is as robust as one could desire, for from (17.1) we have

$$W(\mathbf{A} + \mathbf{E}) \subseteq W(\mathbf{A}) + \overline{\Delta}_{\|E\|},$$

where $\overline{\Delta}_{\|\mathbf{E}\|}$ is the closed disk about 0 of radius $\|\mathbf{E}\|$. Perhaps one can summarize these complementary roles of $\sigma(\mathbf{A})$ and $W(\mathbf{A})$ with the thought that for the question, Where in $\mathbb{C}$ does a matrix $\mathbf{A}$ 'live'?, the smallest reasonable answer is $\sigma(\mathbf{A})$ and the largest reasonable answer is $W(\mathbf{A})$. Other answers, including the ε-pseudospectra for the values of ε of greatest interest, and also the estimates of $\sigma(\mathbf{A})$ that are obtained by various approximate techniques such as Arnoldi iteration (§§27, 28), typically lie in between. See the discussion in §47.

Our discussion up to now has concerned finite-dimensional matrices and the norm $\| \cdot \| = \| \cdot \|_2$. We now turn to the generalization of these ideas to other norms and to linear operators. We follow the formulations of §§4 and 15, assuming that $\mathbf{A}$ is a closed operator with dense domain $\mathcal{D}(\mathbf{A})$ in a complex Banach space X.

Generalizations to Banach space were carried out independently by Lumer in 1961 and Bauer (for matrices) in 1962 [31, 522]; the definition we follow here, now reasonably standard, is Bauer's.[4] Recall from §4 that if $u \in X$ has $\|u\| = 1$, then there exists at least one dual vector $f \in X^*$ with $\|f\| = 1$ such that $(f, u) = 1$. (The expression (f, u) denotes the number that results when f is applied to u.) The *numerical range* of $\mathbf{A}$ is defined

---

[3]It follows that it is simple to compute $\omega(\mathbf{A})$. The task of computing the numerical radius, $\mu(\mathbf{A}) - \sup_{z \in W(\mathbf{A})} |z|$, presents a greater challenge; see [545] for an algorithm.

[4]$W(\mathbf{A})$ as defined by (17.19) is also sometimes called the *Bauer field of values*, the *spatial numerical range*, or the *total spatial numerical range*.

as follows:

$$W(\mathbf{A}) = \{(f, \mathbf{A}u) \colon u \in \mathcal{D}(\mathbf{A}), f \in X^*, \|u\| = \|f\| = 1, (f, u) = 1\}.$$
(17.19)

This definition is a natural generalization of (17.1), but in at least one respect it is genuinely more complicated, for whereas in Hilbert space each vector u has a unique dual vector f satisfying the required conditions, in Banach space f may be nonunique.

For matrices with $\|\cdot\| = \|\cdot\|_2$, we know that $W(\mathbf{A})$ is convex and closed and contains the spectrum. The convexity is a Hilbert space property and is lost in general for norms other than $\|\cdot\|_2$, even for matrices. The conditions of being closed and containing the spectrum are finite-dimensional properties and are lost in general for linear operators, even in Hilbert space. If $\mathbf{A}$ is bounded, however, it is known that $W(\mathbf{A})$ is connected [71] and that $\sigma(\mathbf{A}) \subseteq \overline{W(\mathbf{A})}$ [832]. With the numerical radius defined by (17.5) as before, it is known that for a bounded operator $\mathbf{A}$ (17.6) only holds in Hilbert spaces; in Banach spaces, it must be replaced by

$$\|\mathbf{A}^k\| \le \mathrm{e}(\mu(\mathbf{A}))^k.$$
(17.20)

The constant $\mathrm{e} = 2.718\ldots$ is best possible, even for 2×2 matrices [67, 314].

For an example to illustrate nonconvexity [581], consider the matrix

$$\mathbf{A} = \begin{pmatrix} -\mathrm{i} & \mathrm{i} \\ \mathrm{i} & \mathrm{i} \end{pmatrix}.$$
(17.21)

Since $\mathbf{A}$ is skew-Hermitian with eigenvalues $\pm\sqrt{2}\mathrm{i}$, in the 2-norm we have $W(\mathbf{A}) = [-\sqrt{2}\mathrm{i}, \sqrt{2}\mathrm{i}]$ and $\omega(\mathbf{A}) = 0$. In the 1- or ∞-norms, however, $W(\mathbf{A})$ is the hourglass-shaped figure shown on the left in Figure 17.2, with $\omega(\mathbf{A}) = 1$.[5] To be definite let us take $\|\cdot\| = \|\cdot\|_\infty$. The value $z = \mathrm{i} + \mathrm{i}\mathrm{e}^{\mathrm{i}\theta} \in W(\mathbf{A})$ for any real θ is attained with the choices $u = (\mathrm{e}^{\mathrm{i}\theta}, 1)^\mathrm{T}$ and $f = (0, 1)^\mathrm{T}$ in (17.19):

$$(0 \quad 1) \begin{pmatrix} -\mathrm{i} & \mathrm{i} \\ \mathrm{i} & \mathrm{i} \end{pmatrix} \begin{pmatrix} \mathrm{e}^{\mathrm{i}\theta} \\ 1 \end{pmatrix} = \mathrm{i} + \mathrm{i}\mathrm{e}^{\mathrm{i}\theta}.$$

Another choice of u and f gives $z = 1/2$: if $s = (2 + \sqrt{2})/4$, then

$$(s \quad \mathrm{i}^{1/2}(1 - s)) \begin{pmatrix} -\mathrm{i} & \mathrm{i} \\ \mathrm{i} & \mathrm{i} \end{pmatrix} \begin{pmatrix} 1 \\ \mathrm{i}^{-1/2} \end{pmatrix} = 1/2.$$

---

[5] The hourglass consists of the two closed disks of radius 1 about $\pm\mathrm{i}$, plus some further territory near the origin. These are in fact Gerschgorin disks for this matrix, and in general, for any finite-dimensional matrix $\mathbf{A}$ in $\|\cdot\|_1$ or $\|\cdot\|_\infty$, the convex hull of $W(\mathbf{A})$ is equal to the convex hull of the union of appropriate Gerschgorin disks [581].

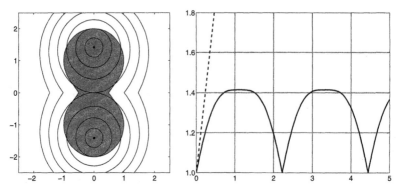

Figure 17.2: On the left, spectrum, ε-pseudospectra ($\varepsilon = 1.5, 1.25, \ldots, 0.25$), and numerical range (shaded) of (17.21) in the $\|\cdot\|_\infty$ or (equivalently for this matrix) $\|\cdot\|_1$ norm. Note that in a Banach space the numerical range need not be convex. The numerical abscissa $\omega(\mathbf{A}) = 1$ is still equal to the initial slope of $\|e^{t\mathbf{A}}\|$, however, as shown in the plot on the right of $\|e^{t\mathbf{A}}\|_\infty$ or $\|e^{t\mathbf{A}}\|_1$ against t. The dashed line has slope 1.

For a first example to illustrate the possibility of 'missing points' in the numerical range, it is enough to take $\mathbf{A}$ to be the shift operator on $\ell^2(\mathbb{Z})$, i.e., the doubly infinite matrix with ones on the first subdiagonal and zeros everywhere else. The spectrum is the unit circle (Theorem 7.1), but $W(\mathbf{A})$ is the open unit disk, i.e., $\sigma(\mathbf{A})$ and $W(\mathbf{A})$ are disjoint.

These ideas bring us to the following very satisfactory situation. We have argued that the main application of the numerical range is to determining the initial slope of $\|e^{t\mathbf{A}}\|$ via the numerical abscissa, which is defined for an operator $\mathbf{A}$ in the usual fashion by

$$\omega(\mathbf{A}) = \sup_{z \in W(\mathbf{A})} \operatorname{Re} z. \tag{17.22}$$

If $\mathbf{A}$ is a linear operator in Banach space, then $W(\mathbf{A})$ may not be closed or convex, but this introduces no difficulties in the definition of $\omega(\mathbf{A})$, which is determined only by the closure of the convex hull of $W(\mathbf{A})$. One might accordingly hope that the results of Theorems 17.1–17.4 would carry over to this more general situation, and if $\mathbf{A}$ is a bounded operator, this is exactly right. The following facts can essentially be found in the original papers of Lumer [522, Lem. 12] and Bauer [31, Thm. 4.3]. For an extended discussion see [72].

Numerical range and bounded operators

Theorem 17.5 *Let $\mathbf{A}$ be a bounded linear operator on a Banach space X. Then all the assertions of Theorems 17.1–17.4 are valid as written, with one modification: In Theorem 17.2, $W(\mathbf{A})$ is replaced by the closure of the convex hull of $W(\mathbf{A})$.*

It remains to record what happens if **A** is unbounded, a subject reviewed in the book by Pazy [606]. Suppose **A** is a densely defined closed operator in X that generates a C_0 semigroup $\{e^{t\mathbf{A}}\}$. From the general theory summarized in §15 we know that there exist real constants M and ω such that $\|e^{t\mathbf{A}}\| \leq Me^{\omega t}$ for all $t \geq 0$. Our concern in this section is the situation where M can be taken to be 1, so that $\mathbf{A} - \omega$ is the generator of a contraction. The central results that characterize semigroups with this property are the Hille–Yosida theorem [395, 606, 845], which relates contractivity to the norm of the resolvent, and the Lumer–Phillips theorem [523, 606], which relates it to the numerical range.

The first thing we note about the case in which **A** is unbounded is that our conditions involving the logarithmic norm $\beta(\mathbf{A})$ can no longer apply, since the numerator of (17.4) is not finite. Dropping this condition, one may hope that the remaining observations concerning the resolvent norm, the norm of the semigroup, and the numerical range may still hold. This is indeed the case, apart from one complication. It may happen that $W(\mathbf{A})$ is 'missing' much more of the complex plane than just some boundary points as in the example above of the shift operator. For example, suppose **A** is the derivative operator in $L^2[0, \infty)$ with domain $\mathcal{D}(\mathbf{A})$ equal to the set of absolutely continuous functions $u \in L^2[0, \infty)$ with $u(0) = 0$. Then the spectrum of **A** is the entire closed right half-plane (see §10), but by integration by parts one can see that the numerical range is just the imaginary axis. Thus, although the numerical abscissa is zero, this operator does not generate a semigroup at all, let alone a contraction semigroup.

Fortunately, large missing patches of the numerical range like this can only occur in a certain special fashion. If $\overline{W(\mathbf{A})}$ is the closure of the numerical range of **A**, let Ω be a connected component of $\mathbb{C} \setminus \overline{W(\mathbf{A})}$. Then Ω may be disjoint from the spectrum $\sigma(\mathbf{A})$, or it may be entirely contained in $\sigma(\mathbf{A})$. No intermediate configuration is possible [606, Thm. 3.9]. As a consequence of such results we obtain the following almost perfect generalization of the results of this section to operators in Banach space.

Numerical range and unbounded operators

Theorem 17.6 *Let* **A** *be a densely defined closed linear operator in a Banach space X. If $\omega(\mathbf{A}) = \infty$, then either* **A** *does not generate a C_0 semigroup or it generates a C_0 semigroup that does not satisfy $\|e^{t\mathbf{A}}\| \leq e^{t\omega}$ for any ω. On the other hand, suppose $\omega(\mathbf{A})$ and $\alpha(\mathbf{A})$ are both finite. Then* **A** *generates a C_0 semigroup with $\|e^{t\mathbf{A}}\| \leq e^{t\omega(\mathbf{A})}$, and all the assertions of Theorems 17.1–17.4 are valid, with two modifications. First, the conditions involving $\beta(\mathbf{A})$ in Theorems 17.3 and 17.4 are dropped. Second, in Theorem 17.2, $W(\mathbf{A})$ is replaced by the closure of the convex hull of $W(\mathbf{A})$, and the additional assumption is made that $\sigma(\mathbf{A})$ contains no unbounded components that are disjoint from $W(\mathbf{A})$.*

18 · The Kreiss Matrix Theorem ⎯⎯⎯⎯⎯⎯

The Kreiss Matrix Theorem, originally published by Heinz-Otto Kreiss in 1962 [465], concerns the characterization of matrices and families of matrices that are *power-bounded*.[1] Thanks in part to its dissemination in the classic monograph by Richtmyer and Morton [639], this theorem quickly came to be regarded as one of the fundamental results of theoretical numerical analysis. The result was given already as (14.28), and its continuous analogue for matrix exponentials as (14.12).

Let $\mathbf{A}$ be an $N \times N$ matrix, let $\|\cdot\|$ denote a vector norm and the matrix norm it induces, and define

$$p(\mathbf{A}) = \sup_{k \geq 0} \|\mathbf{A}^k\|.$$

We say that $\mathbf{A}$ is *power-bounded* if $p(\mathbf{A}) < \infty$. As is well-known, power-boundedness is equivalent to the condition that all the eigenvalues of $\mathbf{A}$ lie in the closed unit disk and that any eigenvalues on the unit circle are nondefective. This eigenvalue condition, however, is not quantitative. It gives no information about how large $p(\mathbf{A})$ may be for a particular choice of $\mathbf{A}$, and consequently, it does not determine whether a family of matrices $\{\mathbf{A}_\nu\}$ satisfies a *uniform* power bound, independently of ν. Such a uniform power bound is needed for the important application to the stability of finite difference discretizations of partial differential equations (§32). The matrices $\{\mathbf{A}_N\}$ considered in §3 are an example of a family that is not uniformly power-bounded, as illustrated in Figure 3.4.

Kreiss's idea was to relate the power bound for $\mathbf{A}$ to what we now call its *Kreiss constant* with respect to the open unit disk, Δ. As in §14, $\mathcal{K}(\mathbf{A})$ is defined as the smallest C for which

$$\|(z - \mathbf{A})^{-1}\| \leq \frac{C}{|z| - 1} \qquad \forall z,\ |z| > 1. \tag{18.1}$$

Equivalently,

$$\mathcal{K}(\mathbf{A}) = \sup_{|z| > 1}\ (|z| - 1)\ \|(z - \mathbf{A})^{-1}\|.$$

$\mathcal{K}(\mathbf{A})$ is a measure of how fast the resolvent norm blows up as z approaches Δ, or equivalently, how far the pseudospectra protrude outside Δ.

Here is the theorem. The analogous result for matrix exponentials is given at the end of this section.

⎯⎯⎯⎯⎯⎯⎯⎯⎯

[1]This section is adapted from [816].

Kreiss Matrix Theorem

Theorem 18.1 *For any $N \times N$ matrix* $\mathbf{A}$,

$$\mathcal{K}(\mathbf{A}) \leq p(\mathbf{A}) \leq eN\mathcal{K}(\mathbf{A}). \qquad (18.2)$$

These bounds are sharp in the sense specified in the proof, below.

In words, though the sequence of norms $\|\mathbf{A}^k\|$ may exhibit an arbitrarily large 'hump' separating its transient from its asymptotic behavior, the height of this hump can be quantified: it is equal to $\mathcal{K}(\mathbf{A})$, up to a factor no greater than eN.

Theorem 18.1 is a long way from the original statement of the Kreiss Matrix Theorem of 1962. At that time, instead of describing the factor eN explicitly, the theorem made the following assertion: A family of $N \times N$ matrices satisfies a uniform power bound if and only if it has a uniform Kreiss constant. Two additional equivalent conditions were also stated, which we shall not discuss, and the equivalence of all four conditions was proved by a cycle of four steps. If one looks at what quantity is implicitly established by this proof in place of the factor eN, one finds a number that is exponentially large as a function of N. In the following years, various authors then made a succession of improvements that reduced this constant. According to Tadmor's remarks in [737] for the earlier developments, the history of progress toward the constant eN involved no fewer than nine steps:

$$\begin{array}{rl}
\text{Kreiss '62:} & \sim c^{N^N} \\
\text{Morton '64:} & \sim 6^N(N+4)^{5N} \\
\text{Miller \& Strang '66:} & \sim N^N \\
\text{Miller '67:} & \sim e^{9N^2} \\
\text{Laptev '75 / Strang '78:} & 32eN^2/\pi \\
\text{Tadmor '81:} & 32eN/\pi \\
\text{LeVeque \& Trefethen '84:} & 2eN \\
\text{Smith '85:} & (1 + \frac{2}{\pi})eN \\
\text{Spijker '91:} & eN
\end{array}$$

For details, see [499, 639, 715, 737, 816] and the references therein.

Most of the remainder of this section is devoted to the proof of Theorem 18.1. This proof is by no means obvious, but it is nonetheless quite simple, depending on a curious mix of a resolvent integral, integration by parts, and some ideas of integral geometry that go back to the famous 'Buffon needle problem' of 1777. This proof can be attributed to the cumulative efforts of Laptev, Strang, Tadmor, LeVeque, Trefethen, Spijker, and Wegert.

Proof of Theorem 18.1. The left-hand inequality of (18.2) asserts that if $\|\mathbf{A}^k\| \leq C$ for all $k \geq 0$, then $\|(z - \mathbf{A})^{-1}\| \leq C/(|z| - 1)$ for all $|z| > 1$. This inequality follows easily by taking the norm of the power series

$$(z - \mathbf{A})^{-1} = z^{-1}\mathbf{I} + z^{-2}\mathbf{A} + z^{-3}\mathbf{A}^2 + \cdots,$$

which is guaranteed to converge for all $|z| > 1$ if $\mathbf{A}$ is power-bounded, since the spectrum of $\mathbf{A}$ then necessarily lies in $\overline{\Delta}$ (Theorem 16.1). We therefore turn to the more difficult right-hand inequality of (18.2).

We begin with the resolvent integral. Following Theorem 16.1, we have

$$\mathbf{A}^k = \frac{1}{2\pi \mathrm{i}} \int_G z^k (z - \mathbf{A})^{-1} \, \mathrm{d}z,$$

where G is any contour enclosing the eigenvalues of $\mathbf{A}$, which themselves must lie in $\overline{\Delta}$ if $\mathcal{K}(\mathbf{A}) < \infty$. In the remainder of this discussion, we assume $\|\cdot\| = \|\cdot\|_2$ and use inner product notation for simplicity, but the same arguments can be generalized to other norms as in §§4, 16, and 17. Let $\mathbf{u}$ and $\mathbf{v}$ be arbitrary N-vectors with $\|\mathbf{u}\| = \|\mathbf{v}\| = 1$. Then

$$\mathbf{v}^*\mathbf{A}^k\mathbf{u} = \frac{1}{2\pi \mathrm{i}} \int_G z^k r(z) \, \mathrm{d}z,$$

where $r(z)$ is the function $\mathbf{v}^*(z - \mathbf{A})^{-1}\mathbf{u}$, which can be shown to be a rational function of order N, that is, a quotient of two polynomials each of degree $\leq N$.

The next step is to integrate by parts, which gives

$$\mathbf{v}^*\mathbf{A}^k\mathbf{u} = \frac{-1}{2\pi \mathrm{i}(k + 1)} \int_G z^{k+1} r'(z) \, \mathrm{d}z.$$

Let the contour of integration be taken as $G = \{z \in \mathbb{C} : |z| = 1 + (k+1)^{-1}\}$. On this contour we have $|z^{k+1}| \leq \mathrm{e}$, giving the bound

$$|\mathbf{v}^*\mathbf{A}^k\mathbf{u}| \leq \frac{\mathrm{e}}{2\pi(k + 1)} \int_G |r'(z)| \, |\mathrm{d}z|.$$

This integral can be interpreted as the arc length of the image of the circle G under the rational function r. By Theorem 18.2 below, this arc length satisfies

$$\int_G |r'(z)| \, |\mathrm{d}z| \leq 2\pi N \sup_{z \in G} |r(z)|.$$

By the definition of $\mathcal{K}(\mathbf{A})$, this supremum is at most $(k + 1)\mathcal{K}(\mathbf{A})$. All together, then, we have

$$|\mathbf{v}^*\mathbf{A}^k\mathbf{u}| \leq \frac{\mathrm{e}}{2\pi(k + 1)} 2\pi N (k + 1)\mathcal{K}(\mathbf{A}) = \mathrm{e}N\mathcal{K}(\mathbf{A}).$$

Since $\|\mathbf{A}^k\|$ is the supremum of $|\mathbf{v}^*\mathbf{A}^k\mathbf{u}|$ over all vectors $\mathbf{u}$ and $\mathbf{v}$ with $\|\mathbf{u}\|_2 = \|\mathbf{v}\|_2 = 1$, this establishes the right-hand inequality of (18.2).

The theorem concludes with an assertion of sharpness of these bounds. We now explain exactly in what sense they are known to be sharp. First, the inequality $\mathcal{K}(\mathbf{A}) \leq p(\mathbf{A})$ is sharp in the sense that for any N, there are $N \times N$ matrices for which $\mathcal{K}(\mathbf{A}) = p(\mathbf{A})$ exactly. In fact we need look no further than the identity matrix, for which we have $\mathcal{K}(\mathbf{A}) = p(\mathbf{A}) = 1$. As for the inequality $p(\mathbf{A}) \leq eN\mathcal{K}(\mathbf{A})$, it is sharp in a weaker sense. In general, given N, there do not necessarily exist matrices for which equality is attained. However, the linear behavior with respect to N and the factor e are both best possible in the sense that if $p(\mathbf{A}) \leq CN^\alpha\mathcal{K}(\mathbf{A})$ for all $\mathbf{A}$ for some constants C and α, then α can be no smaller than 1 and, if $\alpha = 1$, C can be no smaller than e. To prove this, consider the $N \times N$ Jordan block

$$
\mathbf{A} = \begin{pmatrix} 0 & \gamma & & & \\ & 0 & \gamma & & \\ & & \ddots & \ddots & \\ & & & 0 & \gamma \\ & & & & 0 \end{pmatrix}
$$

for some $\gamma = \gamma(N) > 0$. For this matrix one has $p(\mathbf{A}) = \gamma^{N-1}$ and, as shown in [499],

$$
\mathcal{K}(\mathbf{A}) \leq \frac{\gamma^{N-1}}{eN}(1 + \mathcal{O}(N^{-1})),
$$

assuming $N \geq 3$ and $\gamma \geq N$. In the $N \to \infty$ limit, the ratio of these quantities is asymptotic to eN. We thus have $p(\mathbf{A})/\mathcal{K}(\mathbf{A}) \geq eN(1 + \mathcal{O}(N^{-1}))$, and taking the limit $N \to \infty$ finishes the proof. ∎

The proof of the Kreiss Matrix Theorem is complete, except that it made use of Theorem 18.2, 'Spijker's lemma'. This is where the Buffon needle problem comes into the story.

Let r be a rational function of order N. Let $\mathbb{T}$ denote the unit circle $\{z \in \mathbb{C} : |z| = 1\}$, and let $\|\cdot\|_1$, $\|\cdot\|_2$, and $\|\cdot\|_\infty$ denote the 1-, 2-, and ∞-norms on $\mathbb{T}$,

$$
\|f\|_1 = \int_\mathbb{T} |f(z)|\,|dz|, \quad \|f\|_2^2 = \int_\mathbb{T} |f(z)|^2\,|dz|, \quad \|f\|_\infty = \sup_{z \in \mathbb{T}} |f(z)|.
$$

Then the arc length of the curve $r(\mathbb{T})$ in the complex plane, which we denote by $L_\mathbb{C}(r(\mathbb{T}))$, can be represented compactly by the formula

$$
L_\mathbb{C}(r(\mathbb{T})) \equiv \|r'\|_1.
$$

If r is multiplied by a constant α, $L_\mathbb{C}(r(\mathbb{T}))$ changes by the factor $|\alpha|$. However, this scale-dependence can be eliminated by considering the ratio

$$
L_\mathbb{C}(r(\mathbb{T}))\,/\,\|r\|_\infty. \tag{18.3}
$$

In 1984, building upon the earlier work by Laptev, Strang, and Tadmor, LeVeque and Trefethen observed that a bound on (18.3) could be used in the proof of the Kreiss Matrix Theorem [499]. They therefore posed the question, What is the maximum possible value of (18.3)?

It is easy to see that the value $2\pi N$ can be attained: Just take $r(z)$ to be z^N or z^{-N}. If r is restricted to be a polynomial, it follows from Bernstein's inequality that $2\pi N$ is the maximum possible. It is also easy to see that $2\pi N$ is the maximum value for rational functions in the special case $N = 1$, where r is just a Möbius transformation. Based on these facts and on computer experiments, it was conjectured in [499] that $2\pi N$ is the maximum value (18.3) for all rational functions r and all N. However, only the bound $4\pi N$ was proved, and the task of eliminating this gap of a factor of 2 was presented as an Advanced Problem in the *American Mathematical Monthly* [498].

Just one response to the *Monthly* problem was received, from James C. Smith, of the University of South Alabama, who improved the bound to $2(2 + \pi)N$ [702].

Five years later, Marc Spijker of the University of Leiden finally settled the conjecture in the affirmative [715].

Spijker's lemma in the complex plane
Theorem 18.2 $L_{\mathbb{C}}(r(\mathbb{T})) \, / \, \|r\|_\infty \le 2\pi N.$

The proof of Theorem 18.2 below is close to Spijker's, but with more emphasis on geometry. We shall show that Theorem 18.2 is a corollary of a slightly different arc length theorem.

The simplicity of Theorem 18.2 is marred by the need for the normalization by $\|r\|_\infty$. In looking for a cleaner formulation one may ask, What is the analogous result for the Riemann sphere? Let $\mathbb{S}$ denote the Riemann sphere $\{x \in \mathbb{R}^3 : \|\mathbf{x}\|_2 = 1\}$, with the north and south poles corresponding to the points ∞ and 0 in $\mathbb{C}$, respectively, according to the usual stereographic projection, and the equator corresponding to the unit circle $\mathbb{T}$. This identification of $\mathbb{C}$ and $\mathbb{S}$ is discussed in many books on complex analysis [1], and it is readily shown that a unit of arc length $|dz|$ at a position $z \in \mathbb{C}$ is expanded by the factor $2/(1 + |z|^2)$ upon projection onto $\mathbb{S}$. It follows that if $r(\mathbb{T})$ is considered as a closed curve on $\mathbb{S}$, with $L_{\mathbb{S}}(r(\mathbb{T}))$ denoting its arc length on $\mathbb{S}$, then we have

$$L_{\mathbb{S}}(r(\mathbb{T})) \equiv \| 2r'/(1 + |r|^2) \|_1. \tag{18.4}$$

Now the scale-dependence has been eliminated from the problem. It makes sense simply to ask, What is the maximum possible value of $L_{\mathbb{S}}(r(\mathbb{T}))$?

Our second arc length theorem answers this question.

Spijker's lemma on the Riemann sphere

Theorem 18.3 $L_{\mathbb{S}}(r(\mathbb{T})) \leq 2\pi N.$

Note that like Theorem 18.2, Theorem 18.3 is obviously sharp, with equality attained for any r that maps $\mathbb{T}$ with winding number N onto a great circle of $\mathbb{S}$. For example, $r(z) = z^N$ maps $\mathbb{T}$ with winding number N onto the equator, and $r(z) = \mathrm{i}^N(z-1)^N/(z+1)^N$ maps $\mathbb{T}$ with winding number N onto the Greenwich meridian. Note also that for any r with $\|r\|_\infty \leq 1$, we have $L_{\mathbb{C}}(r(\mathbb{T})) \leq L_{\mathbb{S}}(r(\mathbb{T}))$. This follows from (18.4), since $2/(1+|r|^2) \geq 1$ when $|r| \leq 1$. Consequently, Theorem 18.3 implies Theorem 18.2 as a corollary. Thus Spijker's lemma on the Riemann sphere is both simpler and stronger than Spijker's lemma in the complex plane and may be considered the more fundamental result [499].

We now give the proof of Theorem 18.3.

The reader has undoubtedly encountered the Buffon needle problem, published by the Comte de Buffon in 1777. Suppose a needle of length 1 is thrown at random on a plane ruled by parallel lines at a distance 1 apart. What is the probability that the needle will land in a position that crosses a line? Easy calculus shows that the answer is:

$$\text{Probability of intersection} = 2/\pi.$$

Buffon, incidentally, was the leading French naturalist of the eighteenth century and also a translator of Newton. He worked on his 'problème de l'aiguille' long before publishing it as an appendix on 'moral arithmetic' in his forty-four-volume treatise on natural history [113].

The needle problem became well-known, especially among the French, and was generalized. Laplace, without referencing Buffon, solved the analogous problem for a square grid (*Théorie Analytique des Probabilités*, 1812). A more important generalization was to consider the slightly modified question: If the needle has length L, possibly greater than 1, what is the *expected number* of intersections? The answer is easily seen to be:

$$\text{Expected number of intersections} = 2L/\pi. \tag{18.5}$$

And from here it is a small step mathematically, but a big one conceptually, to note that the same formula (18.5) is valid also for a *paper clip*. Various steps in this direction were taken by Cauchy, Lamé, and Barbier, among others [26, 131]. In fact, if any rectifiable curve Γ of arc length L is thrown at random on the parallel grid, the expected number of intersections is (18.5). (A curve is rectifiable if its real and imaginary parts are functions of bounded variation [1].) The idea behind this result is that Γ can be thought of as a concatenation of infinitesimal straight segments, each satisfying

(18.5) for an appropriate infinitesimal value of L. Now it may seem at first that the expected number of intersections for Γ should be more complicated than the sum of the expected numbers for the segments Γ is composed of, since after all, the segments do not fall on the grid independently. However, it is a basic fact of statistics that the expectation of a sum of random variables is equal to the sum of the expectations, regardless of whether they are independent. This observation seems elementary to us now, but its application to the needle problem was evidently not obvious in the nineteenth century.

Bending the paper clip into a circle of radius $1/2$ gives an easy way to remember Buffon's result and its generalization (18.5). For this choice of Γ, L is π and the number of intersections is exactly 2, no matter how the paper clip falls.

We now want to move from the plane to the sphere, a step taken as early as 1860 by Barbier [26]. Consider a 'spherical paper clip'—that is, a curve Γ embeddable in the Riemann sphere. Suppose Γ is placed at random on $\mathbb{S}$. What is the expected number of intersections with the equator? The answer is again essentially a matter of combining calculus with elementary statistics:

$$\text{Expected number of intersections on the sphere} = L/\pi. \qquad (18.6)$$

Or one can skip the calculus and remember this result by thinking of the case in which Γ is itself a great circle. In this case $L = 2\pi$ and the number of intersections is again exactly 2 unless Γ happens to land exactly on the equator, an event of probability zero.

A final development completes this brief history. After Barbier, other mathematicians generalized these results further, including Poincaré, who referenced neither Buffon nor Barbier (*Calcul des Probabilités,* 1896). By this time it was clear that although the needle problem and its generalizations had conventionally been formulated as problems of probability, that interpretation could be dispensed with. Instead of orienting Γ at random on $\mathbb{S}$ and asking for the expected number of intersections with a fixed equator, one can consider Γ to be fixed on $\mathbb{S}$ and compute its arc length $L_{\mathbb{S}}(\Gamma)$ as an integral of the number of intersections with all great circles. To be precise, for any rectifiable curve $\Gamma \subseteq \mathbb{S}$ and any $\mathbf{x} = (x_1, x_2, x_3) \in \mathbb{S}$, let $\nu(\Gamma, \mathbf{x})$ denote the number of points of intersection of Γ with the great circle on $\mathbb{S}$ consisting of points equidistant from the antipodes $\pm\mathbf{x}$. (When this number is infinite, the definition of $\nu(\Gamma, \mathbf{x})$ does not matter, for the set of such points has measure zero.) One obtains the following elegant result, known as *Poincaré's formula*:

$$L_{\mathbb{S}}(\Gamma) = \frac{1}{4} \int_{\mathbb{S}} \nu(\Gamma, \mathbf{x}) \, d\mathbf{x}. \qquad (18.7)$$

The integral is taken with respect to area measure on $\mathbb{S}$.

Poincaré's formula can be expressed in words as follows. To find the arc length of a curve on the Riemann sphere, integrate its numbers of intersections over all great circles, then divide by 4. Or, equivalently, since the sphere has surface area 4π, take the average number of intersections and multiply by π. This latter paraphrase makes plain the equivalence of (18.6) and (18.7).

Poincaré's formula has far-reaching generalizations described in the book by Santaló [658]. It forms a centerpiece of the field known earlier as 'geometric probability' but now as 'integral geometry'.

Is it obvious now how to prove Theorem 18.3? All we need is the following lemma, whose proof can be found in [816] but is an easy exercise. Again, $\nu(r(\mathbb{T}), \mathbf{x})$ denotes the number of intersection points of the curve $r(\mathbb{T})$ with the great circle on the Riemann sphere $\mathbb{S}$ defined by the points $\pm\mathbf{x}$.

> **Lemma 18.4** *If r is a rational function of order N, then $\nu(r(\mathbb{T}), \mathbf{x}) \leq 2N$ for all $\mathbf{x} \in \mathbb{S}$ with the possible exception of a single pair $\mathbf{x} = \pm\mathbf{x}_0$, $\mathbf{x}_0 \in \mathbb{S}$.*

Since the surface area of $\mathbb{S}$ is 4π and since $\frac{1}{4} \cdot 2N \cdot 4\pi = 2\pi N$, Theorem 18.3 is an immediate consequence of this lemma and (18.7). As described above, Theorem 18.3 implies Theorem 18.2, and this completes the proof of the Kreiss Matrix Theorem.

Though the Kreiss Matrix Theorem is most familiar in connection with matrix powers, it has been known since the beginning that essentially the same bound applies to exponentials, too. For this we redefine the Kreiss constant $\mathcal{K}(\mathbf{A})$ with respect to the left half-plane instead of the unit disk, as described on page 138. Further generalizations to other regions and other functions of $\mathbf{A}$ are given in [763].

Kreiss Matrix Theorem for $e^{t\mathbf{A}}$

Theorem 18.5 *For any $N \times N$ matrix $\mathbf{A}$,*

$$\mathcal{K}(\mathbf{A}) \leq \sup_{t \geq 0} \|e^{t\mathbf{A}}\| \leq eN\mathcal{K}(\mathbf{A}), \qquad (18.8)$$

where $\mathcal{K}(\mathbf{A})$ is the Kreiss constant of $\mathbf{A}$ with respect to the left half-plane.

Proof. The proof is essentially as before; we give just a sketch. For any unit vectors $\mathbf{u}$ and $\mathbf{v}$ we have

$$\mathbf{v}^* e^{t\mathbf{A}} \mathbf{u} = \frac{1}{2\pi i} \int_G e^{tz} r(z) \, dz$$

for a rational function r of order N, where G is a contour enclosing $\sigma(\mathbf{A})$. Integration by parts gives

$$\mathbf{v}^* e^{t\mathbf{A}} \mathbf{u} = \frac{-1}{2\pi \mathrm{i} t} \int_G e^{tz} r'(z) \, \mathrm{d}z,$$

and taking $G = t^{-1} + \mathrm{i}\mathbb{R}$ leads to

$$|\mathbf{v}^* e^{t\mathbf{A}} \mathbf{u}| \leq \frac{\mathrm{e}}{2\pi t} \int_G |r'(z)| \, |\mathrm{d}z|,$$

where the integral can again be interpreted as the arc length of the image of G under r. By Theorem 18.2 and the definition of $\mathcal{K}(\mathbf{A})$, this arc length is at most $2\pi N t \mathcal{K}(\mathbf{A})$, from which (18.8) follows. ∎

The question of the sharpness of Theorem 18.5 is addressed in [499].

19 · Growth bound theorem for semigroups _____

If $\mathbf{A}$ is a matrix or bounded operator, we have seen in §15 that the spectrum $\sigma(\mathbf{A})$ is nonempty and determines the growth rate of $\|e^{t\mathbf{A}}\|$ as $t \to \infty$: $\lim_{t\to\infty} t^{-1} \log \|e^{t\mathbf{A}}\| = \alpha(\mathbf{A})$, where α is the spectral abscissa. If $\mathbf{A}$ is unbounded, however, these properties may fail. The spectrum may be empty, and even if it is not, the norms $\|e^{t\mathbf{A}}\|$ may be larger than the spectrum would suggest.

Specifically, let $\mathbf{A}$ be a closed operator, densely defined in a Banach space X, that generates a C_0 semigroup $\{e^{t\mathbf{A}}\}$ as discussed in §15. The *exponential type* or *growth bound* of $\mathbf{A}$ is defined by the formula

$$\omega_0(\mathbf{A}) = \lim_{t\to\infty} t^{-1} \log \|e^{t\mathbf{A}}\|, \tag{19.1}$$

and it is known that this limit always exists (either finite or $-\infty$). Though $\omega_0(\mathbf{A}) = \alpha(\mathbf{A})$ for bounded operators, the most we can say for unbounded operators without further assumptions is that

$$\omega_0(\mathbf{A}) \geq \alpha(\mathbf{A}). \tag{19.2}$$

In the case $\sigma(\mathbf{A}) = \emptyset$, $\alpha(\mathbf{A})$ is defined to be $-\infty$.

The first example of an operator with strict inequality in (19.2) was published by Hille and Phillips in their monograph of 1948 and 1957 [395], and we shall turn to it in a moment. First, however, we mention a simpler example published by Zabczyk in 1975 [847], whose essential idea was implicit in Figure 3.4. We know that a nonnormal matrix $\mathbf{A}$ may exhibit transient behavior in the sense that $\|e^{t\mathbf{A}}\|$ grows faster than $e^{t\alpha(\mathbf{A})}$ (or decays more slowly) for small t. In particular, for the $N \times N$ Jordan block

$$\mathbf{J}_N = \begin{pmatrix} 0 & 1 & & \\ & \ddots & \ddots & \\ & & 0 & 1 \\ & & & 0 \end{pmatrix},$$

$\|e^{t\mathbf{J}_N}\|$ behaves like e^t for $t \ll N$ but is smaller than $e^{\varepsilon t}$ as $t \to \infty$ for any $\varepsilon > 0$, since its spectral abscissa is $\alpha(\mathbf{J}_N) = 0$. Zabczyk proposed an infinite matrix $\mathbf{A}$ in block diagonal form composed of Jordan blocks of increasing dimensions:

$$\mathbf{A} = \begin{pmatrix} i & & & \\ & \mathbf{J}_2 + 2i\mathbf{I}_2 & & \\ & & \mathbf{J}_3 + 3i\mathbf{I}_3 & \\ & & & \ddots \end{pmatrix}. \tag{19.3}$$

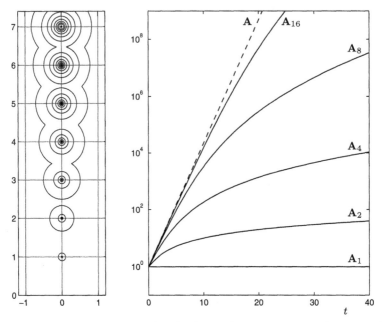

Figure 19.1: Zabczyk example (19.3) of an infinite matrix $\mathbf{A}$ with spectral abscissa $\alpha(\mathbf{A}) = 0$ and growth bound $\omega_0(\mathbf{A}) = 1$. The image on the left shows ε-pseudospectra for $\varepsilon = 10^{-1}, 10^{-2}, \ldots, 10^{-8}$. Further increasingly far-from-normal components of $\sigma_\varepsilon(\mathbf{A})$ lie in an infinite sequence off-scale on the positive imaginary axis. The plot on the right shows $\|e^{t\mathbf{A}}\|$ and some curves $\|e^{t\mathbf{A}_k}\|$ against t, where $\mathbf{A}_N$ is the Nth diagonal block of $\mathbf{A}$, of dimension N.

Here $\mathbf{I}_N$ is the identity of dimension N and we regard $\mathbf{A}$ as an operator on $\ell^2(\mathbb{N})$. The left part of Figure 19.1 confirms what is rather obvious, that for each $\varepsilon > 0$, the ε-pseudospectrum of this infinite matrix is the union of open balls of various radii about the points $i, 2i, 3i, \ldots$, with the radii increasing to $1 + \varepsilon$ as one goes up the imaginary axis. The right part of the figure shows norms $\|e^{t\mathbf{A}_k}\|$ for a few of the diagonal blocks $\mathbf{A}_k$ of $\mathbf{A}$, and it also shows the dashed line corresponding to the envelope of all the blocks, $\|e^{t\mathbf{A}}\|$. Obviously the growth bound is 1 even though the spectral abscissa is 0.

Such examples are also interesting in connection with the *spectral mapping theorem* [221, 248]. If $\mathbf{A}$ is a matrix or bounded operator and f is analytic in a neighborhood of $\sigma(\mathbf{A})$, this theorem asserts that $f(\sigma(\mathbf{A})) = \sigma(f(\mathbf{A}))$. If $\mathbf{A}$ is unbounded, however, then under suitable assumptions the inclusion $f(\sigma(\mathbf{A})) \subseteq \sigma(f(\mathbf{A}))$ still holds, but it may now be strict. In particular this is the case for $f(z) = e^{tz}$, with $e^{t\mathbf{A}}$ and $\mathbf{A}$ defined as a C_0 semigroup and its generator. If $\mathbf{A}$ is the Zabczyk operator, then $e^{t\sigma(\mathbf{A})}$ is a subset of the unit circle, whereas for any value of t that is not a rational multiple of π, $\sigma(e^{t\mathbf{A}})$ is the annulus $e^{-1} \leq |z| \leq e$.

For the Zabczyk example, though the spectral abscissa is only 0, it seems clear why the growth bound has turned out to be larger: Since $\|(\lambda - \mathbf{A})^{-1}\|$ becomes arbitrarily large as λ approaches the line $\operatorname{Re}\lambda = 1$ from the right, it is 'as if' the spectral abscissa were 1. Indeed, if we define the ε-pseudospectral abscissa of $\mathbf{A}$, as always in this book, by $\alpha_\varepsilon(\mathbf{A}) = \sup_{z \in \sigma_\varepsilon(\mathbf{A})} \operatorname{Re} z$, then we note that although $\alpha(\mathbf{A}) = 0$ for this operator, the limit of $\alpha_\varepsilon(\mathbf{A})$ as $\varepsilon \downarrow 0$ is 1. One might imagine that it should be this limit rather than $\alpha(\mathbf{A})$ that determines $\omega_0(\mathbf{A})$. In fact, we can easily prove half of this conjecture,

$$\omega_0(\mathbf{A}) \geq \lim_{\varepsilon \downarrow 0} \alpha_\varepsilon(\mathbf{A}). \tag{19.4}$$

To see this, note that for any $\omega > \omega_0(\mathbf{A})$, the semigroup satisfies $\|e^{t\mathbf{A}}\| \leq Me^{\omega t}$ for all $t \geq 0$, for some M. By (15.14), this implies $\|(z - \mathbf{A})^{-1}\| \leq M/(\operatorname{Re} z - \omega)$ for any z with $\operatorname{Re} z > \omega$, and by taking $\omega \downarrow \omega_0(\mathbf{A})$, we get (19.4).

In Hilbert space, but not in Banach space, the inequality in (19.4) can be replaced by an equality. This result stems from work in the late 1970s and early 1980s, largely independent, by Gearhart [305], Huang [420], and Prüss [615], with related contributions also by Herbst [382], Howland [417], and Greiner [348]. In various sources this theorem is given the names of Gearhart's Theorem, the Gearhart–Prüss Theorem, and Huang's Theorem.

Growth bound theorem

Theorem 19.1 *Let $\mathbf{A}$ be a densely defined closed operator in a Banach space X that generates a C_0 semigroup, and let $\alpha_\varepsilon(\mathbf{A})$ and $\omega_0(\mathbf{A})$ denote the ε-pseudospectral abscissa and the growth bound of $\mathbf{A}$, respectively. The quantity $\alpha_\varepsilon(\mathbf{A}) - \varepsilon$ is a monotonically nondecreasing function of ε, with*

$$\omega_0(\mathbf{A}) \geq \lim_{\varepsilon \downarrow 0} \alpha_\varepsilon(\mathbf{A}). \tag{19.5}$$

If X is a Hilbert space, then

$$\omega_0(\mathbf{A}) = \lim_{\varepsilon \downarrow 0} \alpha_\varepsilon(\mathbf{A}). \tag{19.6}$$

Proof. The monotonicity result is a consequence of the basic properties of pseudospectra; see in particular the final assertion of Theorem 4.3. The inequality (19.5) was established in the text above. For a proof of (19.6), see [248, Thm. V.1.11]. ∎

In the Zabczyk example, the spectrum of $\mathbf{A}$ is nonempty and thus $\alpha(\mathbf{A})$ is finite. Zabczyk pointed out that by adjusting the constants in the example, one can construct an infinite matrix with arbitrary finite $\alpha(\mathbf{A})$ and $\omega_0(\mathbf{A})$ with $\alpha(\mathbf{A}) < \omega_0(\mathbf{A})$. Another possibility, however, would be to go further and construct an example with $\omega_0(\mathbf{A})$ finite but $\sigma(\mathbf{A}) = \emptyset$, hence $\alpha(\mathbf{A}) = -\infty$. The Hille–Phillips example, which appeared in [395, §23.16] (and also in Hille's 1948 first edition of that book), is of this kind.

The context for this example is fractional integrals and fractional derivatives; more generally, one could construct examples from a variety of Volterra integral operators. For a function $u(x)$ defined on $x \in (0,1)$, consider the function $\mathbf{L}^{(\nu)}u$ on $(0,1)$ defined by the *Riemann–Liouville integral*

$$\mathbf{L}^{(\nu)}u(x) = \frac{1}{\Gamma(\nu)} \int_0^x (x-s)^{\nu-1}u(s)\,ds, \qquad (19.7)$$

where Γ is the gamma function. If ν is a positive integer, (19.7) gives the νth indefinite integral of u, and it is natural to take the same formula as a definition of a *fractional integration operator* for arbitrary $\nu > 0$. By differentiating one or more times, one can extend the definition also to *fractional differentiation operators*. These definitions can also be related to Fourier or Laplace transforms, and of course, one can make precise choices about the domains of these operators and their corresponding properties; see [122, §2.6] and the references therein.

For example, taking $\nu = 1/2$ gives the *half-integral operator* on $[0,1]$:

$$\mathbf{L}^{1/2}u(x) = \frac{1}{\sqrt{\pi}} \int_0^x (x-s)^{-1/2}u(s)\,ds. \qquad (19.8)$$

Consideration of the half-integral of $u'(x)$ gives us the corresponding *half-derivative operator*:

$$\mathbf{L}^{-1/2}u(x) = \frac{1}{\sqrt{\pi}} \int_0^x (x-s)^{-1/2}u'(s)\,ds, \qquad (19.9)$$

for sufficiently smooth functions u satisfying $u(0) = 0$. For such functions u, $\mathbf{L}^{1/2}$ and $\mathbf{L}^{-1/2}$ are inverses of one another.

The pseudospectra of $\mathbf{L}^{-1/2}$, interpreted as an operator in $L^2(0,1)$, are shown in Figure 19.2, which is based on a numerical computation by a spectral collocation method as in §43. These images are related to the example presented in §5. There, we considered the operator $\mathbf{L}$ defined by $\mathbf{L}u = u'$ on an interval $[0,d]$. The spectrum of $\mathbf{L}$ was empty, but its resolvent norm $\|(z-\mathbf{L})^{-1}\|$ grew exponentially as $z \to -\infty$, and Figure 5.2 displayed its pseudospectra, half-planes bounded on the right by lines $\mathrm{Re}\,z = \text{constant}$. Now, the boundary condition is at the left rather than the right, so the pseudospectra of the derivative operator would be right half-planes rather than left half-planes. Because of the square root, we instead get pseudospectra filling most of a quadrant in the right half-plane.

Like the derivative operator with a boundary condition, this half-derivative operator with a boundary condition has empty spectrum. Its pseudospectral abscissae are all $+\infty$, so it does not generate a C_0 semigroup,

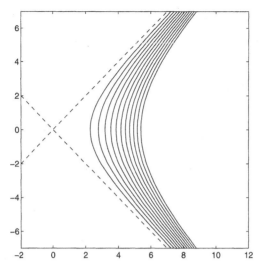

Figure 19.2: ε-pseudospectra of the half-derivative operator (19.9) for $\varepsilon = 10^{-2}$, $10^{-3}, \ldots, 10^{-12}$, from left to right. The spectrum is empty, but the resolvent norms are exponentially large in a quadrant of the complex plane. Multiplying by $e^{\pm 3\pi i/4}$ gives an operator $\mathbf{A}$ with $\alpha(\mathbf{A}) = -\infty$, $\omega_0(\mathbf{A}) = 0$, and $\|e^{t\mathbf{A}}\| = 1$ for all $t \geq 0$.

but we could rotate the picture in the complex plane by defining

$$\mathbf{A} = e^{\pm 3\pi i/4} \mathbf{L}^{-1/2}.$$

Now we have $\alpha_\varepsilon(\mathbf{A}) = \varepsilon$ for all $\varepsilon > 0$, and thus by Theorem 19.1, this is an example of an operator with $\alpha(\mathbf{A}) = -\infty$ and $\omega_0 = 0$. Although the spectrum of $\mathbf{A}$ is empty, the spectrum of $e^{t\mathbf{A}}$ for any $t > 0$ is the closed left half-plane.

The original Hille–Phillips example is closely related. Instead of considering the square root of the derivative, Hille and Phillips considered the 'logarithm' of (19.7). That is, they interpreted the family $\mathbf{L}^{(\nu)}$ as a semigroup with evolution variable ν and took as their example the infinitesimal generator $\mathbf{A}$ of this semigroup. Figure 19.3 shows pseudospectra of this operator. The resolvent norms are *very* large: note the doubly exponential values of ε. (This figure was also computed in part by a spectral collocation method; a more careful calculation is described by Baggett in [18].) Multiplying this example by $\pm i$ gives an example of an operator $\mathbf{A}$ with growth bound $\pi/2$ and spectral abscissa $-\infty$. The spectrum is empty, but at a point at only a distance 10 from 0, the resolvent norm is approximately 10^{10000}.[1] For any $t > 0$, the spectrum of $e^{t\mathbf{A}}$ is the annulus $e^{-\pi/2} \leq |z| \leq e^{\pi/2}$.

---

[1] This is the second-largest resolvent norm mentioned in this book. See Figure 36.6.

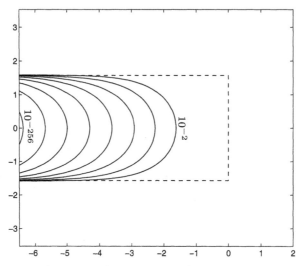

Figure 19.3: ε-pseudospectra of the logarithm $\mathbf{A}$ of the integration operator, i.e., the infinitesimal generator of the fractional integration semigroup $\{\mathbf{L}^{(\nu)}\}$, for $\varepsilon = 10^{-2}, 10^{-4}, 10^{-8}, \ldots, 10^{-256}$. The dashed lines mark the half-strip $\operatorname{Re} z < 0$, $|\operatorname{Im} z| < \pi/2$. By multiplying this example by $\pm i$, we obtain the Hille–Phillips example of an operator $\mathbf{A}$ with $\alpha(\mathbf{A}) = -\infty$ and $\omega_0(\mathbf{A}) = 0$.

The three examples with $\omega_0(\mathbf{A}) > \alpha(\mathbf{A})$ exhibited in this section have a special property: $\|e^{t\mathbf{A}}\| = e^{t\omega_0(\mathbf{A})}$ for all $t > 0$. Their pseudospectral abscissae also have a special property: $\alpha_\varepsilon(\mathbf{A}) = \omega_0(\mathbf{A}) + \varepsilon$ for all $\varepsilon > 0$. For operators in Hilbert space, as pointed out by Baggett [18], these two conditions are equivalent.

Exactly exponential growth or decay

Theorem 19.2 Let $\mathbf{A}$ be a densely defined closed operator in a Hilbert space X that generates a C_0 semigroup, and let $\omega_0(\mathbf{A})$, $\omega(\mathbf{A})$, and $\alpha_\varepsilon(\mathbf{A})$ denote the growth bound, numerical abscissa, and ε-pseudospectral abscissa of $\mathbf{A}$, respectively. The following conditions are equivalent:

$$\omega_0(\mathbf{A}) = \omega(\mathbf{A}), \tag{19.10}$$

$$\|e^{t\mathbf{A}}\| = e^{t\omega_0(\mathbf{A})}, \qquad \forall t > 0, \tag{19.11}$$

$$\alpha_\varepsilon(\mathbf{A}) = \omega_0(\mathbf{A}) + \varepsilon, \qquad \forall \varepsilon > 0, \tag{19.12}$$

$$\sup_{\operatorname{Re}\lambda = \mu} \|(\lambda - \mathbf{A})^{-1}\| = \frac{1}{\mu - \omega_0(\mathbf{A})}, \qquad \forall \mu > \omega_0(\mathbf{A}). \tag{19.13}$$

In a Banach space, (19.10) $\Longleftrightarrow$ (19.11) $\Longleftarrow$ (19.12) $\Longleftrightarrow$ (19.13).

Proof. The equivalence of (19.12) and (19.13) is a matter of definitions. To establish the equivalence of (19.10) and (19.11), we note that by Theorem 17.2, $\omega(\mathbf{\Lambda}) = \lim_{\varepsilon \to \infty} \alpha_\varepsilon(\mathbf{A}) - \varepsilon$. By the definition (19.1) of $\omega_0(\mathbf{A})$ together with the bound $\|e^{t\mathbf{A}}\| \leq e^{t\omega(\mathbf{A})}$ of Theorem 17.1, it follows that $\omega(\mathbf{A})$ and $\omega_0(\mathbf{A})$ are equal if and only if (19.11) holds. It also follows that (19.12) implies (19.10). Finally, in Hilbert space, by Theorem 19.1, $\omega_0(\mathbf{A}) = \lim_{\varepsilon \to 0} \alpha_\varepsilon(\mathbf{A}) - \varepsilon$. If (19.10) holds, then $\alpha_\varepsilon(\mathbf{A}) - \varepsilon$ is a monotonically nondecreasing function of ε with equal limits $\omega_0(\mathbf{A})$ as $\varepsilon \to 0$ and $\varepsilon \to \infty$, implying (19.12). ∎

For examples showing that the conclusions of Theorems 19.1 and 19.2 may not hold in Banach space, see [248].

V. Fluid Mechanics

20 · Stability of fluid flows _____

Together with quantum mechanics and the analysis of acoustic and structural vibrations, the third great area of application of eigenvalues has been the stability of fluid flows. Since the fluids in question are usually either liquids or gases moving slowly enough for compressibility effects to be insignificant, this field goes by the name of *hydrodynamic stability*. The roots of the subject date to Helmholtz, Kelvin, Rayleigh, and Reynolds in the nineteenth century; classic twentieth century books include those of Lin [508], Chandrasekhar [135], Joseph [436], and Drazin and Reid [214]. An important recent book is that of Schmid and Henningson [669].

The basic question of hydrodynamic stability is, Given a fluid at rest or in steady ('laminar') motion, will small perturbations of the flow tend to grow or to decay? Innumerable applications have been investigated over the years, which we may roughly divide into two classes. In one set of problems, the aim is to explain the appearance of regular structures in a fluid (Figure 20.1). Why does a jet of water break into regularly spaced drops, why do clouds form regular rolls or cells, why do ocean waves undulate periodically? The usual mechanism in such problems is that small perturbations of the steady flow grow unstably, and some particular wavelength grows most unstably of all, whereupon nonlinear effects take hold that dampen further growth, leaving a pattern of finite amplitude. For example, Lord Rayleigh analyzed the instability and breakup into droplets of a smooth jet of water [619]. He showed that if such a jet is perturbed sinusoidally in radius at a certain wavelength, then surface tension may amplify the perturbations, with the maximum amplification rate occurring at a wavelength of about 4.51 diameters. In the other set of problems, we are concerned with predicting whether a steady flow will remain steady or break down to an irregular or turbulent flow (Figure 20.2). Here, the notional breakdown mechanism is that small perturbations of the regular flow grow unstably, whereupon nonlinear effects grab hold and lift the flow into some more complicated regime. We see this effect, say, when turbulent eddies form behind a stick in a fast-moving stream.

For both of these classes of hydrodynamic stability problems, the basic mathematical procedure is the same (see Figure 33.3). The equations of fluid mechanics are nonlinear, but by the consideration of infinitesimal perturbations about a steady flow, they are linearized. The linear equations are then examined for eigenvalues in the right half of the complex plane.[1] If such an eigenvalue exists, there is an instability, and the eigenmode asso-

_____

[1]More generally, we consider points of the spectrum, but for simplicity we shall speak here of eigenvalues without mentioning this technicality further.

Figure 20.1: When regular patterns appear in fluid flows, the explanation usually involves eigenvalues. In such cases there is a smooth, pattern-free flow that is theoretically possible, but it is unstable to infinitesimal disturbances, and disturbances at some wavelengths are amplified more than others. These illustrations show capillary instability in a milk-drop coronet (© Harold and Esther Edgerton Foundation, 2005, courtesy of Palm Press, Inc.), the von Karman vortex street in the wake of a cylinder (from an experiment by Maarten A. Rutgers, Xiao-lun Wu, and Walter Goldburg; other photographs from the same series can be found in [647]), and the Kelvin–Helmholtz instability at the interface between two moving fluids (photograph by F. A. Roberts, P. E. Dimotakis, and A. Roshko from [797]).

ciated with the rightmost unstable eigenvalue can be expected to dominate the form of the instability. If all the eigenvalues are in the left half-plane, one expects stability.

Dozens of books and thousands of papers have followed this pattern of analysis over the course of more than a century. There have been many

Figure 20.2: Eigenvalues are also often called upon in an attempt to explain instabilities, typically at higher Reynolds numbers, that end in an irregular or turbulent flow instead of a regular pattern. This image, courtesy of the USGS/Cascades
Volcano Observatory, shows the eruption of Mt. St. Helens on 18 May 1980. Because of the large space scale, the Reynolds number for flow out of a smokestack
or volcano is invariably very high, and these flows are never laminar.

successes, especially for low-speed flows, but there have also been many
failures, especially for high-speed ones. Indeed, many studies have been
published in which predictions based on eigenvalues fail entirely to match
observations. Widely varying explanations of these discrepancies have been
advanced, but in the 1990s a consensus formed that the main cause is
nonnormality.

In this section we describe the three cleanest hydrodynamic stability
problems for which eigenvalue analysis fails: plane Couette flow, plane
Poiseuille flow, and pipe flow. All three go back a century or more and
show a discrepancy between stability predicted by eigenvalue analysis and
instability observed in the laboratory. In each case we consider the idealized problem involving a viscous incompressible fluid in an infinitely long
straight-sided channel or pipe. The mathematical details are complicated,
involving the nonlinear Navier–Stokes equations coupled with an incom-
pressibility condition and no-slip boundary conditions along the walls. We
omit most of these details, which can be found, for example, in [214]
and [669], and simply present certain computational results. In each problem the key parameter is the *Reynolds number R*, a nondimensional measure of the ratio of inertial to viscous effects. A 'high-speed flow' is more
properly characterized as one with high Reynolds number—which might
be achieved by low speed but even lower viscosity.

Plane Couette flow. In the 1880s, Maurice Couette investigated an apparatus for determining the viscosity of a liquid by filling the gap between two concentric cylinders with the liquid and measuring the resistance of this system to relative rotation of the cylinders [166]. At low speeds, the flow is laminar. As the speed is increased, however, ring-shaped vortices with a regular spacing may appear. These *Taylor vortices* were explained in celebrated work by G. I. Taylor in the 1920s [744, 745]. Taylor showed that at a certain speed, the laminar flow becomes unstable to infinitesimal perturbations, and there is a wavelength of maximum unstable growth rate. This is a beautiful example of successful eigenvalue analysis. At still higher speeds, on the other hand, further instabilities cause the Taylor vortices first to become wavy and then to break down to turbulence. Predicting the behavior of these higher speed Taylor–Couette flows is an example of a problem for which eigenvalue analysis is not so successful [738].

We are going to consider not Taylor–Couette flow, but *plane Couette flow.* This is the simplified limit of Taylor–Couette flow in which the radius has been taken to infinity, so there is no curvature present. The geometry is sketched in Figure 20.3. We imagine two infinite solid walls moving in the $+x$ and $-x$ directions. For any Reynolds number, there is a solution to the Navier–Stokes equations in this geometry consisting of a steady parallel flow with velocity depending linearly on the height y (Figure 20.3).

Here is what is seen in the laboratory [175, 753]. For Reynolds number R less than about 350 according to the standard definition (channel half-width times maximum velocity divided by kinematic viscosity), the laminar flow solution is observed. For $R > 350$, irregular perturbations tend to appear instead. For $R \gg 1000$, they almost always appear, and the flow is turbulent. Thus plane Couette flows change from laminar to turbulent as the Reynolds number increases, but there is no sharp transition value.

Attempts at eigenvalue analysis of this problem go back to Orr and Sommerfeld around 1907 [587, 707]. One linearizes the Navier–Stokes equations about the laminar flow and checks for unstable eigenmodes. An important part of this analysis is encapsulated by *Squire's theorem* of 1933 [717], which asserts that if there is an unstable eigenmode that depends upon the

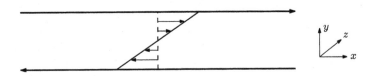

Figure 20.3: Schematic view of plane Couette flow. The flow domain is the infinite three-dimensional region between two parallel plates moving in opposite directions along the x axis. The steady flow is parallel, with a linear velocity profile as a function of y.

cross-steam coordinate z at a Reynolds number R_1, then there is an unstable eigenmode at another Reynolds number $R_2 < R_1$ that is independent of z. Thus for eigenvalue analysis, it is enough to look at two-dimensional $(x\text{–}y)$ perturbations of the laminar flow. The Orr–Sommerfeld equation is a fourth-order ordinary differential eigenvalue equation whose solutions describe the y dependence of these two-dimensional eigenmodes (see §22); the x dependence is sinusoidal.

Such analysis of plane Couette flow shows that no matter how large R is, there are no unstable eigenvalues. This fact was suspected for many years, and in 1973 it was proved by Romanov [642].

Why then is plane Couette flow unstable in practice? A part of the answer can be seen in Figure 20.4 (from [780]) which shows spectra and pseudospectra for the linearized plane Couette flow operator for $R = 350$ and 3500. In each image, the spectrum is a two-dimensional subset of the open left half-plane. It comes close to the imaginary axis, a distance of about $2.47/R$ as $R \to \infty$, but does not touch it or cross it. The ε-pseudospectra, however, tell a different story. For small values of ε, they protrude a distance much greater than ε into the right half-plane. This implies that in various senses, these linear operators will not behave so stably. For example, the second figure shows that for $R = 3500$, the

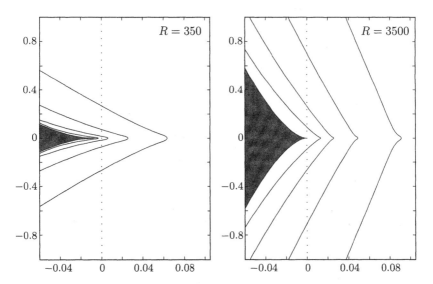

Figure 20.4: Spectra (shaded) and pseudospectra for plane Couette flow at two Reynolds numbers, from [780]. The dotted line is the imaginary axis. Boundaries of ε-pseudospectra are shown, from right to left, for $\varepsilon = 10^{-2}$, $10^{-2.5}$, 10^{-3}, and $10^{-3.5}$. This flow is eigenvalue stable for all R, with spectrum contained strictly inside the left half-plane, but it is unstable in practice for $R \gg 1000$.

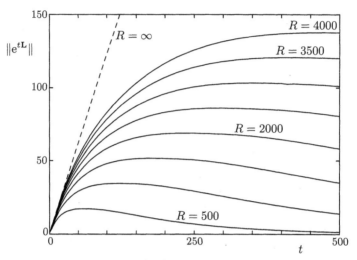

Figure 20.5: Transient growth of infinitesimal perturbations for linearized plane Couette flow, from [780]. Perturbations may be amplified by a factor $\mathcal{O}(R)$ over a time scale $\mathcal{O}(R)$ before viscosity eventually makes them decay. Compare Figure 3.4.

$10^{-3.5}$-pseudospectrum reaches to about $z = 0.012$ along the real axis. Theorem 15.4 thus implies that certain perturbations of plane Couette flow with $R = 3500$ will grow transiently, by purely linear mechanisms, by a factor of at least $0.012 \times 10^{3.5} \approx 38$. Figure 20.5 confirms this prediction. The actual maximum growth factor is about 120, or for general R as $R \to \infty$, $\|e^{t\mathbf{L}}\| \approx R/29.1$ at $t \approx R/8.52$. Here $\mathbf{L}$ is the Navier–Stokes evolution operator linearized about the laminar flow, and $\| \cdot \|$ is the root-mean-square speed of the perturbation.

A potential growth in amplitude by a factor of 120 is of obvious physical significance. In fact, many researchers consider that the proper measure of a flow perturbation is not the amplitude but its square, which has the dimensions of energy. In this measure the amplification factor for plane Couette flow at $R = 3500$ is 14,400.

There is no mystery about the physics underlying this behavior. The linearized Navier–Stokes operator amplifies various perturbations transiently to various degrees, and there is a class of perturbations that it amplifies particularly strongly: *streamwise vortices*, by which we mean velocity fields that approximate vortices oriented in the x direction. Suppose a perturbation of this kind is superimposed on the laminar flow. As t increases, the vorticity will move particles of fluid up and down in the y direction. A high-speed particle moved to a slow region will appear as an anomaly of locally large speed in the $\pm x$ direction, and a low-speed particle moved to a fast region will appear as an anomaly of low speed. We say that a streamwise

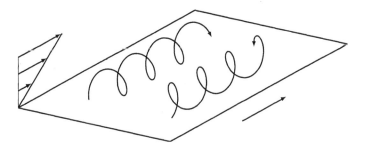

Figure 20.6: Schematic view of a flow perturbation, a pair of streamwise vortices, that achieves transient growth close to the maximum of Figure 20.5.

vortex generates *streamwise streaks* of potentially much greater amplitude (Figure 20.6). This effect goes by various names, including *vortex tilting* and *lift-up*, and it has been recognized for many years; the dashed line marked $R = \infty$ in Figure 20.5 could be derived as a corollary to a 1975 paper by Ellingsen and Palm [238]. For any finite R, the streak will eventually decay due to viscosity, but the higher R is, the longer it will take for this to happen and the more transient growth may occur in the meantime.

This vortex-to-streak mechanism is a consequence of nonnormality. A normal amplification process would be described by an eigenvector input that generated the same eigenvector output at higher amplitude, but here, the output of the amplifier takes a different form from the input. Indeed, the irrelevance of eigenvalues to this process is apparent in the fact that Squires's theorem tells us to look at structures dependent on x and y, whereas streamwise vortices and streaks are orthogonal to this plane, depending on y and z. One could see this by considering a singular value decomposition $e^{t\mathbf{L}} = \mathbf{U\Sigma V}^*$ at some time t (we use matrix notation for simplicity); the dominant right singular vector $\mathbf{v}$ would be approximately a streamwise vortex and the corresponding left singular vector $\mathbf{u}$ approximately a streamwise streak. Alternatively we may obtain a streak of much the same form from an ε-pseudomode of $\mathbf{L}$ for small ε. By contrast, the true eigenvectors of this flow problem give no hint of the importance of streamwise structures.

The history of the growing appreciation of nonnormal effects in fluid mechanics has been complex. For decades, it has been clear from laboratory experiments and computer simulations that high-speed flows are dominated by streamwise structures such as vortices, streaks, and 'hairpins'. What took a long time to develop was a bridge from these observations to the mathematical theory of hydrodynamic stability, which seemed to conclude that two-dimensional structures should be dominant instead. Many researchers assumed that nonlinearity must play an essential part in these effects, since it appeared that linear analysis failed to explain them. However, it was a mistake to confuse linear analysis with eigenvalue analysis.[2] In fact, the essentials of perturbation amplification and generation of streamwise streaks are linear, and the nonlinearities only become important as one tries to track the further evolution of such structures, including possible transition to turbulence. These facts began to become clearer in the fluid mechanics literature around 1990 with landmark papers of Boberg and Brosa [65], Butler and Farrell [121], and Reddy and Henningson [622]. Crucial related work in the preceding years was due to Benney, Landahl, and Gustavsson [360].

If there were nothing more to the dynamics of linearized flows than streamwise vortices and streaks, it might be unnecessary to speak of nonnormality in general; we could focus instead on a 2×2 linear skeleton consisting of just vortex plus streak. Such mechanisms of 'direct resonance' were the subject of some of the work just mentioned by Benney and others during the 1970s and 1980s. However, the reality of linearized flows is more complicated than a single input and a single output. Even for a situation as clean as plane Couette flow, the maximal transient growth is achieved by structures that are not exactly aligned with the x axis, and as soon as one wishes to look at nonmaximal perturbations or more complicated geometries, one needs more general tools. In this respect fluid mechanics is like other fields where nonnormal matrices and operators arise: the study of a 2×2 Jordan matrix or a pair of nearly parallel eigenvectors may capture the essence of the matter in certain situations, but it is insufficiently general to apply to all problems.

We have spoken of transient growth, but other physical implications too can be deduced from the pseudospectra for plane Couette flow shown in Figure 20.4 [780]. One is that although the linearized operator is eigenvalue stable, it takes only a very small perturbation (of norm about $65.9/R^2$) to make it eigenvalue unstable. This suggests that in a laboratory realization of plane Couette flow, with the inevitable imperfections of construction, there is a possibility of eigenvalue instability at higher Reynolds numbers. Another is that a stable flow of this kind may experience great *receptivity* to outside disturbances, acting potentially as an amplifier of distant vibrations of certain forms and frequencies via a process of 'pseudoresonance' [660]. A standard measure of receptivity in fluid mechanics is essentially the maximal norm of the resolvent along the imaginary axis. For plane Couette flow, that quantity is $\mathcal{O}(R^2)$ as $R \to \infty$.

Plane Poiseuille flow. Having discussed plane Couette flow at length, we shall now consider plane Poiseuille and pipe flows much more briefly, for although the details are different, the main features are the same. Fig-

---

[2]In the fluid mechanics literature, the standard term for eigenvalue analysis is 'linear stability analysis'. The justification for this usage is that, apart from borderline cases involving the imaginary axis, the linearized equations are stable as $t \to \infty$ if and only if there are no eigenvalues in the right half-plane. Nevertheless, it is a misleading expression, since it leaves no room for nonnormal linear effects; in this book we avoid it.

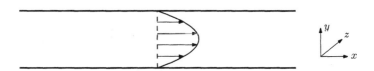

Figure 20.7: Schematic view of plane Poiseuille flow. Again, the flow domain is the infinite 3D region between two parallel plates, but now the plates are stationary and the flow is driven by a pressure gradient. The laminar solution has a parabolic velocity profile.

ure 20.7 shows the geometry of the plane Poiseuille flow problem. As before, there are two infinite plates, but now both plates are fixed and there is a flow between them driven by a pressure gradient. Again there is a laminar solution of the Navier–Stokes equations valid for all R, which is now described by a parabolic velocity profile. Again the laboratory experiments show an indistinct transition from laminar to turbulent flows, with the laminar flow observed always for values of R below about 1000, whereas irregular flows and turbulence may be observed for larger values of R and are almost always observed, say, for $R > 10,000$.

Plane Poiseuille flow has an idiosyncrasy, however, that has muddied the history of this subject: For large enough R, it is eigenvalue unstable. There is a critical Reynolds number at which the instability first appears, which was first calculated accurately by Orszag in 1971 [588], and since this is perhaps the most famous of all numbers arising from eigenvalue analysis, it is worth displaying:

$$R_{\text{crit}} \approx 5772.22.$$

As R increases, this is the value at which the linearized plane Poiseuille flow spectrum first crosses into the right half-plane. Long before Orszag, it was known that this eigenvalue instability was a feature of plane Poiseuille flow, and much attention was given to the physics of the associated eigenmodes, which are two-dimensional $(x\text{-}y)$ structures known as *Tollmien–Schlichting* *(TS) waves*. In experiments, such waves were rarely seen, with turbulence sometimes being observed for $R < R_{\text{crit}}$ and laminar flow sometimes being observed for $R > R_{\text{crit}}$. Nevertheless it was widely assumed that the TS waves must play an important role somehow in these flows, even if in some hidden fashion. In more recent years, such views have faded. It now appears that although TS waves can be seen in the laboratory if they are excited, for example, by a vibrating ribbon, they are not important in most cases in plane Poiseuille flow. The eigenvalue instability that promotes them is weak, associated with long time constants; in the laboratory, the slowly growing perturbation would often be flushed out the downstream end before becoming large enough to be observed. Meanwhile the larger non-

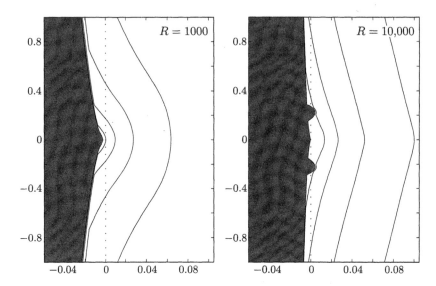

Figure 20.8: Like Figure 20.4 but for plane Poiseuille flow, again from [780]. The solid lines show the boundaries of the ε-pseudospectra for $\varepsilon = 10^{-2}$, $10^{-2.5}$, 10^{-3}, and $10^{-3.5}$. For $R > 5772.22$, two Tollmien–Schlichting bumps in the spectrum protrude into the right half-plane, but in most experiments it is still the 3D effects associated with the pseudospectra that are dominant.

normal three-dimensional effects, though transient in theory, are dominant in practice, and we see streamwise vortices and streaks much as for plane Couette flow. This is the conclusion one would expect from Figure 20.8, which, aside from the small TS bumps, is much like Figure 20.4. These matters are discussed at greater length in §22; see especially Figure 22.3.

Pipe flow. Our third classical flow is the simplest of all conceptually: flow through an infinite circular pipe, sometimes also known as *Poiseuille flow* or *Hagen–Poiseuille flow*. The geometry is suggested by Figure 20.9. We have an infinite circular pipe with flow driven by a pressure gradient. Again there is a laminar solution for all R with a parabolic velocity pro-

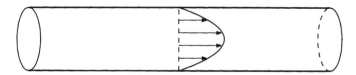

Figure 20.9: Schematic view of circular pipe flow. The flow domain is bounded by a circular pipe of infinite length, and the flow is driven by a pressure gradient. The laminar solution has a parabolic velocity profile.

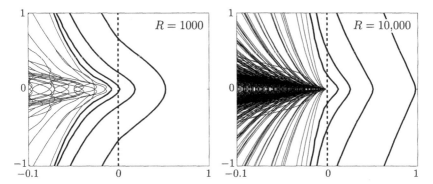

Figure 20.10: Like Figures 20.4 and 20.8 but for pipe flow, from [765]. The light curves represent spectra, which are now curves rather than regions since the pipe has only one unbounded dimension. The heavy curves are the boundaries of the ε-pseudospectra for $\varepsilon = 10^{-2}$, $10^{-2.5}$, 10^{-3}, and $10^{-3.5}$. Like plane Couette flow, this flow is eigenvalue stable for all R. (The small kinks in the pseudospectral boundaries are artifacts of the numerical computation.)

file. Physically and mathematically, the behavior is much the same as for plane Couette flow. The flow is eigenvalue stable for all R (this has not been proved, but the evidence from numerical computations is compelling and there is little disputation of this point), and in practice it is stable for Reynolds numbers less than about 2000 (with R now defined as radius times centerline velocity divided by kinematic viscosity). Reynolds himself observed transition to turbulence at various values of R on the order of 10,000 [636], whereas more recent exceptionally careful experiments have managed to retain laminar flow to values of R as high as 100,000. For large enough R, however, turbulence always appears. The pseudospectra of Figure 20.10 suggest that this is to be expected, since they look much like those of Figures 20.4 and 20.8. The spectra are quite different—curves rather than regions, which is a consequence of there being just one unbounded dimension in the problem rather than two—but it is not clear that this mathematical fact has much physical significance.

In this section we have concentrated on three idealized flows, only one of which, pipe flow, is close to a geometry found in applications. The variety of more complicated flow problems that have been investigated by engineers and mathematicians over the years is enormous. As outlined in §23, there are whole literatures on curved pipes and walls, ribbed walls, boundary layers, swirling flows, transonic flows, non-Newtonian fluids, flows over obstacles, jets, temperature-dependent effects, magnetohydrodynamic effects, and more. Perhaps it is fair to say that in most of these problems, if eigenvalue analysis is possible at all, it reveals unstable eigenmodes in certain parameter ranges that are mixed with other, nonmodal effects, which become more prominent as the flow speed increases.

Readers who compare our plots with those in the papers and books cited will see that most of the latter display individual eigenvalues as dots rather than spectra as curves or regions. Such plots are obtained after restricting attention to a fixed pair of x and z Fourier parameters in the case of plane Couette or Poiseuille flows, or a fixed pair of x and θ parameters in the case of pipe flow. Our plots combine all these modes into one picture.

What about turbulence? Of course this is not a purely linear phenomenon; it depends on an interaction of linear and nonlinear effects. A model of the nonlinear process of transition to turbulence is described in the next section, but this book does not discuss turbulence itself.

21 · A model of transition to turbulence _____

It seems probable, almost certain indeed, that . . . the steady motion is stable for any viscosity, however small; and that the practical unsteadiness pointed out by Stokes forty-four years ago, and so admirably investigated experimentally five or six years ago by Osborne Reynolds, is to be explained by limits of stability becoming narrower and narrower the smaller is the viscosity.
— Lord Kelvin, 1887 [451]

High-speed shear flows, as we have discussed in §20, are usually unstable in practice even though they may be stable in theory. For example, the flow of water through a pipe at high Reynolds number R is usually turbulent, even though in principle the smooth laminar flow solution should be stable to infinitesimal perturbations for any R. In this section we present a simple two-variable model, first published in [780] and studied further in [270], that sheds some light on how such apparently paradoxical behavior is possible. The model blends a nonnormal linear term with an energy-conserving nonlinear term and features a fixed point which, though stable, lies in a basin of attraction whose width shrinks as R increases. The general pattern of thinking that led to this model has roots going back a century to Kelvin and Orr and was first brought into sharp focus in an important paper by Boberg and Brosa in 1988 [65].

Since matters of fluid mechanics tend to be controversial, let us be clear what we do and do not claim for this simple model. We do claim that it illustrates how nonnormality and energy-conserving nonlinearity can combine to render a mathematically stable fixed point unstable to small finite perturbations. We also claim that in certain high speed fluid flows, under some circumstances, practical instability arises from a nonnormal/nonlinear interplay of this general nature. On the other hand we do not claim that the details of any fluids problem, which will be infinite-dimensional, match those of this two-dimensional model, or that the basin of attraction for a fluids problem will shrink as $R \to \infty$ as fast as for this model. Nor do we claim that the model sheds any light on the nature of turbulence, merely on the early stages of transition. Finally, we do not claim that a mechanism of this kind is the only route to practical instability of high-speed flows. On the contrary, it seems clear that practical instability can also come about in other ways. For example, suppose a laboratory apparatus is built that realizes a system that in theory should have no unstable eigenvalues. If the eigenvalues are sensitive to perturbations, then slight imperfections in construction may result in the apparatus corresponding to equations that have unstable eigenvalues after all.

Our model consists of two coupled nonlinear ordinary differential equations. Let R be a parameter, a caricature of the Reynolds number, which will be taken to be moderately large. Here is the system:

$$\frac{d}{dt}\begin{pmatrix} u \\ v \end{pmatrix} = \begin{pmatrix} -R^{-1} & 1 \\ 0 & -2R^{-1} \end{pmatrix}\begin{pmatrix} u \\ v \end{pmatrix} + \sqrt{u^2 + v^2}\begin{pmatrix} 0 & -1 \\ 1 & 0 \end{pmatrix}\begin{pmatrix} u \\ v \end{pmatrix}. \quad (21.1)$$

To make our discussion as simple as possible we shall consider the various terms of (21.1) in three pieces:

$$\frac{d}{dt}\begin{pmatrix} u \\ v \end{pmatrix} = \begin{pmatrix} -R^{-1} & 0 \\ 0 & -2R^{-1} \end{pmatrix}\begin{pmatrix} u \\ v \end{pmatrix} \qquad (21.2)$$

$$+ \begin{pmatrix} 0 & 1 \\ 0 & 0 \end{pmatrix}\begin{pmatrix} u \\ v \end{pmatrix} \qquad (21.3)$$

$$+ \sqrt{u^2 + v^2}\begin{pmatrix} 0 & -1 \\ 1 & 0 \end{pmatrix}\begin{pmatrix} u \\ v \end{pmatrix}. \qquad (21.4)$$

Consider first (21.2). This term is linear and diagonal, hence normal. The diagonal entries are negative and small, corresponding to the small amount of diffusion present in a flow at high Reynolds number. The first part of Figure 21.1 shows the effect of this term in the (u, v) phase plane: it makes an initial state tend exponentially to the origin. The behavior of our model would be much the same if we made the diagonal entries equal, but this would render the first matrix of (21.1) nondiagonalizable, which some find distracting. We keep the eigenvalues separate to emphasize that the behavior of interest does not depend on nondiagonalizability.

Next consider (21.3). This 'shear term' is nonnormal: it adds energy to u without correspondingly reducing v, as shown in the second part of Figure 21.1. In a fluid mechanics application, u and v are sometimes thought of as the amplitudes of a streamwise streak and a streamwise vortex, respectively, but this interpretation is oversimplified. Instead, it is better to think of v as a caricature of all kinds of structures in a flow, including streamwise vortices, that may sustain other structures without being diminished themselves (Boberg and Brosa call them 'mothers' [65]), and of u as a caricature of all kinds of structures, including streamwise streaks, that may be sustained in this fashion ('daughters').

Finally there is (21.4). This term is nonlinear, but it is very simple. Because the matrix is skew-symmetric, it acts orthogonally to the flow, transferring energy from u to v while conserving the total energy. The factor $\sqrt{u^2 + v^2}$ makes the term nonlinear; it 'shuts off' at the origin, growing linearly as u and v increase, as shown in the third part of Figure 21.1. The presence of a square root means that the term is not analytic at $u = v = 0$, unlike the equations of fluid mechanics, but this feature can readily be changed without affecting the main behavior of interest [20].

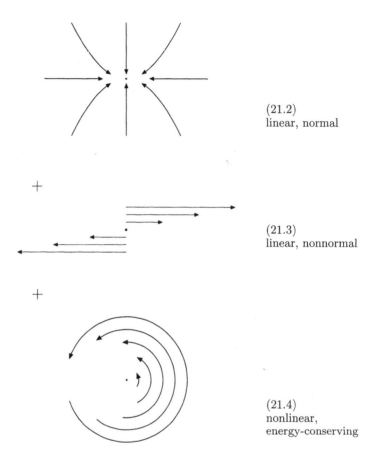

(21.2)
linear, normal

+

(21.3)
linear, nonnormal

+

(21.4)
nonlinear,
energy-conserving

Figure 21.1: Three components that combine in the model (21.1). Each plot shows several trajectories in the (u, v) phase space.

How do the three terms (21.2)–(21.4) act in combination? This is certainly not obvious from a glance at Figure 21.1. First, let us combine the two linear terms, setting

$$\mathbf{A} = \begin{pmatrix} -R^{-1} & 1 \\ 0 & -2R^{-1} \end{pmatrix}.$$

Figure 21.2 shows that the resulting flow is a typical one of the kind analyzed often in this book, with a transient effect unrelated to the eigenvalues. As $R \to \infty$ for this model, both max $\|e^{t\mathbf{A}}\|$ and the time at which this maximum is achieved scale in proportion to R.

Now let us put together the full nonlinear system. Here it becomes necessary to show the axis scales. In the top half of Figure 21.3, ten

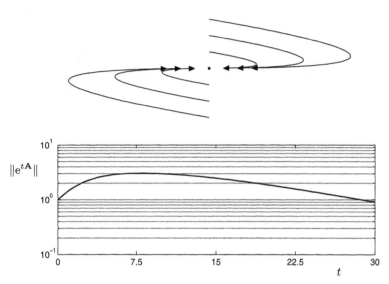

Figure 21.2: Behavior of the linear part of (21.1), that is, the two terms (21.2) and (21.3) in combination, for $R = 12$. The upper plot shows trajectories for $0 \leq t \leq 30$ in the the (u, v) phase space, as in Figure 21.1. The lower plot shows transient growth of this nonnormal linear operator over the same time interval.

trajectories are plotted, five in the upper half-plane and five in the lower half-plane. Three of the trajectories in each half-plane start very close to the origin and then decrease to the origin; these are not visible, appearing just as a small dot. Evidently the origin is a stable fixed point (a sink) of the nonlinear system, as it must be since the linear part of the model has negative eigenvalues. The other two trajectories start a little larger and end up spiraling out to two other stable fixed points located near $(u, v) = (\pm 2/R, \pm 1)$.

This is not a book on dynamical systems, and we shall not go into detail about the geometrical aspects of this flow. Suffice it to say that if there are three sinks in the phase plane, then there must be boundaries separating the regions attracted to each, and along these boundaries must lie saddle points. This is indeed the case; the saddles lie approximately at $(\pm 2/R^2, \pm 2/R^3)$. The full picture appears in Figure 21.4, which is the centerpiece of this section. For $R = 3$ and $R = 4$, the figure shows a spiral region of width about $7/R^3$ in the (u, v) plane that is the basin of attraction of the sink at the origin. The wider spiral regions in between are the basins of attraction of the other two sinks. The sinks and the saddles are visible in the plot, though it is clear that for larger values of R the sink at the origin and the two saddles would quickly become indistinguishable. For further examples of the interplay of nonlinearity and nonnormality in determining basins of attraction of stable states, see [386, 469].

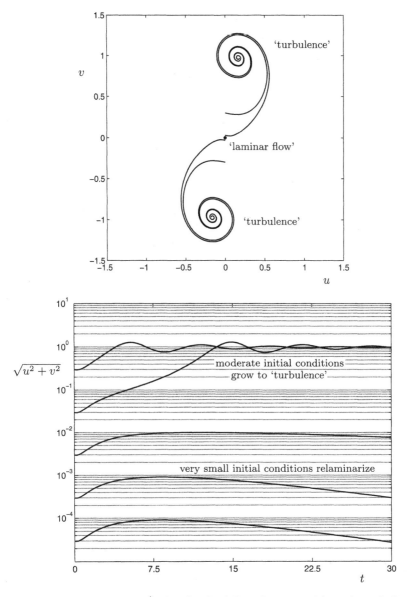

Figure 21.3: Like Figure 21.2 but for the full nonlinear model, again with $R = 12$. Because the equations are nonlinear, the scales now matter. Five trajectories are shown beginning on the v-axis with initial values $v = 3 \times 10^{-1}, 3 \times 10^{-2}, \ldots, 3 \times 10^{-5}$, and the corresponding five trajectories in the lower half of the phase plane. The label 'turbulence' is just suggestive; this system in no way models a turbulent flow.

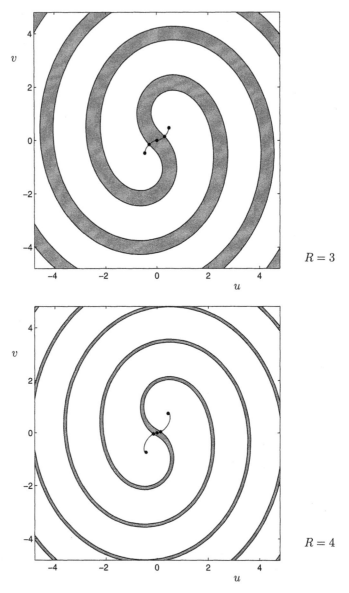

Figure 21.4: Basins of attraction for (21.1); the basin of attraction of the 'laminar' state at the origin is the narrow spiral region bounded by the stable manifolds of the saddle points, colored gray. The five fixed points are shown, lying in the order sink-saddle-sink-saddle-sink along the curve formed by the unstable manifolds of the saddle points. As $R \to \infty$, the width of the basin of attraction shrinks in proportion to R^{-3}. Analogous behavior occurs with plane Poiseuille, plane Couette, and pipe flows, though with vastly more complicated geometry and with exponent closer to -1 than to -3. This suggests how such flows can be stable in theory yet unstable in practice.

Here is our explanation of how certain fluid flows that are stable in theory may be unstable in practice. Consider the basin of attraction of Figure 21.4, but imagine that R is, say, 40 rather than 4. The figure will then look much the same, but the basin will be 1000 times narrower. Now in principle, the fixed point at the origin of such a system is stable, but a trajectory that starts anywhere but the slightest distance from the origin will almost certainly be drawn outward. For some high-speed fluid flows, the geometry will be analogous, though of course far more complicated and infinite-dimensional. For example, in a long circular pipe, the laminar flow may be stable at high Reynolds numbers, but an initial condition that deviates in any but the slightest degree from laminar will almost certainly spiral up to turbulence.

Figure 21.4 contains a curious feature: Some initial conditions very far from the origin lie in the basin of attraction and thus eventually must converge to the origin. This property holds for fluid flows, too. For example, in a pipe or channel flow, any perturbation of the laminar flow that is independent of the streamwise direction, no matter how large the perturbation and no matter how large the Reynolds number, must in principle eventually relaminarize; see §20. Of course one would not expect to be able to observe this in practice at high Reynolds numbers.

Since (21.1) originally appeared in [780], other ODE models of transition to turbulence have also been put forward by various authors. Comparisons of half a dozen of these models are made in [20], and an example of a much more complex system based on the proper orthogonal decomposition appears in [704]. The alternative models that have been proposed usually have more variables, in an attempt to bring them closer to the fluid mechanics; when there are three variables or more, there is the possibility that the state analogous to 'turbulence' can be a chaotic attractor rather than a fixed point [19, 306]. All these models feature basins of attraction of the laminar state that shrink at some rate R^γ as $R \to \infty$; the exponent γ ranges between -3 and -1.

It is interesting to compare the values for the exponent γ suggested by low-dimensional models with the evidence available for actual fluid flows, which comes from three sources: laboratory experiments, numerical simulations of the Navier–Stokes equations, and the landmark theoretical work of Chapman [138, 669]. (Chapman's published paper treats plane Couette and plane Poiseuille flow; his results for pipe flow are unpublished as yet.) In 1993, Trefethen, Trefethen, Reddy and Driscoll raised the question of what the threshold exponent is for these three canonical flows [780], conjecturing that the answer would be strictly less than -1. Chapman's work provides a solution: Assuming his model is correct, the actual exponents are -1 for plane Couette and pipe flow and $-5/4$ for plane Poiseuille flow.[1] This appears to be a tidy resolution, but in fact, the situation is more complicated: Numerical experiments consistently suggest exponents substantially below −1 for these flows. Chapman explains this discrepancy by demonstrating that his analysis predicts convergence with the limiting exponents only for Reynolds numbers on the order of 10^6 or greater, far beyond the usual laboratory or computational range. He argues that measurements based on Reynolds numbers in the realistic range 10^3–10^4 should suggest exponents smaller than −1, just as they do in practice. Thus again we seem to have a tidy resolution of all outstanding problems, though more nuanced than before. Yet there is a further complication, introduced by the most precise laboratory results to date concerning threshold exponents. In an exceptionally careful sequence of experiments involving flow in a long pipe, Hof, Juel, and Mullin [406] have found a very clean R^{-1} relationship in the range where numerical simulations and Chapman's theory both predict an exponent closer to −5/4 or −3/2. Thus a gap in our understanding remains. Perhaps the explanation is that the flow disturbances used to trigger transition in the experiments of [406] are not sufficiently close to 'optimal' in form. If so, this would raise the possibility that perturbations that are optimal in principle may not always be significant in practice.

We can summarize the above complex situation as follows. For the three canonical shear flows introduced in the last section, the width of the basin of attraction of the laminar state shrinks as $R \to \infty$. At laboratory Reynolds numbers, it shrinks faster than R^{-1} for reasons analogous to those at work in the simple model of this section—a mechanism called 'bootstrapping' in [780]. In the limit $R \to \infty$, the bootstrapping effect persists for plane Poiseuille flow but shuts off for plane Couette and pipe flow. The interesting physical mechanisms involved in this behavior are discussed in [138, 623].

Not all researchers agree that the model described in this section has anything to do with transition to turbulence. For a dissident view, see [814].

---

[1] For plane Poiseuille flow one ignores the Tollmien–Schlichting eigenvalue instability that sets in for $R > 5772$ on a time scale too slow to have much importance in practice, as discussed in §22.

22 · Orr–Sommerfeld and Airy operators _____

The most famous of all non-Hermitian eigenvalue problems is the Orr–Sommerfeld equation of hydrodynamic stability. The Orr–Sommerfeld operator was also one of the first differential operators for which pseudospectra were computed numerically, in a pioneering 1993 paper by Reddy, Schmid, and Henningson [624].

The Orr–Sommerfeld problem arises in the analysis of parallel fluid flow in an idealized infinitely long domain [214, 669]. Let us concentrate on the simplest case, *plane Poiseuille flow*, which was introduced in §20. Figure 22.1 shows the configuration, a viscous incompressible flow between infinite parallel flat plates in three dimensions. In Orr–Sommerfeld analysis we begin by restricting attention to flows that are invariant with respect to the unbounded dimension perpendicular to the flow. This restriction yields a two-dimensional problem in variables we shall denote by s (streamwise) and x (spanwise). The governing equations are the incompressible Navier–Stokes equations (not written here) together with zero-velocity ('no slip') boundary conditions. The equations depend on the crucial parameter R, the *Reynolds number*, a nondimensional ratio of inertial to viscous force scales.

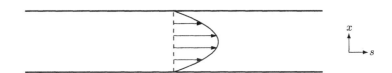

Figure 22.1: Schematic view of plane Poiseuille flow (repeated from Figure 20.7). The parabola represents the laminar flow solution whose stability is at issue.

For any value of R there is a solution to the Navier–Stokes equations, the *laminar solution*, consisting of flow in the s direction with a parabolic velocity flow profile, as suggested in the figure. It is conventional to nondimensionalize the problem by taking the channel to be defined by $-1 \leq x \leq 1$ and the laminar velocity profile to be

$$U(x) = 1 - x^2. \tag{22.1}$$

The classic question about this flow is, Is it stable with respect to infinitesimal perturbations? If not, this might seem to offer an explanation of why high-speed channel flows become turbulent in practice. To find the answer, taking advantage of the Fourier transform in the streamwise direction, we imagine an infinitesimal perturbation that is sinusoidal in s but has an arbitrary shape in x defined by the stream function

$$\Psi(s,x,t) = u(x,t)e^{i\alpha s}, \tag{22.2}$$

where α is a real wave number, with u subject to the boundary conditions $u(\pm 1) = (\partial u/\partial x)(\pm 1) = 0$. Upon inserting this ansatz into the Navier–Stokes equations and defining $\mathbf{D} = \partial/\partial x$ and $u_t = \partial u/\partial t$, we obtain the following equation governing the evolution of $u(x,t)$:

$$\frac{1}{i\alpha}(\mathbf{D}^2 - \alpha^2)u_t = \left[\frac{1}{i\alpha R}(\mathbf{D}^2 - \alpha^2)^2 - (1 - x^2)(\mathbf{D}^2 - \alpha^2) - 2\right]u. \tag{22.3}$$

Equation (22.3) defines a linear autonomous dynamical process that evolves in t—a semigroup (§15). We see that spatial derivatives appear up to the fourth order. In view of the factor $(\mathbf{D}^2 - \alpha^2)$ at the front, however, this dynamical system is not governed simply by a differential operator but is of the generalized form $\mathbf{B}u_t = \mathbf{A}u$, where $\mathbf{B}$ is a second-order differential operator and $\mathbf{A}$ is a fourth-order differential operator (see §45). The *Orr–Sommerfeld operator*, i.e., the generator of the semigroup, is $\mathbf{L} = \mathbf{B}^{-1}\mathbf{A}$:

$$\mathbf{L} = i\alpha(\mathbf{D}^2 - \alpha^2)^{-1}\left[\frac{1}{i\alpha R}(\mathbf{D}^2 - \alpha^2)^2 - (1 - x^2)(\mathbf{D}^2 - \alpha^2) - 2\right]. \tag{22.4}$$

In a rigorous theoretical treatment one would have to be careful to define exactly what function space this expression applies in [677]. In a numerical computation one usually leaves the problem in the generalized form. For computation of pseudospectra and other norm-dependent quantities, one must take care to work in the appropriate energy norm (see §45 and Appendix A of [624]).

We thus find ourselves with a nonnormal operator $\mathbf{L} = \mathbf{L}_{R,\alpha}$ that depends on two parameters: the Reynolds number R and the streamwise wave number α. In the literature, however, the Orr–Sommerfeld problem is almost never presented in this operator form. Instead most authors go directly to the eigenvalue problem $\mathbf{L}u = \lambda u$ by assuming a time dependence $e^{\lambda t}$ for some $\lambda \in \mathbb{C}$. Now (22.4) becomes the famous *Orr–Sommerfeld equation*,

$$\frac{\lambda}{i\alpha}u = (\mathbf{D}^2 - \alpha^2)^{-1}\left[\frac{1}{i\alpha R}(\mathbf{D}^2 - \alpha^2)^2 - (1 - x^2)(\mathbf{D}^2 - \alpha^2) - 2\right]u, \tag{22.5}$$

a fourth-order differential equation containing the eigenvalue λ as an unknown. (It is also common to set $\lambda = -i\alpha c$ so that the left-hand side simplifies to $-cu$.)

Since Orr and Sommerfeld a century ago [587, 707], dozens and indeed probably hundreds of papers have been written about (22.5), treating

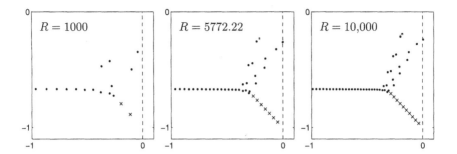

Figure 22.2: The rightmost part of the spectrum of the Orr–Sommerfeld operator for $\alpha = \alpha_{\mathrm{crit}} \approx 1.02$ and three values of R. The crosses mark nearly-degenerate pairs (two eigenvalues closer than 0.01). In the last frame the rightmost eigenvalue is in the interior of the right half-plane, but its real part is only 0.0037.

discreteness of the spectrum, completeness of the eigenmodes [210], numerical range and resolvent, dependence on the flow profile $U(x)$, asymptotic estimates, numerical computation, experimental verification, and numerous other topics [481, 679, 680]. Of course the central question has been, Are there eigenvalues in the right half of the complex plane? It was soon recognized that for large enough R, the answer is yes. A memorable development was the first high-accuracy computation by Orszag in 1971 [588] of the critical parameters for this instability:

$$R_{\mathrm{crit}} \approx 5772.22, \qquad \alpha_{\mathrm{crit}} \approx 1.02055.$$

For $R < R_{\mathrm{crit}}$ and any α, the Orr–Sommerfeld problem is eigenvalue stable, but with $\alpha = \alpha_{\mathrm{crit}}$, an eigenvalue crosses the imaginary axis at $R = R_{\mathrm{crit}}$. The corresponding eigenmode, with vorticity strongly concentrated near the boundaries $x = \pm 1$, is known as a *Tollmien–Schlichting wave*. Orszag's result is illustrated in Figure 22.2.[1]

And here we reach the point where Orr–Sommerfeld eigenvalue analysis fails, although for most of the twentieth century, this fact was not widely appreciated.

A clue that something is wrong is the exceedingly small real part of the rightmost eigenvalue in the third panel of Figure 22.2: just 0.0037, even though the Reynolds number is far above R_{crit}. The determination that this flow is 'unstable' is nothing more than the discovery that it is susceptible to disturbances that grow at the extraordinarily low rate

$$\|u(t)\| \approx e^{0.0037t}. \tag{22.6}$$

---

[1]All the numerical results of this section were obtained by means of a set of MATLAB spectral collocation programs written by Reddy and Henningson in the early 1990s. A version of the these codes appears in Appendix A of [669].

A perturbation growing at this rate would have to travel hundreds of channel widths downstream before it was amplified by even a factor of 10. Probably no laboratory channel has ever been constructed that is so long; yet in actual channels, one regularly sees transition to turbulence at this Reynolds number. Clearly (22.6) is not enough to explain this behavior of real flows.

Figure 22.3 gives some insight into what is overlooked by the eigenvalue analysis. The curves in this figure embody some remarkable physics and are worth studying carefully. First, note that both axes are logarithmic, enabling one to see behavior on multiple time and amplitude scales (cf. Figure 15.3). The time scale of greatest relevance to actual flows would perhaps be $10^1 \leq t \leq 10^2$. Looking at the lowest curve in this part of the plot, we see that the growth experienced by the unstable eigenmode up to time $t = 100$ is not even as great as a factor of 1.5. The middle curve shows the actual norm $\| \exp(t\mathbf{L}) \|$, which corresponds to maximal growth of arbitrary initial perturbations; such calculations were first reported by Farrell [258]. It is a good deal higher, a factor of about 8. Thus it is evident that eigenvalue analysis of the Orr–Sommerfeld operator misses its dominant behavior, which is transient rather than asymptotic. In a moment we shall look at pseudospectra.

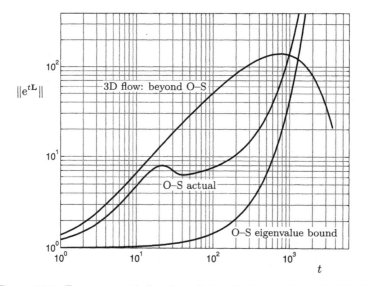

Figure 22.3: Energy growth for plane Poiseuille flow at $R = 10,000$. The lower two curves correspond to the Orr–Sommerfeld problem with $\alpha \approx 1.02$. The top curve corresponds to a three-dimensional flow perturbation with $\alpha = 0$ and spanwise wave number $\beta = 2.04$. This last growth is entirely transient, having nothing to do with eigenvalues, but it is the dominant effect physically.

The third curve in Figure 22.3 tells an even more important story, discussed in §20. This curve shows the values of $\| \exp(t\mathbf{L}) \|$ we would have obtained had we not made the initial assumption that the flow perturbation is two-dimensional. (The mathematics here generalizes from the Orr–Sommerfeld equations to a coupled system of Orr–Sommerfeld and so-called *Squire equations* [214, 669].) The amplification in the range $10^1 \le t \le 10^2$ increases by another order of magnitude, up to a maximum of about 50.

If three-dimensional perturbations are so much more important than two-dimensional perturbations, why does Orr–Sommerfeld analysis limit attention to the latter? There is a clear answer to this question, the 1933 result known as *Squire's theorem*, already mentioned in §20 [717]. Squire began by positing a perturbation with streamwise wave number α and spanwise wave number $\beta \ne 0$. He then showed that the eigenvalue problem associated with these parameters is equivalent to an Orr–Sommerfeld eigenvalue problem with parameters $\beta' = 0$, $\alpha' = \sqrt{\alpha^2 + \beta^2}$, and $R' = R\alpha/\alpha'$, which is smaller than R. It follows that if we are only concerned about the critical Reynolds number as defined by unstable eigenmodes, there is no need to look at three-dimensional perturbations. This mathematically correct but highly misleading observation confused the fluid dynamics literature for many years.

The subject of this section is the Orr–Sommerfeld operator, not fluid mechanics, so we shall now drop the subject of three-dimensional perturbations and return to the matter of the difference between the lower two curves in Figure 22.3. In this book we seek insight into such transient effects by looking at pseudospectra. The pseudospectra of Orr–Sommerfeld operators were considered at length by Reddy, Schmid, and Henningson [624]. Figure 22.4, in the style of their paper, shows pseudospectra of $\mathbf{L}$ for $\alpha = 1.02$ and $R = 10{,}000$.

Perhaps two nonnormal aspects of the Orr–Sommerfeld operator are most apparent in Figure 22.4. One is the rather mild nonnormality in the right part of the spectrum. Physically, this is the important feature, since it is associated with the transient growth that makes the lower two curves of Figure 22.3 differ. The 'hump' scales with amplitude $\mathcal{O}(R^{1/3})$ and time scale $\mathcal{O}(R^{1/3})$ as $R \to \infty$. The other is the more striking nonnormality near the intersection point of the 'Y'. Here the eigenvalue problem is exceedingly ill-conditioned; the figure shows values of ε down to 10^{-8}. The condition number of the infinite-dimensional eigenvector matrix associated with this operator is in fact about 2.03×10^8, a figure that grows approximately in proportion to $e^{\sqrt{R}/5}$ as $R \to \infty$. As a consequence, these eigenvalues are hard to compute numerically, a fact noted by Orszag and others. They are so deep inside the spectrum that there is little physical significance to this ill-conditioning, but there is some analytical significance, for it implies that attempting to expand solutions to an Orr–Sommerfeld flow problem as linear combinations of Orr–Sommerfeld eigenmodes would be a very bad idea.

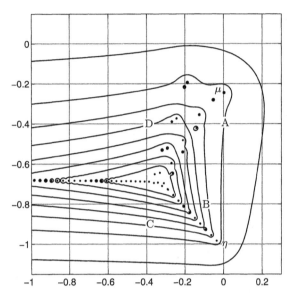

Figure 22.4: Spectrum and ε-pseudospectra of the Orr–Sommerfeld operator (22.4) for $\alpha = 1.02$, $R = 10{,}000$, and $\varepsilon = 10^{-1}, \ldots, 10^{-8}$. Eigenmodes and pseudomodes corresponding to the markers μ, η and A, B, C, D are shown in Figure 22.6.

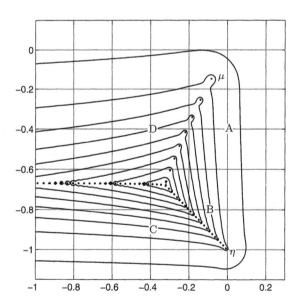

Figure 22.5: Repetition of Figure 22.4, for the approximation (22.8). According to the theory of §11, the pseudospectra fill a half-strip as $R \to \infty$.

The expansion coefficients would be orders of magnitude larger than the functions being expanded, and virtually all the physics would be encoded in the evolving patterns of cancellation among the various nonorthogonal eigenmode components.[2]

It is clear from Figure 22.4 that the Orr–Sommerfeld spectra and pseudo-spectra are complicated, and one might attempt to learn more by analyzing simpler operators with some of the same properties. For example, if we simplify (22.4) by deleting the constant 2 and cancelling the common factor $(D^2 - \alpha^2)$, we obtain the second-order differential operator

$$\mathbf{L} = R^{-1}(\mathbf{D}^2 - \alpha^2) - i\alpha(1 - x^2). \tag{22.7}$$

If we delete the constant α^2 and set $\alpha = 1$ in front of the $1 - x^2$ factor, this becomes

$$\mathbf{L} = R^{-1}\mathbf{D}^2 - i(1 - x^2). \tag{22.8}$$

The spectra and pseudospectra of this operator, shown in Figure 22.5, have much in common with those of the Orr–Sommerfeld operator. This is easily explained, for (22.8) is a variable coefficient differential operator of the kind analyzed in §11. The symbol is

$$f(x, k) = -(k/\sqrt{R})^2 - i(1 - x^2),$$

and the theorems of §11 imply that as $\sqrt{R} \to \infty$, the resolvent norm will grow exponentially throughout the half-strip $\mathrm{Re}\, z < 0$, $-1 < \mathrm{Im}\, z < 0$, with associated pseudomodes in the form of symmetrically positioned pairs of double wave packets; the doubling comes because the two-to-one function $1 - x^2$ has the effect that if (x, k) satisfies the twist condition of §11, so does $(-x, -k)$. This prediction matches the figure nicely and explains why the eigenfunctions for the Orr–Sommerfeld problem have condition numbers scaling exponentially with $\sqrt{R}$. Further explorations, summarized in Figure 22.6, show that the optimal pseudomodes of (22.8) indeed have the form of wave packets, as do the eigenmodes themselves. For the Orr–Sommerfeld operator itself, also shown in the figure, the behavior is similar except for values of z in the upper right corner of the half-strip (marks μ and A).

The operator (22.8) is not the simplest variable coefficient operator with this general kind of behavior. Reddy, Schmid, and Henningson [624] went further and considered

$$\mathbf{L} = R^{-1}\mathbf{D}^2 + ix, \tag{22.9}$$

which in view of the similarity to the Airy equation we call the (complex) *Airy operator* (see p. 107). The same operator has also been investigated

---

[2]Similar spectral instability about a 'Y' branch is observed in magnetohydrodynamics models. In this case the unstable modes and pseudomodes are of particular interest, related to Alfvén waves [74, 795]. See Figure 23.1.

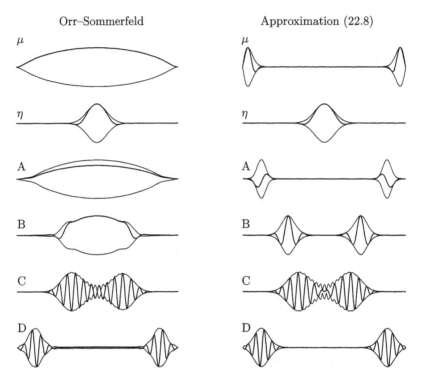

Figure 22.6: Eigenmodes (top two rows) and pseudomodes (bottom four rows) of the Orr–Sommerfeld operator (left) and its approximation (22.5) (right) with $R = 10{,}000$. The Roman markers correspond to A $= -0.4$i, B $= -0.1 - 0.8$i, C $= -0.4 - 0.9$i, D $= -0.4 - 0.4$i. In each case the domain is $-1 \le x \le 1$ and the plot shows the absolute value, the negative absolute value, and the real part of the mode. According to (22.2), these curves represent the stream function; the velocity is associated with the derivative.

by Stoller, Happer, and Dyson [731], Shkalikov [678], and Redparth [628]. (Instead of a further simplification for plane Poiseuille flow, (22.9) can be obtained more directly as an approximation for plane Couette flow.) Again the results of §11 make analysis of pseudospectra easy; the symbol is

$$f(x, k) = -(k/\sqrt{R})^2 + ix,$$

and thus we expect the pseudospectra to approximate the half-strip Re $z < 0$, $-1 < $ Im$z < 1$. Figure 22.7 confirms this expectation. Reddy et al. showed that the numerical range of this operator is contained in the half-strip.

In summary, the Orr–Sommerfeld operator is a beautiful mathematical object, a fascinating example of a nonnormal linear operator. The restriction to two dimensions, however, limits its significance for fluid mechanics.

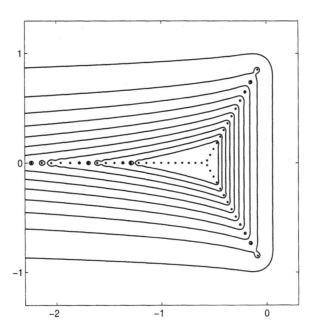

Figure 22.7: Eigenvalues and ε-pseudospectra of the Airy operator (22.9) with $R = 2500$ for $\varepsilon = 10^{-1}, \ldots, 10^{-10}$. Again, the theory of §11 explains why the pseudospectra fill a half-strip in the $R \to \infty$ limit.

23 · Further problems in fluid mechanics _____

Every issue of *Physics of Fluids* or the *Journal of Fluid Mechanics* contains articles on instability and transition to turbulence. For shearing flows at medium or high speeds, a common theme appears again and again: The dominant structures are three-dimensional vortices and streaks roughly aligned with the flow. Experiments show such structures repeatedly. Mathematical theory reveals them too, so long as the analysis admits three-dimensional disturbances and is not confined to eigenvalues and eigenmodes. Sometimes the streaks appear as unstable eigenmodes, but often the linearized operators are highly nonnormal and the streaks are nonmodal.

Sections 20–22 concentrated on the prototypical cases of plane Poiseuille, plane Couette, and pipe flow. In this section we mention a few of the many other problems of fluid mechanics in which linear nonnormal effects have been investigated. Such studies are relatively new, having mostly appeared since the mid-1990s, but there were important earlier theoretical works by Ellingsen and Palm [238] and Landahl [477] and others, as well as a great deal of older experimental evidence. Sometimes old experiments get revisited in the light of new theory, as in the case of Mayer and Reshotko's 1997 paper with the intriguing title, 'Evidence for Transient Disturbance Growth in a 1961 Pipe-Flow Experiment' [539].

Boundary layers and receptivity. A viscous fluid flowing past a solid surface, such as a turbine blade or an aircraft wing, forms a boundary layer in which complicated structures and turbulence may arise [661]. Such flows have much in common with pipe and channel flows, but there are important differences. One difference is that there is often a leading edge that may be roughly independent of the cross-stream dimension; this may be one reason why predictions based on two-dimensional analysis sometimes fare better in boundary layers than in pipes and channels. Another difference is that the domain is infinite in extent, so that the spectrum of the associated operator must be partially or wholly continuous. Also as a result of the unbounded geometry, there can be no laminar solution that is independent of the streamwise direction, so that approximations such as a parallel flow assumption must often be made, and it is common to investigate evolution with respect to space (distance downstream) rather than time. The lack of a second wall further eliminates the length scale provided by the diameter of a pipe or a channel, making the physics more variable and the analysis more challenging. Despite all these facts, boundary layer flows have much in common with confined flows, and in particular, one regularly sees streamwise vortices and streaks (and their folded-over variants known as 'hairpins') as precursors to turbulence. A fascinating aspect of these structures is that they arc often excited by vibrations in the free-stream part of the flow, and since the operators are nonnormal, one finds that to analyze the response in the boundary layer, one must make use of the adjoint of the linearized operator.[1] This process of *receptivity* has been extensively studied [660]. A considerable experimental and numerical literature has accumulated on transient growth and receptivity and their role in transition in boundary layers since the notable early works by Hultgren and Gustavsson [423] and Butler and Farrell [121]; references include [9, 163, 427, 519, 537, 667, 668, 784, 818]. The expression 'bypass transition' (i.e., nonmodal or subcritical transition) was coined in the context of boundary layers by Morkovin in the late 1960s [633]; Reshotko provides an updated view [634].

Taylor–Couette flow. Taylor–Couette flow occurs between two concentric cylinders, the inner one rotating and the outer one, in the simplest case, stationary. If one ignores end effects, then at all rotation rates there is a laminar solution in which the fluid travels in circular orbits. In landmark work of the early 1920s, Taylor showed that at a certain rotation speed, a centrifugal instability develops and the motion evolves into a more complicated pattern known as 'Taylor rolls' [214, 744, 745]. This is a problem in which eigenvalue analysis has been highly successful. Certain variations, however, lead to eigendifficulties. If the outer cylinder is counter-rotating, then one has a configuration that blends the features of the original Taylor problem with those of plane Couette flow. Here nonnormal effects become pronounced, and in some parameter regimes, streamwise streaks may form through transient linear processes and trigger transition to turbulence. Nonnormality and transient effects in Taylor–Couette flow have been investigated in [419] and [547], the former with the aid of pseudospectra.

Curved pipes, walls, and channels. Other flows involving curved walls are also subject to centrifugal instabilities. *Dean flow* involves a curved channel, and *Görtler flow* the boundary layer near a curved wall [192, 214, 280, 329, 736]. In these and other similar configurations, streamwise vortices and streaks usually appear. Sometimes their appearance can be predicted by eigenvalue analysis, and this has led to an interesting pattern of thinking among fluid dynamicists: If curved flows have unstable eigenmodes but straight ones do not, then perhaps straight walls should be regarded as curved walls in the limit of zero curvature? Some fascinating results of this perspective have been developed by Nagata [571]. We believe, however, that some of the appeal of this view is a result of mathematical confusion, for if one goes beyond eigenmodes, the dynamics of a straight pipe or channel can be understood in their own terms, not

---

[1] The same phenomenon of 'adjoint coupling' is of interest to laser engineers (see §60).

as a limit of something else. Indeed, since the associated operators have pseudospectra protruding into the right half-plane, it is hardly surprising that perturbation of the geometry, such as the curving of a boundary wall, frequently produces flows with modal instabilities. In any case, whether or not there are unstable eigenvalues, these flows are often subject to linear nonmodal transient effects, which have received increasing attention in recent years [77, 165, 509].

Compressible and supersonic flow. An 'incompressible' flow is one for which effects of compressibility can be neglected. This includes liquids under most circumstances and also gases in situations where the flow speed is much less than the speed of sound, i.e., $M \ll 1$, where M is the Mach number. Some important flows, however, are strongly compressible, including those involving aircraft or gas turbines. Nonmodal and transient effects for compressible boundary layer flows have been considered in [371, 670], with results in general agreement with those for the incompressible case. Again there is large transient growth, unrelated to eigenmodes, that scales in amplitude in proportion to the Reynolds number R as $R \to \infty$. As with incompressible boundary layers, a crucial aspect of the physics is receptivity to disturbances in the free stream. Even in the supersonic case $M \gg 1$, such effects appear to be involved in transition to turbulence [136, 302, 657]. Reshotko and Tumin have argued that the 'blunt-body paradox' of supersonic flow may have a linear, nonmodal resolution [634, 635].

Non-Newtonian fluids. A non-Newtonian fluid is one whose shear forces are not described by the usual viscosity constant that appears in the Navier–Stokes equations. Important examples are shear-thinning fluids, in which the viscosity diminishes as the fluid is sheared (e.g., paint), viscoplastic fluids, in which there may be no yield at all at low stress (toothpaste), and viscoelastic fluids, which can support tension (egg white). This is a subject of industrial importance, both because some liquids are intrinsically non-Newtonian and also because the flow characteristics of Newtonian liquids may be improved by non-Newtonian additives such as polymers (to inhibit turbulence and reduce drag). All the familiar questions of instability and transition have non-Newtonian counterparts, and just as in the Newtonian case, it appears that nonnormal effects are frequently important at higher speeds. Such effects are studied for viscoelastic fluids in [735], with particular reference to plane Couette flow. An (eigenvalue-based) discussion of the shear-thinning case appears in [293].

Atmospheric flows. The study of atmospheric flows is a huge field, two of whose facets are climate modelling and weather prediction, with their urgent implications for agriculture, hurricane warnings, dispersal of pollutants, global warming, and more. Mathematically, these problems combine the complexities of '$2\frac{1}{2}$-dimensional' spherical geometry, Coriolis forces, periodic thermal forcing, topography, precipitation, interactions with land

and ocean, reflection from clouds, and chemistry. Great financial and computational resources are devoted to these problems, and as in other areas of fluid mechanics, increasing attention has been paid in the past two decades to nonnormal and transient phenomena. This has taken several forms. One is a redirection of attention from atmospheric structures described by eigenmodes (associated with the names Charney and Eady) toward nonmodal structures; the pioneer in this area since the 1980s has been Brian Farrell [256, 257, 259, 260]. Another is a growing emphasis in computational weather prediction on finite time horizons and associated dynamical notions such as finite-time Lyapunov exponents [843] and finite-time 'optimals' [114, 257, 259]. In atmospheric flow problems the doubling time for a perturbation may be on the order of two days, and a factor of 10 may be the difference between a breeze and a hurricane. Clearly the limit $t \to \infty$ is not the only physically important regime under such circumstances; current works give as much attention to singular vectors as to eigenvectors. It was Lorenz who drew attention to finite-horizon issues with analysis of an order-28 generalization of his famous chaotic set of three differential equations [517, 598]. A third new theme has been an emphasis on the stochastic nature of predictions and the use of the technique of 'ensemble forecasting', in which a simulation is run numerous times with various initial data so that not just a single prediction but a cloud of predictions is obtained [28, 266, 283, 444, 515, 556].

Oceanic flows. Oceanic flows are also studied extensively, often as part of coupled ocean-atmosphere models. In recent years linear, nonmodal explanations have been put forward for phenomena involving El Niño [557, 749], large-scale ocean circulation [513, 514, 559], and flows near coasts and past islands and headlands [3, 4], as well as more general discussions of fundamental issues such as numerical modelling, adaptive observations, predictability, oceanic turbulence, transient growth, adjoints, optimal disturbances, and sensitivity to perturbations [137, 267, 558, 599].

Flow control. Advances in electronics, materials, nanotechnology, and fundamental understanding of fluid mechanics have given new prominence in recent years to an old idea: active flow control, for example to suppress turbulence by use of mechanical actuators or blowing and suction in a bounding wall. Some of the recent work in this area builds upon nonmodal analysis of transient growth of disturbances [61, 162, 264, 407, 484, 507].

Magnetohydrodynamics, plasma physics. In an extraordinarily diverse and important set of applications, fluid flows are complicated by the additional feature of electric currents, the magnetic fields they generate, and the forces induced by these magnetic fields. Questions of stability are central to many of these applications. For example, what instabilities are associated with the dynamos that generate the magnetic fields of the earth and the sun [146, 265, 511]? Can a tokamak reactor contain a plasma stably so

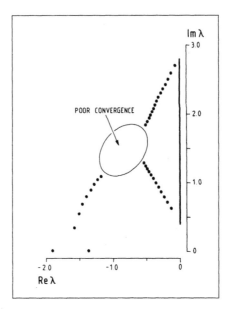

Figure 23.1: A suggestive plot of eigenvalues in the complex plane from Kerner [452].

as to enable controlled fusion [452]? What causes the fluctuations in the ionosphere that disturb satellites and radio waves [278]? How do liquid metals behave [336]? For decades the prevailing technique for addressing such problems has been eigenvalue analysis, despite persistent difficulties of computation and physical interpretation. We cannot resist reproducing in Figure 23.1 an image from a review paper of Kerner [452] that points to some of these difficulties. Readers of this book will suspect that the figure comes from a problem with highly sensitive eigenvalues near an intersection of three branches, like the Orr–Sommerfeld spectrum of Figure 22.4, and this is indeed the case. References pertaining to methods that go beyond eigenvalues in magnetohydrodynamics include [74, 132, 170, 319, 453, 511, 833]. In particular, Borba et al. argue in [74] that pseudospectra provide a resolution of the 'resistive Alfvén paradox'.

Other problems. We have mentioned a number of flow problems, but there are many more. For example, one could consider wakes and jets [27, 164, 191], mixing layers, trailing line vortices [10, 670], thermally driven flows, inclined planes and water tables [59, 585], microscopic flows [675], galactic dynamics [611], compliant or porous walls, rough walls [819], fully developed turbulence [262], aerodynamic flutter [102], two-phase flow, granular flow [671], or combustion [445]. In all of these areas linear nonmodal mechanisms have a place, and there is ongoing activity to elucidate the details of their interplay with nonlinear effects to generate the great complexity of real flows.

VI. Matrix Iterations

24 · Gauss–Seidel and SOR iterations _____

At the time of this writing, the two areas in which pseudospectra have been applied most widely are fluid dynamics and numerical linear algebra. Having considered a variety of fluids applications, we now turn our attention in this part of the book to the convergence behavior of iterative methods for solving linear algebraic equations and matrix eigenvalue problems. Algorithms for both tasks rely on functions of matrices, mostly powers or polynomials, and results from §16 readily lead to convergence bounds that can be descriptive for nonnormal matrices.

First we consider the solution of linear systems $\mathbf{Ax} = \mathbf{b}$, where $\mathbf{A} \in \mathbb{C}^{n \times n}$ is a nonsingular matrix, $\mathbf{b} \in \mathbb{C}^n$ is given, and $\mathbf{x} \in \mathbb{C}^n$ is unknown. We suppose that $\mathbf{A}$ is large and sparse, in which case variants of Gaussian elimination often prove intractable and iterative methods provide an appealing alternative. Though nonnormality has little direct impact on the performance of Gaussian elimination, it manifests itself in important ways in the behavior of iterative algorithms.

In this section and the next, we study classical stationary iterative methods, which include the Jacobi, Gauss–Seidel, and successive over-relaxation (SOR) algorithms. These methods solve the linear system by first splitting $\mathbf{A}$ into two matrices, $\mathbf{A} = \mathbf{M} - \mathbf{N}$, where $\mathbf{M}$ is nonsingular. Then, given some initial guess $\mathbf{x}_0$, they iterate

$$\mathbf{x}_{k+1} = \mathbf{M}^{-1}(\mathbf{Nx}_k + \mathbf{b}). \tag{24.1}$$

Since each iteration requires the solution of a linear system with the coefficient matrix $\mathbf{M}$, it is essential that it be easy to solve equations of the form $\mathbf{Mw} = \mathbf{y}$ for $\mathbf{w}$; for this reason $\mathbf{M}$ is often diagonal or triangular.

If the kth error is denoted by $\mathbf{e}_k = \mathbf{x}_k - \mathbf{x}$, one can show that the iteration (24.1) gives

$$\mathbf{e}_k = (\mathbf{M}^{-1}\mathbf{N})^k \mathbf{e}_0.$$

This error is bounded by powers of the *iteration matrix* $\mathbf{M}^{-1}\mathbf{N}$,

$$\|\mathbf{e}_k\| = \|(\mathbf{M}^{-1}\mathbf{N})^k \mathbf{e}_0\| \le \|(\mathbf{M}^{-1}\mathbf{N})^k\| \|\mathbf{e}_0\|. \tag{24.2}$$

Recall that $\|(\mathbf{M}^{-1}\mathbf{N})^k\| \to 0$ as $k \to \infty$ if and only if $\rho(\mathbf{M}^{-1}\mathbf{N}) < 1$, where ρ is the spectral radius. Furthermore, the spectral radius determines the asymptotic convergence rate [234, 804]:

$$\limsup_{k \to \infty} \left[\sup_{\mathbf{e}_0 \neq 0} \left(\frac{\|\mathbf{e}_k\|}{\|\mathbf{e}_0\|} \right)^{1/k} \right] = \rho(\mathbf{M}^{-1}\mathbf{N}).$$

These formulas indicate that convergence is related to the growth and decay of matrix powers. As we have observed throughout this book and investigated especially in §16, eigenvalues alone are sometimes insufficient to understand this subject, as nonnormality can cause significant transient effects in $\|(\mathbf{M}^{-1}\mathbf{N})^k\|$. For iterative methods, this means a delay in the onset of convergence at the asymptotic rate. In this section and the next, we show that pseudospectra provide a convenient tool for identifying situations where interesting transient convergence behavior can occur. The possibility of transient effects was recognized to some degree years ago; see, e.g., [326, 368, 804]. For example, in his influential 1962 text *Matrix Iterative Analysis*, Varga presents the matrices

$$\mathbf{S}_1 = \begin{pmatrix} \alpha & 4 \\ 0 & \alpha \end{pmatrix}, \qquad \mathbf{S}_2 = \begin{pmatrix} \alpha & 0 \\ 0 & \beta \end{pmatrix},$$

with $0 \ll \alpha < \beta < 1$, and emphasizes that $\|\mathbf{S}_1^k\| > \|\mathbf{S}_2^k\|$ for small values of k, yet for sufficiently large k, $\|\mathbf{S}_1^k\| < \|\mathbf{S}_2^k\|$ [804, §3.2]. Hammarling and Wilkinson note, 'It is not generally appreciated that this concentration on the asymptotic rate of convergence may be extremely misleading as far as the practical behaviour is concerned' [368, p. 2].

Writing $\mathbf{A}$ as the sum of its diagonal, strictly lower, and strictly upper triangular parts, $\mathbf{A} = \mathbf{D} + \mathbf{L} + \mathbf{U}$, we note the following well-known choices for the iteration matrix:

$$\begin{aligned} \text{Jacobi}: \quad & \mathbf{M}^{-1}\mathbf{N} = -\mathbf{D}^{-1}(\mathbf{L} + \mathbf{U}), \\ \text{Gauss–Seidel}: \quad & \mathbf{M}^{-1}\mathbf{N} = -(\mathbf{D} + \mathbf{L})^{-1}\mathbf{U}, \\ \text{SOR}: \quad & \mathbf{M}^{-1}\mathbf{N} = (\omega^{-1}\mathbf{D} + \mathbf{L})^{-1}((\omega^{-1} - 1)\mathbf{D} - \mathbf{U}), \end{aligned}$$

where $\omega \in [0, 2]$ is a parameter chosen to optimize the convergence behavior. (SOR reduces to Gauss–Seidel when $\omega = 1$.)

The Gauss–Seidel and SOR iterations typically give rise to nonnormal iteration matrices even when $\mathbf{A}$ is Hermitian. For example, suppose $\mathbf{A}$ is derived from the standard three-point finite difference discretization of the one-dimensional boundary value problem

$$-u''(x) = f(x) \quad x \in [0, 1], \; u(0) = a, \; u(1) = b. \qquad (24.3)$$

If the interval $[0, 1]$ is discretized with $N + 2$ uniformly spaced grid points, then the coefficient matrix is the $N \times N$ tridiagonal Toeplitz matrix

$$\mathbf{A} = \text{tridiag}(-1, 2, -1).$$

The corresponding Gauss–Seidel iteration matrix takes the form

$$
\mathbf{M}^{-1}\mathbf{N} =
\begin{pmatrix}
0 & \frac{1}{2} & & & \\
0 & \frac{1}{4} & \frac{1}{2} & & \\
\vdots & \vdots & \ddots & \ddots & \\
0 & 2^{-N-1} & \cdots & \frac{1}{4} & \frac{1}{2} \\
0 & 2^{-N} & \cdots & \cdots & \frac{1}{4}
\end{pmatrix}
\in \mathbb{C}^{N \times N}.
\tag{24.4}
$$

All the eigenvalues of this matrix are real. When N is even, zero is an eigenvalue of algebraic multiplicity $N/2$, though it has just a one-dimensional eigenspace. The remaining eigenvalues are, for $k = 1, \dots, N/2$,

$$
\lambda_k = \cos^2(k\pi/(n+1))
$$

with corresponding eigenvectors given entrywise by

$$
[\mathbf{v}_k]_j = \lambda_k^{j/2} \sin\left(\frac{kj\pi}{N+1}\right).
\tag{24.5}
$$

The matrix (24.4) is a Toeplitz matrix, aside from the zeros in the first column. We shall analyze it using the techniques and terminology of §7. The matrices $\mathbf{M}$ and $\mathbf{N}$ are also Toeplitz, with symbols $f_{\mathbf{M}}(z) = 2 - z$ and $f_{\mathbf{N}}(z) = z^{-1}$. Observe, then, that $\mathbf{M}^{-1}\mathbf{N}$ is a modification of the Toeplitz matrix determined by the symbol

$$
f_{\mathbf{M}^{-1}\mathbf{N}}(z) = \frac{f_{\mathbf{N}}(z)}{f_{\mathbf{M}}(z)} = \frac{1}{2z - z^2} = \tfrac{1}{2}z^{-1} + \tfrac{1}{4} + \tfrac{1}{8}z + \tfrac{1}{16}z^2 + \cdots.
\tag{24.6}
$$

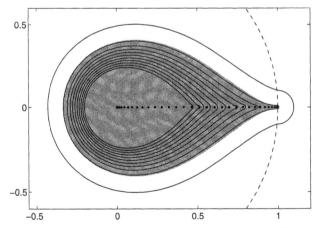

Figure 24.1: Spectrum and ε-pseudospectra ($\varepsilon = 10^{-1}, \dots, 10^{-11}$) of the Gauss–Seidel iteration matrix (24.4) for $N = 64$. The gray region shows the spectrum of $\mathbf{M}^{-1}\mathbf{N}$ in the infinite-dimensional limit, and the dashed line is the unit circle.

Using arguments similar to those found in §§7–9, one can show that the resolvent norm of $\mathbf{M}^{-1}\mathbf{N}$ will grow at least exponentially as $N \to \infty$ at all points z enclosed by the curve $f_{\mathbf{M}^{-1}\mathbf{N}}(\mathbb{T})$, where $\mathbb{T}$ denotes the unit circle. This teardrop-shaped region is shaded gray in Figure 24.1.

One might be tempted to blame the dramatic nonnormality of this example on the fact that $\mathbf{M}^{-1}\mathbf{N}$ has a Jordan block of dimension $N/2$ associated with the eigenvalue $\lambda = 0$. While this certainly contributes to the nonnormality, the distinct nonzero eigenvalues also play a role. As can be deduced from (24.5), the eigenvectors associated with the nonzero eigenvalues are far from orthogonal. This is also evident from Figure 24.2, which shows the pseudospectra of $\mathbf{U}^*\mathbf{M}^{-1}\mathbf{N}\mathbf{U}$, the orthogonal compression of $\mathbf{M}^{-1}\mathbf{N}$ onto the invariant subspace associated with its nonzero eigenvalues, where the columns of $\mathbf{U} \in \mathbb{C}^{N \times N/2}$ form an orthogonal basis for this subspace. Figure 24.3 shows the condition numbers of the nonzero eigenvalues, which are consistent with the nonnormality seen in Figure 24.2.

What influence does such extreme nonnormality have on Gauss–Seidel convergence for this well-known example? Essentially none! Though the boundary of the 10^{-11}-pseudospectrum contains points far from the smallest magnitude eigenvalues in Figure 24.1, it does not extend beyond the unit disk. This fact is highlighted in Figure 24.4, which shows that the pseudospectral radius $\rho_\varepsilon(\mathbf{M}^{-1}\mathbf{N})$ differs negligibly from the spectral radius. As a result, $\|(\mathbf{M}^{-1}\mathbf{N})^k\|$ does not suffer from transient effects, and thus the bound (24.2) ensures that the error converges as predicted by eigenvalue analysis. This is an instance where a geometric understanding

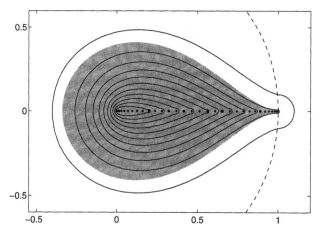

Figure 24.2: Spectrum and pseudospectra of the matrix (24.4) from Figure 24.1, but restricted to the invariant subspace associated with the nonzero eigenvalues. Evidently there is much more to the nonnormality of (24.4) than the defective zero eigenvalue.

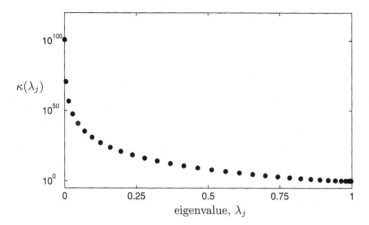

Figure 24.3: Condition numbers $\kappa(\lambda_j)$ of the nonzero eigenvalues of the Gauss–Seidel iteration matrix (24.4) for $N = 64$.

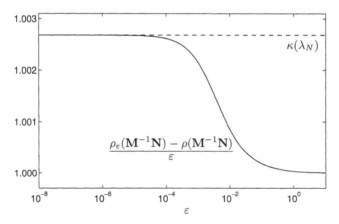

Figure 24.4: Behavior of the pseudospectral radius for the Gauss–Seidel iteration matrix (24.4). Though the problem exhibits strong nonnormality, this does not significantly affect $\rho_\varepsilon(\mathbf{M}^{-1}\mathbf{N})$. From Theorem 52.3 we can deduce that this curve must approach $\kappa(\lambda_{\max}) = 1.00267\ldots$ as $\varepsilon \to 0$; on the other hand, it must approach 1 as $\varepsilon \to \infty$.

of the nonnormality is essential, and the scalar measures of nonnormality addressed in §48 would be wholly insufficient. (For example, the eigenvector matrix for $\mathbf{M}^{-1}\mathbf{N}$ has infinite condition number, since the iteration matrix is nondiagonalizable.)

Though this example is rather elementary because of the small bandwidth of $\mathbf{A}$, similar phenomena arise, for example, when (24.3) is general-

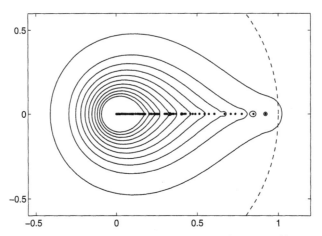

Figure 24.5: Spectrum and ε-pseudospectra ($\varepsilon = 10^{-1}, \ldots, 10^{-11}$) of the Gauss–Seidel iteration matrix corresponding to the three-dimensional problem (24.7) on a grid of $10 \times 10 \times 10$ unknowns.

ized to a multidimensional problem such as

$$-\Delta \mathbf{u} = f \quad \text{on } [0,1] \times [0,1] \times [0,1] \tag{24.7}$$

with Dirichlet boundary conditions. Figure 24.5 shows the pseudospectra for this example.

The pseudospectra of the Gauss–Seidel matrix (24.4) were first investigated by Trefethen in 1992 [772]. Had nonnormality played an important role in the behavior of this classic matrix, research into nonnormal phenomena in numerical linear algebra might have begun in earnest in the 1950s, when Frankel and Young were studying these iterations [289, 846]. There are other important problems, however, where nonnormality does affect the convergence of stationary iterative methods. Such an example is the subject of the next section.

25 · Upwind effects and SOR convergence _____

This section continues the discussion of classical iterative methods begun in the last section. There we studied an example where the Gauss–Seidel iteration matrix exhibited considerable nonnormality that was geometrically concentrated around the origin, having little influence on the pseudospectral radii. Here, we investigate a different one-dimensional boundary value problem where the differential operator, as well as the iteration matrix, is non-self-adjoint.

Consider the advection-diffusion problem

$$-\nu u''(x) + \gamma u'(x) = f(x), \quad x \in [0,1], \; u(0) = \alpha, \; u(1) = \beta \qquad (25.1)$$

for constant 'viscosity' $\nu > 0$ and 'wind speed' $\gamma > 0$. Spectral and pseudospectral properties of advection-diffusion operators were examined in §12. Our present concern is the numerical approximation of solutions to this equation. To begin, we discretize the interval $[0,1]$ with $N+2$ uniformly spaced grid points. The standard centered finite difference approximation gives rise to an $N \times N$ coefficient matrix, and the structure of this matrix depends on how the grid points (and thus the unknowns) are ordered. Two natural choices are to label points from left to right, or from right to left.

$$\rightarrow \quad \gamma > 0 \quad \rightarrow$$

DOWNWIND
$$x_{j-1} \quad x_j \quad x_{j+1}$$

UPWIND
$$x_{j+1} \quad x_j \quad x_{j-1}$$

The terms 'downwind' and 'upwind' arise from the fact that the first ordering follows the direction of the wind γ, while the second is against it. The coefficient matrices induced by these orderings are both tridiagonal and Toeplitz,

$$\text{DOWNWIND} \qquad \mathbf{A} = \text{tridiag}\left(-\nu - \frac{\gamma}{2N}, \; 2\nu, \; -\nu + \frac{\gamma}{2N}\right);$$

$$\text{UPWIND} \qquad \mathbf{A} = \text{tridiag}\left(-\nu + \frac{\gamma}{2N}, \; 2\nu, \; -\nu - \frac{\gamma}{2N}\right).$$

Note that these matrices are transposes of one another.

How does the choice of ordering affect the performance of the iterative method? As described in §24, the asymptotic convergence rate is determined by the spectral radius of the iteration matrix. For convenience, we abbreviate our matrix

$$\mathbf{A} = \text{tridiag}\,(a, b, c)\,.$$

The fact that our discretization is a consistent approximation to the differential equation implies that $a + b + c = 0$, and since $\nu > 0$, we have $b > 0$. As in §24, $\mathbf{A}$ can be decomposed into its diagonal, strictly lower triangular, and strictly upper triangular parts, $\mathbf{A} = \mathbf{D} + \mathbf{L} + \mathbf{U}$.

The Jacobi iteration matrix $\mathbf{S_J} = -\mathbf{D}^{-1}(\mathbf{L} + \mathbf{U})$ is tridiagonal and Toeplitz, but unlike the matrix of the example in §24, where $a = c$, this matrix $\mathbf{S_J}$ is nonnormal. Following the theory for Toeplitz matrices presented in §7, we note that its symbol,

$$f_{\mathbf{S_J}}(z) = -ab^{-1}z - cb^{-1}z^{-1},$$

maps the unit circle $\mathbb{T}$ to an ellipse. The eigenvalues $\{\mu_j\}$ of $\mathbf{S_J}$ fall between the foci of this ellipse and can be computed directly from the formula

$$\mu_j = \frac{2\sqrt{ac}}{b}\,\cos\!\left(\frac{j\pi}{N+1}\right), \quad j = 1, \ldots, N; \qquad (25.2)$$

see, e.g., [701, p. 154ff]. As seen in §§3 and 7, the resolvent norm of $\mathbf{S_J}$ grows exponentially inside the ellipse, so even when n and ε are small, the pseudospectra contain points far from $\{\mu_j\}$ and convergence at the expected asymptotic rate is delayed by transient effects. Interchanging a and c effectively transposes $\mathbf{S_J}$, altering neither the eigenvalues nor the map of the symbol, $f_{\mathbf{S_J}}(\mathbb{T})$. We conclude that the nonnormality of the Jacobi iteration matrix is independent of the downwind or upwind ordering. Figure 25.1 shows the pseudospectra for either ordering for $\nu = 1$, $\gamma = 3N/2$, $N = 32$.

Now consider the Gauss–Seidel iteration matrix $\mathbf{S_{GS}} = -(\mathbf{D} + \mathbf{L})^{-1}\mathbf{U}$. When N is even, $\mathbf{S_{GS}}$ has an eigenvalue at zero with algebraic multiplicity $N/2$, and remaining eigenvalues

$$\lambda_j = \mu_j^2 = \frac{4ac}{b^2}\,\cos^2\!\left(\frac{j\pi}{N+1}\right), \quad j = 1, \ldots, N/2,$$

the squares of those of the Jacobi iteration matrix. These eigenvalues do not change if a and c are interchanged, and so the spectral radius $\mathbf{S_{GS}}$ remains the same for both the upwind and downwind orderings. In fact, if a and c are nonzero, the upwind and downwind iteration matrices are similar. Thus from the asymptotic perspective of eigenvalue analysis, there is nothing to choose between these two ordering schemes.

Various authors have observed over the years, however, that the downwind discretization is actually much better than the upwind one [147, 239, 268, 369, 604]. Unlike the Jacobi case, these two orderings lead to Gauss–Seidel iteration matrices with different nonnormal properties. We begin

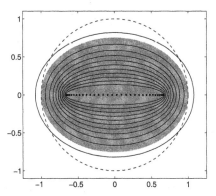

Figure 25.1: Spectrum and ε-pseudospectra ($\varepsilon = 10^{-1}, \ldots, 10^{-11}$) of the Jacobi iteration matrix $\mathbf{S_J}$ for the upwind or downwind ordering with $\nu = 1$, $\gamma = 3N/2$, $N = 32$. The gray region shows the spectrum of $\mathbf{M}^{-1}\mathbf{N}$ in the infinite-dimensional limit.

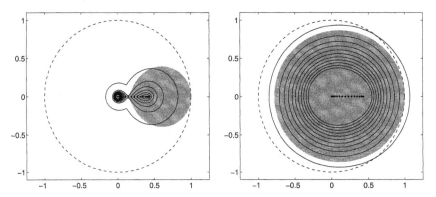

Figure 25.2: Repetition of Figure 25.1, but for the Gauss–Seidel iteration matrices $\mathbf{S_{GS}}$ with the downwind (left) and upwind (right) orderings.

with a simple, extreme example. If $\mathbf{A}$ is lower triangular ($c = 0$ and $a = -b$), then $\mathbf{U} = \mathbf{0}$ and $\mathbf{S_{GS}} = (\mathbf{L} + \mathbf{D})^{-1}\mathbf{U} = \mathbf{0}$. Thus the Gauss–Seidel method must converge in a *single iteration*. On the other hand, if $\mathbf{A}$ is upper triangular ($a = 0$ and $c = -b$), then $\mathbf{S_{GS}} = \text{tridiag}(0, 0, 1)$, a Jordan block. Again, the spectrum consists only of the eigenvalue $\lambda = 0$, and hence $\mathbf{S_{GS}}$ is nilpotent and $\|\mathbf{S_{GS}^N}\| = 0$. For the upwind ordering, however, $\|\mathbf{S_{GS}^{N-1}}\| = 1$, while for the downwind ordering, $\|\mathbf{S_{GS}^k}\| = 0$ for all $k \geq 1$.

When c is small but nonzero, the iteration matrix will exhibit similar nonnormality. One can develop some understanding of this nonnormality by analyzing the symbol of the Toeplitz matrices associated with the iteration matrices. Since $f_{\mathbf{M}}(z) = b + az$ and $f_{\mathbf{N}}(z) = -cz^{-1}$, the symbol for the infinite-dimensional iteration matrix is

$$f_{\mathbf{S}_{GS}}(z) = \frac{c}{az^2 + bz}.$$

We call the maximum modulus of $\{f_{\mathbf{S}_{GS}}(z) : z \in \mathbb{T}\}$ the *symbol radius*, in this case given by

$$\max_{z \in \mathbb{T}} |f_{\mathbf{S}_{GS}}(z)| = \frac{|c|}{||a| - |b||}.$$

Recalling that $a+b+c = 0$, one can see that when $|a|$ is small, $|f_{\mathbf{S}_{J}}(z)| \approx 1$, while when $|c|$ is small, $|f_{\mathbf{S}_{J}}(z)| \ll 1$ for $z \in \mathbb{T}$. This is confirmed in Figure 25.2 for $a = -7/4$, $b = 2$, $c = -1/4$ (i.e., $\nu = 1$ and $\gamma = 3N/2$), where we plot the spectrum of $\mathbf{S}_{GS}$ in the infinite-dimensional limit in gray, superimposing the spectrum and pseudospectra for $N = 32$. Though these downwind and upwind orderings share the same eigenvalues, the corresponding symbols and pseudospectra are very different. In particular, the eigenvalue of largest modulus is much more ill-conditioned in the upwind case, as illustrated in Figure 25.3. Compare this plot to Figure 24.4, the analogous image for $a = c = -1$ and $b = 2$ with $N = 64$. In that case, the pseudospectral radii exceed the spectral radius only modestly. In the present example, both orderings depart significantly from normality, though Figure 25.3 makes it clear that this deviation is much more acute for the upwind ordering.

What is the effect of these orderings in practice? Figure 25.4 illustrates $\|\mathbf{S}_{GS}^k\|$ for the example described in the last paragraph, along with the convergence curve for the Gauss–Seidel method applied to $\mathbf{A}\mathbf{x} = \mathbf{b}$ where $\mathbf{b} = (1, 1, \ldots, 1)^{\mathrm{T}}$ and $\mathbf{x}_0 = \mathbf{0}$.

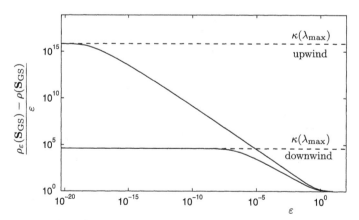

Figure 25.3: Pseudospectral radii for the Gauss–Seidel iteration matrix for the advection-diffusion problem with upwind and downwind ordering ($N = 32$). Both matrices are nonnormal, but the nonnormality is far more pronounced in the upwind case.

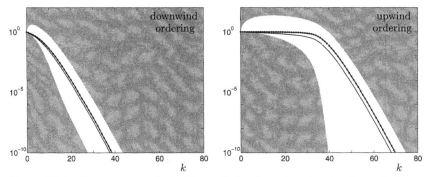

Figure 25.4: Convergence of the Gauss–Seidel iteration for the downwind and upwind example, $N = 32$. The dotted curves denote $\|\mathbf{S}_{\text{GS}}^k\|$, while the solid lines just below them show the actual Gauss–Seidel error $\|\mathbf{e}_k\|/\|\mathbf{e}_0\|$ for a particular $\mathbf{b}$ and $\mathbf{x}_0$. The gray regions represent the upper and lower bounds (16.6) and (16.24) for $\|\mathbf{S}_{\text{GS}}^k\|$ based on the pseudospectral radius.

Further nonnormal effects appear when we examine the SOR iteration, based on the iteration matrix

$$\mathbf{S}_\omega = \mathbf{M}_\omega^{-1}\mathbf{N}_\omega = (\omega^{-1}\mathbf{D} + \mathbf{L})^{-1}((\omega^{-1} - 1)\mathbf{D} - \mathbf{U}).$$

The parameter ω is selected to yield the most rapid asymptotic convergence factor, i.e., to minimize the spectral radius of $\mathbf{S}_\omega$. For the simple advection-diffusion example, we can explicitly determine the spectrum of $\mathbf{S}_\omega$ for any ω. Again using the Jacobi eigenvalues $\{\mu_j\}$ given by (25.2), we find that the eigenvalues of the SOR iteration matrix are

$$\lambda_j = 1 - \omega + \tfrac{1}{2}\omega^2\mu_j^2 \pm \tfrac{1}{2}\omega\mu_j\sqrt{\omega^2\mu_j^2 + 4\mu_j(1 - \omega)}, \qquad (25.3)$$

for $j = 1, \ldots, N$. The optimal choice of ω can be written explicitly for the present example [804, 846],

$$\omega_{\text{opt}} = \frac{2}{1 + \sqrt{1 - \max_j |\mu_j|^2}}.$$

Since $\mathbf{M}_\omega$ and $\mathbf{N}_\omega$ are both Toeplitz, we can generalize the Gauss–Seidel symbol analysis. The symbols for $\mathbf{M}_\omega$ and $\mathbf{N}_\omega$ take the form

$$f_{\mathbf{M}_\omega}(z) = \omega^{-1}b + az, \qquad f_{\mathbf{N}_\omega}(z) = \omega^{-1}b - b - cz^{-1}.$$

From these we can get some idea of the behavior of $\mathbf{S}_\omega$ in the infinite-dimensional limit,

$$f_{\mathbf{S}_\omega}(z) = \frac{(1 - \omega)bz - \omega c}{(a\omega z + b)z}. \qquad (25.4)$$

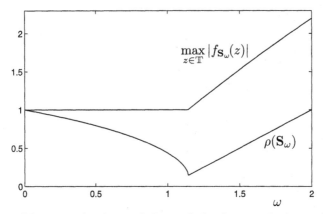

Figure 25.5: The spectral radius and the symbol radius as a function of the SOR parameter ω for a tridiagonal Toeplitz matrix arising from an upwind discretization.

It is interesting to examine how the spectral radius and the symbol radius behave as functions of ω. It can be shown that both are minimized for the same value of ω [604]. For other values of ω, however, the symbol radius tells us more. For example, take $\nu = 1$ and $\gamma = 3N/2$, but now with $N = 64$. In Figure 25.5, we see that for the upwind ordering ($a = -1/4$, $b = 2$, $c = -7/8$), the symbol radius warns of significant transient growth if ω is taken larger than the optimal value. This expectation is confirmed in Figure 25.6, which shows norms of powers of the iteration matrix $\mathbf{S}_\omega$ for three values of ω. In addition to the optimal value $\omega_{\mathrm{opt}} = 1.14241291\ldots$

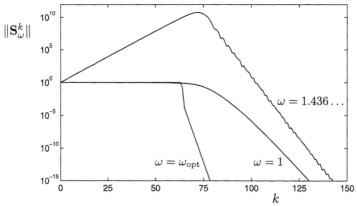

Figure 25.6: Norms of powers of the SOR iteration matrix $\mathbf{S}_\omega$ for three different values of ω. The asymptotic convergence rates for $\omega = 1$ and $\omega = 1.436\ldots$ are the same, but the transient behaviors differ. The chatter in the upper curve is genuine, not a numerical artifact.

suggested by the above analysis, we take $\omega = 1$ (Gauss–Seidel iteration) and $\omega = 1.436478795\ldots$, a value chosen to yield the same asymptotic convergence rate as $\omega = 1$. This latter value of ω leads to transient growth of more than ten orders of magnitude before the asymptotic convergence rate is realized. As pointed out by Böttcher [80], one can prove that exponential transient growth must take place by combining theorems of §§7 and 16.

We have concentrated on exhibiting upwind and downwind effects in a one-dimensional model problem so simplified that it could hardly pose any difficulties in practice. The same ordering effects, however, arise in more complicated problems in multiple space dimensions, and coping with them effectively can be a challenge; see, e.g., [369].

26 · Krylov subspace iterations _____

Since the mid-1980s, powerful new iterative methods for solving large, sparse systems of equations $\mathbf{Ax} = \mathbf{b}$ have become popular for many scientific computing applications. These non-Hermitian descendants of the conjugate gradient algorithm, such as GMRES, bi-conjugate gradients, QMR, and Bi-CGSTAB, are collectively known as Krylov subspace iterations [17, 338, 654, 791]. These methods often converge quickly, but they are hard to analyze; their behavior depends on eigenvalues but also on further properties of a matrix associated with nonnormality. In this section, we describe some of the key convergence phenomena, focusing on GMRES as the iteration of this kind with the cleanest mathematical properties.

Krylov iterations construct approximate solutions of the form $\mathbf{x}_k = \mathbf{x}_0 + \mathbf{q}_k$, where $\mathbf{x}_0$ is an initial guess and $\mathbf{q}_k$ is drawn from the *Krylov subspace*

$$\mathcal{K}_k(\mathbf{A}, \mathbf{r}_0) = \mathrm{span}\{\mathbf{r}_0, \mathbf{A}\mathbf{r}_0, \ldots, \mathbf{A}^{k-1}\mathbf{r}_0\}. \tag{26.1}$$

Here $\mathbf{r}_0$ denotes the initial residual, $\mathbf{r}_0 = \mathbf{b} - \mathbf{A}\mathbf{x}_0$. The design and analysis of such methods is closely linked to the polynomial structure of the Krylov subspaces. Any vector in $\mathcal{K}_k(\mathbf{A}, \mathbf{r}_0)$ can be written as a polynomial in $\mathbf{A}$ times $\mathbf{r}_0$, and so $\mathbf{q}_k = q_k(\mathbf{A})\mathbf{r}_0$ for some $q_k \in \mathcal{P}_{k-1}$, where $\mathcal{P}_{k-1}$ denotes the set of polynomials of degree $k - 1$ or less. We shall measure convergence through the residual

$$\begin{aligned} \mathbf{r}_k &= \mathbf{b} - \mathbf{A}\mathbf{x}_k = \mathbf{r}_0 - \mathbf{A}q_k(\mathbf{A})\mathbf{r}_0 \\ &= p_k(\mathbf{A})\mathbf{r}_0, \end{aligned} \tag{26.2}$$

where the *residual polynomial* $p_k(z) = 1 - zq_k(z)$ has degree $\leq k$ and satisfies the normalization $p_k(0) = 1$. Krylov subspace algorithms differ in their choices for p_k, balancing the need for effective polynomials against the expense of computing them.

The GMRES algorithm of Saad and Schultz [655] determines optimal iterates, in the sense that they minimize the 2-norm of the residual:

$$\|\mathbf{r}_k\| = \min_{\substack{p_k \in \mathcal{P}_k \\ p_k(0)=1}} \|p_k(\mathbf{A})\mathbf{r}_0\|. \tag{26.3}$$

GMRES must terminate exactly in no more than N steps for $\mathbf{A} \in \mathbb{C}^{N \times N}$, since once k reaches the degree of the minimal polynomial of $\mathbf{A}$, there exists some $p_k \in \mathcal{P}_k$, $p_k(0) = 1$, that annihilates $\mathbf{A}$. For GMRES to be effective, it must produce small residual norms for $k \ll N$.

The core of the GMRES iteration is the *Arnoldi process*, a mechanism for building an orthonormal basis for the Krylov subspace. The first vector in this basis is $\mathbf{u}_1 = \mathbf{r}_0/\|\mathbf{r}_0\|$, and at the kth iteration the Arnoldi process determines a vector $\mathbf{u}_{k+1}$ so that

$$\text{span}\{\mathbf{u}_1, \ldots, \mathbf{u}_{k+1}\} = \mathcal{K}_{k+1}(\mathbf{A}, \mathbf{r}_0).$$

This new basis vector is formed by applying a step of the Gram–Schmidt algorithm to orthogonalize $\mathbf{A}\mathbf{u}_k \in \mathcal{K}_{k+1}(\mathbf{A}, \mathbf{r}_0)$ against the basis vectors for $\mathcal{K}_k(\mathbf{A}, \mathbf{r}_0)$:

$$\mathbf{u}_{k+1} = \mathbf{A}\mathbf{u}_k - \sum_{j=1}^{k} (\mathbf{A}\mathbf{u}_k, \mathbf{u}_j)\mathbf{u}_j. \tag{26.4}$$

This procedure becomes increasingly costly as k grows, which explains the need for more efficient suboptimal methods. The convergence of such algorithms is incompletely understood; for a discussion of some peculiarities that arise, see [245, 569].

To analyze the Arnoldi process, it is convenient to organize the orthogonalization steps into matrix form. Let the coefficients of the Gram–Schmidt process (26.4) form the entries of the upper Hessenberg matrix $\mathbf{H}_k \in \mathbb{C}^{k \times k}$, i.e., $h_{jk} = (\mathbf{A}\mathbf{u}_k, \mathbf{u}_j)$ for $j > k+1$, and let $\widetilde{\mathbf{H}}_k \in \mathbb{C}^{(k+1) \times k}$ denote its extension by one row. If the basis vectors $\mathbf{u}_1, \ldots, \mathbf{u}_k$ are arranged into the columns of $\mathbf{U}_k$, the first k steps of (26.4) take the form

$$\begin{aligned} \mathbf{A}\mathbf{U}_k &= \mathbf{U}_k \mathbf{H}_k + h_{k+1,k}\mathbf{u}_{k+1}\mathbf{e}_k^* \\ &= \mathbf{U}_{k+1}\widetilde{\mathbf{H}}_k. \end{aligned} \tag{26.5}$$

Since the columns of $\mathbf{U}_k$ are orthonormal, premultiplying by $\mathbf{U}_k^*$ gives

$$\mathbf{H}_k = \mathbf{U}_k^* \mathbf{A} \mathbf{U}_k. \tag{26.6}$$

The matrix $\mathbf{H}_k$ is the restriction of $\mathbf{A}$ to the degree-k Krylov subspace, and its eigenvalues $\{\theta\}_{j=1}^{k}$, called *Ritz values*, approximate those of $\mathbf{A}$ and are an important element of iterative eigenvalue computations, as discussed further in §28.

We can use this description of the Arnoldi process to derive the form of the optimal polynomial p_k that satisfies (26.3). Let $\{\nu_j\}_{j=1}^{k}$ denote the roots of p_k, and factor out the ℓth root: $p_k(z) = (1 - z/\nu_\ell)q(z)$, where $q \in \mathcal{P}_{k-1}$. Thus $q(\mathbf{A})\mathbf{r}_0 \in \mathcal{K}_k(\mathbf{A}, \mathbf{r}_0)$, and so there exists $\mathbf{y} \in \mathbb{C}^k$ such that $q(\mathbf{A})\mathbf{r}_0 = \mathbf{U}_k \mathbf{y}$. The least-squares optimality property (26.3) implies that the residual $\mathbf{r}_k$ must be orthogonal to the approximating subspace $\mathbf{A}\mathcal{K}_k(\mathbf{A}, \mathbf{r}_0)$, and hence

$$0 = (\mathbf{A}\mathbf{U}_k)^*\mathbf{r}_k = \mathbf{U}_k^*\mathbf{A}^*(\mathbf{I} - \nu_\ell^{-1}\mathbf{A})\mathbf{U}_k\mathbf{y}.$$

Substituting the identity $\mathbf{A}\mathbf{U}_k = \mathbf{U}_{k+1}\widetilde{\mathbf{H}}_k$ twice into the above equation gives

$$\nu_\ell \widetilde{\mathbf{H}}_k^* \mathbf{U}_{k+1}^* \mathbf{U}_k \mathbf{y} = \widetilde{\mathbf{H}}_k^* \mathbf{U}_{k+1}^* \mathbf{U}_{k+1} \widetilde{\mathbf{H}}_k \mathbf{y},$$

and hence ν_ℓ solves the generalized eigenvalue problem

$$\widetilde{\mathbf{H}}_k^* \widetilde{\mathbf{H}}_k \mathbf{y} = \nu_\ell \, \mathbf{H}_k^* \mathbf{y}.$$

Equivalently, by (26.5), ν_ℓ is an eigenvalue of a rank-1 update of $\mathbf{H}_k$,

$$(\mathbf{H}_k + h_{k+1,k}^2 \mathbf{H}_k^{-*} \mathbf{e}_k \mathbf{e}_k^*) \mathbf{y} = \nu_\ell \, \mathbf{y}.$$

Like the Ritz values $\{\theta_j\}$, the roots of p_k are eigenvalue estimates for $\mathbf{A}$, known as *harmonic Ritz values* [38, 128, 292, 526, 595]. These roots are reciprocals of eigenvalue estimates for $\mathbf{A}^{-1}$ obtained by restricting $\mathbf{A}^{-1}$ to the subspace $\mathbf{A}\mathcal{K}_k(\mathbf{A}, \mathbf{r}_0)$. When $\mathbf{H}_k$ is singular, one of these roots ν_k must be infinite, implying that $\deg(p_k) < k$. Consequently, GMRES makes no progress at the kth step: $\|\mathbf{r}_k\| = \|\mathbf{r}_{k-1}\|$. At the other extreme, when $h_{k+1,k} = 0$, the harmonic Ritz values and the Ritz values coincide and are equal to eigenvalues of $\mathbf{A}$. This case is known as 'lucky breakdown', for it also implies that the iteration terminates with the exact solution, $\mathbf{r}_k = \mathbf{0}$.

The easiest way to quantify the accuracy of Ritz and harmonic Ritz values is to show that they are pseudoeigenvalues, as described in the following theorem. The first result is well-known [694], [494, §4.6]; the second is due to Simoncini and Gallopoulos [693].

Theorem 26.1 *The Ritz and harmonic Ritz values are pseudoeigenvalues of* $\mathbf{A}$:

$$\{\theta_j\}_{j=1}^k \subseteq \sigma_\varepsilon(\mathbf{A}) \qquad \forall \varepsilon > |h_{k+1,k}|;$$

$$\{\nu_j\}_{j=1}^k \subseteq \sigma_\varepsilon(\mathbf{A}) \qquad \forall \varepsilon > |h_{k+1,k}| + |h_{k+1,k}^2|/s_{\min}(\mathbf{H}_k),$$

where $s_{\min}(\cdot)$ *denotes the minimal singular value.*

Proof. Both results are proved by constructing specific perturbation matrices $\mathbf{E}$ such that $\sigma(\mathbf{A} + \mathbf{E})$ contains the desired pseudoeigenvalues. For the Ritz values, set $\mathbf{E} = -h_{k+1,k} \mathbf{u}_{k+1} \mathbf{u}_k^*$, for which $\|\mathbf{E}\| = |h_{k+1,k}|$. Then using (26.5),

$$(\mathbf{A} + \mathbf{E})\mathbf{U}_k = \mathbf{A}\mathbf{U}_k - h_{k+1,k} \mathbf{u}_{k+1} \mathbf{e}_k^* = \mathbf{U}_k \mathbf{H}_k,$$

so the columns of $\mathbf{U}_k$ span an invariant subspace of $\mathbf{A} + \mathbf{E}$, and thus $\{\theta_j\}_{j=1}^k = \sigma(\mathbf{H}_k) \subseteq \sigma(\mathbf{A} + \mathbf{E})$. The result for harmonic Ritz values follows similarly, using the perturbation

$$\mathbf{E} = h_{k+1,k}^2 \mathbf{H}_k^{-*} \mathbf{e}_k \mathbf{e}_k^* - h_{k+1,k} \mathbf{u}_{k+1} \mathbf{e}_k^*,$$

which has norm bounded by $|h_{k+1,k}^2|/s_{\min}(\mathbf{H}_k) + |h_{k+1,k}|$. ∎

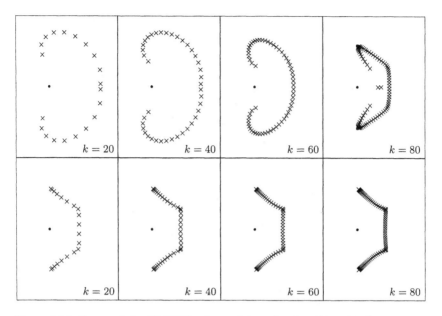

Figure 26.1: Roots of the GMRES polynomials p_k for the 100×100 Grcar matrix (top) and a normal matrix with the same spectrum (bottom). Each plot shows the rectangle $-2 \leq \mathrm{Re}\, z \leq 4$, $-4 \leq \mathrm{Im}\, z \leq 4$ of the complex plane. Compare Figure 7.5.

The bounds given in this theorem can be rather slack. For one thing, they are based on constructing perturbation matrices $\mathbf{E}$ that make all the roots $\{\theta_j\}$ or $\{\nu_j\}$ eigenvalues of $\mathbf{A} + \mathbf{E}$ simultaneously, whereas smaller perturbations may make an individual θ_j or ν_j an eigenvalue. Still, in many cases the Ritz and harmonic Ritz values reflect the pseudospectra of $\mathbf{A}$, as suggested in Figure 26.1, which illustrates GMRES behavior for the Grcar matrix (7.14) of dimension $N = 100$. The top row of plots shows the roots $\{\nu_j\}$ of the optimal GMRES polynomial p_k for various degrees k, which roughly follow the boundaries of ε-pseudospectra for decreasing values of ε, as can be seen by comparison with Figure 7.5. We repeat this experiment in the second row of plots, but now for a normal matrix with the same spectrum as the Grcar matrix. In this case, the roots of p_k closely approximate the eigenvalues even at early iterations.

Can we be more precise about how the nonnormality of $\mathbf{A}$ affects the convergence behavior of GMRES? This has been a significant research question for more than a decade. The initial residual $\mathbf{r}_0$ complicates the analysis, but often has little effect on the convergence behavior, so it is typically removed from the optimization via the bound[1]

---

[1]There are some nonnormal matrices for which no $\mathbf{r}_0$ attains equality in (26.7) [252, 758], but such examples are thought to be rare in practice [757]. For an example of GMRES analysis that incorporates $\mathbf{r}_0$, see [503].

$$\|\mathbf{r}_k\| = \min_{\substack{p_k \in \mathcal{P}_k \\ p_k(0)=1}} \|p_k(\mathbf{A})\mathbf{r}_0\| \leq \min_{\substack{p_k \in \mathcal{P}_k \\ p_k(0)=1}} \|p_k(\mathbf{A})\|\,\|\mathbf{r}_0\|. \qquad (26.7)$$

Throughout the rest of this section we study the *Ideal GMRES* problem [344],

$$\min_{\substack{p_k \in \mathcal{P}_k \\ p_k(0)=1}} \|p_k(\mathbf{A})\|, \qquad (26.8)$$

which is closely related to the Chebyshev polynomials of a matrix discussed in §29. Since the minimal polynomial of $\mathbf{A}$ annihilates $\mathbf{A}$, one might expect the roots of $p_k(\mathbf{A})$ to approximate the eigenvalues of $\mathbf{A}$ (i.e., the roots of the minimal polynomial). When $\mathbf{A}$ is normal, $\mathbf{A} = \mathbf{U\Lambda U}^*$ for unitary $\mathbf{U}$, this is essentially true:

$$\min_{\substack{p_k \in \mathcal{P}_k \\ p_k(0)=1}} \|p_k(\mathbf{A})\| = \min_{\substack{p_k \in \mathcal{P}_k \\ p_k(0)=1}} \|\mathbf{U}p_k(\mathbf{\Lambda})\mathbf{U}^*\| = \min_{\substack{p_k \in \mathcal{P}_k \\ p_k(0)=1}} \max_{\lambda \in \sigma(\mathbf{A})} |p_k(\lambda)|. \quad (26.9)$$

The quantity on the right of (26.9) is a lower bound on the convergence of Ideal GMRES for nonnormal $\mathbf{A}$: For any $p \in \mathcal{P}_k$, the spectral mapping theorem gives $p(\sigma(\mathbf{A})) = \sigma(p(\mathbf{A}))$, and so

$$\max_{\lambda \in \sigma(\mathbf{A})} |p_k(\lambda)| = \rho(p_k(\mathbf{A})) \leq \|p_k(\mathbf{A})\|, \qquad (26.10)$$

where $\rho(\cdot)$ denotes the spectral radius [422, 575].

Several different approaches can be used to develop upper bounds on (26.8). When $\mathbf{A}$ is diagonalizable, $\mathbf{A} = \mathbf{V\Lambda V}^{-1}$,

$$\min_{\substack{p_k \in \mathcal{P}_k \\ p_k(0)=1}} \|p_k(\mathbf{A})\| = \min_{\substack{p_k \in \mathcal{P}_k \\ p_k(0)=1}} \|p_k(\mathbf{V\Lambda V}^{-1})\|$$

$$\leq \min_{\substack{p_k \in \mathcal{P}_k \\ p_k(0)=1}} \kappa(\mathbf{V})\,\|p_k(\mathbf{\Lambda})\|$$

$$= \kappa(\mathbf{V}) \min_{\substack{p_k \in \mathcal{P}_k \\ p_k(0)=1}} \max_{\lambda \in \sigma(\mathbf{A})} |p_k(\lambda)|, \qquad (26.11)$$

where $\kappa(\mathbf{V}) = \|\mathbf{V}\|\,\|\mathbf{V}^{-1}\|$ is the 2-norm condition number of $\mathbf{V}$ [236, 655]. Like (26.10), this analysis reduces the GMRES convergence question to a discrete approximation problem in the complex plane. For nonnormal problems where $\kappa(\mathbf{V})$ is large, Ideal GMRES will typically fall between the extremes of these eigenvalue-based upper and lower bounds, depending upon the nature of the nonnormality. Pseudospectra provide an alternative approach that can be more descriptive when $\kappa(\mathbf{V})$ is large or infinite. Using a Dunford–Taylor integral (see §14),

$$
\min_{\substack{p_k \in \mathcal{P}_k \\ p_k(0)=1}} \| p_k(\mathbf{\Lambda}) \| - \min_{\substack{p_k \in \mathcal{P}_k \\ p_k(0)=1}} \left\| \frac{1}{2\pi i} \int_\Gamma p_k(z)(z - \mathbf{A})^{-1} \, dz \right\|
$$

$$
\leq \min_{\substack{p_k \in \mathcal{P}_k \\ p_k(0)=1}} \frac{1}{2\pi} \int_\Gamma |p_k(z)| \, \|(z - \mathbf{A})^{-1}\| \, |dz|.
$$

Taking $\Gamma = \Gamma_\varepsilon$ to be a contour or union of contours enclosing $\sigma_\varepsilon(\mathbf{A})$ and coarsely approximating the integral yields the following bound from [771]. Very roughly speaking, this bound indicates that for highly nonnormal problems, convergence of GMRES depends on approximation by polynomials not on the spectrum, but on the pseudospectra.

Pseudospectral bound for GMRES

Theorem 26.2 *Let Γ_ε be a union of contours enclosing $\sigma_\varepsilon(\mathbf{A})$. Then*

$$
\frac{\|\mathbf{r}_k\|}{\|\mathbf{r}_0\|} \leq \min_{\substack{p_k \in \mathcal{P}_k \\ p_k(0)=1}} \| p_k(\mathbf{A}) \| \leq \frac{L_\varepsilon}{2\pi\varepsilon} \min_{\substack{p_k \in \mathcal{P}_k \\ p_k(0)=1}} \max_{z \in \Gamma_\varepsilon} |p_k(z)|, \qquad (26.12)
$$

where L_ε is the arc length of Γ_ε.

As with the pseudospectral bounds presented in §§15 and 16, the parameter ε is an essential part of Theorem 26.2. Large values of ε yield small constants $L_\varepsilon/2\pi\varepsilon$, but with approximation problems over large domains (potentially including the normalization point $z = 0$), while small values lead to larger constants but approximation problems over reduced domains. In one respect this is a drawback of the pseudospectral approach, but at the same time the ε-dependence provides a mechanism for describing different phases of GMRES convergence: for nonnormal problems one often observes stagnation at early iterations (where bounds with large ε are descriptive) followed by more rapid convergence later (described by small ε). Figuratively speaking, this corresponds to GMRES starting out with only a hazy impression of the spectral properties of $\mathbf{A}$ that comes into focus with successive iterations.

In the large ε limit, (26.12) becomes the trivial upper bound $\|\mathbf{r}_k\| \leq \|\mathbf{r}_0\|$, and in the small ε limit, the asymptotic characterization of the resolvent norm described in §52 implies that (26.12) captures the exact termination that must occur when k reaches the degree of the minimal polynomial of $\mathbf{A}$. Of course the important matter is the performance of Theorem 26.2 between these large-ε and small-ε extremes of stagnation and termination. We investigate several computational examples. First, let $\mathbf{A}$ be the scaled

Jordan block

$$\mathbf{A} = \begin{pmatrix} 1 & \beta & & \\ & 1 & \ddots & \\ & & \ddots & \beta \\ & & & 1 \end{pmatrix} \in \mathbb{C}^{N \times N}. \qquad (26.13)$$

Theorem 26.2 works out very cleanly in this example, as the pseudospectra are disks (see §51) whose radii depend upon the magnitude of β and the matrix dimension. For a fixed β, denote this radius by r_ε, so that

$$\sigma_\varepsilon(\mathbf{A}) = 1 + \Delta_{r_\varepsilon},$$

where Δ_{r_ε} is the open disk about 0 of radius r_ε. Theorem 26.2 reduces to

$$\min_{\substack{p_k \in \mathcal{P}_k \\ p_k(0)=1}} \|p_k(\mathbf{A})\| \leq \left(\frac{2\pi r_\varepsilon}{2\pi\varepsilon}\right) \max_{z \in 1 + \Delta_{r_\varepsilon}} |1 - z|^k = \frac{r_\varepsilon^{k+1}}{\varepsilon}, \qquad (26.14)$$

where we have used the fact that $(1 - z)^k$ is the minimizing polynomial for the disk $1 + \Delta_{r_\varepsilon}$. Figure 26.2 illustrates this bound for all $\varepsilon > 0$.

GMRES appears to converge in a nearly linear fashion for the matrix (26.13). The bound (26.12) captures this behavior, reflecting the fact that in the large-N limit, the pseudospectra of this matrix are disks of radius $\beta + \varepsilon$ (see §7). The next example, an 'integration matrix' [215, 632],

$$\mathbf{A} = \begin{pmatrix} 1 & \gamma & & & \\ & 1 & \frac{1}{2}\gamma & & \\ & & \ddots & \ddots & \\ & & & 1 & \frac{1}{N-1}\gamma \\ & & & & 1 \end{pmatrix} \in \mathbb{C}^{N \times N}, \qquad (26.15)$$

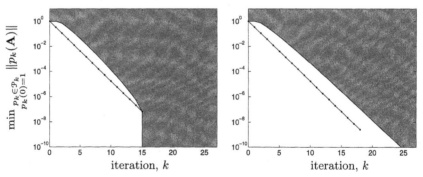

Figure 26.2: Convergence of Ideal GMRES for the Jordan block (26.13) with $\beta = 1/3$ for $N = 16$ (left) and $N = 128$ (right). The gray regions illustrate the envelope of pseudospectral convergence bounds taken over all values of ε. The Ideal GMRES curves were computed using the SDPT3 package [759]; due to numerical constraints, we only show values up to $k = 18$ for $N = 128$.

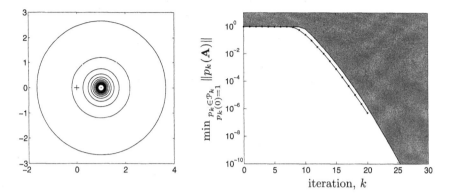

Figure 26.3: Spectra and ε-pseudospectra (left) and convergence of Ideal GMRES (right) for (26.15) with $\gamma = 4$ for dimension $N = 128$. The pseudospectral boundaries are for $\varepsilon = 10^{-12}, 10^{-11}, \ldots, 10^{-1}$; the cross marks the origin, where the optimizing polynomials are normalized. The gray regions on the right illustrate the envelope of pseudospectral convergence bounds taken over all $\varepsilon > 0$.

is more interesting. Again the pseudospectra are circular disks, as can be shown in the same way as for (26.13). Thus the formula (26.14) still holds, though the radii r_ε differ markedly from the Jordan block case: They are no longer large discs in the large-N limit. As a result, Ideal GMRES initially stagnates for this problem and then converges at a much improved rate. Pseudospectra capture this transition very well, as seen in Figure 26.3.

For our final example, we turn to a finite-element discretization of a two-dimensional advection-diffusion problem that has become a popular model problem for GMRES; see [276, 244, 249, 504] and also Figure 51.2. Here, we study a particular example of dimension $N = 169$. As with the integration matrix, Ideal GMRES initially stagnates due to nonnormality, then converges more rapidly. Pseudospectra for this matrix are shown in Figure 26.4 along with convergence bounds from Theorem 26.2 for a range of values of ε. Not only do the pseudospectra of $\mathbf{A}$ give an indication about the convergence of GMRES; quantities derived from the GMRES iteration shed light on the pseudospectra of $\mathbf{A}$. The bottom left plot in this figure shows pseudospectra of the rectangular upper Hessenberg matrix $\widetilde{\mathbf{H}}_{24}$, which approximate $\sigma_\varepsilon(\mathbf{A})$, at least for large values of ε; see §40 for details. The final plot shows the harmonic Ritz values $\{\nu_j\}$ at step $k = 24$, with level curves of the residual polynomial, $p_{24}(z)$. Indeed, this residual polynomial is small not just over the eigenvalues of $\mathbf{A}$, but also over the ε-pseudospectra for moderate values of ε.

The predictions of Theorem 26.2 are not always as accurate as shown in the examples here, and several factors are to blame. First, this theorem does not use the fact that the spectrum is a discrete point set, a property

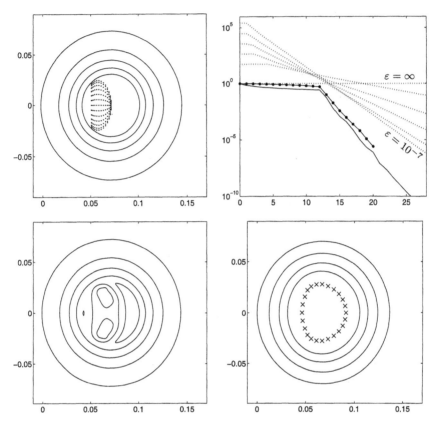

Figure 26.4: Pseudospectra and GMRES convergence data for an advection-diffusion model problem of dimension $N = 169$. The top left plot shows $\sigma(\mathbf{A})$ and $\sigma_\varepsilon(\mathbf{A})$ for $\varepsilon = 10^{-2}, 10^{-3}, \ldots, 10^{-6}$. These pseudospectra are well approximated by the corresponding pseudospectra of $\widetilde{\mathbf{H}}_{24}$, shown in the bottom left plot. The top right plot illustrates $\|\mathbf{r}_k\|/\|\mathbf{r}_0\|$ for a particular right hand side (solid line), the convergence of Ideal GMRES (dotted line), and bounds from Theorem 26.2 for $\varepsilon = \infty, 10^{-3}, 10^{-4}, \ldots, 10^{-7}$. The bottom right plot shows the harmonic Ritz values at step $k = 24$ and level curves $|p_{24}(z)| = 10^0, 10^{-2}, \ldots, 10^{-6}$, read from the outside in.

exploited by the GMRES iteration when $\mathbf{A}$ has a small number of isolated eigenvalues [792]. Moreover, if $\mathbf{A}$ is highly nonnormal only because several eigenvalues are ill-conditioned, GMRES will initially concentrate on those eigenvalues, effectively eliminating them from consideration at future iterations. Theorem 26.2 lacks the flexibility to capture such behavior, and as a result one must take ε to be very small to get a tractable approximation problem, which in turn gives an unacceptably large constant. As a remedy, one can use different values of ε on different disjoint pseudospectral components. One can also use eigenvalue and eigenvector information more carefully than in (26.11); see [244, 339].

Problems from applications are rarely as clean as those shown here, and often the convergence of Krylov subspace iterations in their pure forms is unacceptably slow. The performance is usually improved by *preconditioning* the equations, replacing $\mathbf{A}\mathbf{x} = \mathbf{b}$ by one or the other of the formulas

$$(\mathbf{B}^{-1}\mathbf{A})\mathbf{x} = \mathbf{B}^{-1}\mathbf{b}, \qquad (\mathbf{A}\mathbf{B}^{-1})(\mathbf{B}\mathbf{x}) = \mathbf{b},$$

where $\mathbf{B}$ is some easily invertible matrix and $\mathbf{B}^{-1}\mathbf{A}$ or $\mathbf{A}\mathbf{B}^{-1}$ has more favorable properties than $\mathbf{A}$ itself. From the perspective of eigenvalues, it makes no difference if one preconditions by multiplying $\mathbf{A}$ with $\mathbf{B}^{-1}$ on the left or right, as $\mathbf{B}^{-1}\mathbf{A}$ and $\mathbf{B}^{-1}\mathbf{A}$ are similar. Yet we saw in the last section that because of nonnormality, two matrices that are similar may have very different iterative properties. Manteuffel and Parter investigated the great differences between left and right preconditioning for a given matrix $\mathbf{A}$ derived from discretization of an elliptic partial differential equation and related this to boundary conditions and adjoints [527], and Lee conducted an interesting computational study of an example arising from the discretization of a differential algebraic equation from general relativity [490]. A general strategy for preconditioning matrices by similarity transformations has been suggested by Gutknecht and Loher—an idea that would make no sense from the point of view of eigenvalues alone [362].

27 · Hybrid iterations

The GMRES algorithm discussed in the last section is an optimal iterative method for $\mathbf{A}\mathbf{x} = \mathbf{b}$: It computes the approximate solution $\mathbf{x}_k$ that minimizes the norm of the residual $\mathbf{r}_k = \mathbf{b} - \mathbf{A}\mathbf{x}_k$ over all vectors in the Krylov subspace:[1]

$$\|\mathbf{r}_k\| = \min_{\substack{p \in \mathcal{P}_k \\ p(0)=1}} \|p(\mathbf{A})\mathbf{b}\|.$$

To compute this optimal iterate, GMRES builds and stores an orthonormal basis for the Krylov subspace—an amount of work per step that grows linearly with the step number k. This cost becomes prohibitive for problems that require many steps, and one must turn to alternative algorithms that compute suboptimal iterates, at a cheaper cost per iteration. There are a number of options: One can restart GMRES every m steps, or apply algorithms derived from the non-Hermitian Lanczos process such as BiCGSTAB and QMR; see, e.g., [338, 654, 791]. Another alternative is to use *hybrid iterative methods*, which consist of three components:

Step 1. Estimate the spectrum $\sigma(\mathbf{A})$,

Step 2. Find a sequence of polynomials $\{p_k(z)\}$ with $p_k(0) = 1$
that are small in magnitude on the estimated spectrum,

Step 3. Perform a polynomial iteration with $\{p_k(z)\}$.

Readers of this book will suspect this sequence to be of dubious merit when $\mathbf{A}$ is nonnormal. Before turning to this aspect of the problem, however, let us survey the literature on these hybrid iterative algorithms.

First, we must consider why hybrid algorithms are potentially advantageous. Alternatives to the full GMRES method attempt to make each iteration cheaper without increasing their total number too much. Restarted GMRES and methods based on the non-Hermitian Lanczos process are popular, though their convergence properties are not fully understood. At each iteration these methods compute multiple inner products, which require 'all-to-all' communication on distributed memory parallel computers. Hybrid algorithms largely avoid these operations, which explains the attention these methods attracted as new computer architectures became popular during the late 1980s and early 1990s. Hybrid methods are particularly appealing in situations where matrix-vector multiplications do not dominate the total cost of computation, e.g., when $\mathbf{A}$ has sparsity or other exploitable structure, inner products are expensive, or when full or restarted GMRES requires many operations to construct an orthonormal

---

[1] This section is adapted from [570].

basis for the Krylov subspace.[2] Ideas related to hybrid iterations also arise in polynomial preconditioning [21, 120, 654, 652] and augmented Krylov subspace methods [233, 560]. Similar methods apply to the iterative computation of matrix eigenvalues [126, 383, 403, 650].

Most of the hybrid algorithms published in the past twenty-five years are summarized in Table 27.1. The starting point of this literature is Manteuffel's algorithm of the late 1970s.[3] In this method a number of the extreme eigenvalues of **A** are estimated by a modified power iteration, these eigenvalue estimates are surrounded by an ellipse not enclosing the origin, and a Chebyshev iteration is then carried out with parameters corresponding to that ellipse. In brief,

> *Step* 1: Modified power iteration,
>
> *Steps* 2 *and* 3: Chebyshev iteration based on ellipse enclosing the
> eigenvalue estimates.

Manteuffel's algorithm has had considerable influence over the years, and it was implemented by Ashby in a Fortran package called ChebyCode [15].

Subsequent hybrid algorithms have modified all of the steps of Manteuffel's algorithm. First, beginning with Smolarski and Saylor around 1981, Step 3 was generalized to a Richardson or Horner iteration corresponding to an arbitrary polynomial $p_k(z)$ rather than just a scaled and shifted Chebyshev polynomial. This eliminates the restriction to a spectrum contained in an ellipse, and Step 2 thus changes too. Algorithms in this class typically approximate $\sigma(\mathbf{A})$ by estimating some outlying eigenvalues and then forming the polygon that is the convex hull of these eigenvalue estimates, or sometimes, for matrices with real entries, the union of the convex hulls of the eigenvalue estimates in the upper and lower half-planes. A polynomial approximation problem is then solved on that polygon via some combination of methods that may include linear programming [241], least-squares fitting [651, 664], numerical conformal mapping [501, 720], Faber or Faber–CF approximation [231, 255, 361], and interpolation in Féjer points [277, 742], among other techniques [235, 333, 586].

Beginning with Elman, Saad, and Saylor around 1986, Step 1 of Manteuffel's algorithm was also changed. The modified power iteration was replaced by an Arnoldi iteration, which is the standard Krylov subspace method for estimating eigenvalues (discussed next in §28). One advantage of the Arnoldi process is that it also forms the basis of the GMRES algo-

---

[2]In the symmetric case, GMRES reduces to the MINRES iteration, which can be implemented by a three-term recurrence, requiring an amount of work per step independent of k [274, 596]. Therefore matrix-vector multiplications usually do dominate in the symmetric case, at least on serial computers.

[3]Manteuffel [524] attributes earlier work in this direction to Wachspress (1962), Wrigley (1963), Hageman (1972), and Kincaid (1974). Another early contributor was Rutishauser. There is also an early Russian literature that anticipates semi-iterative algorithms based on ellipses and more general conformal maps; see [253, Chap. 9].

Table 27.1: Non-Hermitian hybrid iterative algorithms since 1977. Except for
the Hybrid GMRES algorithm [570], all are based on estimates of eigenvalues or
related quantities in the complex plane.

Manteuffel 1977 [524, 525, 15]:
modified power iteration → eigenvalue estimates → ellipse
                                                        → Chebyshev iteration

Smolarski and Saylor 1981 [705, 706]:
modified power iteration → eigenvalue estimates → polygon
                                → L^2-optimal $p_k(z)$ → Richardson iteration

Elman, Saad, and Saylor 1986 [240]:
Arnoldi/GMRES → eigenvalue estimates → ellipse → Chebyshev iteration

Elman and Streit 1986 [241]:
Arnoldi/GMRES → eigenvalue estimates → polygon
                                → L^∞-optimal $p_k(z)$ → Horner iteration

Saad 1987 [651]:
Arnoldi/GMRES → eigenvalue estimates → polygon
   → Chebyshev basis → L^2-optimal $p_k(z)$ → 2nd-order Richardson iteration

Saylor and Smolarski 1991 [663, 664]:
Arnoldi/GMRES → eigenvalue estimates → polygon
                                → L^2-optimal $p_k(z)$ → Richardson iteration

Li 1992 [501]:
Arnoldi/GMRES → eigenvalue estimates → polygon
        → conformal map → rational approximation → (k, ℓ)-step iteration

Nachtigal, Reichel, and Trefethen 1992 [570]: Hybrid GMRES
GMRES → $p_k(z)$ → Richardson iteration

Starke and Varga 1993 [720]:
Arnoldi/GMRES → eigenvalue estimates → polygon
        → conformal map → Faber polynomials → Richardson iteration

Manteuffel and Starke 1996 [528]:
Arnoldi/GMRES → numerical range and reciprocal estimates → polygon
        → conformal map → Faber polynomials → Richardson iteration

---

rithm for solving $\mathbf{Ax} = \mathbf{b}$, so that a hybrid algorithm that uses the Arnoldi
iteration in Step 1 can begin the process of convergence toward the solution
of $\mathbf{Ax} = \mathbf{b}$ simultaneously at little additional cost. All of the post-1985
algorithms take advantage of this possibility, and this is the meaning of the
indications 'Arnoldi/GMRES' in Table 27.1. Manteuffel and Starke [528]
use the Arnoldi information not just to estimate eigenvalues but to estimate
the numerical ranges of $\mathbf{A}$ and $\mathbf{A}^{-1}$.

In summary, eigenvalue-based hybrid iterations follow this template:

Step 1: Arnoldi/GMRES for eigenvalues or related estimates,

Step 2: Enclose the eigenvalue estimates by a polygon and solve an approximation problem on that polygon,

Step 3: Richardson or Horner iteration.

Of course this sequence is only a bare outline of how these algorithms are assembled in practice. A robust code will monitor convergence as the computation proceeds and may loop back to Step 1 to acquire more spectral information if necessary. Another issue of crucial practical importance that we shall not consider here is the efficient and stable implementation of Step 3; see [631] for a discussion of Leja ordering of polynomial roots.

This completes our survey of conventional hybrid algorithms, based on eigenvalues. We now come to the reason for including this topic here: the limitations, partly practical and partly conceptual, of using eigenvalues and estimated eigenvalues as a basis for matrix algorithms.

Suppose that one is lucky enough to know the spectrum of **A** *exactly*. If **A** is normal, this is all it takes to design an optimal polynomial iteration, at least in principle, via a problem in complex Chebyshev approximation. If **A** is far from normal, however, knowledge of the spectrum may be of little use. Two matrices with the same eigenvalues may have very different convergence rates for iterative methods, as seen in the previous two sections. On the other hand, in practice one does not know the spectrum of **A** exactly, but must settle for estimates. Obviously this is a further source of error, which may apply regardless of whether **A** is normal.

Figure 27.1 presents an example that can be interpreted as illustrating either of these sources of error. The shaded region is bounded by the lemniscate

$$\{z \in \mathbb{C} : |p(z)| = 4\}, \qquad p(z) = (z-1)(z-5). \qquad (27.1)$$

Suppose **A** is a normal matrix whose spectrum is approximately the boundary of this region. For example, **A** might be a diagonal matrix with eigenvalues closely spaced along the lemniscate. Or, equally well for the purposes of this example, let **A** be a nonnormal matrix with eigenvalues 1 and 5 and pseudospectra closely approximating the lemniscatic region for some small value of ε. A block Toeplitz variant of Theorem 7.2 (see [632]) implies that a matrix with this property can be constructed in the form of a large bidiagonal block Toeplitz matrix with 2 on the superdiagonal and alternating numbers $1, 5, 1, 5, \ldots$ on the diagonal. It is such a matrix with dimension $N = 300$ that is used for Figure 27.2; for pseudospectra of this matrix, look ahead to Figure 27.5.

Since **A** 'lives' on a lemniscate defined by $p(z)$, its optimal iteration polynomials will be normalized powers of $p(z)$, with asymptotic convergence factor $2/\sqrt{5} \approx 0.89$. Figure 27.2 confirms that this is the linear

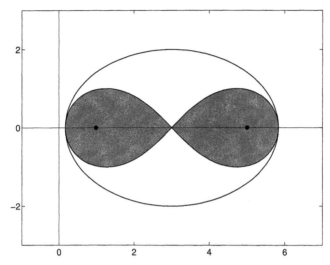

Figure 27.1: If the spectrum or ε-pseudospectra for small ε fill the shaded region (bounded by the lemniscate $\{z \in \mathbb{C} : |(z-1)(z-5)| = 4\}$), ...

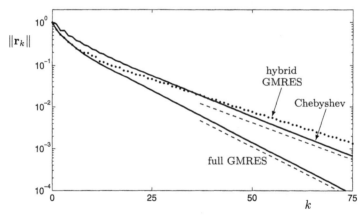

Figure 27.2: ... then a Chebyshev iteration based on the eigenvalue estimates $\{1,5\}$ will converge at a suboptimal rate. The dashed lines show the theoretical linear convergence rates for the Chebyshev iteration and full GMRES (optimal). The solid curves show actual convergence of these algorithms in a numerical experiment involving a 300×300 bidiagonal matrix $\mathbf{A}$ and a random vector $\mathbf{b}$. The dotted curve corresponds to the Hybrid GMRES algorithm, discussed below, starting from the degree-6 GMRES polynomial.

rate at which the full GMRES algorithm converges for this problem, as described in §26.[4] On the other hand suppose than the actual eigenval-

---

[4]To derive this asymptotic rate, we divide the lemniscate level 4 of (27.1) by $|p(0)|$, then take the square root of the result because $p(z)$ is of degree 2.

ues of $\mathbf{A}$ are 1 and 5 and that a hybrid algorithm manages to find these exactly; or suppose that $\mathbf{A}$ is normal with actual eigenvalues along the lemniscate but that the hybrid algorithm estimates these eigenvalues as $\{1, 5\}$, as in principle it might do. (For reasons related to the phenomenon known as 'balayage' in potential theory, it would never do so in practice.) Suppose the hybrid algorithm then uses the estimates $\{1, 5\}$ as the basis of a Step 2/Step 3 iteration. Most of the algorithms in Table 27.1 will take the interval $[1, 5]$ as a set on which to construct iteration polynomials. The Chebyshev iteration is optimal for this set, as it is a real interval. Its asymptotic convergence factor is determined by the smallest ellipse with foci $\{1, 5\}$ that encloses the shaded region in the figure. This number is ≈ 0.92.[5]

We conclude that an eigenvalue-based hybrid iteration for the matrix of Figure 27.1, under certain circumstances, may be expected to be about 40% slower than optimal, as reflected in Figure 27.2. If $\mathbf{A}$ is normal with eigenvalues along the lemniscate, this poor performance can be blamed on imperfect eigenvalue estimates. If it is nonnormal with eigenvalues exactly 1 and 5, it can be blamed on the nonnormality.

These effects may be more pronounced. In Figure 27.3 the shaded region corresponds to the lemniscate

$$\{z \in \mathbb{C} : |p(z)| = \tfrac{256}{27}\}, \qquad p(z) = (z-1)(z-5)^2. \qquad (27.2)$$

Since this lemniscate is better separated from the origin than the last one, the asymptotic convergence factor for the optimal iteration based on powers of the polynomial $p(z)$ improves to ≈ 0.72. However, a hybrid iteration based on the estimate $\sigma(\mathbf{A}) \approx [1, 5]$ in Step 1 will now diverge, because the ellipse that circumscribes the shaded region lies outside the origin. The asymptotic convergence factor is ≈ 1.15, as illustrated in Figure 27.4. (The matrix used to generate this figure has entries 1, 5, and 5 repeated on the main diagonal, with $(256/27)^{1/3}$ on all the first superdiagonal entries; its pseudospectra are shown in Figure 27.5.)

These examples show that there is reason to doubt whether hybrid iterative algorithms ought to work very well in practice. After all, they are based on imperfect estimates of information that is inappropriate to begin with. Nevertheless, they often do work well, and one part of the explanation is a particularly intriguing fact. To a degree, these two sources of error tend to cancel. *Eigenvalue estimates are better than exact eigenvalues!* In practice an Arnoldi process for a matrix like that of Figure 27.1 would virtually never deliver eigenvalue estimates close to $\{1, 5\}$.

---

[5]This figure is obtained by calculating that under the conformal map that carries the exterior of the ellipse with foci $\{1, 5\}$ that passes through the origin onto the exterior of the unit disk, the ellipse that circumscribes the shaded region in Figure 27.1 is carried to the circle of radius approximately 0.92 [215]. The exact number is $(2 + 2\sqrt{2})/(3 + \sqrt{5})$.

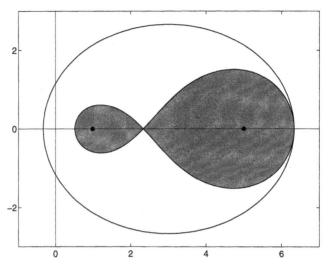

Figure 27.3: If the spectrum or ε-pseudospectra for small ε fill the shaded region (bounded by the lemniscate $\{z \in \mathbb{C} : |(z-1)(z-5)^2| = 256/27\}$), ...

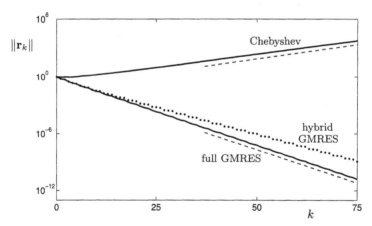

Figure 27.4: ...then a Chebyshev iteration based on the eigenvalues estimates $\{1, 5\}$ will not converge at all. Again, the dashed lines are theoretical estimates and the other curves come from numerical experiments. The Hybrid GMRES curve is again based on the degree-6 GMRES polynomial.

The explanation of this curious phenomenon is related to the robustness of typical eigenvalue estimators. An estimate as simple as Gerschgorin's theorem, for example, may be crude, but it has the virtue that it is insensitive to small perturbations in the matrix [247]. The same conclusion applies to other related estimates, such as the Eneström–Kakeya theorem [7]. In other words, what is conventionally thought of as an estimate of the spectrum is often also a good estimate of pseudospectra. As described in the previous section, knowledge of pseudospectra may be sufficient for the design of good polynomial iterations even when knowledge of the spectrum alone is insufficient. Consequently hybrid iterative algorithms sometimes work better than they 'ought' to, and most of the algorithms listed in Table 27.1 are better than the eigenvalue ideas they are based upon.

Suppose one tried to design a hybrid iterative algorithm from the start without making hidden assumptions of near-normality. In Step 1, one might aim to get a good estimate of an ε-pseudospectrum for some suitable value of ε. One way to do this would be to use a *Chebyshev lemniscate* as the estimated pseudospectral boundary—a level curve of the polynomial $p_k(z)$ implicitly constructed by the Arnoldi iteration (§29). In Step 2, one would then want to compute polynomials that are as small as possible on the region bounded by this lemniscate. But these polynomials are simply $p_k(z)$ itself and its powers! In other words, the lemniscate and indeed the complex plane drop out of the discussion and one is left with an unexpectedly simple algorithm: Construct a polynomial in Step 1; use it over and over again in Step 3.

This is the *Hybrid GMRES algorithm* [570], the only algorithm in Table 27.1 that is not based on estimating eigenvalues or related sets in the complex plane.[6] One point was omitted from the description above: because the iteration polynomials must be normalized by $p_k(0) = 1$, the appropriate polynomial to estimate in Step 1 and then apply in Step 3 is the GMRES polynomial, not the Arnoldi polynomial. Here is the algorithm:

Step 1: GMRES (giving a polynomial $p_k(z)$),

Step 3: Richardson iteration (with the same $p_k(z)$).

Eigenvalues, eigenvalue estimates, and approximation problems have vanished from consideration; there is no Step 2. For details, including a crucial principle of how many GMRES steps to take before switching to Richardson iteration, see [570]. The efficacy of this approach is seen in Figures 27.2 and 27.4. The Hybrid GMRES algorithm, based in these experiments on iterating the degree-6 GMRES polynomial, converges not much more slowly than the more expensive full GMRES method, even in the case where Chebyshev iteration fails. The roots of the Hybrid GMRES polynomials are shown in Figure 27.5 along with the pseudospectra of the associated matrices.

A lesson put forward by the authors of the Hybrid GMRES algorithm is that sometimes the correct response to the limitations of eigenvalue tech-

---

[6]Gene Golub has pointed out to us that, in an article years ahead of its time, I. M. Khabaza proposed a similar algorithm in 1963 [454]. This method explicitly computes the coefficients of what we would now call the GMRES polynomial. Though in some respects this work anticipated the GMRES algorithm that was to follow twenty years later, the numerical implementation is unstable for high-degree polynomials. For whatever reason, Khabaza's paper was overlooked.

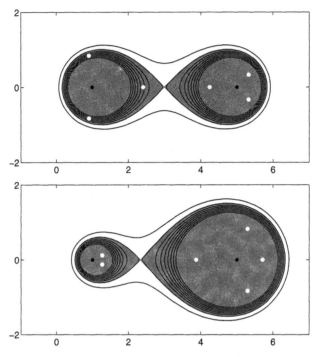

Figure 27.5: Spectra and ε-pseudospectra of the 300×300 bidiagonal matrices used for the computations shown in Figures 27.2 and 27.4; $\varepsilon = 10^{-1}$, 10^{-2}, 10^{-4}, ..., 10^{-16}. The gray regions are bounded by the lemniscates from Figures 27.1 and 27.2. The six white dots in each plot show the roots of a degree-6 GMRES polynomial, which are repeatedly applied to give the Hybrid GMRES convergence curves in the previous figures. The roots are poor eigenvalue estimates, but they lead to fine convergence.

niques may be not to use the complex plane more skillfully, by pseudo-spectra or other means, but to bypass it entirely. But is this method more effective in practice than the other hybrid algorithms summarized in Table 27.1? Unfortunately, practical experience on which to decide this question is lacking. Despite the sizable academic literature on hybrid algorithms, practitioners use restarted GMRES and methods based on Lanczos biorthogonalization much more often.

28 · Arnoldi and related eigenvalue iterations _____

The preceding sections have described how nonnormality can complicate the convergence of iterative methods for solving linear systems of equations. These algorithms are closely related to methods for computing eigenvalues of large, sparse matrices, so one naturally expects nonnormality to play a similar role there too. In fact, eigenvalue algorithms introduce a few new wrinkles.

Over the past dozen years large-scale non-Hermitian eigenvalue computations have become increasingly routine due to improvements in algorithms, software, and computers. Like iterative methods for linear systems, these algorithms construct approximate solutions from low-dimensional subspaces. Methods vary in the subspaces they use, which should contain good approximations to the eigenspace or invariant subspace of interest, but must also be efficiently computable.

The Arnoldi method, a non-Hermitian generalization of the Lanczos algorithm, is the cornerstone of this subject. It draws its approximations from the Krylov subspace

$$\mathcal{K}_k(\mathbf{A}, \mathbf{u}_1) = \text{span}\{\mathbf{u}_1, \mathbf{A}\mathbf{u}_1, \dots, \mathbf{A}^{k-1}\mathbf{u}_1\} \tag{28.1}$$

generated by $\mathbf{A} \in \mathbb{C}^{N \times N}$ and a starting vector $\mathbf{u}_1 \in \mathbb{C}^N$. Proposed in 1951 [12], the Arnoldi algorithm only gained popular attention as an iterative method following an important 1980 paper by Saad [648]. A decade later, Sorensen [709] proposed several key improvements that made it sufficiently robust to allow for implementation in black-box software. The resulting ARPACK subroutine library [494] has enjoyed widespread use among physicists and engineers; it is incorporated in MATLAB via the `eigs` command.

As explained below, the Arnoldi algorithm typically has greatest success in locating eigenvalues on the periphery of the spectrum; in most cases, interior eigenvalues can only be found after much more effort. Thus, an important class of eigenvalue algorithms works not with (28.1), but with a Krylov subspace generated by $(\mathbf{A} - \mu)^{-1}$ for some $\mu \in \mathbb{C}$ near the desired eigenvalues. The eigenvalues of $\mathbf{A}$ nearest to μ are mapped to the largest magnitude eigenvalues of $(\mathbf{A} - \mu)^{-1}$. The simplest method of this type is the *shift-and-invert Arnoldi algorithm*, which applies the usual Arnoldi iteration to $(\mathbf{A} - \mu)^{-1}$. The matrix $\mathbf{A} - \mu$ is often difficult to invert due to the size of $\mathbf{A}$ and its spectral properties; alternative algorithms avoid factorization of $\mathbf{A} - \mu$ in favor of various approximations, which yield subspaces that are not precisely Krylov. One approach is to use 'inexact methods', which iteratively solve equations involving $\mathbf{A} - \mu$ to a prescribed tolerance. Preconditioners can also be applied; see [23, Chap. 11]. Among the most prominent such algorithms are the Davidson [171, 176] and Jacobi–Davidson methods [697]. For detailed information about these and other non-Hermitian eigensolvers, see the surveys [23, 653, 790].

Before beginning our analysis, we must first address a philosophical question: Doesn't the desire to compute eigenvalues of matrices that are far from normal run counter to this book's prevailing theme?

There can be good reasons to seek eigenvalues of nonnormal matrices. First, the eigenvalues significant to the motivating application may be well-conditioned, even though eigenvalues in a distant region of the complex plane are not. One finds such localized nonnormality in applications such as the Gauss–Seidel iteration for the standard discretization of the Laplacian (§24), stable oscillators in laser theory (§60), and transition matrices in Markov chains (§§56 and 57). Alternatively, a set of eigenvalues may be individually ill-conditioned, while the associated invariant subspace is well-conditioned, i.e., nearly orthogonal to the complementary invariant subspace [729]. In such circumstances, the well-conditioned invariant subspace should be computable to good accuracy, even if the individual eigenvalues are not. Even more fundamentally, in many cases of practical interest the nonnormality of a non-Hermitian matrix is strong enough to affect convergence of algorithms but not strong enough to make the eigenvalues useless for applications.

Here we shall focus our attention on the convergence of the standard Arnoldi algorithm; more practical algorithms add further subtleties. We begin by studying the most elementary eigenvalue algorithm of all, the power method. Even in this simple setting, we shall see a number of the primary issues that arise in the Arnoldi convergence theory to follow.

The power method approximates the eigenvector of $\mathbf{A} \in \mathbb{C}^{N \times N}$ associated with the largest magnitude eigenvalue by repeatedly applying $\mathbf{A}$ to a starting vector $\mathbf{u}_1$, then normalizing to prevent under- or overflow. Given the unit-length starting vector $\mathbf{u}_1$, the kth iterate takes the form $\mathbf{u}_k = \mathbf{A}^{k-1}\mathbf{u}_1 / \|\mathbf{A}^{k-1}\mathbf{u}_1\|$. Provided there is a simple eigenvalue λ_1 whose magnitude is larger than all others, and $\mathbf{u}_1$ has a component in the direction of the associated eigenvector $\mathbf{v}_1$, $\mathrm{span}\{\mathbf{u}_k\}$ will converge to $\mathrm{span}\{\mathbf{v}_1\}$ as $k \to \infty$. For simplicity, suppose $\mathbf{A}$ is diagonalizable with eigenvalues $\{\lambda_j\}$ and associated eigenvectors $\{\mathbf{v}_j\}$. Order the eigenvalues by magnitude,

$$|\lambda_1| > |\lambda_2| \geq |\lambda_3| \geq \cdots \geq |\lambda_N|.$$

Decomposing the starting vector in the eigenvector basis,

$$\mathbf{u}_1 = c_1 \mathbf{v}_1 + \cdots + c_N \mathbf{v}_N,$$

one readily sees that

$$\mathbf{A}^k \mathbf{u}_1 = \lambda_1^k c_1 \left(\mathbf{v}_1 + \frac{c_2}{c_1}\left(\frac{\lambda_2}{\lambda_1}\right)^k \mathbf{v}_2 + \cdots + \frac{c_N}{c_1}\left(\frac{\lambda_N}{\lambda_1}\right)^k \mathbf{v}_N \right). \qquad (28.2)$$

This simple expression reveals three critical influences on the convergence behavior: the ratio of the two dominant eigenvalues, $|\lambda_2|/|\lambda_1|$; the bias of the starting vector toward $\mathbf{v}_1$, described by c_1; and nonnormality, reflected by the possibility that $|c_j| \gg \|\mathbf{u}_1\|$ due to cancellation effects.

To illustrate the influence of nonnormality, we consider the matrix

$$\mathbf{A} = \begin{pmatrix} 1 & \alpha & 0 \\ 0 & 3/4 & \beta \\ 0 & 0 & -3/4 \end{pmatrix} \tag{28.3}$$

with eigenvalues $\lambda_1 = 1$ and $\lambda_{2,3} = \pm 3/4$ and eigenvectors

$$\mathbf{v}_1 = \begin{pmatrix} 1 \\ 0 \\ 0 \end{pmatrix}, \qquad \mathbf{v}_2 = \begin{pmatrix} -4\alpha \\ 1 \\ 0 \end{pmatrix}, \qquad \mathbf{v}_3 = \begin{pmatrix} 8\alpha\beta/21 \\ -2\beta/3 \\ 1 \end{pmatrix}.$$

The parameters α and β control nonnormality. We seek $\mathbf{v}_1$ using the power method with starting vector $\mathbf{u}_1 = (1, 1, 1)^{\mathrm{T}}/\sqrt{3}$. When $\alpha \gg 1$, $\mathbf{v}_1$ and $\mathbf{v}_2$ are nearly aligned; when $|\alpha| \ll 1$ and $\beta \gg 1$, $\mathbf{v}_2$ and $\mathbf{v}_3$ are nearly aligned, but nearly orthogonal to $\mathbf{v}_1$. Figure 28.1 illustrates the early iterates of the power method for three different choices of α and β. One observes a curious phenomenon, already apparent from (28.2): When the desired eigenvector $\mathbf{v}_1$ forms a small angle with one of the other eigenvectors, say $\mathbf{v}_2$, the convergence actually seems to *accelerate*. This is because the components of $\mathbf{u}_1$ in these two directions will be exceptionally large in such a case, but nearly cancelling; steps of the power method reduce the cancellation effect, leaving large vector components in these directions. In contrast, when $\mathbf{A}$ is nonnormal due to a small angle between $\mathbf{v}_2$ and $\mathbf{v}_3$, convergence slows. Now it is the components in directions $\mathbf{v}_2$ and $\mathbf{v}_3$ that are large and nearly cancelling, so that when steps of the power method reduce the cancellation, one is left with large components in unwanted directions.

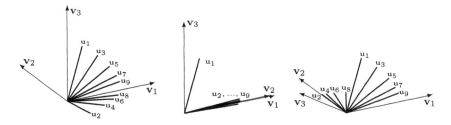

Figure 28.1: Power method iterates $\{\mathbf{u}_1, \ldots, \mathbf{u}_9\}$ for the matrix (28.3) with starting vector $\mathbf{u}_1 = (1, 1, 1)^{\mathrm{T}}/\sqrt{3}$. On the left, $\alpha = \beta = 0$: The problem is normal, and convergence is steady but gradual. In the middle, $\alpha = 10$ and $\beta = 0$: The eigenvectors $\mathbf{v}_1$ and $\mathbf{v}_2$ are nearly aligned, resulting in rapid initial convergence. On the right, $\alpha = 0$ and $\beta = 10$: Now the eigenvectors $\mathbf{v}_2$ and $\mathbf{v}_3$ are nearly aligned, and the power method is biased toward them at early iterations.

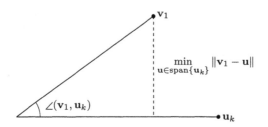

Figure 28.2: The angle between $\mathbf{v}_1$ and $\mathbf{u}_k$.

To obtain general convergence bounds, we shall measure the accuracy of the eigenvector approximation $\mathbf{u}_k$. Assuming that $\|\mathbf{v}_1\| = 1$, we have

$$\sin \angle(\mathbf{v}_1, \mathbf{u}_k) = \min_{\mathbf{u}\in\text{span}\{\mathbf{u}_k\}} \frac{\|\mathbf{v}_1 - \mathbf{u}\|}{\|\mathbf{v}_1\|} = \min_{\gamma\in\mathbb{C}} \|\mathbf{v}_1 - \gamma\mathbf{u}_k\|;$$

see Figure 28.2. To separate the desired eigenvector from the complementary invariant subspace, we shall repeatedly use the *spectral projector* $\mathbf{P}$ associated with λ_1. Note that $\mathbf{P}$ projects onto span$\{\mathbf{v}_1\}$, with Ker $\mathbf{P}$ equal to the complementary invariant subspace. Hence $\mathbf{P}\mathbf{v}_1 = \mathbf{v}_1$, while $\mathbf{P}\mathbf{v}_j = \mathbf{0}$ for $j \neq 1$; further details are provided in §52. Though $\mathbf{P}$ is not in general an orthogonal projector, it always commutes with $\mathbf{A}$, a property we shall repeatedly use. For the 3×3 example (28.3),

$$\mathbf{P} = \begin{pmatrix} 1 & 4\alpha & 16\alpha\beta/7 \\ 0 & 0 & 0 \\ 0 & 0 & 0 \end{pmatrix}.$$

Applying $\mathbf{P}$ to the starting vector $\mathbf{u}_1$ reveals the component of $\mathbf{u}_1$ in the direction $\mathbf{v}_1$: $\mathbf{P}\mathbf{u}_1 = c_1\mathbf{v}_1$, and by our previous assumption, $c_1 \neq 0$. Thus follows an alternative characterization of the convergence angle:

$$\begin{aligned}
\sin \angle(\mathbf{v}_1, \mathbf{u}_k) &= \min_{\gamma\in\mathbb{C}} \|\mathbf{v}_1 - \gamma\mathbf{u}_k\| \\
&= \min_{\gamma\in\mathbb{C}} \|\mathbf{v}_1 - \gamma\mathbf{P}\mathbf{u}_k - \gamma(\mathbf{I} - \mathbf{P})\mathbf{u}_k\| \\
&= \min_{\gamma\in\mathbb{C}} \|\mathbf{v}_1 - \gamma\mathbf{P}\mathbf{A}^{k-1}\mathbf{u}_1 - \gamma(\mathbf{I} - \mathbf{P})\mathbf{A}^{k-1}\mathbf{u}_1\| \\
&= \min_{\gamma\in\mathbb{C}} \|\mathbf{v}_1 - \gamma c_1\lambda_1^{k-1}\mathbf{v}_1 - \gamma(\mathbf{I} - \mathbf{P})\mathbf{A}^{k-1}\mathbf{u}_1\|.
\end{aligned}$$

Taking $\gamma = (c_1\lambda_1^{k-1})^{-1}$ gives

$$\sin \angle(\mathbf{v}_1, \mathbf{u}_k) \leq \frac{\|(\mathbf{I} - \mathbf{P})\mathbf{A}^{k-1}\|}{|c_1||\lambda_1|^{k-1}}. \tag{28.4}$$

Since $\mathbf{P}$ is a spectral projector, $\mathbf{I} - \mathbf{P}$ is the spectral projector for the invariant subspace complementary to $\mathrm{span}\{\mathbf{v}_1\}$. Because this latter matrix commutes with $\mathbf{A}$,

$$(\mathbf{I} - \mathbf{P})\mathbf{A}^{k-1} = \mathbf{A}^{k-1}(\mathbf{I} - \mathbf{P}) = \mathbf{A}^{k-1}(\mathbf{I} - \mathbf{P})^{k-1} = \left(\mathbf{A}(\mathbf{I} - \mathbf{P})\right)^{k-1},$$

provided $k > 1$, and we thus obtain the following convergence bound.

Convergence of the power method

Theorem 28.1 *Let the power method be applied to the matrix $\mathbf{A}$ with starting vector $\mathbf{u}_1$. Suppose λ_1 is a simple eigenvalue of $\mathbf{A}$ with associated eigenvector $\mathbf{v}_1$, and $|\lambda_1| > |\lambda|$ for all other $\lambda \in \sigma(\mathbf{A})$. Then, in the above notation,*

$$\sin \angle(\mathbf{v}_1, \mathbf{u}_k) \le \frac{1}{|c_1|} \frac{\|\mathbf{A}^{k-1}(\mathbf{I} - \mathbf{P})\|}{|\lambda_1|^{k-1}},$$

and, provided $\mathbf{v}_1^* \mathbf{u}_k \ne 0$, the eigenvalue estimate $\theta = \mathbf{u}_k^* \mathbf{A} \mathbf{u}_k$ satisfies

$$|\lambda_1 - \theta| \le 4 \frac{\delta \|\mathbf{A}\|}{\sqrt{1 - \delta^2}},$$

where $\delta = \sin \angle(\mathbf{v}_1, \mathbf{u}_k)$.

Results similar to the first part of this theorem can be found in Saad [653, Thm. 5.2] (in the context of the more general subspace iteration algorithm) and Stewart [727, p. 57]. The second part follows from Jia and Stewart [432, Cor. 4.2].

Bounding convergence of the power method reduces, then, to bounding the norms of the powers of a matrix, a subject addressed at length in §16. (The same problem of decay of powers of $\mathbf{A}(\mathbf{I} - \mathbf{P})$ is fundamental to the convergence of Markov chains, as detailed in §§56 and 57.) Note that the matrix $\mathbf{A}(\mathbf{I} - \mathbf{P})$ has the same spectrum as $\mathbf{A}$, except that λ_1 is moved to zero $(\mathbf{A}(\mathbf{I} - \mathbf{P})\mathbf{v}_1 = \mathbf{A}\mathbf{v}_1 - \mathbf{A}\mathbf{P}\mathbf{v}_1 = 0)$. In fact, if $\mathbf{A}$ is diagonalizable, $\mathbf{A} = \mathbf{V}\mathbf{\Lambda}\mathbf{V}^{-1}$, then $\mathbf{A}(\mathbf{I} - \mathbf{P})$ is the matrix obtained by replacing λ_1 on the diagonal of $\mathbf{\Lambda}$ by zero. For the 3×3 example (28.3),

$$\mathbf{A}(\mathbf{I} - \mathbf{P}) = \begin{pmatrix} 0 & -3\alpha & -16\alpha\beta/7 \\ 0 & 3/4 & \beta \\ 0 & 0 & -3/4 \end{pmatrix}.$$

When $\alpha = 0$, the desired eigenvector $\mathbf{v}_1$ is orthogonal to the other two, and for behavioral purposes $\mathbf{A}(\mathbf{I} - \mathbf{P})$ reduces to a 2×2 matrix. When $|\alpha| \gg 0$, the zero eigenvalue of $\mathbf{A}(\mathbf{I} - \mathbf{P})$ is ill-conditioned, with implications for $\|[\mathbf{A}(\mathbf{I} - \mathbf{P})]^{k-1}\|$.

Bounds on matrix powers transform Theorem 28.1 into a more concrete bound. For diagonalizable $\mathbf{A}$, (16.10) implies

$$\sin \angle(\mathbf{v}_1, \mathbf{u}_k) \leq \frac{\kappa(\mathbf{V})}{|c_1|} \left| \frac{\lambda_2}{\lambda_1} \right|^{k-1}.$$

For any $\mathbf{A}$, we can apply (16.6) to obtain

$$\sin \angle(\mathbf{v}_1, \mathbf{u}_k) \leq \frac{1}{\varepsilon |c_1|} \frac{\rho_\varepsilon (\mathbf{A}(\mathbf{I} - \mathbf{P}))^k}{|\lambda_1|^{k-1}}$$

for all $\varepsilon > 0$, where the pseudospectral radius $\rho_\varepsilon(\mathbf{A}(\mathbf{I} - \mathbf{P}))$ approaches $|\lambda_2|$ as $\varepsilon \to 0$. For a normal matrix, the convergence rate depends upon the relative sizes of the two largest magnitude eigenvalues. For nonnormal problems, these bounds allow for potential stagnation. As observed in §16, the latter bound has the ability to model a convergence rate that improves to $|\lambda_2/\lambda_1|$ as k increases. (Though the bounds can predict transient growth, note that $\sin \angle(\mathbf{v}_1, \mathbf{u}_k) \in [0, 1]$ for all k. The apparent potential for growth reflects a poor choice for γ in (28.4); in those cases, $\gamma = 0$ would be superior.)

The power method serves as a prototype for understanding the convergence behavior of more powerful eigenvalue algorithms. In practice it often fails to be effective, as the desired eigenvalue may be part of a cluster ($|\lambda_2/\lambda_1| \approx 1$) or may not be the largest in magnitude. For this reason one must often turn to more sophisticated algorithms that construct larger approximating subspaces.

One can readily appreciate why the Arnoldi algorithm, with its approximating subspace

$$\mathcal{K}_k(\mathbf{A}, \mathbf{u}_1) = \text{span}\{\mathbf{u}_1, \mathbf{A}\mathbf{u}_1, \ldots, \mathbf{A}^{k-1}\mathbf{u}_1\},$$

should yield an improvement. This Krylov subspace consists of the span of the power method's first k iterates, and thus should be rich in eigenvectors corresponding to the largest magnitude eigenvalues. Furthermore, Krylov subspaces are invariant with respect to shifts in $\mathbf{A}$,

$$\mathcal{K}_k(\mathbf{A}, \mathbf{u}_1) = \mathcal{K}_k(\mathbf{A} + \mu, \mathbf{u}_1),$$

for any $\mu \in \mathbb{C}$; see, e.g., [601]. Hence they should also yield good approximations to any eigenvalues that can be made largest in magnitude through some shift.

As described in §26, the Arnoldi process constructs an orthonormal basis $\{\mathbf{u}_j\}$ for $\mathcal{K}_k(\mathbf{A}, \mathbf{u}_1)$, stored as columns of the matrix $\mathbf{U}_k \in \mathbb{C}^{N \times k}$. Eigenvalue approximations, called *Ritz values*, are obtained by restricting the domain and range of $\mathbf{A}$ to $\mathcal{K}_k(\mathbf{A}, \mathbf{u}_1)$. Specifically, the Ritz values are the eigenvalues $\{\theta_j\}_{j=1}^k$ of the upper Hessenberg matrix

$$\mathbf{H}_k = \mathbf{U}_k^* \mathbf{A} \mathbf{U}_k \in \mathbb{C}^{k \times k}.$$

When $k = N$, this is a unitary similarity transformation, and the Ritz values match the eigenvalues of $\mathbf{A}$ exactly. For large scale problems, one hopes some Ritz values provide good approximations to the desired eigenvalues for $k \ll N$. One can gain some intuition about the location of the Ritz values by noting that they are roots of the polynomial that satisfies the approximation problem

$$\min_{\substack{p \in \mathcal{P}_k \\ p \text{ monic}}} \|p(\mathbf{A})\mathbf{u}_1\|,$$

as described in [344]. For typical $\mathbf{u}_1$ we may expect that the optimal polynomial p will be close to that associated with the 'ideal Arnoldi minimization problem' discussed in the next section,

$$\min_{\substack{p \in \mathcal{P}_k \\ p \text{ monic}}} \|p(\mathbf{A})\|;$$

when the Arnoldi method obtains eigenvalue information about $\mathbf{A}$, it is at least in part because a polynomial that minimizes $\|p(\mathbf{A})\|$ must necessarily have roots that come close to certain eigenvalues. Just as we expect the Ritz values $\{\theta_j\}$ to approximate eigenvalues of $\mathbf{A}$, so we can use the pseudospectra of $\mathbf{H}_k$ or $\widetilde{\mathbf{H}}_k = \mathbf{U}_{k+1}^* \mathbf{A} \mathbf{U}_k \in \mathbb{C}^{(k+1) \times k}$ to approximate those of $\mathbf{A}$ [761]. Figure 28.3 illustrates Ritz values and approximate pseudospectra for a sample nonnormal matrix of dimension $N = 2000$. For further details, see §§40 and 46. Careful non-Hermitian eigenvalue trappers can look to these approximate pseudospectra for insight into the accuracy of their computations, in the same way as one can check a condition number estimate following Gaussian elimination.

If $\mathbf{z}_j$ is the eigenvector of $\mathbf{H}_k$ associated with θ_j, then $\mathbf{y}_j = \mathbf{U}_k \mathbf{z}_j \in \mathbb{C}^N$ is the associated *Ritz vector*. From the fundamental Arnoldi relationship (26.5),

$$\mathbf{A}\mathbf{U}_k = \mathbf{U}_k \mathbf{H}_k + h_{k+1,k} \mathbf{u}_{k+1} \mathbf{e}_k^*,$$

follows a compact formula for the residual norm:

$$\begin{aligned}
\|\mathbf{A}\mathbf{y}_j - \theta_j \mathbf{y}_j\| &= \|(\mathbf{A}\mathbf{U}_k - \mathbf{U}_k \mathbf{H}_k)\mathbf{z}_j\| \\
&= \|h_{k+1,k} \mathbf{u}_k \mathbf{e}_k^* \mathbf{z}_j\| = |h_{k+1,k}||\mathbf{e}_k^* \mathbf{z}_j|.
\end{aligned}$$

With the normalization $\|\mathbf{z}_k\| = 1$, we conclude that $\theta_j \in \sigma_\varepsilon(\mathbf{A})$ for $\varepsilon = |h_{k+1,k}||\mathbf{e}_k^* \mathbf{z}_j| \leq |h_{k+1,k}|$; cf. Theorem 26.1. A small value of $|h_{k+1,k}|$ reflects the fact that $\operatorname{Ran}\mathbf{U}_k$ is nearly an invariant subspace of $\mathbf{A}$, and all the Ritz values may be regarded as accurate, at least in the sense of backward error (see §53). On the other hand, a small value of $|\mathbf{e}_k^* \mathbf{z}_j|$ implies that the Ritz vector $\mathbf{y}_j$ received minimal benefit from the most recent Krylov direction $\mathbf{A}^{k-1}\mathbf{u}_1$. The associated Ritz value is an accurate eigenvalue approximation, though the other Ritz values may be far from accurate.

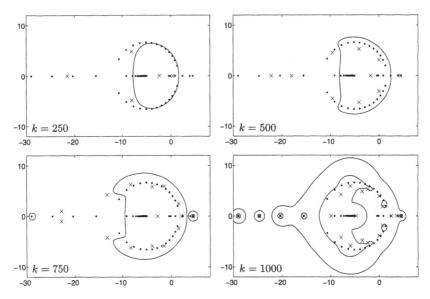

Figure 28.3: Eigenvalues ($\cdot$), Ritz values ($\times$), and approximate ε-pseudospectra ($\varepsilon = 10^0, 10^{-0.5}$) from the Arnoldi iteration for the dimension $N = 2000$ Olmstead matrix from [22, 68]. The spectrum of $\mathbf{A}$ extends far into the left half-plane; these plots show only the rightmost eigenvalues. For approximate pseudospectra, we show $\sigma_\varepsilon(\widetilde{\mathbf{H}}_k)$, guaranteed to approximate $\sigma_\varepsilon(\mathbf{A})$ from below, as described in §§40 and 46. In practice, one would use the restarted Arnoldi iteration to obtain greater accuracy with much smaller subspace dimensions.

Residual norms are the most accessible means for an eigenvalue algorithm to monitor its progress. For example, when ARPACK [494] reports an eigenvalue θ accurate to the user-specified tolerance TOL, a user actually knows that $\theta \in \sigma_\varepsilon(\mathbf{A})$ for all $\varepsilon > \text{TOL}\|\mathbf{A}\|$.

When $\mathbf{A}$ is Hermitian, the Ritz values obey an interlacing property that ensures monotone convergence. For general non-Hermitian matrices, this property is lost, and convergence can be irregular. Defective eigenvalues lead to further complications, with multiple Ritz values converging to a single eigenvalue. Figure 28.4 illustrates how nonnormality can affect the convergence of Ritz values. We seek the rightmost eigenvalue, a simple eigenvalue located in the left half-plane; there are 100 other simple eigenvalues on a vertical line in the complex plane. For a normal matrix, the Ritz values contain excellent approximations to the rightmost eigenvalue for iteration $k = 20$ and beyond. Now suppose the eigenvalues on the vertical line are made nonnormal, though the desired eigenvalue remains perfectly conditioned. The convergence behavior changes completely; though all the eigenvalues are in the left half-plane, the rightmost Ritz value is in the right half-plane for all four iterations shown in the figure. Moreover, one would struggle to identify a particular Ritz value that was converging toward the desired eigenvalue: The Ritz values markedly change from iteration to iteration. Admittedly, this is an extreme example, one in which the rightmost eigenvalue would be of little physical importance due to the nonnormality of the undesired eigenvalues. In practice, one most often sees behavior somewhere between these normal and nonnormal extremes, as in Figure 28.3.

Though the behavior of the Ritz values can be erratic, the convergence of the Arnoldi algorithm is more easily described by the angle between the desired eigenvector and the Krylov subspace. This approach has the advantage that it applies not only to the Arnoldi iteration, but to any method that draws approximations from Krylov subspaces, such as the non-Hermitian Lanczos algorithm. The derivation follows the same tack that led to Theorem 28.1, though now the angle of interest is between the desired eigenvector and an approximating subspace:

$$\sin \angle(\mathbf{v}_1, \mathcal{K}_k(\mathbf{A}, \mathbf{u}_1)) = \min_{\mathbf{u} \in \mathcal{K}_k(\mathbf{A}, \mathbf{u}_1)} \frac{\|\mathbf{v}_1 - \mathbf{u}\|}{\|\mathbf{v}_1\|}.$$

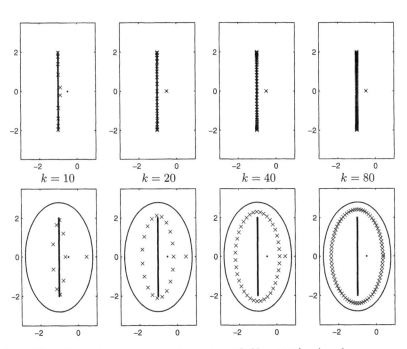

Figure 28.4: Ritz values for a normal matrix with $N = 101$ (top) and a nonnormal variant (bottom) for $k = 10$, 20, 40, and 80. True eigenvalues are shown as small dots, and the Ritz values are marked by ×. The contours on the bottom plots show the boundary of $\sigma_\varepsilon(\mathbf{A})$ for $\varepsilon = |h_{k+1,k}|$. (In all the cases shown, $|h_{k+1,k}| \approx 2$.)

Assume again that λ_1 is a simple eigenvalue with associated unit-length eigenvector $\mathbf{v}_1$ and that the starting vector $\mathbf{u}_1$ is not deficient in the desired eigenvector $\mathbf{v}_1$, i.e., $\mathbf{P}\mathbf{u}_1 = c_1\mathbf{v}_1$ for some $c_1 \neq 0$. We utilize the polynomial characterization of Krylov subspaces,

$$\mathcal{K}_k(\mathbf{A}, \mathbf{u}_1) = \{p(\mathbf{A})\mathbf{u}_1 : p \in \mathcal{P}_{k-1}\},$$

to obtain

$$
\begin{aligned}
\sin \angle(\mathbf{v}_1, \mathcal{K}_k(\mathbf{A}, \mathbf{u}_1)) &= \min_{\mathbf{u} \in \mathcal{K}_k(\mathbf{A}, \mathbf{u}_1)} \|\mathbf{v}_1 - \mathbf{u}\| \\
&= |c_1|^{-1} \min_{\mathbf{u} \in \mathcal{K}_k(\mathbf{A}, \mathbf{u}_1)} \|c_1\mathbf{v}_1 - \mathbf{u}\| \\
&= |c_1|^{-1} \min_{p \in \mathcal{P}_{k-1}} \|c_1\mathbf{v}_1 - p(\mathbf{A})\mathbf{u}_1\| \\
&= |c_1|^{-1} \min_{p \in \mathcal{P}_{k-1}} \|c_1\mathbf{v}_1 - p(\lambda_1)c_1\mathbf{v}_1 - p(\mathbf{A})(\mathbf{I} - \mathbf{P})\mathbf{u}_1\|.
\end{aligned}
$$

We may eliminate terms in this last formula by requiring that $p(\lambda_1) = 1$. This restriction leads to the following characterization due to Saad [648, Prop. 2.1]; see also [727, Thm. 3.10]. (Again, the Ritz value estimate was established in [432, Cor. 4.2]. For further convergence results for Arnoldi Ritz values, see [456].)

Convergence of the Arnoldi iteration

Theorem 28.2 *Let the Arnoldi iteration be applied to the matrix $\mathbf{A}$ with starting vector $\mathbf{u}_1$. Suppose λ_1 is any simple eigenvalue of $\mathbf{A}$ with associated eigenvector $\mathbf{v}_1$. Then, in the above notation,*

$$\sin \angle(\mathbf{v}_1, \mathcal{K}_k(\mathbf{A}, \mathbf{u}_1)) \leq \frac{1}{|c_1|} \min_{\substack{p \in \mathcal{P}_{k-1} \\ p(\lambda_1)=1}} \|p(\mathbf{A})(\mathbf{I} - \mathbf{P})\|, \qquad (28.5)$$

and, provided $\mathbf{v}_1$ is not orthogonal to $\mathcal{K}_k(\mathbf{A}, \mathbf{u}_1)$, there is a Ritz value $\theta \in \sigma(\mathbf{H}_k)$ such that

$$|\lambda_1 - \theta| \leq 4 \left(\frac{\delta\|\mathbf{A}\|}{\sqrt{1-\delta^2}} \right)^{1/k} \left(2\|\mathbf{A}\| + \frac{\delta\|\mathbf{A}\|}{\sqrt{1-\delta^2}} \right)^{1-1/k},$$

where $\delta = \sin \angle(\mathbf{v}_1, \mathcal{K}_k(\mathbf{A}, \mathbf{u}_1))$.

First compare this result to the GMRES optimization problem (26.3). In the present setting, the polynomial is normalized at the eigenvalue λ_1 of $\mathbf{A}$ rather than the origin. This normalization makes it impossible to drive $\|p(\mathbf{A})\|$ to zero, but the presence of the projector $\mathbf{I} - \mathbf{P}$ following the

$p(\mathbf{A})$ in (28.5) removes this obstacle. Also compare Theorem 28.2 to the power method bound in Theorem 28.1. In both bounds, the influence of the starting vector is incorporated via the constant $1/|c_1|$. While this quantity scales the bounds, it does not affect the asymptotic convergence rate. Just as in Theorem 28.1, that rate is governed by a function of $\mathbf{A}(\mathbf{I} - \mathbf{P})$: now a polynomial rather than a simple power.[1] As demonstrated in §26 and elsewhere, the problem of making a polynomial of a matrix small depends on both the spectrum and nonnormality. When $\mathbf{A}$ is diagonalizable, $\mathbf{A} = \mathbf{V}\mathbf{\Lambda}\mathbf{V}^{-1}$,

$$\sin \angle(\mathbf{v}_1, \mathcal{K}_k(\mathbf{A}, \mathbf{u}_1)) \leq \frac{\kappa(\mathbf{V})}{|c_1|} \min_{\substack{p \in \mathcal{P}_{k-1} \\ p(\lambda_1)=1}} \max_{\lambda \in \sigma(\mathbf{A}) \setminus \lambda_1} |p(\lambda)|.$$

Pseudospectral bounds also follow simply. One can write $p(\mathbf{A}(\mathbf{I} - \mathbf{P}))$ as the integral

$$p(\mathbf{A})(\mathbf{I} - \mathbf{P}) = \frac{1}{2\pi i} \int_\Gamma p(z)(z - \mathbf{A})^{-1}\, dz,$$

where Γ is a finite union of positively oriented Jordan curves containing in their collective interior all the eigenvalues of $\mathbf{A}$ *except* λ_1. Suppose that ε is sufficiently small that the component of the pseudospectrum $\sigma_\varepsilon(\mathbf{A})$ containing λ_1 is disjoint from those components containing the other eigenvalues, and suppose that Γ encloses these other components but not λ_1. Our usual procedure for bounding resolvent integrals (see §14) gives

$$\|p(\mathbf{A})(\mathbf{I} - \mathbf{P})\| \leq \frac{L_\varepsilon}{2\pi\varepsilon} \max_{z \in \Gamma} |p(z)|,$$

where L_ε denotes the arc length of Γ. Substituting this bound into Theorem 28.2 leads to a pseudospectral bound on the angle the desired eigenvector makes with the Krylov subspace:

$$\sin \angle(\mathbf{v}_1, \mathcal{K}_k(\mathbf{A}, \mathbf{u}_1)) \leq \frac{L_\varepsilon}{2\pi\varepsilon|c_1|} \min_{\substack{p \in \mathcal{P}_{k-1} \\ p(\lambda_1)=1}} \max_{z \in \Gamma} |p(z)|.$$

The ability of the Krylov subspace to capture the eigenvector $\mathbf{v}_1$ depends on how small polynomials can be on the rest of the (pseudo)spectrum, while remaining normalized at the desired eigenvalue. This indicates that $\mathbf{v}_1$ may be difficult to find when λ_1 is part of a tight cluster, or otherwise located in the interior of the spectrum. The leading constant $|c_1|^{-1} = \|\mathbf{P}\mathbf{u}_1\|^{-1}$ describes bias in the starting vector and depends implicitly on the conditioning of the eigenvalue λ_1. Its presence confirms the notion that convergence should be quicker the richer the starting vector is in the desired eigenvector $\mathbf{u}_1$, though the starting vector does not generally influence the

---

[1]Since $\mathbf{I} - \mathbf{P}$ commutes with $\mathbf{A}$, we can write $p(\mathbf{A})(\mathbf{I} - \mathbf{P}) = p(\mathbf{A}(\mathbf{I} - \mathbf{P}))(\mathbf{I} - \mathbf{P})$.

asymptotic convergence behavior. The eigenvalue condition number $\kappa(\lambda_1)$, discussed in §52, is a measure of the angle formed between left and right eigenvectors:

$$\kappa(\lambda_1) = \frac{\|\mathbf{v}_1\|\|\widehat{\mathbf{v}}_1\|}{|\widehat{\mathbf{v}}_1^*\mathbf{v}_1|},$$

for the left eigenvector $\widehat{\mathbf{v}}_1$ associated with λ_1. One can show that $\|\mathbf{P}\| = \kappa(\lambda_1)$, and thus

$$|c_1| = \|\mathbf{P}\mathbf{u}_1\| \le \|\mathbf{P}\|\|\mathbf{u}_1\| = \kappa(\lambda_1)\|\mathbf{u}_1\|.$$

This expression suggests that the bounds in Theorems 28.1 and 28.2 may improve as λ_1 becomes increasingly ill-conditioned. We observed such counterintuitive behavior in the middle plot of Figure 28.1. Eigenvalue ill-conditioning implies that several eigenvectors are nearly linearly dependent. The Krylov subspace is initially rich in this over-represented direction, leading to a small angle at early iterations.

Our present setting remains several steps removed from practical eigenvalue computations. One often seeks a collection of eigenvalues (and the associated invariant subspace), rather than one particular eigenpair. This is particularly important when the eigenvalue of practical interest is part of a cluster, defective, or otherwise ill-conditioned. In this context the angle we previously studied can be generalized to the *containment gap* between the desired invariant subspace $\mathcal{V}$ and the Krylov subspace:

$$\delta(\mathcal{V}, \mathcal{K}_k(\mathbf{A}, \mathbf{u}_1)) = \max_{\mathbf{v} \in \mathcal{V}} \min_{\mathbf{u} \in \mathcal{K}_k(\mathbf{A}, \mathbf{u}_1)} \frac{\|\mathbf{v} - \mathbf{u}\|}{\|\mathbf{v}\|}.$$

Let $m = \dim(\mathcal{V})$. The analysis in [39, 40] leads to several bounds for this situation. For example,

$$\delta(\mathcal{V}, \mathcal{K}_k(\mathbf{A}, \mathbf{u}_1)) \le C_1 \left(\frac{L_\varepsilon}{2\pi\varepsilon}\right) \min_{p \in \mathcal{P}_{k-2m}} \max_{z \in \Gamma} |1 - \alpha_g(z)p(z)|, \qquad (28.6)$$

where $\alpha_g \in \mathcal{P}_m$ consists of the factors of the minimal polynomial of $\mathbf{A}$ associated with the desired eigenvalues, and the constant C_1 describes the bias of the starting vector:

$$C_1 := \max_{\psi \in \mathcal{P}_{m-1}} \frac{\|\psi(\mathbf{A})\mathbf{P}\mathbf{u}_1\|}{\|\psi(\mathbf{A})(\mathbf{I} - \mathbf{P})\mathbf{u}_1\|}.$$

Here Γ encircles the components of $\sigma_\varepsilon(\mathbf{A})$ associated with the undesired eigenvalues, omitting the eigenvalues associated with $\mathcal{V}$. Again in (28.6), nonnormality associated with the good eigenvalues alone does not appear to influence the bound (aside from an incidental impact on C_1). On the other hand, ill-conditioning of unwanted eigenvalues impedes the predicted convergence, even if the desired eigenvalues are perfectly conditioned.

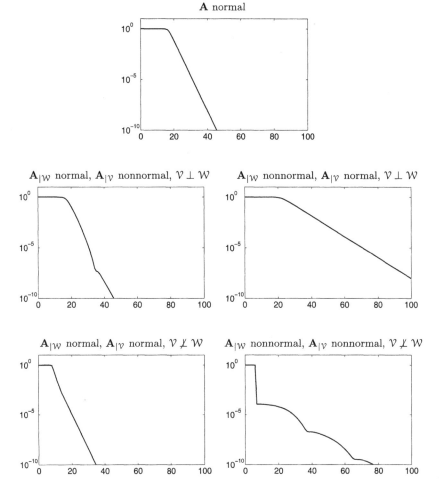

Figure 28.5: Gap convergence to a six-dimensional invariant subspace for five matrices with the same eigenvalues but varying nonnormality. Internal ill-conditioning of the desired eigenvalues (associated with the eigenspace $\mathcal{V}$) does not much affect convergence, but ill-conditioning of the unwanted eigenvalues (associated with the eigenspace $\mathcal{W}$) does. Convergence improves by a constant factor when $\mathcal{V}$ is nearly aligned with $\mathcal{W}$. In each plot, the vertical axis represents $\delta(\mathcal{V}, \mathcal{K}_k(\mathbf{A}, \mathbf{u}_1))$, and the horizontal axis shows the iteration number, k.

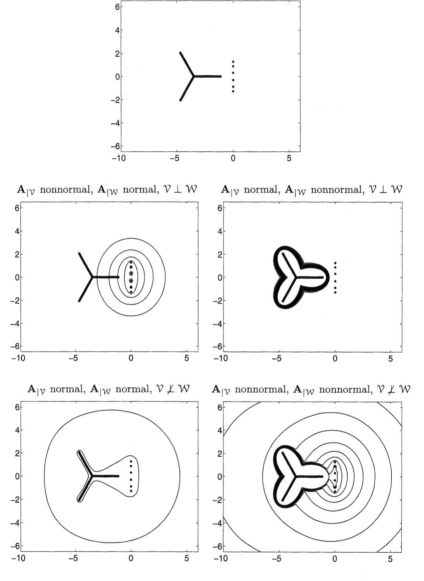

Figure 28.6: Spectra and ε-pseudospectra of the five matrices of Figure 28.5, with $\varepsilon = 10^{-2}, \ldots, 10^{-10}$. (In three of the five plots, the lower level curves are not visible.) We seek the six eigenvalues with zero real part, corresponding to the invariant subspace $\mathcal{V}$.

Figure 28.5 confirms these predictions, illustrating convergence of the containment gap $\delta(\mathcal{V}, \mathcal{K}_k(\mathbf{A}, \mathbf{u}_1))$ for a six-dimensional subspace associated with the rightmost eigenvalues of a matrix of dimension $N = 200$. We create five matrices with the same spectrum, but with other factors varied: the nonnormality of the desired eigenvalues (associated with $\mathcal{V}$), the nonnormality of the remaining eigenvalues (associated with the complementary invariant subspace, $\mathcal{W}$), and the angle between $\mathcal{V}$ and $\mathcal{W}$. Pseudospectra of these matrices are shown in Figure 28.6. As expected, the nonnormality associated with $\mathcal{V}$ has little effect on the convergence, while nonnormality associated with the unwanted eigenvalues slows the rate considerably. When $\mathcal{V}$ and $\mathcal{W}$ form a small angle with one another, the convergence improves by a constant factor, as predicted by a decreased value of C_1 in (28.6). The pseudospectra of $\mathbf{A}$ alone do not reveal everything about convergence; one must also understand how $\mathbf{A}$ behaves on the subspaces $\mathcal{V}$ and $\mathcal{W}$.

Practical eigenvalue algorithms require further refinements still. Construction and storage of the orthogonal basis for $\mathcal{K}_k(\mathbf{A}, \mathbf{u}_1)$ become increasingly expensive as k grows, so one typically fixes a maximal dimension for the approximating subspace and restarts the method with an improved starting vector [648, 709]. Convergence analysis for restarted Krylov subspaces is presented in [39, 40]. Alternative algorithms based on $(\mathbf{A} - \mu)^{-1}$ introduce further complications, not all of which are yet understood.

29 · The Chebyshev polynomials of a matrix _____

The theme throughout this part of the book has been norms of polynomials of matrices, $\|p_k(\mathbf{A})\|$, where k is the polynomial degree. Iterative methods for solving linear systems of equations depend on convergence of $\|p_k(\mathbf{A})\|$ to zero as $n \to \infty$, and as we have seen in the last section, iterative methods for finding eigenvalues also depend on such convergence, typically extracting eigenvalue approximations as roots of p_k. In fact, the Arnoldi iteration for finding eigenvalues can be precisely characterized as a method that finds the sequence of monic polynomials p_k that minimize the successive norms $\|p_k(\mathbf{A})\mathbf{u}_1\|$, where $\mathbf{u}_1$ is a starting vector [344, 649]; Arnoldi reduces to Lanczos if $\mathbf{A}$ is Hermitian [476, 601]. The purpose of this section is to consider the same problem, except minimizing $\|p_k(\mathbf{A})\|$ instead of $\|p_k(\mathbf{A})\mathbf{u}_1\|$. This modified problem is elegant theoretically, and it captures some of the fundamental properties of the convergence of the Arnoldi iteration, which is typically insensitive to the choice of $\mathbf{u}_1$. It also sheds further light on the links between nonnormality of matrices and their behavior.

Chebyshev polynomials of matrices were defined in [344] (under the name 'ideal Arnoldi polynomials') and considered at length in [757, 762].

Chebyshev polynomial of a matrix

Let $\mathbf{A}$ be an $N \times N$ matrix, let k be a nonnegative integer, and let $\|\cdot\|$ denote the matrix 2-norm. The degree-k *Chebyshev polynomial of* $\mathbf{A}$ is the degree-k monic polynomial ϕ_k for which

$$\|\phi_k(\mathbf{A})\| = \text{minimum} \qquad (29.1)$$

or, equivalently,

$$\|\mathbf{A}^k - (c_0\mathbf{I} + c_1\mathbf{A} + \cdots + c_{k-1}\mathbf{A}^{k-1})\| = \text{minimum}. \qquad (29.2)$$

The term 'Chebyshev polynomial' is widely familiar in connection with polynomials of minimal norm on an interval [535], and in approximation theory one also finds this idea generalized to the Chebyshev polynomials of a compact set K in the complex plane, defined as monic polynomials of specified degree with minimax norm on K [656, 815, 822]; this generalization originated with Faber in 1920 [250]. The extension (29.1)–(29.2) to matrices could be regarded as a further generalization, for it is easily shown that if $\mathbf{A}$ is normal, then the degree-k Chebyshev polynomial of $\mathbf{A}$

is the same as the degree-k Chebyshev polynomial in Faber's sense of the spectrum $\sigma(\mathbf{A})$.

The wording of our definition suggests that Chebyshev polynomials of matrices exist and are unique. This is indeed the case, as is spelled out in the following theorem from [344].

Existence and uniqueness of Chebyshev polynomials of a matrix

Theorem 29.1 *A Chebyshev polynomial ϕ_k exists for any $\mathbf{A}$ and k. It is unique if and only if k is less than or equal to the degree of the minimal polynomial of $\mathbf{A}$, hence in particular if $\|\phi_k(\mathbf{A})\| > 0$.*

Proof. The proof of existence is routine if we note that we can describe (29.2) in an abstract vector space as a problem of finding a best approximation of a vector $\mathbf{w}$ (namely $\mathbf{A}^k$) by a vector $\mathbf{v}_k$ in a finite-dimensional subspace $\mathcal{V}$ (spanned by $\mathbf{I}, \mathbf{A}, \ldots, \mathbf{A}^{k-1}$). The existence of such a vector $\mathbf{v}_k$ follows by a standard compactness argument that can be found in any book on approximation theory; see, for example, [144, p. 20] or [516, p. 17].

Since the matrix norm $\|\cdot\|$ is not strictly convex, the proof of uniqueness is not routine. First we note that by the assumption on k, the matrices $\mathbf{I}, \mathbf{A}, \ldots, \mathbf{A}^{k-1}$ are linearly independent. Therefore any $\mathbf{v} \in \mathcal{V}$ has a unique expression as a linear combination of these matrices, and so, proving the uniqueness of ϕ_k is the same as proving uniqueness of $\mathbf{v}_k$. We now argue by contradiction. Suppose that $\widehat{\phi}$ and $\widetilde{\phi}$ are two distinct solutions to (29.1)–(29.2), and let the minimal norm they attain be $\|\widehat{\phi}(\mathbf{A})\| = \|\widetilde{\phi}(\mathbf{A})\| = C$. If we define $\phi = (\widehat{\phi} + \widetilde{\phi})/2$, then $\|\phi(\mathbf{A})\| \leq C$, so we have $\|\phi(\mathbf{A})\| = C$ since $\widehat{\phi}$ and $\widetilde{\phi}$ are minimal. Let $\mathbf{w}_1, \ldots, \mathbf{w}_J$ be a set of maximal right singular vectors for $\phi(\mathbf{A})$, i.e., a set of orthonormal vectors with

$$\|\phi(\mathbf{A})\mathbf{w}_j\| = C, \quad 1 \leq j \leq J,$$

with J as large as possible. For each $\mathbf{w}_j$ we have $\|\widehat{\phi}(\mathbf{A})\mathbf{w}_j\| = \|\widetilde{\phi}(\mathbf{A})\mathbf{w}_j\| = C$ and $\widehat{\phi}(\mathbf{A})\mathbf{w}_j = \widetilde{\phi}(\mathbf{A})\mathbf{w}_j$, for otherwise, by the strict convexity of the norm $\|\cdot\|$ applied to vectors, we would have $\|\phi(\mathbf{A})\mathbf{w}_j\| < C$. Thus

$$(\widetilde{\phi} - \widehat{\phi})(\mathbf{A})(\mathbf{w}_j) = \mathbf{0}, \quad 1 \leq j \leq J.$$

Now since $(\widetilde{\phi} - \widehat{\phi})(z)$ is not identically zero, we can multiply it by a scalar and a suitable power of z to obtain a monic polynomial $\Delta\phi$ of degree k such that $\Delta\phi(\mathbf{A})\mathbf{w}_j = \mathbf{0}$ for $1 \leq j \leq J$. For $\varepsilon \in (0,1)$, consider now the monic polynomial ϕ_ε defined by

$$\phi_\varepsilon(z) = (1 - \varepsilon)\phi(z) + \varepsilon\Delta\phi(z).$$

If $\mathbf{w}_{J+1}, \ldots, \mathbf{w}_N$ denote the remainder of a set of N singular vectors of $\phi(\mathbf{A})$, with corresponding singular values $C > \sigma_{J+1} \geq \cdots \geq \sigma_N \geq 0$, then we have

$$\|\phi_\varepsilon(\mathbf{A})\mathbf{w}_j\| \leq \begin{cases} (1-\varepsilon)C, & 1 \leq j \leq J, \\ (1-\varepsilon)\sigma_{J+1} + \varepsilon\|\Delta\phi(\mathbf{A})\|, & J+1 \leq j \leq N. \end{cases}$$

The first case is $<C$ for arbitrary ε, and the second row is $<C$ for sufficiently small ε, since $\sigma_{J+1} < C$. Since the singular vectors $\mathbf{w}_1, \ldots, \mathbf{w}_N$ form an orthonormal basis, this implies that $\|\phi_\varepsilon(\mathbf{A})\| < C$ for sufficiently small ε, contradicting the assumption that $\widehat{\phi}$ and $\widehat{\phi}$ are minimal.   ∎

Chebyshev polynomials of the familiar kind on $[-1, 1]$ are defined by explicit formulas, and Faber's Chebyshev polynomials of sets $K \subseteq \mathbb{C}$ can be computed by various methods related to the Remez algorithm or to linear or semi-infinite programming [275, 489, 743]. What about computing the Chebyshev polynomials of a matrix $\mathbf{A}$? A powerful technique in this case is *semidefinite programming*, a term that refers to solution of optimization problems in which the objective function or the constraints involve eigenvalues of Hermitian matrices [101, 573, 755]. In (29.1)–(29.2), the aim is to find parameters $c_0, \ldots, c_{k-1}$ to minimize the largest singular value of the matrix $\|p_k(\mathbf{A})\|$ of dimension N; this singular value is the square root of the largest eigenvalue of

$$\begin{pmatrix} \mathbf{0} & p_k(\mathbf{A}) \\ p_k(\mathbf{A})^* & \mathbf{0} \end{pmatrix},$$

a Hermitian matrix of dimension $2N$. Determination of the Chebyshev polynomial of $\mathbf{A}$ thus reduces to a semidefinite programming problem, and powerful primal-dual interior point iterative methods have been developed for solving it numerically. The details are presented in [762], and a MATLAB program called chebymat is included as part of the SDPT3 semidefinite programming package of Toh, Todd, and Tütüncü [759].

For example, chebymat informs us that the Chebyshev polynomials of

$$\mathbf{A} = \begin{pmatrix} 1 & 2 & 3 \\ 4 & 5 & 6 \\ 7 & 8 & 9 \end{pmatrix}$$

are

$$\phi_0(z) = 1, \quad \phi_1(z) = z - 15/2, \quad \phi_2(z) = z^2 - 15z - 9,$$

with

$$\|\phi_0(\mathbf{A})\| = 1, \quad \|\phi_1(\mathbf{A})\| \approx 11.4077, \quad \|\phi_2(\mathbf{A})\| = 9.$$

By looking at matrices that are normal or reasonably close to normal, we can get an indication of how the Arnoldi and Lanczos iterations locate eigenvalues. For example, suppose $\mathbf{A}$ is the 32×32 (normal) diagonal matrix with diagonal entries 1 and $20, 21, \ldots, 50$. For $k = 1, 2, \ldots, 10$, we find that the smallest roots of the Chebyshev polynomials ϕ_k of $\mathbf{A}$ are approximately as follows:

25.5000, 8.1723, 3.6779, 1.7032, 1.1668,

1.0386, 1.0090, 1.0021, 1.0005, 1.0001.

Evidently we have geometric convergence to the smallest eigenvalue, 1. Similarly, suppose we have a 100×100 matrix whose entries are independent, normally distributed real random numbers. Figure 29.1 shows the eigenvalues of a matrix $\mathbf{A}$ of this kind and also the roots of its Chebyshev polynomial of degree 30. Again there is evidence of convergence of the outer roots to the outer eigenvalues of $\mathbf{A}$.

In both of the examples just given, a true Arnoldi iteration, built on a starting vector $\mathbf{q}$, would converge faster, for Chebyshev polynomials of matrices give only partial insight into the convergence of Arnoldi iterations. They do better at explaining failure of convergence in cases that are highly nonnormal.

For an example of this kind, let $\mathbf{A}$ be the Grcar matrix, a Toeplitz matrix with symbol (7.14), of dimension $N = 60$. Figure 29.2 shows spectra and pseudospectra of this matrix in the first panel, and in the other panels it shows roots and lemniscates of Chebyshev polynomials. We must explain what this means. For any polynomial p, a *lemniscate* of p is the curve or union of curves in the complex plane defined by the condition

$$|\phi(z)| = C, \tag{29.3}$$

where C is a constant. Given $\mathbf{A}$, k, and the degree-k Chebyshev polynomial

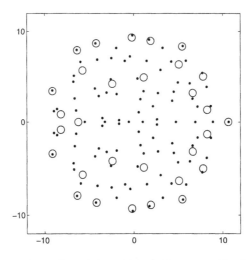

Figure 29.1: Eigenvalues of a random 100×100 matrix (dots) and roots of the corresponding Chebyshev polynomial ϕ_k of degree 30 (circles). The outer roots approximate some of the outer eigenvalues.

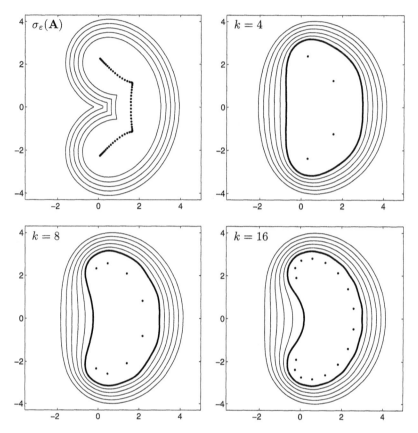

Figure 29.2: Spectra and ε-pseudospectra (first plot) and roots and ρ-lemniscates (other plots) of Chebyshev polynomials for the Grcar matrix of dimension $N = 60$. The dark curves are the Chebyshev lemniscates defined by (29.5), and the light curves are the additional lemniscates given by $|\phi_k(z)|^{1/k} = \rho_k(\mathbf{A}) + \varepsilon$. In both cases ε takes values $\frac{1}{5}, \frac{2}{5}, \frac{3}{5}, \frac{4}{5}, 1$.

ϕ_k of $\mathbf{A}$, we define the *k-capacity* of $\mathbf{A}$ to be the number

$$\rho_k(\mathbf{A}) \equiv \|p_k(\mathbf{A})\|^{1/k}, \tag{29.4}$$

and we define the *degree-k Chebyshev lemniscate* of $\mathbf{A}$ to be the lemniscate given by the condition

$$|\phi_k(z)|^{1/k} = \|p_k(\mathbf{A})\|^{1/k} = \rho_k(\mathbf{A}). \tag{29.5}$$

The Chebyshev lemniscate is the one shown as a dark curve in Figure 29.2. In addition, further lemniscates are shown corresponding to

$$|\phi_k(z)|^{1/k} = \rho_k(\mathbf{A}) + \varepsilon, \quad \varepsilon = \tfrac{1}{5}, \tfrac{2}{5}, \tfrac{3}{5}, \tfrac{4}{5}, 1.$$

We see from the images that the Chebyshev polynomials of $\mathbf{A}$ have little to do with the spectrum, but quite a lot to do with the pseudospectra. The agreement of pseudospectra and lemniscates is not exact in any sense, but it is nevertheless quite striking.

In [341], Greenbaum presents a figure much like our Figure 29.2, except with lemniscates for true Arnoldi polynomials based on $\|p_k(\mathbf{A})\mathbf{u}_1\|$ rather than Chebyshev or ideal Arnoldi polynomials based on $\|p_k(\mathbf{A})\|$.

Figure 29.3 repeats this experiment for the Scottish flag matrix of Figure 9.4, which also is strongly nonnormal. Again we have a remarkable correspondence between lemniscates and pseudospectra, with the eigenvalues nowhere in evidence.

These two examples are exceptionally clean. To illustrate that the connection between pseudospectra and lemniscates is not always so close, our

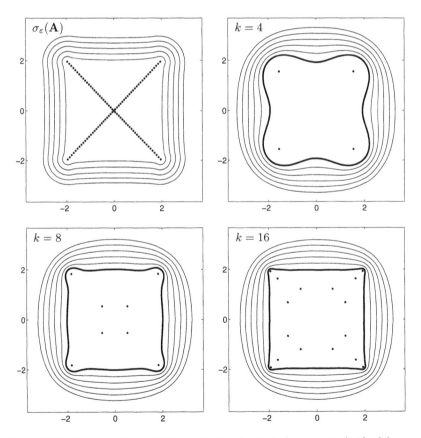

Figure 29.3: Same as Figure 29.2 but for the Scottish flag matrix (9.7) of dimension $N = 101$.

next example is generated with the MATLAB command

$$A = \texttt{triu(rand(60)+1i*rand(60),-2)}. \qquad (29.6)$$

This is a matrix whose entries in the second subdiagonal and above are random complex numbers with real and imaginary parts each uniformly distributed in $[0, 1]$, with zeros below the second subdiagonal. The result in Figure 29.4 shows much greater complexity than before.

A connection can be made between the Chebyshev lemniscates of Figures 29.2–29.4 and the sets known as the *polynomial numerical hulls* of a matrix, defined originally by Nevanlinna [575, 576] and considered subsequently by Greenbaum et al. [251, 340, 341]; see especially [341] for an extensive discussion. We shall take $\| \cdot \| = \| \cdot \|_2$, though other norms can also be used (as indeed is true of Chebyshev polynomials, too).

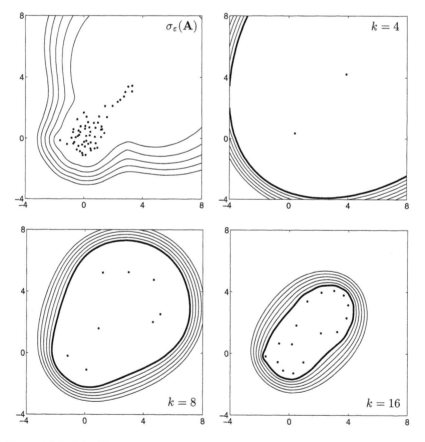

Figure 29.4: Like Figures 29.2 and 29.3 but for the random matrix of dimension $N = 60$ defined by (29.6).

Polynomial numerical hull

Let $\mathbf{A}$ be an $N \times N$ matrix, let k be a nonnegative integer, and let $\|\cdot\|$ denote the matrix 2-norm. The *degree-k polynomial numerical hull* $\mathcal{H}_k(\mathbf{A})$ is the set of all $z \in \mathbb{C}$ such that for every polynomial p of degree k, $|p(z)| \leq \|p(\mathbf{A})\|$.

The results of the following theorem can be found in [341].

Chebyshev lemniscates and polynomial numerical hulls

Theorem 29.2 *Given a matrix* $\mathbf{A}$ *and a nonnegative integer* k, *let* $\mathcal{L}_k(\mathbf{A})$ *be the region bounded by the degree-k Chebyshev lemniscate, let* $\mathcal{H}_k(\mathbf{A})$ *be the degree* k *polynomial numerical hull of* $\mathbf{A}$, *and let* $\sigma(\mathbf{A})$ *denote the spectrum of* $\mathbf{A}$ *as usual. Then*

$$\sigma(\mathbf{A}) \subseteq \mathcal{H}_k(\mathbf{A}) \subseteq \mathcal{L}_k(\mathbf{A}). \tag{29.7}$$

Proof. The first inclusion is a consequence of the spectral mapping theorem. For the second, note that the region $\mathcal{L}_k(\mathbf{A})$ is defined by the condition $|p(z)| \leq \|p(\mathbf{A})\|$ for the particular polynomial $p = \phi_k$, whereas $\mathcal{H}_k(\mathbf{A})$ is defined by the same inequality for *all* polynomials of degree k. The latter is thus a stricter condition, and therefore $\mathcal{H}_k(\mathbf{A}) \subseteq \mathcal{L}_k(\mathbf{A})$. ∎

One might wonder if perhaps $\mathcal{H}_k(\mathbf{A}) = \mathcal{L}_k(\mathbf{A})$ for any $\mathbf{A}$ and k, but this is not so. Indeed, $\mathcal{L}_1$ is always a disk, whereas $\mathcal{H}_1(\mathbf{A}) = W(\mathbf{A})$, the numerical range, is a polygon if $\mathbf{A}$ is normal.

Chebyshev polynomials of matrices have in common with polynomial numerical hulls the property that unlike pseudospectra, they are specially tied to polynomials of $\mathbf{A}$ rather than more general functions. As of this writing, they are easier to compute.[1] We do not know which concept will prove more useful in the long run.

Figure 29.5, in the style of the previous three, presents an example considered by Greenbaum in [341]. The matrix here is the decay matrix $\mathbf{A} = \mathbf{P} - \mathbf{P}^\infty$ associated with the riffle shuffle of a deck of $N = 52$ cards, described in §57. The figure resembles corresponding figures in [341], though those show hulls instead of lemniscates and are computed in the 1-norm (appropriate for the shuffling problem) rather than the 2-norm. For $k = 1, 2, \ldots, 12$ the maximal moduli of all points on the Chebyshev lemniscates $\mathcal{L}_k(\mathbf{A})$ for this matrix are approximately

2.717, 1.390, 1.175, 1.043, .997, .958, .826, .685, .584, .516, .501, .500.

---

[1]In fact, the primary method for computing polynomial numerical hulls requires the *ideal GMRES polynomial* of $z_{j\ell} - \mathbf{A}$ on a grid of points $\{z_{j\ell}\}$ in the complex plane [340]. The ideal GMRES polynomial is similar to the Chebyshev polynomial of a matrix, and its computation is a problem of similar complexity [761].

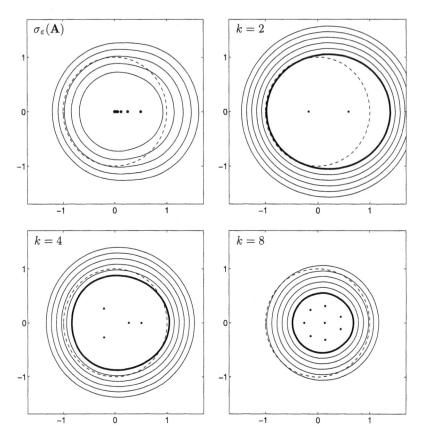

Figure 29.5: Plots as in the previous three figures, but for a riffle shuffle matrix with $N = 52$, following Greenbaum [341]. The dashed curve is the unit circle.

Eventually these numbers converge to $1/2$, the largest eigenvalue of $\mathbf{A}$. But not until $n = 7$ do they fall much below 1, and this is related to the Bayer–Diaconis phenomenon that 'it takes seven shuffles to randomize a deck of cards', discussed in §57.

VII. Numerical Solution
of Differential Equations

30 · Spectral differentiation matrices _____

Much of scientific computing depends on processes of numerical differentiation, in which a derivative of a function is replaced by some kind of discrete approximation. Specifically, in the field of spectral collocation methods,[1] derivatives may be computed numerically by multiplying a vector of data by a *differentiation matrix* [99, 127, 282, 775]. If the underlying grid is not periodic, these differentiation matrices are typically nonnormal, and the departure from normality may grow exponentially as a function of the number of grid points. In such cases the nonnormality may have a big effect on numerical stability and behavior of the methods. This was one of the applications that contributed to the rise of interest in nonnormality and pseudospectra among numerical analysts in the early 1990s [625, 770, 782].

Let us consider the classic example of Chebyshev spectral differentiation on the interval $[-1, 1]$. Let N be a positive integer, and let $x_0, \ldots, x_N$ be the $N + 1$ *Chebyshev points*

$$x_j = \cos(j\pi/N), \qquad j = 0, 1, \ldots, N. \tag{30.1}$$

Note that the points are numbered from right to left for convenience and that they cluster near ± 1. For example, the Chebyshev points for $N = 8$ and 24 are shown in Figure 30.1.

Figure 30.1: Chebyshev points in $[-1, 1]$ for $N = 8$ and 24.

Spectral differentiation proceeds as follows. Given a function u defined on the grid $\{x_j\}$, we obtain a discrete derivative w by differentiating a polynomial interpolant:

- *Let p be the unique polynomial of degree $\leq N$ with $p(x_j) = u_j$, $0 \leq j \leq N$.*
- *Set $w_j = p'(x_j)$.*

We do not construct $p(x)$ explicitly, but instead note that since the differentiation operator is linear, it can be represented by multiplication by an

_____
[1]Spectral collocation methods are also known, confusingly, as *pseudospectral methods*, the idea being to distinguish them from the alternative *spectral Galerkin* methods, based on integrals rather than point evaluations. There is no connection between that use of the term pseudospectral and the one that is the basis of this book.

$(N + 1) \times (N + 1)$ matrix $\mathbf{D}_N$:

$$\mathbf{w} = \mathbf{D}_N \mathbf{u}.$$

The entries of $\mathbf{D}_N$ can be determined analytically, and spectral methods manipulate these matrices explicitly to solve problems of ODEs, PDEs, and integral equations.[2]

For example, for $N = 1$, the interpolation points are $x_0 = 1$ and $x_1 = -1$, and the differentiation process takes the form

$$\begin{pmatrix} w_0 \\ w_1 \end{pmatrix} = \begin{pmatrix} \frac{1}{2} & -\frac{1}{2} \\ \frac{1}{2} & -\frac{1}{2} \end{pmatrix} \begin{pmatrix} u_0 \\ u_1 \end{pmatrix}.$$

For $N = 2$ the points are $x_0 = 1$, $x_1 = 0$, and $x_2 = -1$, and we have

$$\begin{pmatrix} w_0 \\ w_1 \\ w_2 \end{pmatrix} = \begin{pmatrix} \frac{3}{2} & -2 & \frac{1}{2} \\ \frac{1}{2} & 0 & -\frac{1}{2} \\ -\frac{1}{2} & 2 & -\frac{3}{2} \end{pmatrix} \begin{pmatrix} u_0 \\ u_1 \\ u_2 \end{pmatrix}.$$

Note that these matrices are neither symmetric nor skew-symmetric. The formulas for general N can be found on page 410 and are embodied in the following MATLAB function adapted from [775]:

```
function [D,x] = cheb(N)
x = cos(pi*(0:N)/N)';
c = [2; ones(N-1,1); 2].*(-1).^(0:N)';
X = repmat(x,1,N+1); dX = X-X';
D = (c*(1./c)')./(dX+(eye(N+1)));
D = D - diag(sum(D,2));
```

A simple theorem establishes that there is a departure from normality.[3]

Theorem 30.1 *For any N, $\|\mathbf{D}_N\| > N^2/3$ but $(\mathbf{D}_N)^{N+1} = 0$.*

Proof. The upper left corner entry of $\mathbf{D}_N$ is $(2N^2 + 1)/6$, and as this is greater than $N^2/3$, the same must be true of $\|\mathbf{D}_N\|$. As for the nilpotency, the definition of $\mathbf{D}_N$ implies that for any vector $\mathbf{u}$, $(\mathbf{D}_N)^{N+1}\mathbf{u}$ is the vector obtained by interpolating $\mathbf{u}$ by $p(x)$, differentiating $N + 1$ times, and evaluating the result on the grid. Since $p(x)$ is a polynomial of degree at most N, the result will always be zero, regardless of $\mathbf{u}$. Thus $(\mathbf{D}_N)^{N+1}$ is the zero matrix. ∎

---

[2] Alternatively, the implementation sometimes involves the Fast Fourier Transform rather than explicit matrix manipulation.

[3] Though mathematically nilpotent, these matrices do not behave as such on a computer in finite precision arithmetic; see page 165.

In fact, the norms grow approximately like $\|\mathbf{D}_N\| \sim 0.5498 N^2$ as $N \to \infty$ [770]. This scaling suggests that the cleanest results will be obtained if we consider the matrices $\mathbf{A}_N = N^{-2}\mathbf{D}_N$. Figure 30.2 shows pseudospectra of these matrices for $N = 8$ and $N = 24$. The departure from normality is pronounced. As $N \to \infty$, such figures appear to converge to a limit in which $\|(x - \mathbf{A}_\infty)^{-1}\|$ grows roughly in proportion to $(1.8)^{1/x}$ as $x \to 0$.

Theorem 30.1 and Figure 30.2 describe a spectral differentiation process with no boundary conditions. In applications, however, boundary conditions are usually present. In the simplest case, suppose the condition $u_0 = 0$ is imposed at the point $x = 1$, which would be the 'inflow' point for the PDE $u_t = u_x$. This condition can be imposed in the spectral method by deleting the first row and the first column of $\mathbf{D}_N$, producing a square matrix $\widetilde{\mathbf{D}}_N$ of dimension N instead of $N + 1$. Figure 30.3 shows the effect on the spectra and pseudospectra of $\widetilde{\mathbf{A}}_N = N^{-2}\widetilde{\mathbf{D}}_N$. The matrix is no longer nilpotent; the eigenvalues are all nonzero. Most of them shrink to the origin as $N \to \infty$ in this normalization, but as they shrink, they leave behind certain 'outliers' of size $\mathcal{O}(1)$. This limit was considered in [625]. The largest of the outliers has magnitude approximately 0.0886 as $N \to \infty$, corresponding to $0.0886 N^2$ before normalization.

In applications, spectral differentiation aims to provide a 'spectrally accurate' approximation of exact differentiation—meaning that for smooth functions, the errors decrease faster than any power of N as $N \to \infty$, and for analytic functions, they decrease exponentially. An aspect of the spectral accuracy is revealed graphically in the portion in Figure 30.3 near the origin. As far as the eye can see, the boundaries of the pseudospectra are exactly vertical in this fascinating corner of the plot, as the enlargement of Figure 30.4 confirms. Now for the differentiation operator d/dx on $[-1, 1]$ with boundary condition $u(1) = 0$, as shown in §5, the pseudospectra are precisely half-planes bounded by vertical lines. Thus we see here a

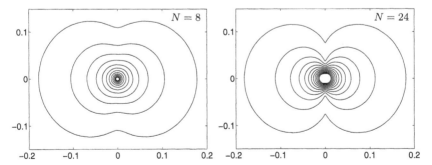

Figure 30.2: ε-pseudospectra of the normalized Chebyshev spectral differentiation matrices $N^{-2}\mathbf{D}_N$ for $N = 8$ and 24, $\varepsilon = 10^{-2}$, 10^{-4}, 10^{-6}, ..., 10^{-14}. These matrices are nilpotent, so their eigenvalues are all zero.

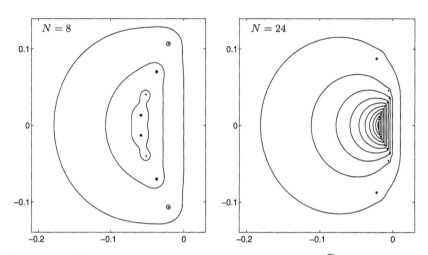

Figure 30.3: Same as Figure 30.2, but for the matrices $N^{-2}\widetilde{\mathbf{D}}_N$ obtained from $N^{-2}\mathbf{D}_N$ by removing a row and a column to impose a boundary condition.

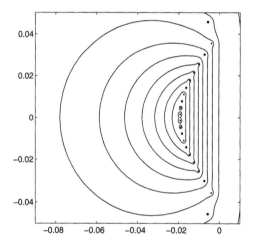

Figure 30.4: Closeup of the second plot of Figure 30.3.

beautiful example of 'spectrally accurate pseudospectra'. The eigenvalues, by contrast, do not approximate anything in the original problem at all; they are an epiphenomenon.

In 1986, before the significance of the nonnormality of spectral differentiation matrices had been recognized, Tal-Ezer published an intriguing paper based in part on ideas of Dubiner [741]. Tal-Ezer proposed that the

$\mathcal{O}(N^2)$ eigenvalues of the matrices $\widetilde{\mathbf{D}}_N$ could be shrunk to $\mathcal{O}(N)$ by making a small change: replacing the Chebyshev grid (30.1) by a *Legendre grid* consisting of the point $x = 1$ together with the zeros of the degree N Legendre polynomial $P_N(x)$. This reduction in eigenvalues would have huge advantages in applications, for it would permit an increase in stable time step sizes for time-dependent PDE simulations from $\mathcal{O}(N^{-2})$ to $\mathcal{O}(N^{-1})$.

Figure 30.5 repeats Figure 30.3 for matrices $\mathbf{A}_N = N^{-2}\widetilde{\mathbf{D}}_N$ based on these Legendre points. We see that Legendre grids do indeed have smaller eigenvalues: The outliers are gone. Their pseudospectra, however, are no smaller than for Chebyshev grids. Thus here is a case in which the scaling of the spectra differs from that of the pseudospectra by a whole factor of N. Which of the two is it that matters in applications? Do the Legendre grids really permit increased time steps? As we shall see in the next two sections, in general the answer is no.

Figure 30.6, a closeup of Figure 30.5, reveals much the same behavior as before: vertical lines that appear perfectly straight. Evidently, although Chebyshev and Legendre grids differ in their eigenvalues, their pseudospectra near the origin agree to spectral accuracy.

The comparison of the nonnormal properties of Chebyshev and Legendre spectral differentiation matrices began with a paper in 1987 by Trefethen and Trummer [782], who noted that the eigenvalues of these matrices were so sensitive to perturbations that it could make an important difference whether they were computed in single precision or double precision. Trefethen and Trummer did not investigate pseudospectra, and as a re-

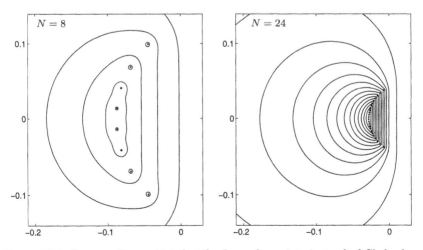

Figure 30.5: Same as Figure 30.3, but for Legendre points instead of Chebyshev points.

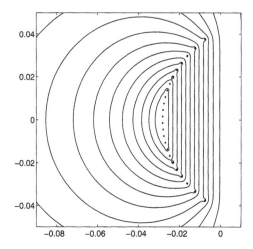

Figure 30.6: Closeup of the second plot of Figure 30.5.

sult overestimated the significance of machine precision. Soon afterward it became clear that the computed eigenvalues in question were lining up nicely near the boundary of the ε-pseudospectrum, with ε on the order of machine epsilon, but that deeper aspects of the matrices in question might be necessary to explain their behavior in applications.

We close by mentioning an interesting theoretical problem where much work remains to be done. Do pseudospectra of spectral approximations really converge to those of corresponding differential or integral operators? How fast? Under what conditions? For general studies of spectral approximation for differential and integral equations, see [2, 139]. Several theorems have been published that describe the convergence of the pseudospectra of operator approximations [375, 836], though it remains to be seen if spectral discretizations fit into such frameworks. For many operators, the numerical evidence of convergence is convincing; see §43.

31 · Nonmodal instability of PDE discretizations _____

In this section we describe the phenomenon of nonmodal instability of certain discretizations of time-dependent partial differential equations. In the next, we present theorems to characterize discretizations that are free of this effect.

As computers began to appear in the late 1940s and early 1950s, they were employed almost immediately for the solution of time-dependent PDEs by finite difference discretizations. It was quickly discovered that some discretizations were *stable* and others were *unstable*, giving useless results. For example, consider the first-order wave equation

$$\frac{\partial u}{\partial t} = \frac{\partial u}{\partial x}, \qquad x \in (-1, 1), \quad t \geq 0, \tag{31.1}$$

together with initial data

$$u(x, 0) = \begin{cases} \cos^2(\pi(x - \frac{1}{4})), & |x - \frac{1}{4}| \leq \frac{1}{2}, \\ 0, & \text{otherwise,} \end{cases}$$

and boundary data $u(1, t) = 0$ for all t. Suppose this initial value problem is approximated numerically on a regular Δx–Δt grid by centered differences in x and the third-order Adams–Bashforth formula in t; see, e.g., [426]. If u_j^n denotes the discrete approximation at $x = -1 + j\Delta x$, $t = n\Delta t$, then the spatial discretization is

$$\frac{\partial u}{\partial x}(j\Delta x, n\Delta t) \approx \mathbf{D} u_j^n, \tag{31.2}$$

where $\mathbf{D}$ is the tridiagonal matrix defined by[1]

$$\mathbf{D} u_j^n = \frac{u_{j+1}^n - u_{j-1}^n}{2\Delta x}, \tag{31.3}$$

and the full space-time discretization is

$$u_j^{n+1} = u_j^n + \Delta t\, \mathbf{D}(\tfrac{23}{12}u_j^n - \tfrac{16}{12}u_j^{n-1} + \tfrac{5}{12}u_j^{n-2}). \tag{31.4}$$

To be specific, let us take $N = 60$, $\Delta x = 2N^{-1}$, $u_N^n = 0$, $u_0^n = u_1^n$, and initial values u_j^0, u_j^1, u_j^2 ($1 \leq j \leq N - 1$) taken from the exact solution $u(x, t) = u(x + t, 0)$. Figure 31.1 shows computed results with $\Delta t = 0.6\Delta x$ and $\Delta t = 1.2\Delta x$. For the first choice, the computation is stable, generating the correct solution, a wave propagating left at speed 1. For the second, it is unstable. The discretization errors introduced at each step are not

_____
[1]In principle one should write $(\mathbf{D}u^n)_j$, but the notation $\mathbf{D}u_j^n$ is standard in the field of finite difference methods.

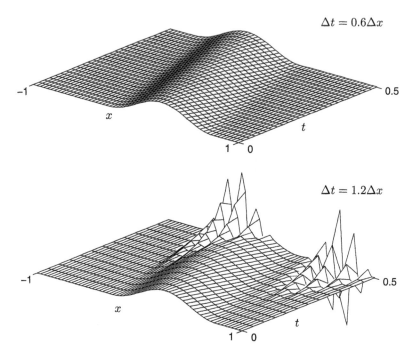

Figure 31.1: Stable and unstable finite difference approximations of (31.1). This is a modal instability, associated with a sawtoothed eigenvector (four grid points per wavelength) with eigenvalue of modulus > 1. At $t = 1$, the amplitude exceeds 10^5.

much bigger than before, but they compound exponentially in successive time steps until the wave is obliterated by a sawtooth oscillation. If this computation is continued for a longer time, the values u_j^n are greater than 10^5 at $t = 1$ and close to 10^{15} at $t = 2$.

The standard technique for explaining the instability of finite difference formulas was developed by von Neumann and others and described in a 1951 paper of O'Brien, Hyman, and Kaplan [583].[2] 'Von Neumann analysis'

---

[2]With hindsight, it was seen that the roots of von Neumann analysis lie earlier, before the invention of computers. A 1928 paper of Courant, Friedrichs, and Lewy showed that a discrete approximation to a wave equation cannot converge to the correct solution in general as the mesh is refined unless Δt and Δx satisfy a certain inequality [167]. This 'CFL condition' was justified on the basis of numerical and mathematical domains of dependence, not instability, but it was later realized that when it is not satisfied, the failure of the numerical scheme does indeed take the form of instability. Even earlier, L. F. Richardson solved PDEs by finite differences around 1910 [637] and foresaw the era of numerical weather prediction [638]. Among other schemes, Richardson proposed a 'leap frog' formula for the heat equation that we now recognize as unstable. Fornberg [282, pp. 58–60] points out that Richardson computed just five time steps with this scheme and saw no problems, but if he had been able to take 25 steps, he would have seen a result much like the lower plot of Figure 31.1.

is another term for discrete Fourier analysis. One begins by noting that if we ignore the complication of boundary conditions and imagine that the domain is unbounded, then any initial condition for the finite difference formula can be written as a superposition of waves

$$u_j^n = \lambda^n \mathrm{e}^{\mathrm{i}jk\Delta x} \tag{31.5}$$

for real *wave numbers* k and corresponding *amplification factors* (i.e., eigenvalues) $\lambda = \lambda(k)$. If $|\lambda| > 1$ for some k, we have exponential growth and instability. In the case of the formula (31.4), inserting (31.5) and cancelling the common factor $\mathrm{e}^{\mathrm{i}jk\Delta x}$ yields the cubic equation

$$\lambda - 1 = \frac{\mathrm{i}(23 - 16\lambda^{-1} + 5\lambda^{-2})}{12} \frac{\Delta t}{\Delta x} \sin(k\Delta x).$$

It can be shown that for $\Delta t/\Delta x = 0.6$, or indeed for any value of $\Delta t/\Delta x$ less than about 0.724, all solutions of this equation satisfy $|\lambda| \leq 1$, whereas for $\Delta t/\Delta x = 1.2$, this is no longer true, and the numerical solutions blow up. The wave number that blows up fastest is $k = \frac{1}{2}\pi/\Delta x$, i.e., four grid points per wavelength, and the corresponding eigenvalue is $\lambda \approx 0.28 + 2.09\mathrm{i}$, with $|\lambda| \approx 2.11$.

By now it should be clear why the subject of stability of discretizations of PDEs is a part of this book. Discretizations of linear time-dependent PDEs are discrete time dynamical systems, and the problem of numerical stability is a problem of ensuring that these dynamical systems cannot amplify rounding or truncation errors explosively. This much was evident to experts by the end of the 1940s.

Then, in the 1950s, a general mathematical theory was developed by Lax and Richtmyer, now often called the theory of *Lax-stability* [488, 639]. Lax and Richtmyer observed that the problem of stability is the problem of bounding the norms of powers of certain discrete solution operators. They developed conditions that characterize *convergence* of discrete approximations to the correct solution as the mesh is refined (ignoring rounding errors, and assuming the underlying problem is linear and well-posed). The main result is the famous *Lax Equivalence Theorem*, which states that if the discrete approximation is *consistent*, meaning that it approximates the right PDE as $\Delta x \to 0$ and $\Delta t \to 0$, then

$$\text{convergence} \iff \text{stability}.$$

Here 'stability' means that the solution operators are uniformly bounded as the time and space grid sizes approach zero. For example, if $\mathbf{S}_{\Delta t}$ is an operator that advances the numerical solution for a time-independent PDE from one time step to the next on a grid associated with time step Δt, then stability is the condition

$$\|(\mathbf{S}_{\Delta t})^n\| \leq C(n\Delta t) \tag{31.6}$$

for some fixed function $C(t)$, uniformly for all n as $\Delta t, \Delta x \to 0$.

All this is a bit vague, omitting various details necessary for a mathematically precise statement. One detail is that (31.6) does not involve a single function space or norm, but a set of spaces and norms that must converge to a continuous limit as $\Delta t \to 0$; the solution operator $\{\mathbf{S}_{\Delta t}\}$ will involve matrices of growing dimensions as $\Delta t \to 0$. Another is that a multistep numerical process such as (31.4) does not simply advance from one step to the next, but combines several steps together. For a full treatment of such matters, see [639]. The essential point, however, is clear: Convergence of PDE discretizations depends on norm-boundedness of families of matrices. If the powers are bounded, the computation is stable. The problem of characterizing power-bounded families of matrices led to the Kreiss matrix theorem of 1962 (§18) [465].

We can now explain how von Neumann analysis fits into the general theory of Lax-stability. A priori, the question of stability requires the analysis of families of matrices, and eigenvalue analysis alone could never give bounds on norms of powers of arbitrary families of matrices. In the special case of constant-coefficient problems on regular grids, however, the Fourier transform takes what would be families of matrices of unbounded dimensions in space into families of matrices of a fixed dimension, indexed over wave numbers. The transformation is unitary, and as a consequence, eigenvalue analysis of the resulting matrices is enough to ensure stability. For practical problems involving boundaries or variable coefficients, further theorems have been proved to show that von Neumann analysis still gives the correct results provided certain additional assumptions are satisfied such as smoothness of coefficients [639, 748].

On the other hand, there are some discretizations of PDEs that are fundamentally not translation-invariant. For these, von Neumann analysis is inapplicable, and instabilities may appear that are nonmodal in nature. A vivid example of such effects was presented in a paper by Parter from 1962 [603]. More extreme examples of nonmodal instabilities arise in the field of spectral methods for the solution of PDEs [127, 775], introduced in the last section. The remainder of this section is devoted to exploring an example of this kind, taken from [770, 782]. We shall look at the same example from another angle in the next section.

Let us consider (31.1) again, but now discretize x by spectral differentiation in the unevenly spaced grid of N *Legendre points* $\{x_j\}$, i.e., roots of the Legendre polynomial of degree N, as described in the last section. Such discretizations were proposed in [217] and [740]. Given N data points u_j on this grid, we construct their approximate spatial derivative by interpolating these data, as well as the boundary value 0 at $x = 1$, by a single global polynomial p_N of degree N, and then differentiating this interpolant.

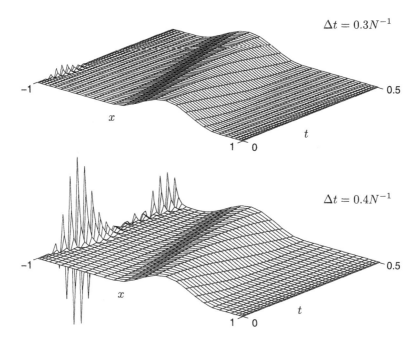

Figure 31.2: Legendre spectral approximation of the same problem (31.1) on a spatial grid with $N = 30$. Now there is a nonmodal, transient instability associated with a pseudoeigenvector localized at the boundary. For slightly larger time steps the solution becomes rapidly worse than shown here.

Thus (31.2) is replaced by

$$\frac{\partial u}{\partial x}(x_j, n\Delta t) \approx \mathbf{D} u_j^n = p_N'(x_j), \qquad (31.7)$$

where $\mathbf{D}$ is the Legendre spectral differentiation matrix. Figure 31.2 now reveals quite a different kind of instability than we saw in Figure 31.1. As $t \to \infty$, on a fixed grid, things do not look too bad. For finite t, however, large transient errors appear near the boundary $x = -1$. This is a perfect example of transient growth and eventual decay of a highly nonnormal linear, time-independent dynamical system.

We can reduce this multistep Adams–Bashforth formula to a one-step process by forming the $3N \times 3N$ block matrix

$$\mathbf{S} = \begin{pmatrix} \mathbf{I} & \mathbf{0} & \mathbf{0} \\ \mathbf{I} & \mathbf{0} & \mathbf{0} \\ \mathbf{0} & \mathbf{I} & \mathbf{0} \end{pmatrix} + \frac{\Delta t}{12} \begin{pmatrix} 23\mathbf{D} & -16\mathbf{D} & 5\mathbf{D} \\ \mathbf{0} & \mathbf{0} & \mathbf{0} \\ \mathbf{0} & \mathbf{0} & \mathbf{0} \end{pmatrix}, \qquad (31.8)$$

which maps $(\mathbf{u}^n, \mathbf{u}^{n-1}, \mathbf{u}^{n-2})^{\mathrm{T}}$ to $(\mathbf{u}^{n+1}, \mathbf{u}^n, \mathbf{u}^{n-1})^{\mathrm{T}}$. Stability now becomes a matter of norms of powers of $\mathbf{S}$, and these are plotted in Fig-

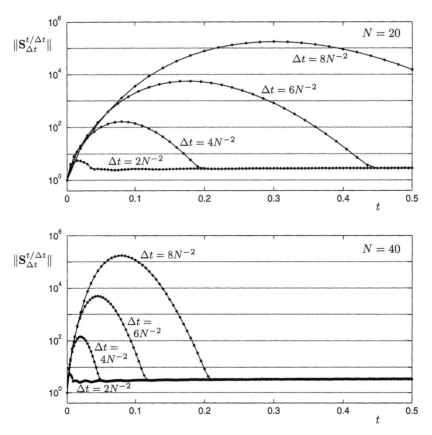

Figure 31.3: Norms of powers of the compound matrix $\mathbf{S}$ of (31.8) for the Legendre spectral discretization of (31.1) with a nonmodal instability. The scheme is Lax-stable if $\Delta t = \mathcal{O}(N^{-2})$ as $N \to \infty$, otherwise Lax-unstable.

ure 31.3. This figure reveals that if $\Delta t = \mathcal{O}(N^{-2})$ as $N \to \infty$, then the maximal norm, though possibly large, is uniformly bounded for all N. The discretization is Lax-stable, and the numerical solution will converge to the exact solution in the absence of rounding errors. If $\Delta t \neq \mathcal{O}(N^{-2})$ as $N \to \infty$, on the other hand, there will be Lax-instability and no convergence. In particular, a choice such as $\Delta t = 0.4 N^{-1}$ will be catastrophic, even though the eigenvalues in that case remain inside the unit disk for all N.

Figure 31.4 shows the pseudospectra of $\mathbf{S}$ for the particular choice $N = 20$ and $\Delta t = 0.4 N^{-1} = 8 N^{-2}$ (thus $\mathbf{S}$ has dimension 60). Around most of the unit circle, the resolvent norm is of modest size, but in the region $z \approx -1$ it takes values beyond 10^{6}, making it clear that there must be large transient growth. Since the boundary of the 10^{-6}-pseudospectrum crosses

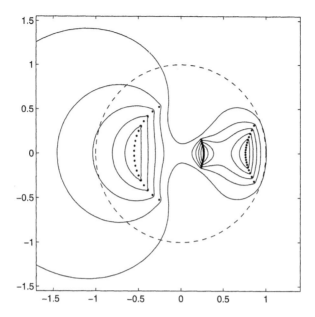

Figure 31.4: ε-pseudospectra of the matrix $\mathbf{S}$ of (31.8) with $N = 20$, $\Delta t = 0.4N^{-1}$ for $\varepsilon = 10^{-2}, 10^{-4}, \ldots, 10^{-10}$. The dashed curve is the unit circle. The large resolvent norms for $z \approx -1$ imply that the powers of this matrix must grow transiently by a factor greater than 10^4 before eventually decaying, as seen in Figure 31.3.

the real axis near $z = -1.035$, Theorem 16.4 implies that there must be transient growth of the powers $\|\mathbf{S}^n\|$ by a factor of at least $10^6 \times 0.035 = 3.5 \times 10^4$. From Figure 31.3 it is evident that the actual growth is about six times greater than this.

Could one use a discretization of this kind for large-t simulations, since the instability is transient and dies away eventually? At a glance it might seem so, but as Kreiss emphasized in 1962 [465], the instability can only be expected to be transient for a purely constant-coefficient linear problem in the absence of rounding errors. As soon as variable coefficients or nonlinearities or other perturbations are introduced, the loss of convergence is likely to become global. Just as Richardson failed to recognize his modal instability in 1910, so Dubiner [217], Tal-Ezer [740], and Trefethen and Trummer [782] all failed to recognize this nonmodal instability during 1986–87.

32 · Stability of the method of lines _____

The last section showed that for discretizations of time-dependent PDEs, the proper location of eigenvalues is not enough in general to ensure stability, and hence convergence to the correct solution, as the mesh is refined. Here we show that for many such problems, the correct condition for stability is that *the pseudospectra of the spatial discretization operator must be contained in the stability region of the time-stepping formula.* More precisely, the ε-pseudospectrum must lie within a distance $\mathcal{O}(\varepsilon)$ of the stability region as $\varepsilon \to 0$. For stability over $t \in [0, T]$ instead of $[0, \infty)$, this condition loosens to $\mathcal{O}(\varepsilon) + \mathcal{O}(\Delta t)$ as $\Delta t \to 0$ and $\varepsilon \to 0$.

The problems in question are finite difference, finite element, or spectral discretizations of time-dependent PDEs for which one can separate the treatments of space and time. Suppose we have a time-dependent PDE and discretize it with respect to the spatial variables. The result is a system of coupled ODEs, one for each grid point for a simple finite-difference discretization of a scalar problem. Suppose we then discretize this system in time by a numerical formula for ODEs such as an Adams or Runge–Kutta formula. This two-step process is called the *method of lines*, an allusion to the idea that after discretization in space, one has a system of coupled problems whose solutions are defined along 'lines' in the t direction [424].

Here is an example. Suppose we have the first-order wave equation on the domain $[-\pi, \pi]$ with periodic boundary conditions,

$$\frac{\partial u}{\partial t} = \frac{\partial u}{\partial x}, \qquad x \in [-\pi, \pi], \tag{32.1}$$

together with initial data $u(x, 0) = f(x)$. If we discretize by a one-sided 'upwind' approximation in x, the result is the system of ODEs

$$\frac{du_j(t)}{dt} = \frac{u_{j+1}(t) - u_j(t)}{\Delta x} \tag{32.2}$$

indexed over appropriate values of j. One way to solve a system of ODEs $du/dt = f(u)$ numerically is by the forward Euler formula,

$$\mathbf{u}^{n+1} = \mathbf{u}^n + \Delta t\, f(\mathbf{u}^n). \tag{32.3}$$

For our particular choice of f defined by (32.2), we obtain

$$u_j^{n+1} = u_j^n + \frac{\Delta t}{\Delta x}(u_{j+1}^n - u_j^n). \tag{32.4}$$

This discretization in x and t can be described as the method of lines discretization of (32.1) obtained by combining forward ('upwind') differencing in x with forward ('Euler') differencing in t.

At a practical computational level, it is common to use the method of lines to reduce a PDE to a set of ODEs that can then be treated by black-box adaptive software. At a theoretical level, one may hope that to determine stability and convergence of a fully discrete PDE discretization such as (32.4), one may be able to use the well-established theory of discretizations of ODEs worked out by Dahlquist and others beginning in the 1950s. That is the theme of this section.

The key idea is the notion of a stability region [365, 473]. Suppose a discrete ODE formula with time step Δt is applied to the linear scalar constant-coefficient model problem

$$\frac{du}{dt} = \lambda u, \tag{32.5}$$

where $\lambda \in \mathbb{C}$ is a constant. This process amounts to a linear recurrence relation whose order depends on the number of steps or stages of the ODE scheme. The *stability region* for the ODE formula is the set of points in the $\lambda \Delta t$-plane for which this recurrence is stable in the sense that all its solutions are bounded as $n \to \infty$. Figure 32.1 shows stability regions for three of the most important families of methods. In the field of numerical solution of ODEs, stability regions are used to assess whether a discretization is likely to behave successfully or not, especially for stiff ODEs, i.e., those with widely disparate time scales (see §33). Many method of lines discretizations of PDEs lead to stiff systems of ODEs, because the space derivatives lead to low- and high-wave number components that evolve on different time scales. For such problems, the standard rule of thumb is that *a discretization is stable if all eigenvalues of the spatial discretization operator (scaled by Δt) lie in the stability region of the time discretization operator.*

Let us work out the example (32.4). By considering Fourier modes

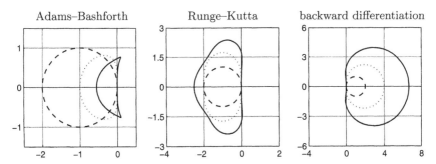

Adams–Bashforth   Runge–Kutta   backward differentiation

Figure 32.1: Stability regions in the $\lambda \Delta t$-plane for ODE formulas of orders 1 (dashed), 2 (dotted), and 3 (solid). For the Adams–Bashforth and Runge–Kutta formulas, the stability regions are the interiors of the curves drawn. For the backward differentiation formulas, they are the exteriors.

$u_j^n = \lambda^n \exp(\mathrm{i}kj\Delta x)$ for arbitrary wave numbers k, we find that the space discretization is a normal operator with eigenvalues

$$\lambda = \frac{e^{\mathrm{i}k\Delta x} - 1}{\Delta x}.$$

(More precisely, since the domain $[-\pi, \pi]$ is bounded, the wave numbers are restricted to the integers, but as $\Delta x \to 0$ this makes little difference.) Multiplying by Δt gives the scaled eigenvalues

$$\lambda = \frac{\Delta t}{\Delta x}(e^{\mathrm{i}k\Delta x} - 1),$$

a set of numbers that lie along the circle of radius $\Delta t / \Delta x$ centered at $-\Delta t / \Delta x$. On the other hand, the stability region for the forward Euler formula (32.3) is the disk of radius 1 centered at -1. Comparing these computations, we find that (32.4) can be expected to be stable provided that $\Delta t / \Delta x \leq 1$, i.e., $\Delta t \leq \Delta x$; see Figure 32.2. A more rigorous analysis confirms that this is indeed the correct stability condition for (32.4).

Figure 32.3 considers a second example, the finite difference discretization of $\partial u/\partial t = \partial u/\partial x$ with nonperiodic boundary conditions described at the beginning of §31. The time discretization is now carried out by the third-order Adams–Bashforth formula, which has a more interesting stability region. The spatial difference is centered rather than upwind, leading to eigenvalues on an imaginary interval (approximately) rather than a circle.

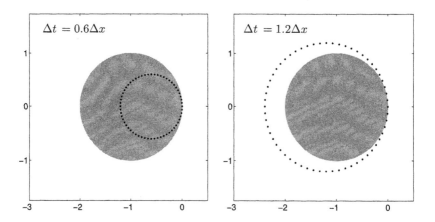

Figure 32.2: Eigenvalues of Δt times the spatial difference operator (32.2) superimposed on the stability region (shown in gray) of the forward Euler formula (32.3) for the finite difference approximation (32.4), for two different choices of Δt. On the left, the eigenvalues fit in the stability region and the computation is stable. On the right it is unstable.

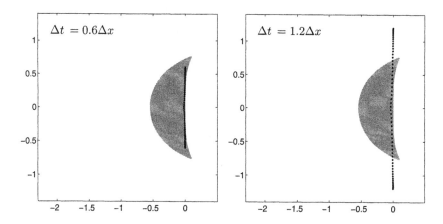

Figure 32.3: Eigenvalues of Δt times the spatial difference operator $\mathbf{D}$ of (31.3) superimposed on the stability region of the third-order Adams–Bashforth formula for the finite difference approximation (31.2) that generated Figure 31.1. Again, the computation is stable on the left and unstable on the right.

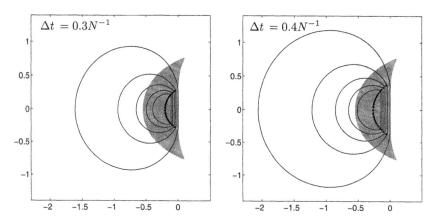

Figure 32.4: Like Figure 32.3 but for the Legendre spectral discretization (31.7) that generated Figure 31.2. Now ε-pseudospectra of $\Delta t \mathbf{D}$ are shown for $\varepsilon = 10^{-2}, 10^{-4}, \ldots, 10^{-10}$ as well as eigenvalues, since this discretization is far from normal. The slight splitting of the eigenvalues is caused by rounding errors, which were investigated in [782].

In the figure we see that the choice $\Delta t = 0.6\Delta x$ appears stable, whereas $\Delta t = 1.2\Delta x$ appears unstable, confirming the results of Figure 31.3. The reason our prediction is correct is that although this spatial discretization matrix is not normal, it is close to normal.

Figure 32.4, by contrast, shows the situation for the highly nonnormal

Legendre spectral discretization of Figure 31.2. We see now that for both choices $\Delta t = 0.3N^{-1}$ and $\Delta t = 0.4N^{-1}$, the eigenvalues are well within the stability region. However, the pseudospectra protrude decisively outside, and it is hardly surprising that there should be some significant transient instability, just as observed in §31.

In a pair of papers 1990 and 1992, Reddy and Trefethen proved a set of theorems that make these connections precise [625, 626], and related results have been published by many authors. The remainder of this section summarizes the results of [625, 626], together with indications of how these results have been sharpened by later authors. Some of our wording is taken directly from [626]. For a different presentation of much the same material, see [425].

Consider a time-independent linear evolution equation

$$\frac{\partial u}{\partial t} = \mathbf{L}u, \quad u(x,0) = f(x), \quad t \in [0,T] \text{ or } [0,\infty), \qquad (32.6)$$

where $\mathbf{L}$ is a linear differential operator, which may incorporate boundary conditions, $T > 0$, and u is a scalar or vector function of t and of one or more space variables x. (A time-dependent forcing term could also be included, but this drops out in the analysis of propagation of errors.) Equation (32.6) is first approximated with respect to the space variables by finite differences, finite elements, or spectral methods on a discrete grid, transforming the PDE into the system of ODEs,

$$\frac{\partial \mathbf{u}}{\partial t} = \mathbf{L}_{\Delta t}\mathbf{u}, \quad \mathbf{u}(0) = \mathbf{f}_{\Delta t}, \qquad (32.7)$$

where $\mathbf{u}(t)$ is a vector of dimension $N_{\Delta t} \leq \infty$ and $\mathbf{L}_{\Delta t}$ is a matrix or bounded linear operator. At this stage the subscript Δt is an arbitrary positive parameter that determines the spatial grid in an unspecified manner. The semidiscretization (32.7) is then approximated with respect to t by a linear multistep, Runge–Kutta, or more general one-step formula with time step Δt. If we write $\mathbf{u}^n \approx \mathbf{u}(n\Delta t)$, the resulting full discretization becomes

$$\mathbf{v}^{n+1} = \mathbf{A}_{\Delta t}\mathbf{v}^n = G(\Delta t\,\mathbf{L}_{\Delta t})\mathbf{v}^n, \qquad (32.8)$$

with appropriate initial conditions. For a one-step time integration formula, $\mathbf{v}^n = \mathbf{u}^n$, while for an s-step formula we define

$$\mathbf{v}^n = \begin{pmatrix} \mathbf{u}^n \\ \mathbf{u}^{n-1} \\ \vdots \\ \mathbf{u}^{n-s+1} \end{pmatrix}. \qquad (32.9)$$

The function G characterizes the time integration formula. For a linear multistep method, $G(w)$ is a companion matrix. Its entries are affine and rational functions of w for explicit and implicit multistep methods, respectively. For Runge–Kutta or one-step methods, $G(w)$ is a polynomial or rational function that approximates e^w for $w \approx 0$.

The full discretization (32.8) is defined to be *Lax-stable* if

$$\|\mathbf{A}_{\Delta t}^n\| \le C(n\Delta t) \quad \text{for all } n \text{ and } \Delta t \text{ with } 0 \le n\Delta t \le T \qquad (32.10)$$

for some fixed function $C(\Delta t)$ and all sufficiently small Δt. The Lax Equivalence Theorem states that (32.10) is a necessary and sufficient condition for convergence of the discrete approximation as $\Delta t \to 0$, assuming that the initial-value problem (32.6) is well-posed and that the discretization (32.8) is consistent [639]. Throughout this discussion, $\|\cdot\|$ denotes the weighted vector 2-norm defined by a nonsingular weight matrix $\mathbf{W}$, $\|\mathbf{x}\| = \|\mathbf{x}\|_{\mathbf{W}} = \|\mathbf{W}\mathbf{x}\|_2$, and also the corresponding matrix norm $\|\mathbf{E}\| = \|\mathbf{E}\|_{\mathbf{W}} = \|\mathbf{W}\mathbf{E}\mathbf{W}^{-1}\|_2$. The matrix $\mathbf{W}$ depends on the grid, and hence on Δt, in a fashion that in principle is arbitrary. In applications, if $\mathbf{W}$ is not the identity, it will typically be a discrete diagonal approximation to a smooth weight function such as a Jacobi or Laguerre weight; see §43.

We shall state two theorems from [626], one for $t \in [0, \infty)$ and one for $t \in [0, T]$. These theorems apply to both one-step and multistep formulas, but the details of the assumptions made in the two cases differ.

First consider the one-step case. If a one-step formula is applied to the model problem (32.5), then at each step the result is multiplied by $\phi(\lambda \Delta t)$, where ϕ is a rational function. The formula can be written

$$\mathbf{v}^{n+1} = \phi(\Delta t \, \mathbf{L}_{\Delta t})\mathbf{v}^n = \mathbf{A}_{\Delta t}\mathbf{v}^n, \quad \mathbf{v}^0 = \mathbf{f}_{\Delta t}. \qquad (32.11)$$

We assume that ϕ is analytic in a neighborhood of the spectrum of $\Delta t \, \mathbf{L}_{\Delta t}$, which ensures that $\phi(\Delta t \, \mathbf{L}_{\Delta t})$ is well defined. The stability region S is defined by

$$S = \{w \in \mathbb{C} : |\phi(w)| \le 1\}.$$

We make the following assumption about S (but will discuss later how such assumptions can be loosened).

Assumption A1. S is bounded and $\phi'(w) \ne 0$ for $w \in \partial S$.

This condition excludes many common implicit one-step methods with unbounded stability regions, such as the first-order backward Euler formula and other A-stable methods.

Assumption A2. There exists a nonempty domain $V \subseteq \mathbb{C}$ and a constant $M < \infty$ such that $\|(\mu - \Delta t \, \mathbf{L}_{\Delta t})^{-1}\| \le M$ for all $\mu \in V$ and all Δt.

Uniform boundedness of $\{\Delta t \, \mathbf{L}_{\Delta t}\}$ is enough to imply this condition.

Next consider multistep formulas, whether explicit or implicit. An s-step linear multistep approximation to the semidiscretization (32.2) can be written in the form

$$\sum_{j=0}^{s} \alpha_j \mathbf{v}^{n+j} - \Delta t \sum_{j=0}^{s} \beta_j \mathbf{L}_{\Delta t}\, \mathbf{v}^{n+j} = 0 \tag{32.12}$$

and is characterized by the polynomials

$$\rho(z) = \sum_{j=0}^{s} \alpha_j z^j, \quad \sigma(z) = \sum_{j=0}^{s} \beta_j z^j \tag{32.13}$$

with the convention

$$\alpha_s = 1, \quad |\alpha_0| + |\beta_0| > 0.$$

By introducing the vector $\mathbf{v}$ in (32.9), the full discretization (32.12) can be written in the compact form (32.8) with

$$\mathbf{A}_{\Delta t} = G(\Delta t\, \mathbf{L}_{\Delta t}) = \begin{pmatrix} \mathbf{a}_{s-1} & \cdots & \mathbf{a}_1 & \mathbf{a}_0 \\ \mathbf{I} & & & \\ & \ddots & & \\ & & \mathbf{I} & \end{pmatrix}. \tag{32.14}$$

Here

$$\mathbf{a}_j = (\mathbf{I} - \beta_s \Delta t\, \mathbf{L}_{\Delta t})^{-1}(\beta_j \Delta t\, \mathbf{L}_{\Delta t} - \alpha_j \mathbf{I}) \in \mathbb{C}^{N_{\Delta t} \times N_{\Delta t}}$$

for $0 \le j \le s - 1$, and $\mathbf{I}$ is the identity operator of dimension $N_{\Delta t}$.

The stability region S of the linear multistep formula is the set of $w \in \mathbb{C}$ for which all roots z_j of the stability polynomial $\pi_w(z) = \rho(z) - w\sigma(z)$ satisfy $|z_j| \le 1$, and any root with $|z_j| = 1$ is simple. This region can be characterized in terms of the image of the unit circle under the rational function

$$r(z) = \frac{\rho(z)}{\sigma(z)}.$$

Again we need some assumptions.

Assumption B1. S is bounded, with $r(z) \ne \infty$ for $|z| = 1$;

Assumption B2. $r'(z) \ne 0$ for $|z| = 1$.

For explicit formulas the condition $r(z) \ne \infty$ for $|z| = 1$ implies that S is bounded, but this does not hold for implicit methods. Assumption B2 implies that the stability region does not have cusps; generalizations for regions with cusps are given in [626].

Here are our fundamental stability theorems.

Stability of the method of lines on $[0, \infty)$

Theorem 32.1 *Let* (32.8) *be the method of lines discretization of* (32.6) *based upon a time integration formula with stability region S satisfying Assumptions* A1–A2 *(one-step formula) or* B1–B2 *(multistep). If*

$$\|\mathbf{A}_{\Delta t}^n\| \leq C_1 \quad \forall n \geq 0 \qquad (32.15)$$

for all sufficiently small Δt, then the ε-pseudoeigenvalues μ_ε of the operators $\Delta t\, \mathbf{L}_{\Delta t}$ satisfy

$$\mathrm{dist}\,(\mu_\varepsilon, S) \leq C_2 \varepsilon \quad \forall \varepsilon \geq 0. \qquad (32.16)$$

Conversely, (32.16) *implies*

$$\|\mathbf{A}_{\Delta t}^n\| \leq C_3 \min\{N_{\Delta t}, n\} \quad \forall n > 0 \qquad (32.17)$$

for all sufficiently small Δt. Here C_1, C_2, and C_3 are constants independent of Δx and Δt.

Stability of the method of lines on $[0, T]$

Theorem 32.2 *Let* (32.8) *be the method of lines discretization of* (32.6) *based upon a time integration formula with stability region S satisfying Assumptions* A1–A2 *or* B1–B2. *If*

$$\|\mathbf{A}_{\Delta t}^n\| \leq C_1, \qquad 0 \leq n\Delta t \leq T, \qquad (32.18)$$

then the ε-pseudoeigenvalues μ_ε of the operators $\Delta t\, \mathbf{L}_{\Delta t}$ satisfy

$$\mathrm{dist}\,(\mu_\varepsilon, S) \leq C_2 \varepsilon + C_3 \Delta t \quad \forall \varepsilon \geq 0. \qquad (32.19)$$

Conversely, (32.19) *implies*

$$\|\mathbf{A}_{\Delta t}^n\| \leq C_4 \min\{N_{\Delta t}, n\}, \qquad 0 < n\Delta t \leq T. \qquad (32.20)$$

Here C_1, C_2, C_3, and C_4 are constants independent of Δx and Δt.

We shall not give the proofs of these theorems, which are complicated. However, let us indicate the kinds of arguments involved. The key idea is to transplant the ODE formula and stability region to a problem of powers of matrices and the unit disk, where the Kreiss Matrix Theorem (§18) and the estimates described in §16 can be employed. The factor $\min\{N_{\Delta t}, n\}$ is thus the same one that we saw in (14.27) and (14.28). The crucial matter in this transplantation is the relationship between two resolvent norms:

$$\|(\mu - \Delta t \, \mathbf{L}_{\Delta t})^{-1}\| \quad \leftrightarrow \quad \|(\lambda - \mathbf{A}_{\Delta t})^{-1}\|.$$

On the left, μ is a point outside the stability region S and $\mathbf{L}_{\Delta t}$ is the operator defining the semidiscretization (32.7). On the right, λ is a point outside the unit disk and $\mathbf{A}_{\Delta t}$ is the operator defining the equivalent one-step process (32.8).

In words, if $\{\Delta t \, \mathbf{L}_{\Delta t}\}$ satisfies the Kreiss condition with respect to S (i.e., at most inverse-linear blow-up of the resolvent norm as $\mu \to S$), does $\{\mathbf{A}_{\Delta t}\}$ satisfy the Kreiss condition with respect to the unit disk—and conversely? To prove the theorems, we need to bound each of these resolvents in terms of the other. In [625, 626] this is done with arguments involving contour integrals and partial fractions. Various further results and estimates are also presented in these papers that do not appear in Theorems 32.1–32.2, including more information about the constants C_j.

We have emphasized the results of [625] and [626] rather than those of many other authors because we know them best, and because they are very general. Briefly, however, here are a few remarks about some of these other contributions up to 1992, which on the whole tend to concentrate on sufficient rather than necessary conditions for stability, based, for example, on some variety of contractivity. Brenner and Thomée use the Hille–Phillips operational calculus to prove a stability result for A-stable one-step formulas applied to a special class of operators $\mathbf{L}_{\Delta t}$ [108]. Spijker, Lenferink, and Kraaijevanger consider more general one-step time integration formulas, but restrict attention to operators satisfying a *circle condition*: $\|\Delta t \, \mathbf{L}_{\Delta t} + \rho \mathbf{I}\| \le \rho$ for some $\rho > 0$ [463, 495, 714]. Sanz-Serna and Verwer derive sufficient conditions for stability based on contractivity and *C-stability* for more general PDEs, both linear and nonlinear [659, 805]. Di Lena and Trigiante focus on the notion of the *spectrum of a family of matrices* (§6), showing that a necessary condition for stability is that the spectrum of the family $\{\Delta t \, \mathbf{L}_{\Delta t}\}$ must be contained in the stability region S [204]. Lenferink and Spijker obtain several results based on the convex set known as the *M-numerical range* of $\Delta t \, \mathbf{L}_{\Delta t}$ [496]. Lubich and Nevanlinna show stability for A-stable one-step and multistep formulas applied to problems where the pseudospectra of $\{\Delta t \, \mathbf{L}_{\Delta t}\}$ lie within a distance $\mathcal{O}(\varepsilon)$ of the left half-plane [518]. Kreiss and Wu derive sufficient conditions for stability of one-step and multistep formulas in norms exponentially weighted in t [470]. They show that the full method of lines discretization is stable if the semidiscretization is stable and the full discretization is *locally stable* in the sense that the open half-disk $\{w : \mathrm{Re} \, w < 0, \, |w| < \|\Delta t \, \mathbf{L}_{\Delta t}\|\}$ is contained in the stability region.

Since 1992, refinements of Theorems 32.1 and 32.2 have appeared, especially in work by Spijker and coauthors [425, 716]. Spijker and Straetemans point out that these theorems are not restricted to a weighted 2-norm; any matrix norm induced by a vector norm will do. In the direction of sufficient conditions for stability, i.e., (32.16)⇒(32.17) and (32.19)⇒(32.20), Theorem 3.2 of [716] shows that Assumptions A1 and A2 can be dispensed with for one-step schemes, and Theorem 3.2 of [425] shows that Assumptions B1 and B2 can be dispensed with for multistep schemes, provided that S is a closed subset of $\mathbb{C} \cup \{\infty\}$. Theorems 2.2 and 3.4 of [425] also show that the factor $\min\{N_{\Delta t}, n\}$ of (32.17) and (32.20) cannot be eliminated except in trivial cases. Sharper estimates are also given in these papers of the relationships, usually linear, between the constants C_1, C_2, C_3, C_4.

According to (32.10), the presence of the factor $\min\{N_{\Delta t}, n\}$ implies that the estimates (32.17) and (32.20) fall short of Lax-stability. By the Lax Equivalence Theorem, it follows that (32.17) and (32.20) are not enough to imply convergence of these discretizations as $\Delta t \to 0$. This sounds worrisome, but since the factor $\min\{N_{\Delta t}, n\}$ is so modest, one still gets convergence for a large subclass of problems defined by data with a small degree of smoothness (i.e., one more derivative than usually assumed). Thus as a practical matter the factor $\min\{N_{\Delta t}, n\}$ is usually not a concern; the instabilities that cause trouble in practice are generally exponential, not algebraic.

For another approach to transplantation of the Kreiss Matrix Theorem from the unit disk to a general region in the complex plane, see [763].

Let us close with one more example, the simplest and best-known example of a nonnormal method of lines discretization, which has been mentioned, among other places, in [564]. (The example is so simple that its explanation certainly does not need the generality of Theorems 32.1 and 32.2.) Suppose we consider again the problem of (31.1),

$$\frac{\partial u}{\partial t} = \frac{\partial u}{\partial x}, \qquad x \in (-1, 1), \quad t \geq 0, \tag{32.21}$$

again with boundary data $u(1, t) = 0$ and the same initial data as in §31. We discretize by the same upwind/Euler formula of (32.4); what is new is that now the domain is $(-1, 1)$ and nonperiodic rather than $[-\pi, \pi]$ and periodic. It is easily seen that this discretization leads to (32.8) taking the form

$$\mathbf{v}^{n+1} = \mathbf{A}\mathbf{v}^n,$$

where $\mathbf{A}$ is the Jordan block

$$\mathbf{A} = \begin{pmatrix} 1-\sigma & \sigma & & & \\ & 1-\sigma & \sigma & & \\ & & \ddots & \ddots & \\ & & & 1-\sigma & \sigma \\ & & & & 1-\sigma \end{pmatrix},$$

where $\sigma = \Delta t / \Delta x$. Readers of this book will see immediately that whereas the eigenvalues of $\mathbf{A}$ satisfy $|\lambda| < 1$ for $\sigma < 2$, the powers of $\mathbf{A}$ will grow exponentially before eventually decaying for any $1 < \sigma < 2$. As a consequence, the discretization is strongly unstable for values of σ in this range.

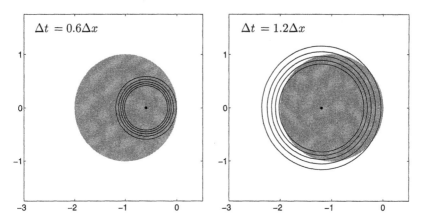

Figure 32.5: Like Figure 32.2 but for the problem (32.21) with the Dirichlet boundary condition $u(1) = 0$ instead of periodic boundary conditions. The spatial discretization is now upper triangular and far from normal. Together with eigenvalues and the stability region, the plots show ε-pseudospectra for $\varepsilon = 10^{-2}, 10^{-4}, \ldots, 10^{-10}$.

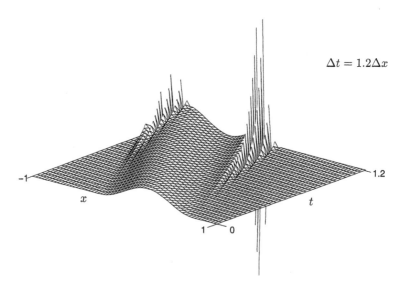

Figure 32.6: Nonmodal instability of the second discretization of Figure 32.5. For larger t the unstable oscillation grows rapidly.

Figure 32.5 illustrates the application of Theorem 32.1 to this problem. For both $\Delta t / \Delta x = 0.6$ and $\Delta t / \Delta x = 1.2$, the eigenvalue of the forward difference spatial discretization operator lies in the stability region for the Euler ODE formula. In the latter case the pseudospectra, however, extend significantly outside the stability region, and the discretization is unstable, as is confirmed in Figure 32.6.

33 · Stiffness of ODEs

A central problem in the numerical solution of ordinary differential equations is the phenomenon known as stiffness. For the purposes of this book, one might regard this as a specialized topic: If you are not interested in discretizations of ODEs, you could bypass this section. Yet at the same time, the issues that arise here are at the very heart of our subject, for stiffness is a phenomenon of transient behavior of dynamical systems that blends together the universal themes of discrete and continuous time, linear and nonlinear equations, and constant and variable coefficients.

What is a stiff ODE? The following are the symptoms most often mentioned:

1. *The problem contains widely varying time scales.*
2. *Stability is more of a constraint on the time step than accuracy.*
3. *Explicit methods do not work.*

Each of these statements has been used as a characterization of stiffness by some authors.[1] Here we attempt to explain how they relate to one another and to spectra and pseudospectra.

We begin with a scalar example; matrices and their eigenvalues will appear on page 318. The linear initial-value problem

$$u'(t) = -100(u(t) - \cos t) - \sin t, \quad u(0) = 1 \qquad (33.1)$$

has the unique solution $u(t) = \cos t$. We note that for this solution, the first term on the right-hand side of (33.1) is zero and thus the large coefficient -100 drops out of the equation. That coefficient has a dominant effect on *nearby* solutions of the ODE corresponding to different initial data, however, as is illustrated in Figure 33.1. A typical trajectory $u(t)$ of a solution to this ODE begins by shooting rapidly toward the curve $\cos t$ on a time scale $\approx 1/100$. This is the hallmark of stiffness: rapidly changing components that are present in an ODE but absent from the particular solution being tracked.

---

[1] Though the roots of the study of stiffness go back to a paper by Curtiss and Hirschfelder in 1952 [172], stiffness was not a part of the 'classic' theory of stability and convergence developed by Dahlquist in the 1950s [173] and disseminated in the 1962 book of Henrici [381], because this theory focused upon convergence in the limit $\Delta t \to 0$ rather than on behavior for finite Δt. Later in the 1960s, following another landmark theoretical paper by Dahlquist [174] and the work on stiff solvers by Gear and others [304], stiffness came to be seen as crucial. For an extensive discussion of every corner of this subject, see [365]. The actual term 'stiff' may have been coined by the statistician John Tukey [172].

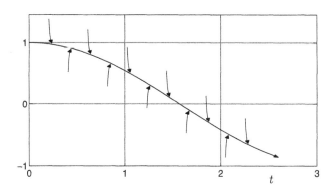

Figure 33.1: Example of a stiff ODE: $u(t) = \cos t$ and some nearby solutions of (33.1).

Table 33.1 shows the remarkable influence that this property has on numerical computations. For six values of the time step Δt, the table compares the results at $t = 1$ computed by the second-order Adams–Bashforth and backward differentiation formulas, abbreviated by AB2 and BD2 (both computations start from exact initial data). BD2 behaves beautifully, converging smoothly and quadratically to the correct answer, but AB2 generates enormous incorrect solutions. Yet when Δt becomes small enough, it settles down to be just as accurate as BD2. We have here a prototypical stiff equation, and BD2 is a prototypical *stiff solver*. The effect is shown graphically in Figure 33.2.

In practice, ODE software employs adaptive time-stepping to obtain results accurate to a user-specified tolerance [364, 365, 673]. For this problem, we can see that if the user asks for, say, three digits of accuracy at $t = 1$, an adaptive solver built on BD2 could achieve this with $\Delta t \approx 0.2$, whereas a solver built on AB2 would need $\Delta t \approx 0.01$. That is on the order of 20 times more computation, a factor that could grow to thousands or millions if the constant 100 in (33.1) were increased. It is for this reason that stiffness is such an important practical issue in scientific computing. Clearly we would want to use BD in preference to AB formulas for certain

Table 33.1: Computed values for $u(1)$ for the ODE (33.1).

Δt	AB2	BD2
0.2	14.40	0.5404
0.1	-5.70×10^4	0.54033
0.05	-1.91×10^9	0.540309
0.02	-5.77×10^{10}	0.5403034
0.01	0.54030196	0.54030258
0.005	0.54030222	0.54030238
$\vdots$	$\vdots$	$\vdots$
0	$0.540302306\ldots$	$0.540302306\ldots = \cos(1)$

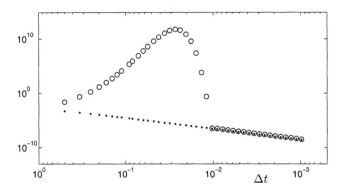

Figure 33.2: Errors $|u_{1/\Delta t} - \cos(1)|$ as a function of step size Δt for the same problem (log-log scale). The AB formula is useless until the time step becomes very small, whereas the BD formula gives good results for all time steps.

equations. For other equations, on the other hand, the AB formulas work perfectly well and the BD formulas are much more expensive because they are implicit—requiring an iterative solution at each time step if the ODE is nonlinear.

The computational challenge is clear, and so is the related theoretical challenge, which is the subject of this section. How can we characterize ODEs that are subject to the effect shown in Figure 33.2?

Our example (33.1) can be analyzed completely. If $u(t)$ is any solution to $u' = -100(u - \cos t) - \sin t$, then $w(t) = u(t) - \cos t$ satisfies the scalar, linear, constant-coefficient ODE

$$w' = \lambda w, \quad \lambda = -100. \tag{33.2}$$

The AB2 formula applied to this equation takes the form of the three-term recurrence relation

$$u_{n+2} - u_{n+1} = \lambda \Delta t \left(\tfrac{3}{2} u_{n+1} - \tfrac{1}{2} u_n \right),$$

where u_n denotes the approximate solution at time step n; that is,

$$u_{n+2} - \left(\tfrac{3}{2} \lambda \Delta t + 1 \right) u_{n+1} + \tfrac{1}{2} \lambda \Delta t \, u_n = 0.$$

The characteristic polynomial of this recurrence is

$$p(z) = z^2 - \left(\tfrac{3}{2} \lambda \Delta t + 1 \right) z + \tfrac{1}{2} \lambda \Delta t.$$

For $\lambda \Delta t < -1$, one of the two roots of this polynomial lies in the negative real interval $(-\infty, -1)$. This means that for $\Delta t > -1/\lambda = 1/100$, successive steps of the AB2 formula will amplify any truncation errors in the solution exponentially, as observed in Table 33.1 and Figure 33.2. For $\Delta t \leq 1/100$, on the other hand, the troublesome root crosses into the interval $[-1, 0)$ and there is no amplification. This is why $\Delta t = -1/\lambda = 1/100$

is the critical value for the AB2 formula applied to this problem, as is evident in Figure 33.2. A similar analysis of the BD2 formula would reveal no unstable roots of the recurrence relation, regardless of Δt.

The explanation just given can be recast in the language of *stability regions*. As was described in the previous section, the stability region of an ODE formula is defined by considering discretizations of the scalar model problem (33.2) for various values of λ. It is the set of points in the complex $\lambda\Delta t$-plane for which the recurrence relation obtained by such discretization is stable in the sense that all of its solutions are bounded as $n \to \infty$. For AB2, Figure 32.1 shows that the stability region is a bounded subset of the left half-plane whose intersection with the negative real axis is $[-1, 0]$. For $w' = -100w$, it follows that AB2 is good for $\Delta t \leq 1/100$. By contrast, Figure 32.1 shows that the stability region for BD2 includes the entire left half-plane, including the negative real axis, and that is why this formula has no difficulty with stiff equations.

We are now ready to turn to the general problem: a system of ordinary differential equations

$$\mathbf{u}' = \mathbf{f}(\mathbf{u}, t) \tag{33.3}$$

where $\mathbf{u}(t)$ is an N-vector for each t and $\mathbf{f}$ is in general nonlinear. Systems of ODEs arise routinely in ODE applications and also in the treatment of PDEs by the method of lines, discussed in the previous section. Suppose we are interested in a particular solution $\mathbf{u}^*(t)$ for values of t near a particular time t^*. When will the problem of computing $\mathbf{u}^*(t)$ for $t \approx t^*$ be stiff? The

$$
\begin{array}{lll}
\mathbf{u}' \;=\; \mathbf{f}(\mathbf{u}, t) & & \text{(i)} \\[4pt]
\quad\downarrow & \textit{linearize} & \\[4pt]
\mathbf{u}' \;=\; \mathbf{A}(t)\mathbf{u} & & \text{(ii)} \\[4pt]
\quad\downarrow & \textit{freeze coefficients} & \\[4pt]
\mathbf{u}' \;=\; \mathbf{A}\mathbf{u} & & \text{(iii)} \\[4pt]
\quad\downarrow & \textit{diagonalize} & \\[4pt]
u' \;=\; \lambda u & & \text{(iv)}
\end{array}
$$

Figure 33.3: Outline of how eigenvalues arise in the analysis of stiffness and of dynamical processes across the mathematical sciences. We begin with a time-dependent nonlinear equation and reduce it by a sequence of three approximations to a set of scalar, constant-coefficient linear problems. The first two approximations are valid for short times, and the last is valid for long times. If there is no overlap between these ranges of validity, eigenvalue analysis is unlikely to give useful predictions.

simplest approach to this question, going back many years, is to look at eigenvalues. The argument that may be used to justify such an analysis is summarized in Figure 33.3.

We begin with (i), a nonlinear system of N first-order ODEs. If we make the substitution

$$\mathbf{u}(t) = \mathbf{u}^*(t) + \mathbf{w}(t),$$

as before, then stability and stiffness depend on the evolution of $\mathbf{w}(t)$. The first step is to *linearize* the equation by assuming $\mathbf{w}$ is small. If $\mathbf{f}$ is differentiable with respect to each component of $\mathbf{u}$, we have

$$\mathbf{f}(\mathbf{u}, t) = \mathbf{f}(\mathbf{u}^*, t) + \mathbf{A}(t)\mathbf{w}(t) + o(\|\mathbf{w}\|),$$

where $\mathbf{A}(t)$ is the $N \times N$ *Jacobian matrix* of partial derivatives of $\mathbf{f}$ with respect to $\mathbf{u}$:

$$[\mathbf{A}(t)]_{jk} = \frac{\partial f_j}{\partial u_k}(\mathbf{u}^*(t), t), \qquad 1 \leq j, k \leq N.$$

This means that if $\mathbf{w}$ is small, the ODE can be accurately approximated by a linear problem:

$$\mathbf{u}'(t) = \mathbf{f}(\mathbf{u}^*, t) + \mathbf{A}(t)\mathbf{w}(t).$$

If we subtract (33.3) from this equation, we obtain

$$\mathbf{w}' = \mathbf{A}(t)\mathbf{w}.$$

One can think of this result as approximate, if $\mathbf{w}$ is small, or exact, if $\mathbf{w}$ is infinitesimal. Rewriting $\mathbf{w}$ as a new variable $\mathbf{u}$ gives (ii).

The second step is to *freeze coefficients* by setting

$$\mathbf{A} = \mathbf{A}(t^*).$$

The idea here is that stability and stiffness are local phenomena, which may appear at some times t^* and not others. The result is the constant-coefficient linear problem (iii).

Finally, assuming $\mathbf{A}$ is diagonalizable, we *diagonalize* it. This gives us the set (iv) of N scalar, linear, constant-coefficient model problems $u' = \lambda u$.

The traditional 'eigenvalue view' of stability and stiffness of a numerical formula can now be summarized as follows (these are rough conditions, not precise statements).

> *Eigenvalue characterization of stability.* A numerical ODE formula is stable for computing $\mathbf{u}^*(t)$ for $t \approx t^*$ if Δt is small enough that for each eigenvalue λ of $\mathbf{A}(t^*)$, $\lambda \Delta t$ lies inside the stability region (or at least within a distance $\mathcal{O}(\Delta t)$).

Eigenvalue characterization of stiffness. An ODE is stiff for the solution $\mathbf{u}^*(t)$ for $t \approx t^*$ if the largest eigenvalue modulus $|\lambda|$ of $\mathbf{A}(t^*)$ is much greater than the rate of change of $\mathbf{u}^*(t)$ itself.

For our scalar example (33.1), the Jacobian is the 1×1 matrix $\mathbf{A}(t) = A = -100$, whose entry and eigenvalue have nothing to do with the time scale $\mathcal{O}(1)$ of the solution $u(t) = \cos t$. Since $100 \gg 1$, the problem is stiff. For many other ODE problems, however, the solution $\mathbf{u}(t)$ changes at a rate also determined by $\mathbf{A}(t)$—typically by its smaller eigenvalues. And from here one derives the notion of a *stiffness ratio* for a solution of an ODE. Depending on the author and the context, this may be a ratio of the absolute values of the eigenvalues of $\mathbf{A}$, or of their real parts. For example, if $\mathbf{A}$ is a Hermitian matrix with eigenvalues ranging from -10^6 to -1, the stiffness ratio would be 10^6 and one would conclude that the problem is highly stiff.

Readers of this book will quickly recognize that since eigenvalues need not be closely coupled to matrix behavior, attempts to characterize stiffness by eigenvalues or stiffness ratios cannot fully succeed. This fact has been investigated from various angles in the literature of numerical solution of ODEs, and it was discussed from the point of view of pseudospectra by D. J. Higham and Trefethen in 1993 [387], who wrote:

> Stiffness cannot properly be characterized in terms of the eigenvalues of the Jacobian, because stiffness is a transient phenomenon whereas the significance of eigenvalues is asymptotic.

An extensive discussion appears in [387] of links between pseudospectra and stiffness, with a variety of numerical experiments and some interesting quotations from the literature about eigenvalues.

Here is a numerical illustration following the pattern of Figure 33.2. Consider the linear constant-coefficient matrix equation

$$\mathbf{u}' = \mathbf{A}\mathbf{u}, \quad \mathbf{u}(0) = \mathbf{u}_0, \tag{33.4}$$

where $\mathbf{u}_0$ is the N-vector of all ones and $\mathbf{A}$ is the $N \times N$ triangular matrix with -1 on the main diagonal and -2 everywhere below. The pseudospectra of this highly nonnormal matrix are shown in Figure 33.4, and they show that the matrix 'lives' in a region of size $\mathcal{O}(N)$ in the left half-plane that extends much beyond the single eigenvalue -1. It is hardly surprising that the figure shows explosive behavior of AB2 just as in Figure 33.2 for time steps $\Delta t = \mathcal{O}(1)$, a transient instability that settles down only for $\Delta t = \mathcal{O}(10^{-1})$. In the light of such examples, Higham and Trefethen proposed these alternative characterizations in terms of pseudospectra:

Pseudospectral characterization of stability. A numerical ODE formula is stable for computing $\mathbf{u}^*(t)$ for $t \approx t^*$ if Δt is small enough

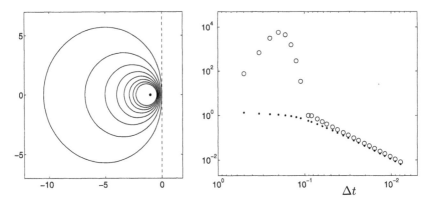

Figure 33.4: On the left, eigenvalue and ε-pseudospectra of the 40×40 matrix $\mathbf{A}$ of (33.4) for $\varepsilon = 10^{-2}, 10^{-4}, \ldots, 10^{-16}$; the dashed line is the imaginary axis. On the right, the analogue of Figure 33.2 for this problem. Clearly an equation controlled by this matrix may be stiff, though the eigenvalues are all equal to -1.

that the ε-pseudospectra of $\Delta t\, \mathbf{A}(t^*)$ lie within a distance $\mathcal{O}(\varepsilon + \Delta t\,)$ of the stability region.

Pseudospectral characterization of stiffness. An ODE is stiff for the solution $\mathbf{u}^*(t)$ for $t \approx t^*$ if the pseudospectra of $\mathbf{A}(t^*)$ extend far into the left half-plane as compared with the time scale of the solution $\mathbf{u}^*(t)$ itself.

While it would be good for theorists to reach a consensus on the characterization of stiffness, the business of practical scientific computing does not depend on this event. Adaptive ODE codes are rarely misled by harmless-looking eigenvalue distributions into taking dangerously large time steps.[2] For the example (33.4), the adaptive Runge–Kutta MATLAB code `ode23` samples the ODE at fully 274 grid points and obtains an accurate result, whereas if the numbers -2 below the diagonal are changed to -0.2, which has no effect on the eigenvalue, the number of sample points shrinks to 82. The subject of adaptive step size selection in ODE software is a fascinating one, and in recent years it has been linked more explicitly to various areas of dynamics and control theory [358, 359, 365, 366, 614].

The theoretical literature on the numerical solution of stiff ODEs, exemplified by the books [193, 365], gives more attention to coping with stiffness than to characterizing it. In these and related works one instead finds discussion of topics such as *B-stability* and *G-stability*, *contractivity* and *dissipativity*, *Lipschitz* and *one-sided Lipschitz constants*, and *logarithmic norms*. The aim of such discussions is to derive sufficient conditions for certain schemes to behave successfully when applied to certain prob-

---

[2]For an analogous observation about adaptive algorithms in iterative numerical linear algebra, see page 261.

lems. All of the tools just mentioned are employed to ensure that when a problem has some kind of contractivity property, nothing can go wrong. Because this is a local condition, it applies to nonlinear as well as linear problems, and thus such analysis is widely called *nonlinear stability analysis*. The same term is used in exactly the same way in the field of fluid mechanics [436]. However, the essential idea of this 'nonlinear' analysis, as emphasized in §17, is linear: If the numerical range of a Jacobian matrix lies in the open left half-plane, then certain desirable properties follow. From here, attractive corollaries for nonlinear problems can be derived, but these are always sufficient conditions for good behavior, never necessary and sufficient.

In this section we have discussed ordinary differential equations, the field in which the subject of stiffness was first investigated, while hardly mentioning the broader world of partial differential equations—and indeed without making much reference to the previous three sections. In fact, however, the most important applications of the notion of stiffness are in the solution of PDEs, where one so often encounters contrasts between fast time scales (e.g., diffusion or dispersion) and slow ones (e.g., convection or reaction). Roughly speaking, any time-dependent PDE for which it is advantageous to use an implicit or semi-implicit discretization in time is likely to correspond to a stiff system of ODEs if it is discretized in space. Mathematically, the observations of this section are much the same as those of §32, and ultimately it is results like Theorems 32.1 and 32.2 that must be appealed to if one wants to make rigorous connections between stiffness and pseudospectra.

34 · GKS-stability of boundary conditions _____

The last four sections have focused on nonmodal instabilities of a potentially explosive kind in the numerical solution of differential equations. In this final section of this part of the book we turn to a milder kind of instability that is perhaps the best known example of nonmodal behavior in numerical analysis. At issue here are numerical boundary conditions for finite difference discretizations of linear hyperbolic PDEs. In the late 1960s and early 1970s, Kreiss, Osher, and others showed that poorly chosen boundary conditions for such discretizations are susceptible to a special kind of instability. The standard name for their work became the theory of *GKS-stability*, after an important but difficult 1972 paper of Gustafsson, Kreiss, and Sundström [356, 357]. This theory can be interpreted in terms of the group velocities of certain waves propagating dispersively on the finite difference grid [768, 769].

As an example, consider the linear scalar hyperbolic constant-coefficient one-dimensional initial boundary value problem

$$u_t = u_x, \quad u(x,0) = u_0(x), \quad u(1,t) = 0 \tag{34.1}$$

defined for $t > 0$ on the interval $0 < x < 1$, where u_0 represents the initial data. The analytical solution is

$$u(x,t) = \begin{cases} u_0(x+t) & \text{for } x+t < 1, \\ 0 & \text{for } x+t \ge 1, \end{cases}$$

i.e., a wave propagating leftward at speed 1 until it is absorbed at the boundary. To solve the problem numerically, we set up a regular grid in the x-t plane with space step $\Delta x = 1/N$ for a positive integer N and time step $\Delta t = \sigma \Delta x$ for a constant $\sigma < 1$, and compute approximations

$$v_j^n \approx u(j\Delta x, n\Delta t)$$

on this grid. For $t = n = 0$, we take initial values $v_j^0 = u_0(j\Delta x)$. To march forward to later time steps, we use the Crank–Nicolson-type implicit finite difference formula

$$\frac{v_j^{n+1} - v_j^n}{\Delta t} = \frac{\frac{1}{2}(v_{j+1}^n - v_{j-1}^n)}{2\Delta x} + \frac{\frac{1}{2}(v_{j+1}^{n+1} - v_{j-1}^{n+1})}{2\Delta x} \tag{34.2}$$

for $1 \le j \le N-1$ together with the inflow boundary condition

$$v_N^{n+1} = 0.$$

One would not usually use an implicit formula like this for a wave equation, but it is convenient for illustration.

So far we have specified $N - 1$ equations involving the N unknowns $v_0^{n+1}, \dots, v_{N-1}^{n+1}$. We need one more equation to define the numerical method fully, a *numerical boundary condition* involving the outflow point v_0^{n+1}. A standard approach is to define v_0^{n+1} by some kind of extrapolation of values from the interior. For example, here are two possibilities:

$$\text{(a)} \ v_0^{n+1} = v_1^n, \qquad \text{(b)} \ v_0^{n+1} = v_2^n. \tag{34.3}$$

Each of these choices gives us a complete numerical method, a well-defined procedure for marching from step n to $n+1$. Since the procedure is linear, it must be equivalent to multiplication by a matrix:

$$(v_0^{n+1}, v_1^{n+1}, \dots, v_{N-1}^{n+1})^{\mathrm{T}} = \mathbf{A}(v_0^n, v_1^n, \dots, v_{N-1}^n)^{\mathrm{T}}.$$

These matrices can be readily computed since they take the form $\mathbf{A} = \mathbf{A}_1^{-1}\mathbf{A}_2$ for appropriate tridiagonal finite differencing matrices $\mathbf{A}_1$ and $\mathbf{A}_2$, suitably modified in their first rows to impose the boundary conditions.

Have we got a stable numerical method? According to the Lax Equivalence Theorem, stability is the condition that $\|\mathbf{A}^n\| \leq C$ for all $n \leq N$ for some constant C independent of N, and this is necessary and sufficient for convergence to the solution of the PDE as $N \to \infty$. As usual in this book, we seek insight into these norms of powers by examining pseudospectra. Figure 34.1 shows ε-pseudospectra for the matrices $\mathbf{A}$ corresponding

(a) $v_0^{n+1} = v_1^n$ (b) $v_0^{n+1} = v_2^n$

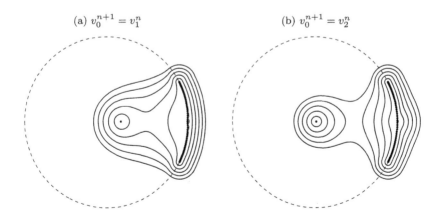

Figure 34.1: Eigenvalues and ε-pseudospectra of the Crank–Nicolson matrices $\mathbf{A}$ corresponding to (34.2) with $N = 60$ and $\sigma = 1/2$ for the outflow boundary conditions (a) and (b) of (34.3), for $\varepsilon = 0.4, 0.8, 1.2, 1.6, 2.0$; the dashed curve is the unit circle. The bulge in the pseudospectra in case (b) reveals the instability of this boundary condition.

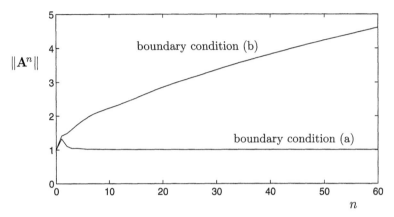

Figure 34.2: Norms $\|\mathbf{A}^n\|$ for the two discretizations, confirming instability in case (b).

to boundary conditions (a) and (b), with linearly spaced values of ε. In case (a), we see that the boundaries of these pseudospectra are likewise linearly spaced outside the unit disk. In case (b), by contrast, there is a bulge near $z = 1$. These observations suggest that (a) may be stable, whereas (b) may be unstable.

Figure 34.2 shows that these predictions are right on target. Condition (a) is abundantly stable, but it is clear that (b) is not. The instability is mild, of magnitude $\|\mathbf{A}^n\| = \mathcal{O}(n^{1/2})$, but this will be enough to prevent convergence in general to the correct solution as $N \to \infty$. Figure 34.3 gives a physical feeling for what is going on by showing the evolution up to $t = 1$ resulting from the initial data

$$u_0(x) = e^{-100(x-1/2)^2}. \tag{34.4}$$

For boundary condition (a), the computed solution for $t \gg 1/2$ is close to the correct result—zero—whereas for (b), we see a spurious sawtoothed solution on the grid of amplitude $\mathcal{O}(1)$. Here the initial Gaussian propagates leftward with the correct velocity -1, but upon hitting the boundary, it excites a sawtoothed reflected wave of similar amplitude traveling right with group velocity $+1$. This numerical artifact will not go away as the mesh is refined. The standard GKS stability criterion can be interpreted as a test for the possibility that the finite difference formula with homogeneous boundary conditions, as defined on a semi-infinite grid bounded by a single left-hand boundary, admits a nonzero wave solution like this one with group velocity ≥ 0. In this example the unstable wave is

$$v_j^n - (-1)^j,$$

which satisfies both the interior formula (34.2) and the boundary condition

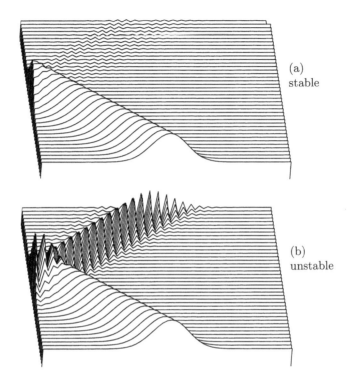

(a)
stable

(b)
unstable

Figure 34.3: Results for $0 \leq x \leq 1$ (horizontal axis) and $0 \leq t \leq 1$ (vertical axis) for the two boundary conditions starting from Gaussian initial data. Every third time step is shown. The instability of boundary condition (b) takes the form of a sawtoothed reflected wave.

(b) of (34.3). By differentiation of a numerical dispersion relation, its group velocity can be found to be $+1$. See [768] and [769] for theorems and numerical examples.

For another example of stable and unstable boundary conditions, we can discretize the same problem (34.1) by the leap frog approximation

$$\frac{v_j^{n+1} - v_j^{n-1}}{2\Delta t} = \frac{v_{j+1}^n - v_{j-1}^n}{2\Delta x}. \tag{34.5}$$

This formula is explicit rather than implicit, meaning that no linear algebra is needed for its implementation. Since it couples three levels of data, however, we must analyze it by a matrix of dimension $2N$ rather than N:

$$(v_0^{n+1}, \ldots, v_{N-1}^{n+1}, \ v_0^n, \ldots, v_{N-1}^n)^{\mathrm{T}} = \mathbf{A}(v_0^n, \ldots, v_{N-1}^n, \ v_0^{n-1}, \ldots, v_{N-1}^{n-1})^{\mathrm{T}}.$$

Figure 34.4 is an analogue of Figure 34.1 for this new discretization, now with the boundary conditions

$$\text{(a)} \ v_0^{n+1} = v_1^n, \qquad \text{(b)} \ v_0^{n+1} = v_1^{n+1}. \tag{34.6}$$

Again, the second boundary condition is unstable. This time the unstable wave takes the form $v_j^n = (-1)^n$, again with group velocity $+1$, but now the instability is of order $\|\mathbf{A}^n\| = \mathcal{O}(n)$ (because of an infinite reflection coefficient, in the terminology of [768]).

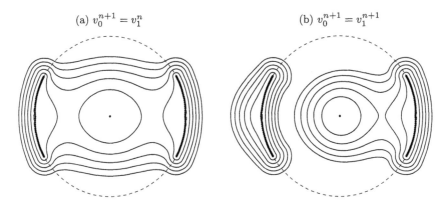

(a) $v_0^{n+1} = v_1^n$ (b) $v_0^{n+1} = v_1^{n+1}$

Figure 34.4: Same as Figure 34.1, but for the leap frog discretization (34.5) with boundary conditions (34.6). Now boundary condition (b) has an unstable bulge at $z = -1$, indicating the existence of an unstable mode that is sawtoothed with respect to t.

We shall not pursue this leap frog example further but return to Crank–Nicolson. In practice, the GKS instabilities that appear usually share the following features of Figures 34.1–34.3: a bulge in the pseudospectra near some troublesome point z_0 on the unit circle; growth of the norms $\|\mathbf{A}^n\|$ at a rate $\mathcal{O}(n^{1/2})$ or higher; and existence of a wave, often sawtoothed in x or t, that propagates from the boundary into the interior. We can make a more precise connection in this example between the norm of the resolvent near $z = 1$ and the norms $\|\mathbf{A}^n\|$. Define the function

$$K(x) = (x - 1)\|(x - \mathbf{A})^{-1}\| \tag{34.7}$$

for $x > 1$. If $\mathbf{A}$ were normal with eigenvalue of largest magnitude at $z = 1$, we would have $\|\mathbf{A}^n\| = 1$ for all $n \geq 0$ and $K(x) = 1$ for all $x > 1$. Here, however, the resolvent norms are larger. Figure 34.5 plots $K(x)$ as a function of $x - 1$ on a log-log scale for matrices with $N = 60$, 120, and 240. We see that whereas $K(x)$ is close to 1 for large x, it grows like $(x - 1)^{-1/2}$ for smaller x. By condition (16.15) of Theorem 16.4, it follows that some norm $\|\mathbf{A}^n\|$ with n of order $(x - 1)^{-1}$ must be at least as great as

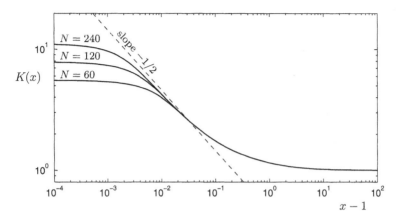

Figure 34.5: Resolvent norms $\|(x - \mathbf{A})^{-1}\|$ for x near the unit circle, as revealed by the function K of (34.7).

$K(x) \approx (x - 1)^{-1/2}$. This closely matches the $\mathcal{O}(n^{1/2})$ instability actually observed.

We can construct a simpler system that has much the same behavior. Consider the $N \times N$ 'GKS-instability matrix'

$$\mathbf{A} = \begin{pmatrix} 1 & & & & & \\ 1 & 0 & & & & \\ & 1 & 0 & & & \\ & & 1 & 0 & & \\ & & & \ddots & \ddots & \\ & & & & 1 & 0 \end{pmatrix}, \tag{34.8}$$

where all entries not listed are zero; thus $\mathbf{A}$ is a lower bidiagonal Jordan block except for the special value $a_{11} = 1$. When $\mathbf{A}$ acts on a vector it shifts it down, duplicating the top entry. Thus, for example, we have

$$\mathbf{A}^3(1, 0, 0, \ldots, 0)^{\mathrm{T}} = (1, 1, 1, 1, 0, 0, \ldots, 0)^{\mathrm{T}}.$$

Successive powers $\mathbf{A}^n$ generate waves propagating rightward from the boundary, just as in the case of a GKS instability, and the norms are $\|\mathbf{A}^n\| = \sqrt{n+1}$ (exactly) for $n < N$. It is no surprise that Figure 34.6 shows a familiar bulge in the pseudospectra.

The discussion up to this point has been straightforward. It would seem that GKS instabilities correspond to small bulges in the pseudospectra, to algebraic growth of norms of powers, and to propagation of waves into the domain from a boundary. They can even be modeled by taking powers of a matrix that differs in just one entry from a Jordan block. In view

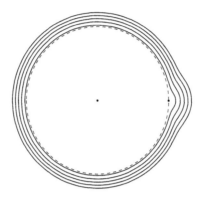

Figure 34.6: Pseudospectra for the GKS-instability matrix (34.8) for $N = 60$. The bulge near $z = 1$ looks much the same as in Figure 34.1; the same values of ε are plotted. This simple model captures some of the 'physics' of typical GKS-instabilities.

of this pretty picture, why is it widely thought that the GKS theory and paper [357] are difficult?

Unfortunately, there are good reasons for these perceptions. Suppose one tries to expand the observations we have made into a general theory of stability of boundary conditions for finite difference approximations of PDEs. To keep the project in bounds we assume that the domain is one-dimensional and that the equations are linear and hyperbolic. The trouble is, there are still many further simplifying assumptions that it would be helpful to make, such as:

(1) constant coefficients (no explicit dependence on x);

(2) scalar (e.g., wave equation, not linearized Navier–Stokes);

(3) initial data only (no interior or boundary forcing terms);

(4) two-level formula (e.g., Crank–Nicolson, not leap frog);

(5) dissipative formula (e.g., Lax–Wendroff, not Crank–Nicolson);

(6) explicit formula (e.g., Lax–Wendroff, not Crank–Nicolson).

Our Crank–Nicolson example has the first four of of these convenient properties, and the earlier, pre-GKS literature on stability of boundary conditions also depended on such assumptions. The principal papers in this era were by Strang, Kreiss, and Osher [466, 467, 468, 589, 590, 733, 734]. Each of these articles, in different contexts, obtained theorems to the effect that under appropriate assumptions, stability could be assured if there were no rightgoing waves of the kind we have described, known as 'generalized eigensolutions'.

Kreiss and his colleagues, however, were determined to find a general theory. A hint that difficulties must arise in such a theory can be found in the theorems of §§16, 18, and 32. We know from the Kreiss Matrix Theorem that although a bulge in pseudospectra guarantees instability, the lack of a bulge cannot guarantee stability under all circumstances; there is a gap between upper and lower bounds of a factor of either n, the time step, or N, the matrix dimension. Thus a complete connection between resolvent norms or pseudospectra and stability would necessarily be delicate. Assumptions such as those above can be used to shrink the gap. For example, the dissipativity assumption (5) is utilized in [467] to preclude the possibility of bulges near points of the unit circle other than $z = 1$, making it possible to obtain an upper bound, independent of n and N, from a Cauchy integral over a contour that passes outside the unit circle only near that single point.

In the general setting, Gustafsson, Kreiss, and Sundström were unable to keep the theory as clean as in the previous, more specialized publications. It is generally agreed that their central result (stated in [357] as Corollary 10.3 and the sentence that follows it!) is a theorem of necessity and sufficiency:

$$GKS\text{-}stable \iff \text{no rightgoing waves.}$$

This looks simple, and indeed, the condition of no rightgoing waves or generalized eigensolutions (called the 'determinant condition') is essentially what we have described above. However, the term *GKS-stable* is quite complicated. This is a special definition of stability, given as Definition 3.3 of [357], that involves exponential decay factors with respect to t and other algebraic terms that remove it significantly from the familiar stability notion of bounded norms of powers. Gustafsson et al. argued that GKS-stability is the 'right' definition in important senses, and Gustafsson showed in [355] that under suitable assumptions it is equivalent to a certain kind of convergence. Yet the details remain technical.

There are sizeable literatures of the stability of boundary conditions and of pseudospectra, but the two are nearly disjoint. (So far as we know, the intersection consists of the single paper [849], together with some unpublished notes communicated to us by Niles Pierce in 2002.) In particular, there is no tradition of looking for GKS instabilities by plotting pseudospectra as in Figures 34.1 and 34.4. Part of the reason for this situation is undoubtedly that the routine computation of pseudospectra is a more recent development than GKS stability theory, popular among a younger generation of researchers. It will be interesting to see if the use of pseudospectra for analyzing boundary conditions catches on in the future.

VIII. Random Matrices

35 · Random dense matrices

Random matrices are of interest in condensed matter physics [544], number theory [58], statistics [567], and numerical analysis [226]; their literature includes hundreds of published articles and a number of books. What do their pseudospectra look like? Are random matrices sufficiently close to normal that one need only consider the eigenvalues? This section will suggest answers to these questions. Dense random matrices are indeed close to normal, but subsequent sections will show the distance from normality increases from algebraic to exponential if the matrix is not dense but has triangular or other nonsymmetric sparsity structure.

Let us say that an N-dimensional (dense) *real random matrix* is a matrix with independent entries drawn from the normal distribution of mean 0 and standard deviation $N^{-1/2}$. That is, each entry has the form

$$a_{jk} = \frac{x}{\sqrt{N}}, \tag{35.1}$$

where x is a sample from the standard normal distribution $N(0,1)$. This normalization by $\sqrt{N}$ turns out to be the natural one for obtaining regular behavior as $N \to \infty$. A *complex random matrix*, similarly, has independent entries from the complex normal distribution of mean 0 and standard deviation $N^{-1/2}$, which means that each entry has the form

$$a_{jk} = \frac{x}{\sqrt{2N}} + \frac{iy}{\sqrt{2N}}, \tag{35.2}$$

where x and y are independent samples from $N(0,1)$.

Figures 35.1 and 35.2 illustrate the beautiful behavior of eigenvalues of real and complex random matrices: They are *uniformly distributed in the unit disk*. Of course, for each finite N, this statement cannot be exactly correct. For complex matrices, the probability density function of the eigenvalues is a continuous function that is nonzero throughout the complex plane. For real matrices, not only is the probability density nonzero throughout the plane, but it has a delta function singularity along the real axis, since for any finite N, there is a positive probability that some eigenvalues will be purely real.[1] In the limit $N \to \infty$, however, these complexities fade away and the distribution converges to the constant π^{-1} inside the unit disk and zero outside. This fact, known as the *circular law*,

---

[1] The problem of what proportion of the eigenvalues can be expected to be real has been solved by Edelman, Kostlan, and Shub [227]: The answer is asymptotic to $(2/\pi N)^{1/2}$. For $N = 10$, the expected fraction of real eigenvalues is exactly $67843\sqrt{2}/327680 \approx 0.2928$.

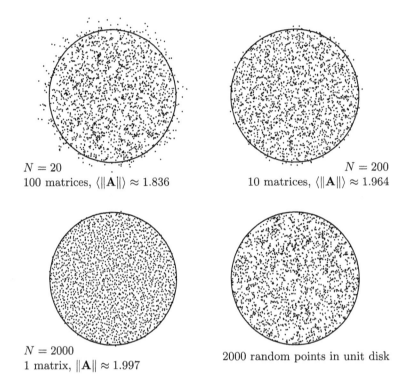

$N = 20$
100 matrices, $\langle \|\mathbf{A}\| \rangle \approx 1.836$

$N = 200$
10 matrices, $\langle \|\mathbf{A}\| \rangle \approx 1.964$

$N = 2000$
1 matrix, $\|\mathbf{A}\| \approx 1.997$

2000 random points in unit disk

Figure 35.1: Eigenvalues of random complex matrices of various dimensions. The first three plots show superpositions of eigenvalues of $2000/N$ matrices of the indicated dimensions, hence 2000 dots altogether in each image; the solid curve is the unit circle. Though the spectral radii of these matrices are ≈ 1, their norms are ≈ 2. Note that eigenvalues of random matrices tend to avoid one another: They are more smoothly distributed than random points in the disk. (This effect is concealed in the top images by the superposition of eigenvalues of different matrices.)

was first established by Ginibre and Mehta and has been generalized by Girko [310, 311, 313, 544].

Like the eigenvalues, the singular values of random matrices behave simply in the limit $N \to \infty$: The probability distribution converges to a quarter-circle over the interval $[0, 2]$. This fact is known as the *quarter-circle law*, established by Marčenko and Pastur in 1967 [529], and it is closely related to the *Wigner semicircle law* [824] for eigenvalues of random Hermitian matrices. For related results, as well as applications in physics and multivariate statistics, see [159, 226, 313, 544, 567, 824].

We summarize these facts in the following theorem. For details and proofs, including precise definitions of the probabilistic notions of convergence underlying all the statements of this section, see the above references.

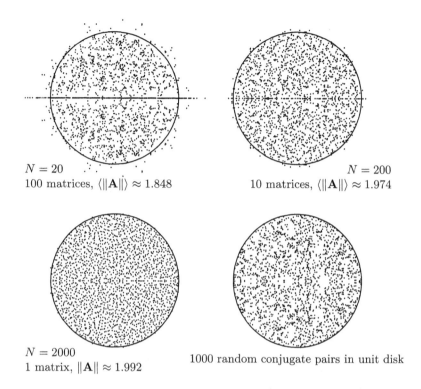

$N = 20$
100 matrices, $\langle \|\mathbf{A}\| \rangle \approx 1.848$

$N = 200$
10 matrices, $\langle \|\mathbf{A}\| \rangle \approx 1.974$

$N = 2000$
1 matrix, $\|\mathbf{A}\| \approx 1.992$

1000 random conjugate pairs in unit disk

Figure 35.2: Same as Figure 35.1, but for real instead of complex matrices. For finite N, there is a positive fraction of purely real eigenvalues, but as $N \to \infty$ the distribution is the same as before.

Eigenvalues and singular values of random matrices

Theorem 35.1 *Let the distribution of dense random matrices $\mathbf{A}$ be defined as above, either real or complex. As $N \to \infty$, in both the real and the complex cases, the probability density functions for the eigenvalues and singular values approach the following limits:*

(i) *Eigenvalues:*
$$\lim_{N\to\infty} \frac{p(\lambda)}{N} = \begin{cases} \pi^{-1} & \text{for } |\lambda| \leq 1, \\ 0 & \text{for } |\lambda| > 1; \end{cases} \tag{35.3}$$

(ii) *Singular values:*
$$\lim_{N\to\infty} \frac{p(\sigma)}{N} = \begin{cases} \frac{2}{\pi}\sqrt{1 - \frac{\sigma^2}{4}} & \text{for } 0 \leq \sigma \leq 2, \\ 0 & \text{for } \sigma > 2. \end{cases} \tag{35.4}$$

Though we shall not prove Theorem 35.1, we mention an elegant technique that can be used in its proof, due originally to Silverstein [691], which provides an intuitive explanation of where such results come from. In numerical linear algebra, the standard algorithm for computing eigenvalues first reduces a matrix to Hessenberg form by a sequence of unitary similarity transformations involving Householder matrices, while the standard algorithm for computing singular values first reduces a matrix to bidiagonal form by multiplying on the left and right by sequences of Householder matrices; see, e.g., [327, 776]. Silverstein's idea was to apply such reductions to random matrices. For example, suppose an $N \times N$ random matrix is bidiagonalized in the standard fashion. Since each unitary multiplication preserves norms, we find that the resulting matrix will be approximately

$$
\begin{pmatrix}
1 & \sqrt{\frac{N-1}{N}} & & & \\
& \sqrt{\frac{N-1}{N}} & \sqrt{\frac{N-2}{N}} & & \\
& & \ddots & \ddots & \\
& & & \sqrt{\frac{2}{N}} & \sqrt{\frac{1}{N}} \\
& & & & \sqrt{\frac{1}{N}}
\end{pmatrix}.
$$

The argument sounds heuristic, but in fact it can easily be made precise, leading to the conclusion that the distribution of singular values of a dense random matrix is exactly that of a bidiagonal random matrix with independent entries from certain chi-squared distributions. For large N the bidiagonal matrix begins with entries approximately equal to 1,

$$
\begin{pmatrix}
1 & 1 & & & \\
& 1 & 1 & & \\
& & 1 & 1 & \\
& & & \ddots & \ddots
\end{pmatrix},
$$

and from here it is clear that the norm is approximately 2, as is implicit in (35.4). Indeed, an intriguing conclusion suggested by Theorem 35.1 is that for a large random matrix, the norm is approximately twice the spectral radius. Geman has proved that this is true [307, 308]:

$$
\lim_{N\to\infty} \|\mathbf{A}_N\| \overset{\text{a.s.}}{=} 2, \qquad \lim_{N\to\infty} \rho(\mathbf{A}_N) \overset{\text{a.s.}}{=} 1, \tag{35.5}
$$

where $\overset{\text{a.s.}}{=}$ means 'almost surely', i.e., with probability 1 if we consider a sequence $\{\mathbf{A}_N\}$ of increasing dimensions N (see, e.g., [350]).

The two parts of Theorem 35.1 can be viewed as boundary points of a continuum. Girko and Sommers et al. have derived a result that interpolates between them by considering matrices $\mathbf{A}$ for which the correlation of $\mathbf{A}$ and $\mathbf{A}^*$ is a prescribed constant between -1 and 1 [312, 313, 708]. The resulting spectra are uniformly distributed in an ellipse, and as the correlation coefficient approaches 1, the ellipse squashes to a real interval with a semicircular density function.

The limits (35.5) are a good starting point for a discussion of nonnormality. If random matrices were normal, $\|\mathbf{A}\|$ and $\rho(\mathbf{A})$ would be equal. Evidently, then, random matrices deviate from normality to a degree. However, the deviation is mild. As a first illustration of this fact, consider Figure 35.3, which presents pseudospectra of a single random matrix of dimension 256. From this plot it is clear that the eigenvalues of this matrix are much less sensitive to perturbations than those of most other examples presented in this book; the most sensitive eigenvalue in this figure has condition number (defined in §52) less than 100. An analysis of the eigenvectors of random matrices has been carried out by Mehlig and Chalker [543], whose results imply that for dense random matrices, the eigenvalue condition numbers scale as $\mathcal{O}(\sqrt{N})$ as $N \to \infty$. (Mehlig and Chalker show much more than this, establishing details of the probability distributions of eigenvectors and their correlations.)

Figure 35.4 illustrates the mild nonnormality of random matrices from another point of view. For each dimension N from 1 to 100, the figure plots average values of $\kappa(\mathbf{A}_N)$ from an experiment involving 100 random complex matrices. To be precise, it is not exactly $\kappa(\mathbf{A}_N)$ that is averaged in making the figure, since this distribution has a heavy tail (in fact an unintegrable tail, in the case of real matrices). Instead, following Smale and Edel-

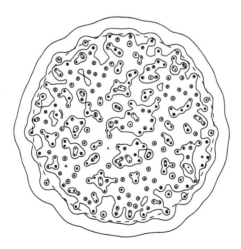

Figure 35.3: Spectrum and ε-pseudospectra of a complex random matrix of dimension $N = 256$, for $\varepsilon = 10^{-1}, 10^{-1.5}, 10^{-2}, 10^{-2.5}, 10^{-3}$. The dashed curve is the unit circle. The degree of nonnormality is mild, of order $\mathcal{O}(\sqrt{N})$, increasing gently as one moves in toward the origin.

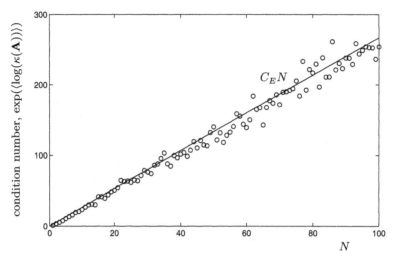

Figure 35.4: Condition numbers of complex random matrices $\mathbf{A}$ (circles). Each data point represents a sample mean $\exp(\langle\log(\kappa(\mathbf{A}))\rangle)$ over 100 matrices of the given dimension N. The condition numbers grow linearly with N. The asymptotic slope for $\kappa(\mathbf{A})$ is $C_E = 2e^{\gamma/2} \approx 2.669$, as indicated by the solid line.

man [225, 699], the logarithms of these condition numbers are averaged and the average is then exponentiated. Thus the plot shows $\exp(\langle\log(\kappa(\mathbf{A}_N))\rangle)$, where $\langle\cdot\rangle$ signifies a mean.

These experiments suggest that the condition number of a random dense matrix grows linearly with the dimension, and indeed this has been proved by Edelman, who found the asymptotic slope to be $C_E = 2e^{\gamma/2} \approx 2.669$, where $\gamma \approx 0.5772$ is Euler's constant [225]. Similar experiments indicate that the condition number of the associated eigenvector matrix also grows linearly. This hardly indicates a high degree of nonnormality; it suggests, as a practical matter, that eigenvalue analysis of random dense matrices is unlikely to lead to difficulties.

Perhaps the modest degree of nonnormality of dense random matrices can best be summarized as follows. From Edelman's theorems we know the condition number scales as $\mathcal{O}(N)$. This rate can be interpreted as the product of two square roots:

$$\mathcal{O}(N) = \mathcal{O}(\sqrt{N}) \times \mathcal{O}(\sqrt{N}).$$

One factor $\sqrt{N}$ comes from the eigenvalues, whose absolute values range from $\mathcal{O}(N^{-1/2})$ to $\mathcal{O}(1)$, and the other comes from their condition numbers. To put it another way, if we take a point z in the unit disk and consider the probability distributions for its distance to the nearest eigenvalue and for its resolvent norm, we find the former scales as $\mathcal{O}(N^{-1/2})$ and the latter as $\mathcal{O}(N)$.

36 · Hatano–Nelson matrices and localization _____

One of the most influential scientific papers of the twentieth century was 'Absence of Diffusion in Certain Random Lattices', published by Nobel laureate Philip Anderson in 1958 [8]. At the heart of Anderson's analysis of the quantum mechanics of disordered systems is a non-Hermitian random matrix problem now known as the *Anderson model*. In its simplest form, the Anderson model is a tridiagonal matrix with 1 on the sub- and superdiagonals and independent samples from a fixed random variable X on the main diagonal:

$$
\mathbf{A} = \begin{pmatrix} \ddots & \ddots & & \\ \ddots & \text{\textit{random}} & 1 & \\ & 1 & \ddots & \ddots \\ & & \ddots & \ddots \end{pmatrix}.
\tag{36.1}
$$

The dimension N is large—in principle, on the order of 10^8, the cube root of Avogadro's number. Mathematically, such matrices are fascinating because in the limit $N \to \infty$, with probability 1, each eigenvector is localized around a central point, decaying exponentially to either side. Physically, this implies that one-dimensional disordered media tend to be insulators, since electrons cannot propagate [751]. (In two or three dimensions, whether or not localization occurs is a more complicated matter.)

Beginning in 1996, Hatano, Nelson, and Shnerb published several papers based on a non-Hermitian analogue of the Anderson model [377, 378, 572, 681], which attracted a great deal of attention [109, 110, 182, 271, 272, 324, 325, 379, 428, 778]. The *non-Hermitian Anderson* or *Hatano–Nelson model* takes the form

$$
\mathbf{A} = \begin{pmatrix} \ddots & \ddots & & e^{-g} \\ \ddots & \text{\textit{random}} & e^{g} & \\ & e^{-g} & \ddots & \ddots \\ e^{g} & & \ddots & \ddots \end{pmatrix},
\tag{36.2}
$$

where g is a fixed real parameter. Again, the diagonal entries are independent samples from a fixed random variable X. Note that Hatano–Nelson matrices differ from Anderson matrices in two respects: They are

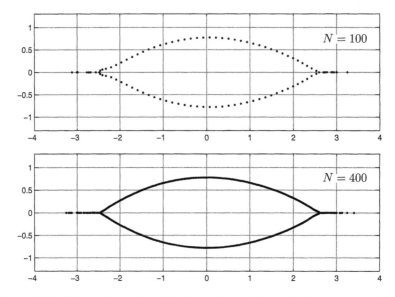

Figure 36.1: Eigenvalues of realizations of the Hatano–Nelson matrix (36.2), based on uniform $[-2, 2]$ random numbers and $g = 1/2$.

non-Hermitian, and they contain corner entries that make the structure periodic. The parameter g controls the degree of asymmetry.

Hatano and Nelson discovered that the eigenvalues of these matrices have a very interesting behavior, illustrated in Figure 36.1 for the case in which X is the uniform distribution on $[-2, 2]$ and $g = 1/2$. We see that most of the eigenvalues lie along a complex 'bubble' extending from about -2.5 to 2.5, and the rest lie on two real 'wings'. As the dimension N is increased, the pattern stays in essentially the same place and becomes more regular. Hatano and Nelson observed this behavior, and theorems explaining it mathematically and quantifying the locations of the curves were published subsequently by Goldsheid and Khoruzhenko [324, 325], Brezin and Zee [109], and Brouwer, Silvestrov, and Beenakker [110].

The behavior of the eigenvectors of (36.2) is even more interesting. The real eigenvalues in Figure 36.1 correspond to localized eigenvectors, as in the Anderson model (although no longer symmetrical). The complex eigenvalues, however, correspond to global, 'delocalized' eigenvectors. Figure 36.2 illustrates these different structures. Roughly speaking, for $g \approx 0$ we have the Anderson situation, with pinned eigenvectors, but as g increases, the 'wind blows stronger' (e.g., in one application of Shnerb and Nelson [681], the magnetic field strength increases); it becomes harder for an eigenvector to remain pinned. Hatano and Nelson speak of a *delocalization transition* as g is increased. For sufficiently large g, all the eigenvalues

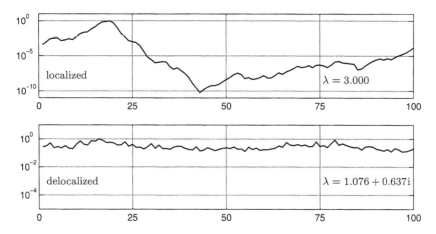

Figure 36.2: Two eigenvectors of the matrix with $N = 100$ of Figure 36.1. The first is exponentially localized, and the second is delocalized. Note the log scales, and also the asymmetry of the localized eigenvector.

move into the complex plane and all the eigenvectors are global.

The Hatano–Nelson matrices (36.2) are not normal, but their nonnormality is mild, and it has little to do with their behavior. If a physical system were governed by one of these matrices, one could expect the eigenvalues and eigenvectors to have physical significance. However, everything changes if the corner entries e^g and e^{-g} are replaced by zero so that we have a *nonperiodic Hatano–Nelson matrix*:

$$\mathbf{A} = \begin{pmatrix} & & & & 0 \\ & & & e^g & \\ & random & & \\ & e^{-g} & & \\ 0 & & & & \end{pmatrix}. \tag{36.3}$$

Now we have a symmetrizable matrix: if $\mathbf{D} = \mathrm{diag}(1, e^g, e^{2g}, \ldots, e^{(N-1)g})$, then $\mathbf{DAD}^{-1}$ is Hermitian. (Physicists call this a gauge transformation.) Thus the eigenvalues of $\mathbf{A}$ must be real, and indeed, they are identical to those for the Anderson model. However, the condition number of $\mathbf{D}$ is exponentially large, so we may expect pronounced effects of nonnormality, and Figure 36.3 reveals just this. The eigenvalues are no longer interesting, but in their place, the pseudospectra trace the very same bubble as before [778]. For further illustrations of pseudospectra for such matrices, see [379, 842].

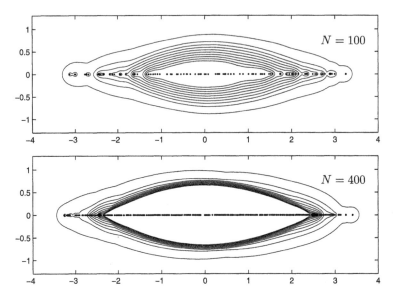

Figure 36.3: Eigenvalues and ε-pseudospectra of nonperiodic Hatano–Nelson matrices (36.3) of dimensions $N = 100$ and 400, for $\varepsilon = 10^{-1}, 10^{-2}, \ldots, 10^{-10}$. Though the eigenvalues are real, the pseudospectra reveal the Hatano–Nelson 'bubble'.

Looking more closely at the pseudospectra reveals an interesting feature of these matrices. In the lower plot of Figure 36.3, the boundaries of the illustrated pseudospectra cluster tightly at the inner edge. What is happening is that as $N \to \infty$, the resolvent norm is diverging to ∞ at an exponential rate inside the bubble, but at a much slower algebraic rate in a certain domain outside the bubble. As $N \to \infty$, the complex plane divides neatly into three regions of distinct behavior, as shown in Figure 36.4: exponential growth of $\|(z - \mathbf{A})^{-1}\|$, algebraic growth, and no growth. Figure 36.5 illustrates this trichotomy by considering one point in each region.

These observations can be explained as follows. For a fixed $z \in \mathbb{C}$, consider the resolvent $\mathbf{R} = (z - \mathbf{A})^{-1}$ for the nonperiodic Hatano–Nelson matrix (36.3). If $\mathbf{y}$ denotes column k of $\mathbf{R}$, then $\mathbf{y}$ satisfies the equation

$$(z - \mathbf{A})\mathbf{y} = \mathbf{e}_k,$$

where $\mathbf{e}_k$ is the kth column of the $N \times N$ identity matrix. Thus the entries y_j of $\mathbf{y}$ are related by a three-term recurrence relation

$$-e^g y_{j-1} + (z - x_j)y_j - e^{-g} y_{j+1} = \begin{cases} 1, & \text{if } j = k, \\ 0, & \text{if } j \neq k, \end{cases} \qquad (36.4)$$

with boundary conditions $y_0 = y_{N+1} = 0$, where x_j is the jth indepen-

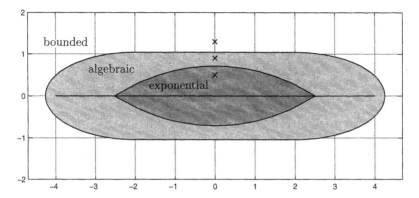

Figure 36.4: Behavior of the nonperiodic Hatano–Nelson matrices (36.3) as $N \to \infty$. At any point z in the dark region ($\Omega_{\mathrm{I}} \cup \Omega_{\mathrm{II}}$), the resolvent norm grows exponentially at a rate determined by the Lyapunov constant of a random system. At any point z in the light region (Ω_{III}), it grows at a far slower algebraic rate determined by rare events. At any point z in the region exterior to these (Ω_{IV}), it is uniformly bounded. The crosses mark the special values $z = 0.5\mathrm{i}, 0.9\mathrm{i}, 1.3\mathrm{i}$ considered in Figure 36.5.

dent sample from the random variable X. Now a three-term recurrence relation has in general a two-dimensional space of solutions. Suppose for a moment that the coefficients x_j are constant. Then generically, this space is spanned by the particular solutions $y_j = \lambda_1^j$ and $y_j = \lambda_2^j$ for two numbers λ_1 and λ_2, each of which is either <1 or >1 in absolute value. Depending on these inequalities, we may say that the two fundamental solutions fit the pattern growth/decay, growth/growth, or decay/decay. If it is growth/decay, (36.4) will generally have a solution of size $\mathcal{O}(1)$ localized at $j = k$, decaying exponentially to either side. In the growth/growth case we get an exponentially large solution with a boundary layer at $j \approx N$, and in the decay/decay case, an exponentially large solution with a boundary layer at $j \approx 0$. (These are precisely the phenomena underlying the behavior of pseudospectra of tridiagonal Toeplitz matrices discussed in §7, where growth/growth, growth/decay, and decay/decay correspond to winding numbers 1, 0, and -1 of z with respect to the symbol curve.)

For our present application, the coefficients are random because of the numbers x_j. The behavior remains much as just described, except that the roles of λ_1 and λ_2 are now played by two real numbers known as the *Lyapunov constants* of this stochastic recurrence relation, each of which generically will be >1 or <1. Again we have $\mathcal{O}(1)$ solutions localized at $j = k$ if the pattern is growth/decay, but exponentially large solutions with a boundary layer if the pattern is growth/growth or decay/decay. In addition there is a further complication: Depending on z, g, and X, these

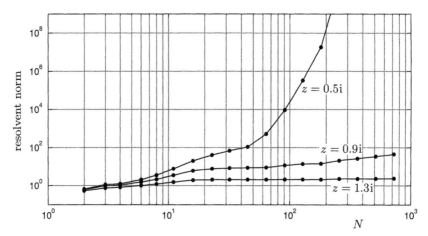

Figure 36.5: Resolvent norm $\|(z - \mathbf{A})^{-1}\|$ as a function of dimension N for the nonperiodic Hatano–Nelson matrices of (36.3), for three values of z.

conclusions may hold surely (i.e., for all possible sequences $\{x_j\}$) or *almost surely*, abbreviated *a.s.* (i.e., with probability 1—for all but a measure 0 subset of sequences $\{x_j\}$).

The different regions of Figure 36.4 are explained by these various possibilities. We can distinguish four cases, corresponding to four regions Ω_{I}, Ω_{II}, Ω_{III}, Ω_{IV} in the complex plane:

Ω_{I}: If solutions to (36.4) grow/grow or decay/decay,
    $\|(z - \mathbf{A})^{-1}\|$ grows exponentially as $N \to \infty$.

Ω_{II}: If solutions to (36.4) grow/grow or decay/decay (a.s.),
    $\|(z - \mathbf{A})^{-1}\|$ grows exponentially as $N \to \infty$ (a.s.).

Ω_{III}: If solutions to (36.4) grow/decay (a.s. but not surely),
    $\|(z - \mathbf{A})^{-1}\|$ grows algebraically as $N \to \infty$ (a.s.).

Ω_{IV}: If solutions to (36.4) grow/decay (surely),
    $\|(z - \mathbf{A})^{-1}\|$ is bounded as $N \to \infty$.

Both Ω_{I} and Ω_{II} contribute to the region of exponential growth observed in practice, i.e., the region inside the bubble in Figure 36.4. Goldsheid and Khoruzhenko have shown that the bubble itself, i.e., the curve separating Ω_{II} and Ω_{III}, can be characterized by an equipotential condition [324, 325].

The observation that there may be distinct regions of exponential and algebraic resolvent norm growth comes from [778], where the division into regions $\Omega_{\mathrm{I}}, \Omega_{\mathrm{II}}, \Omega_{\mathrm{III}}, \Omega_{\mathrm{IV}}$ is introduced. A rigorous treatment of these distinctions is presented in that paper for a bidiagonal or 'one-way' analogue of the nonperiodic Hatano–Nelson matrices,

$$
\mathbf{A} = \begin{pmatrix} \ddots & \ddots & & \\ & \textit{random} & 1 & \\ & & \ddots & \ddots \\ & & & \ddots \end{pmatrix},
\tag{36.5}
$$

where as always, the diagonal entries are samples from a random variable X. (A periodic version of this matrix had previously been considered in [109, 271, 272].) The behavior of this matrix is essentially the same as that of (36.3), but much easier to analyze. Since (36.5) is bidiagonal, its resolvent can be computed immediately: If r_{jk} is the (j,k) entry of $(z-\mathbf{A})^{-1}$ for $k \geq j$, we have $r_{jk} = \prod_{\ell=j}^{k} (z - x_\ell)^{-1}$ and therefore

$$
\log |r_{jk}| = -\sum_{\ell=j}^{k} \log |z - x_\ell|.
\tag{36.6}
$$

(If $z = x_\ell$ in any of these factors, we define $|r_{jk}| = \infty$.) This formula reveals that $\log |r_{jk}|$ is the sum of independent random variables, so it is readily analyzed by the Central Limit Theorem and related tools of probability theory.[1] Assume that $\mathrm{supp}(X)$, the support of X, is compact. Define

$$
d_{\mathsf{min}}(z) = \min_{x \in \mathrm{supp}(X)} |z - x|, \qquad d_{\mathsf{max}}(z) = \max_{x \in \mathrm{supp}(X)} |z - x|,
$$

and

$$
d_{\mathsf{mean}}(z) = \exp(\langle \log |z - X| \rangle),
$$

where $\langle \cdot \rangle$ denotes expected value. Thus $d_{\mathsf{mean}}(z)$ is the geometric mean distance of z to $\mathrm{supp}(X)$, weighted appropriately by the probability measure of X. Then we have $0 \leq d_{\mathsf{min}}(z) \leq d_{\mathsf{mean}}(z) \leq d_{\mathsf{max}}(z) < \infty$, and we can define the four subsets of the complex plane by

$$
\begin{aligned}
\Omega_{\mathrm{I}} &: \quad d_{\mathsf{max}}(z) < 1, \\
\Omega_{\mathrm{II}} &: \quad d_{\mathsf{mean}}(z) < 1 \leq d_{\mathsf{max}}(z), \\
\Omega_{\mathrm{III}} &: \quad d_{\mathsf{min}}(z) \leq 1 \leq d_{\mathsf{mean}}(z), \\
\Omega_{\mathrm{IV}} &: \quad 1 < d_{\mathsf{min}}(z).
\end{aligned}
$$

(Either or both of Ω_{I} and Ω_{II} may be empty, but Ω_{III} and Ω_{IV} are always nonempty.) The sets are disjoint, with $\Omega_{\mathrm{I}} \cup \Omega_{\mathrm{II}} \cup \Omega_{\mathrm{III}} \cup \Omega_{\mathrm{IV}} = \mathbb{C}$. (Any or all of Ω_{I}, Ω_{II}, and Ω_{III} may contain a portion of $\mathrm{supp}(X)$, but Ω_{IV}

---

[1]For a similar analysis of a related problem of stochastic differential equations, see [385].

lies at a distance 1 from supp(X).) Lemma 3.1 of [778] asserts that for z in Ω_I, Ω_{II}, Ω_{III}, and Ω_{IV} we have guaranteed exponential growth, almost sure exponential growth, almost sure exponential decay, and guaranteed exponential decay of $|r_{j-i}|$ as $j - i \to \infty$, respectively. From this point the following theorem is then proved. As always, we write $\|(z - \mathbf{A})^{-1}\| = \infty$ if $(z - \mathbf{A})^{-1}$ does not exist, and the definition of the limit of sets at the end is that $K_N \to K$ as $N \to \infty$ if for every $\varepsilon > 0$, there exists an integer N_0 such that for all $N \geq N_0$, K_N and K are each contained in the ε-neighborhood of the other. The notation conv($\cdot$) denotes the convex hull.

Pseudospectra of nonperiodic bidiagonal random matrices

Theorem 36.1 *Let A be an $N \times N$ matrix of the form (36.5), let $z \in \mathbb{C}$ be fixed, and let $d_{\mathsf{min}}(z)$, $d_{\mathsf{mean}}(z)$, $d_{\mathsf{max}}(z)$ and Ω_I, Ω_{II}, Ω_{III}, Ω_{IV} be defined as above.*

(i) *If $z \in \Omega_I$, then $\|(z - \mathbf{A})^{-1}\| \geq d_{\mathsf{max}}(z)^{-N}$ (exponential growth).*

(ii) *If $z \in \Omega_I \cup \Omega_{II}$, then $\|(z - \mathbf{A})^{-1}\| \to \infty$ a.s., and if in addition $z \notin \mathrm{supp}(X)$, then $\|(z - \mathbf{A})^{-1}\|^{1/N} \to d_{\mathsf{mean}}(z)^{-1}$ a.s. as $N \to \infty$ (almost sure exponential growth).*

(iii) *If $z \in \Omega_{III}$, then $\|(z - \mathbf{A})^{-1}\| \to \infty$ a.s., and if in addition $z \notin \mathrm{supp}(X)$, then $\|(z - \mathbf{A})^{-1}\|^{1/N} \to 1$ a.s. as $N \to \infty$ (almost sure subexponential growth).*

(iv) *If $z \in \Omega_{IV}$, then $\|(z - \mathbf{A})^{-1}\| < (d_{\mathsf{min}}(z) - 1)^{-1}$ (boundedness).*

The spectrum satisfies $\sigma(\mathbf{A}) \subseteq \mathrm{supp}(X)$, with $\sigma(\mathbf{A}) \to \mathrm{supp}(X)$ a.s. as $N \to \infty$. The numerical range satisfies $W(\mathbf{A}) \subseteq \mathrm{conv}(\Omega_{III})$, with $W(\mathbf{A}) \to \mathrm{conv}(\Omega_{III})$ a.s. as $N \to \infty$.

The variable d_{mean} is essentially a potential, and it is the appearance of this variable in the theorem that corresponds to the use of potential theory by Goldsheid and Khoruzhenko [324]. In [778], analogous theorems are proved for periodic bidiagonal random matrices, singly infinite bidiagonal random matrices ('stochastic Toeplitz operators'), and doubly infinite bidiagonal random matrices ('stochastic Laurent operators'). In the case of periodic bidiagonal random matrices, the eigenvalues lie almost surely along the bubble and wings as in the Hatano–Nelson example, defined by

$$S_{\mathsf{bubble}} = \{z \in \mathbb{C} : d_{\mathsf{mean}}(z) = 1\}, \quad S_{\mathsf{wings}} = \{z \in \mathrm{supp}(X) : d_{\mathsf{mean}}(z) > 1\}.$$

For the infinite matrices, the spectrum is $\Omega_I \cup \Omega_{II} \cup \Omega_{III}$ a.s. in the singly infinite case and $\Omega_{II} \cup \Omega_{III}$ a.s. in the doubly infinite case. These large spectra of the infinite matrices result from the fact that at a point $z \in \Omega_{III}$, although the resolvent entries generally decay away from the diagonal, unusual random patterns in the coefficients will cause localized pockets of growth. For a finite matrix, these pockets account for the algebraic growth in condition (iii) of Theorem 36.1, and for an infinite matrix, since arbitrarily large pockets will almost surely appear, the resolvent norm is almost surely ∞ throughout Ω_{III}. These matters of spectra of infinite random matrices are investigated in detail in [182], but although it is a mathematical fact that Ω_{III} is part of the spectrum of the infinite matrix (a.s.), it would be a rare application in which the finite-dimensional matrices would be sufficiently large for Ω_{III} to behave like part of the spectrum in practice.

Figure 36.6 illustrates Theorem 36.1 with the first pseudospectra of a matrix of dimension in the millions ever computed, from [778]. Here $\mathbf{A}$ is a single $10^6 \times 10^6$ realization of (36.5), where X is again the uniform distribution on $[-2, 2]$.

Of course, this is not the only possible choice of X. Figure 36.7, also from [778], shows pseudospectra for bidiagonal matrices of dimensions 10^2 and 10^4, with X taken as the uniform discrete distribution on the points $\{1, -1\}$. Here Ω_{I} is again empty, Ω_{II} is the open set bounded by the lemniscate $|z^2 - 1| = 1$, and Ω_{III} is the region defined by $|z^2 - 1| \geq 1$ and $|z - 1| \leq 1$ or $|z + 1| \leq 1$.

In this section we have considered several classes of non-Hermitian random matrices, some periodic and others nonperiodic. The two types of matrices appear quite different, and it is natural to ask, which one might be a realistic model of a physical system? We believe the answer is simply that the periodic matrices might be good models of certain periodic

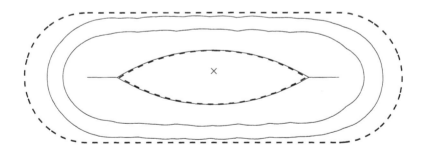

Figure 36.6: Numerically computed ε-pseudospectra for a single $N \times N$ nonperiodic bidiagonal matrix (36.5) with $N = 10^6$ for $\varepsilon = 10^{-1}$, 10^{-2}, and 10^{-100}, from [778], where X is the uniform distribution on $[-2, 2]$. The inner dashed curve is the bubble separating Ω_{II} and Ω_{III}, defined by $d_{\mathsf{mean}}(z) = 1$, and the outer dashed curve is the boundary of Ω_{III} and Ω_{IV}, defined by $d_{\mathsf{min}}(z) = 1$. Inside the bubble, the resolvent norm is very large: at the point $z = 0.1\mathrm{i}$ marked by the cross, about 10^{99698}. The eigenvalues, 10^6 random numbers in $[-2, 2]$, are not shown.

systems and the nonperiodic ones might be good models of certain nonperiodic systems. In the papers by Hatano and Nelson and others that made
these problems famous, only the periodic matrices appear. It was argued
in [778] that the ultimate reason for this may be that these papers rely on
eigenvalue analysis, which fails in the nonperiodic case. In [377, 378], the
matrix (36.2) is proposed as a model of a type II superconductor in the
shape of a cylinder, and in [572, 681], the same matrix models a biological
system in a circular petri dish.

This raises the question of whether a delocalization effect occurs in the
nonperiodic case, and if so, how it can be examined if not by eigenvectors.
Indeed, what is the meaning of localization if not as a statement about
eigenvectors?

The philosophy throughout this book is that what matters about a
matrix or operator $\mathbf{A}$ is ultimately its *behavior* as measured by functions
such as $\mathbf{A}^{-1}$, $\mathbf{A}^k$, or $e^{t\mathbf{A}}$ and their norms (see §47). Let us first ask, if
$\mathbf{A}$ is Hermitian and all its eigenvectors are localized, as in the Anderson
case, what does this imply about functions of $\mathbf{A}$? One way to answer

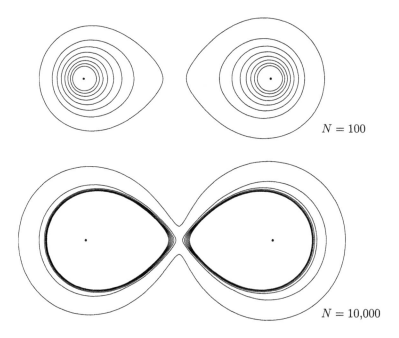

$N = 100$

$N = 10,000$

Figure 36.7: Spectra and ε-pseudospectra of two nonperiodic bidiagonal matrices (36.5), where X is the uniform discrete distribution on $\{1, -1\}$, for
$\varepsilon = 10^{-2}, 10^{-6}, 10^{-10}, \ldots, 10^{-30}$, from [778]. The solid dots, the eigenvalues,
are the points $z = \pm 1$. The resolvent norm increases exponentially with N
throughout the lemniscatic region defined by $|z^2 - 1| < 1$.

the question is to diagonalize $\mathbf{A}$: $\mathbf{A} = \mathbf{Q}\mathbf{\Lambda}\mathbf{Q}^*$, where $\mathbf{Q}$ is unitary, that is, $\mathbf{A} = \sum_k \lambda_k \mathbf{q}_k \mathbf{q}_k^*$, where $\{\mathbf{q}_k\}$ are the normalized eigenvectors. If the eigenvectors are localized, then the significant entries in each matrix $\mathbf{q}_k \mathbf{q}_k^*$ are concentrated near the diagonal. Since $f(\mathbf{A}) = \sum_k f(\lambda_k) \mathbf{q}_k \mathbf{q}_k^*$ for any function f analytic on $\sigma(\mathbf{A})$, we see that the same is true of $f(\mathbf{A})$, and thus here is our answer: For any f analytic on the spectrum of $\mathbf{A}$, $f(\mathbf{A})$ maps inputs localized at a particular point to outputs localized at the same point.

This argument breaks $\mathbf{A}$ into its various eigenspaces and then puts them together. But we can argue without eigenvectors if we break $\mathbf{A}$ into its actions at various complex frequencies z and then put *these* together. For any z, consider the ODE system

$$\mathbf{u}'(t) = \mathbf{A}\mathbf{u}(t) + e^{zt}\mathbf{v}, \tag{36.7}$$

where $\mathbf{v}$ is a fixed vector. The solution is

$$\mathbf{u}(t) = e^{zt}\mathbf{u}_0, \qquad \mathbf{u}_0 = (z - \mathbf{A})^{-1}\mathbf{v}. \tag{36.8}$$

Thus the response to $\mathbf{v}$ at frequency z is determined by the resolvent (or *Green's function*) $(z - \mathbf{A})^{-1}$. Now suppose that $(z - \mathbf{A})^{-1}$, like the matrices $\mathbf{q}_k \mathbf{q}_k^*$ in the last paragraph, has its entries concentrated near the diagonal. Then the response to a localized input at frequency z will be localized. Combining frequencies, we can write a function $f(\mathbf{A})$ as a Cauchy integral

$$f(\mathbf{A}) = \frac{1}{2\pi i}\int_\Gamma (z - \mathbf{A})^{-1} f(z)\, dz, \tag{36.9}$$

where Γ is any Jordan curve enclosing the spectrum of $\mathbf{A}$ but no singularities of f; see §14. From here we reach the same conclusion as before: for any f, $f(\mathbf{A})$ maps localized inputs to localized outputs.

Next we may ask, what if $\mathbf{A}$ is a matrix of the Hatano–Nelson type (36.2), non-Hermitian but not too far from normal, for which some of the eigenvectors are localized but not all of them? Does $f(\mathbf{A})$ map localized inputs to localized outputs? The answer will depend on f. Supposing $\mathbf{A}$ is diagonalizable, we can argue with eigenvectors again. Decompose $\mathbf{A}$ as $\mathbf{A} = \mathbf{X}\mathbf{\Lambda}\mathbf{X}^{-1} = \sum_k \lambda_k \mathbf{x}_k \mathbf{y}_k^*$, where $\mathbf{x}_k$ denotes the kth column of $\mathbf{X}$ and $\mathbf{y}_k^*$ denotes the kth row of $\mathbf{X}^{-1}$. We can assume that since $\mathbf{A}$ is not too far from normal, the left eigenvectors $\{\mathbf{y}_k^*\}$ are localized like the right eigenvectors $\{\mathbf{x}_k\}$. Now suppose that for any eigenvalue of $\mathbf{A}$ whose eigenvector is not localized, $f(\lambda)$ is negligible, such as $f(z) = z^k$ for $|z| < 1$ and large k or $f(z) = e^{tz}$ for $\mathrm{Re}\, z < 0$ and large t. Then $f(\mathbf{A}) = \sum_k f(\lambda_k)\mathbf{x}_k \mathbf{y}_k^*$ will again have its significant entries near the diagonal and map localized inputs to localized outputs. On the other hand, if f takes significant values at some eigenvalues with delocalized eigenvectors, this will not be so.

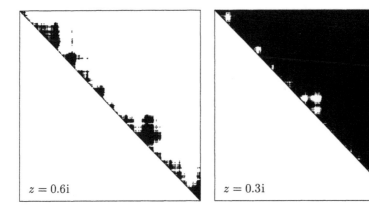

$z = 0.6\text{i}$ $z = 0.3\text{i}$

Figure 36.8: Delocalization as seen in the resolvents $(z - \mathbf{A})^{-1}$ for a nonperiodic bidiagonal matrix (36.5) of dimension $N = 400$, from [778]. A dot is printed at each position of the matrix with $|r_{jk}| > 1/2$. For z outside the bubble (first plot), the resolvent is concentrated near the diagonal, while for z inside the bubble (second plot), it is not concentrated. As z approaches the bubble, local exceptions of larger and larger scale appear.

The Cauchy integral (36.9) enables us to broaden this conclusion to an arbitrary $\mathbf{A}$, which may or may not be diagonalizable or close to normal. Suppose finally that we have a matrix like (36.3) or (36.5) with the property that for some values of z, the nonnegligible entries of $(z - \mathbf{A})^{-1}$ are concentrated near the main diagonal, while for other values of z, they are not. Suppose that for those values of z where $(z - \mathbf{A})^{-1}$ is not localized, $f(z)$ is negligible. Then again, from the Cauchy integral, we conclude that $f(\mathbf{A})$ will map localized inputs to localized outputs.

Figure 36.8 illustrates these ideas for a nonperiodic bidiagonal matrix (36.5) as in Figure 36.6 but of dimension 400. We see that a stimulus at $z = 0.6\text{i}$, outside the bubble, excites a response only near the diagonal (exponentially decaying away from the diagonal), whereas $z = 0.3\text{i}$, inside the bubble, excites a global response (exponentially growing). It follows that $\exp(t(-\text{i}\mathbf{A} - 0.6))$, for example, would tend to have localized behavior for large t, whereas $\exp(t(-\text{i}\mathbf{A} - 0.3))$ would not. Both images show pockets of exceptional behavior near the diagonal, whose scale will diverge to ∞ as z approaches the bubble, a phenomenon familiar to physicists interested in the behavior of condensed matter near critical points.

37 · Random Fibonacci matrices

In the last section we saw that the resolvents of certain random bidiagonal and tridiagonal matrices are related to recurrence relations with random coefficients. Here we consider related matrices in which the second super-diagonal may also be nonzero, and in particular, make a connection with random Fibonacci sequences and related problems.

We concentrate on what we shall call the *random Fibonacci matrix*,

$$\mathbf{A} = \begin{pmatrix} 0 & \pm 1 & \pm 1 & & & & \\ & 0 & \pm 1 & \pm 1 & & & \\ & & \ddots & \ddots & \ddots & & \\ & & & 0 & \pm 1 & \pm 1 & \\ & & & & 0 & \pm 1 & \\ & & & & & 0 \end{pmatrix}, \tag{37.1}$$

of dimension N. This matrix has two nonzero diagonals along which the numbers 1 and -1 appear at each position randomly with equal proba-bility $1/2$. We say that these numbers are independent samples from the $\{\pm 1\}$ distribution. Figure 37.1 shows pseudospectra of a matrix of this kind of dimension $N = 300$. We see that the pseudospectra are approximately disks about the origin and that the resolvent norm $\|(z - \mathbf{A})^{-1}\|$ is very large for smaller values of $|z|$. Figure 37.2 shows norms of powers of the same matrix. Since $\mathbf{A}$ is strictly upper triangular, it is nilpotent, but as one would expect from the pseudospectra, a great deal of transient growth in the powers occurs before the eventual collapse to zero.

A closer look at Figure 37.1 raises some questions. Several of the con-tours show irregularities, and one might wonder whether these are caused by some kind of computational mistake. In fact, they are genuine, as in-dicated by the closeups shown in Figure 37.3. We are observing that, at various points $z \in \mathbb{C}$, $\|(z-\mathbf{A})^{-1}\|$ has a local minimum. If we took a differ-ent matrix $\mathbf{A}$ from the same class, the locations of the local minima would be different, but the general pattern would typically remain the same.

Figure 37.4 shows the norm of the resolvent $\|(z - \mathbf{A})^{-1}\|$ for values of z along the positive real axis for matrices $\mathbf{A}$ of various dimensions. We see that above a certain number $z \approx 1.231$, the norms are modest, while for smaller z, they are exponentially large. The increase as $z \to 0$ is not monotonic, however. There are little pockets of contrary behavior, another reflection of the local minima of Figure 37.3.

Although the behavior revealed in Figures 37.1–37.4 is intricate, it is not hard to analyze, at least the main features. We shall look at resolvents

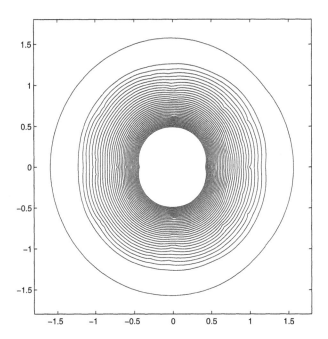

Figure 37.1: ε-pseudospectra of a random Fibonacci matrix (37.1) of dimension 300, for $\varepsilon = 10^{-1}, 10^{-4}, 10^{-7}, \dots, 10^{-100}$. The irregularities are genuine.

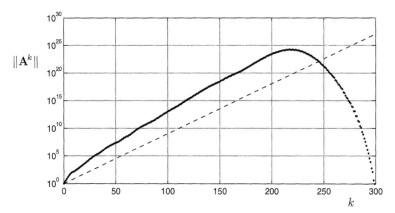

Figure 37.2: Norms of powers of a matrix of the form (37.1) of dimension 300. The norm of $\mathbf{A}^{299}$ is exactly 1, and for $k \geq 300$, $\mathbf{A}^k = \mathbf{0}$. The dashed line represents 1.231^k. Compare Figure 14.4.

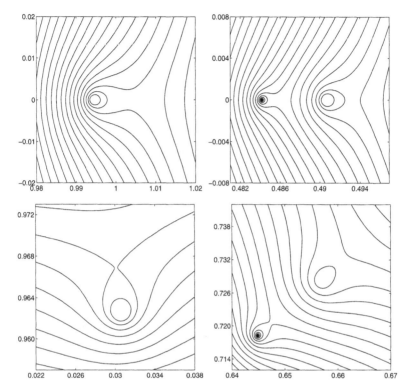

Figure 37.3: Closeups of Figure 37.1 near four points z where $\|(z - \mathbf{A})^{-1}\|$ has a local minimum. The largest values of ε (i.e., those nearest the minima) in the four plots are (a) $10^{-15.2}$, (b) $10^{-81.2}$, (c) $10^{-23.6}$, and (d) $10^{-21.2}$, with the additional contours in each case corresponding to successive reductions of ε by powers of $10^{0.2}$.

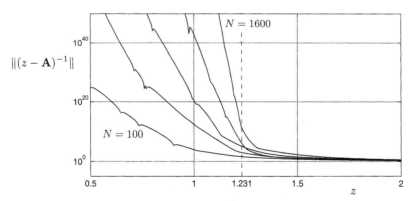

Figure 37.4: Resolvent norms $\|(z - \mathbf{A})^{-1}\|$ along the $z > 0$ axis for random Fibonacci matrices (37.1) of dimensions $N = 100, 200, 400, 800, 1600$.

and pseudospectra; the analysis of the the matrix powers themselves would be much the same. Taking first $z = 1$ for simplicity, consider

$$\mathbf{I} - \mathbf{A} = \begin{pmatrix} 1 & -a_1 & -b_2 & & \\ & 1 & -a_2 & -b_3 & \\ & & 1 & -a_3 & \ddots \\ & & & & \ddots & \ddots \end{pmatrix}, \tag{37.2}$$

where $\{a_j\}$ and $\{b_j\}$ are independent samples from the $\{\pm 1\}$ distribution. Since $\mathbf{I} - \mathbf{A}$ is upper triangular, so is $(\mathbf{I} - \mathbf{A})^{-1}$. If we write

$$(\mathbf{I} - \mathbf{A})^{-1} = \begin{pmatrix} r_{11} & r_{12} & r_{13} & r_{14} & \cdots \\ & r_{22} & r_{23} & r_{24} & \cdots \\ & & r_{33} & r_{34} & \cdots \\ & & & & \ddots & \ddots \end{pmatrix}, \tag{37.3}$$

then the product of (37.3) and (37.2) must be the identity, and from the top row of this product $(\mathbf{I} - \mathbf{A})^{-1}(\mathbf{I} - \mathbf{A})$ we find that

$$\begin{aligned} r_{11} &= 1, \\ r_{12} &= a_1 r_{11}, \\ r_{13} &= a_2 r_{12} + b_2 r_{11}, \\ r_{14} &= a_3 r_{13} + b_3 r_{12}, \\ r_{15} &= a_4 r_{14} + b_4 r_{13}, \\ &\quad\cdots\cdots \end{aligned}$$

In other words, the first row of $(\mathbf{I} - \mathbf{A})^{-1}$ contains numbers r_{1j} generated by a three-term recurrence relation with random coefficients, the *random Fibonacci recurrence*,

$$x_{j+1} = \pm x_j \pm x_{j-1}, \tag{37.4}$$

where each coefficient is an independent sample from the $\{\pm 1\}$ distribution. (The other rows also contain numbers generated by the same formula.) This recurrence relation was analyzed in a beautiful paper by Viswanath [809], who proved that any sequence generated by (37.4) almost surely satisfies

$$\lim_{j \to \infty} |x_j|^{1/j} = V = 1.13198824\ldots; \tag{37.5}$$

V is *Viswanath's constant*. From (37.5) it follows that

$$\lim_{N \to \infty} \|(\mathbf{I} - \mathbf{A}_N)^{-1}\|^{1/N} = V \tag{37.6}$$

for a sequence of $N \times N$ matrices (37.1) with probability 1. For $N = 300$ we have $V^N \approx 1.42 \times 10^{16}$; this is consistent with the fact that $z = 1$ is between the $\varepsilon = 10^{-15.6}$ and $\varepsilon^{-15.8}$ contours in the first panel of Figure 37.3.

All this concerns the resolvent $\|(z - \mathbf{A})^{-1}\|$ at $z = 1$. For general z, the diagonal of (37.2) changes from 1 to z and the equations take the form

$$
\begin{aligned}
r_{11} &= z^{-1}, \\
r_{12} &= z^{-1} a_1 r_{11}, \\
r_{13} &= z^{-1} a_2 r_{12} + z^{-1} b_2 r_{11}, \\
r_{14} &= z^{-1} a_3 r_{13} + z^{-1} b_3 r_{12}, \\
r_{15} &= z^{-1} a_4 r_{14} + z^{-1} b_4 r_{13}, \\
&\cdots
\end{aligned}
$$

corresponding to the generalization of (37.4),

$$
x_{j+1} = \pm z^{-1} x_j \pm z^{-1} x_{j-1}. \tag{37.7}
$$

The investigation of multistep random recurrences like these is often pursued by formulating them as one-step matrix processes,

$$
\begin{pmatrix} x_j \\ x_{j+1} \end{pmatrix} = \begin{pmatrix} 0 & 1 \\ \pm z^{-1} & \pm z^{-1} \end{pmatrix} \begin{pmatrix} x_{j-1} \\ x_j \end{pmatrix}. \tag{37.8}
$$

By iterating this step, we see that the random recurrence is a problem of *random matrix products*, a subject initiated by Bellman in 1954 and studied by Furstenberg and many others since then [45, 94, 295]. For the particular recurrence (37.7), we find that for larger values of z, solutions decay exponentially with probability 1, whereas for smaller values of z they grow exponentially with probability 1; the exponential growth or decay constant is the *Lyapunov constant* for the recurrence (more precisely, the larger of its two Lyapunov constants). The transition value at which the Lyapunov constant is 1, so that there is neither exponential growth nor exponential decay, is ≈ 1.231; we do not know this number exactly. Figure 37.5 illustrates these three situations. The irregularity in the curves reflects the same statistical fluctuations that gave rise to the irregularities in Figures 37.1, 37.3, and 37.4 discussed above.

The growth or decay of solutions to the recurrence (37.7) affects the resolvent norm $\|(z - \mathbf{A})^{-1}\|$ for a random Fibonacci matrix (37.1) in the same manner as we saw for Hatano–Nelson and related matrices in the last section. Within a region Ω_{I} close to the origin, $\|(z - \mathbf{A})^{-1}\|$ must grow exponentially as $N \to \infty$, and within a somewhat larger region Ω_{II} it will almost surely grow exponentially (i.e., with probability 1); in a typical experiment Ω_{I} and Ω_{II} will be indistinguishable. Further from the origin

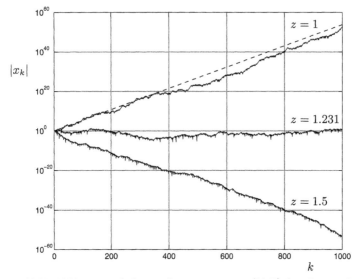

Figure 37.5: 1000 steps of the random recurrence (37.7) for three values of the parameter z. The dashed line represents V^k, where V is Viswanath's constant $1.13198824\ldots$.

lies a region Ω_{III} of algebraic but not exponential growth (almost sure), and further out still, for $|z| > 2$, a region Ω_{IV} in which the resolvent norm is bounded by $1/(|z| - 2)$, independently of N. This behavior is summarized in Figure 37.6, which is analogous to Figure 36.4. Theorems to establish these behaviors rigorously have not yet been proved, and indeed, we do not know whether the boundary between Ω_{II} and Ω_{III} is smooth or irregular, as might be suggested by the fractal dependences of random recurrences on their defining parameters investigated in [246].

Random Fibonacci matrices can be generalized in all kinds of ways. An empty generalization would be to eliminate the random signs on one of the diagonals, for the matrices

$$
\begin{pmatrix}
0 & 1 & \pm 1 & & & \\
& \ddots & \ddots & \ddots & & \\
& & 0 & 1 & \pm 1 \\
& & & 0 & 1 \\
& & & & 0
\end{pmatrix},
\quad
\begin{pmatrix}
0 & \pm 1 & 1 & & & \\
& \ddots & \ddots & \ddots & & \\
& & 0 & \pm 1 & 1 \\
& & & 0 & \pm 1 \\
& & & & 0
\end{pmatrix}
$$

have the same properties as (37.1), as is readily proved by unitary diagonal similarity transformations. A more substantial change would be to modify the $\{\pm 1\}$ distribution, replacing it, for example, by a normal distribution or by a distribution of complex numbers on the unit circle; some of the

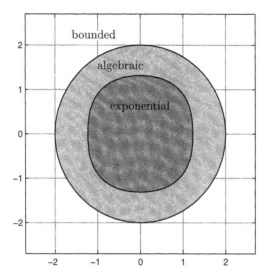

Figure 37.6: Behavior of random Fibonacci matrices (37.1) as $N \to \infty$ (schematic); compare Figure 36.4. In the dark region, the resolvent norm grows exponentially (surely or almost surely), in the light region it grows subexponentially (almost surely), and in the region with $|z| > 2$ it is bounded by $1/(|z| - 2)$.

effects that arise from such generalizations have been studied in [246, 839]. Alternatively, we could retain the $\{\pm 1\}$ distribution but modify the selection of diagonals on which it is applied. Figure 37.7 shows four variations of this type. In the first, 0 is replaced on the main diagonal by $\{\pm 1\}$; the result is again a strongly nonnormal matrix, but now with two eigenvalues ± 1 instead of the single eigenvalue 0. In the second, it is the first subdiagonal that is replaced by $\{\pm 1\}$; now the eigenvalues are complex and the matrix is again far from normal, though not so far as before. In the third, we have a matrix in which random $\{\pm 1\}$ entries appear solely on the first subdiagonal and the first superdiagonal. Note the two axes of symmetry and the nonzero fraction of eigenvalues that are pure imaginary. Such 'sign model' matrices, which are not strongly nonnormal, have been analyzed by Feinberg and Zee and others [152, 271]. Finally, the fourth panel returns to the random Fibonacci matrix (37.1), with the single change that the structure is made periodic by including random ± 1 entries in positions $(N-1, 1)$, $(N, 1)$, and $(N, 2)$. As with the Hatano–Nelson matrices of the last section, this periodicity changes the picture completely inside the 'bubble' that separates Ω_{II} and Ω_{III}, where the resolvent is delocalized, but has little effect outside, where it is localized. To physicists interested in random matrices, this is a familiar principle: if the boundary conditions matter, the eigenvectors are delocalized [751].

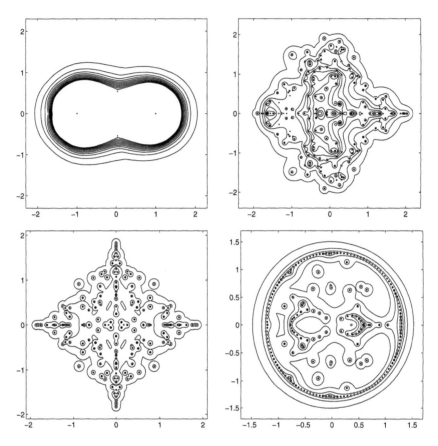

Figure 37.7: Variations on the theme of the random Fibonacci matrix (37.1) with $N = 200$ and $\varepsilon = 10^{-1}, 10^{-1.5}, \dots, 10^{-6}$. (a) Main diagonal $\{\pm 1\}$ instead of 0. (b) First subdiagonal $\{\pm 1\}$ instead of 0. (c) $\{\pm 1\}$ entries on first sub- and superdiagonals. (d) Same as (37.1) but with nonzero entries in lower left corner to make the structure periodic.

All the matrices we have mentioned share a pattern: Along each diagonal, the entries are independent samples from a fixed distribution. They might accordingly be called *stochastic Toeplitz matrices*, or in the periodic case, *stochastic circulant matrices*. Clearly the possible structures of the spectra and pseudospectra of such matrices are extremely varied. A further example we shall investigate, in the next section, is a case in which nonzero entries appear throughout the upper triangular block.

38 · Random triangular matrices

In this final section of this part of the book, we turn our attention to another class of random matrices with exponentially strong nonnormality: matrices of the strictly upper triangular form

$$
\mathbf{A} = \begin{pmatrix}
0 & a_{12} & a_{13} & \cdots & a_{1N} \\
 & 0 & a_{23} & \cdots & a_{2N} \\
 & & 0 & \ddots & \vdots \\
 & & & \ddots & a_{N-1,N} \\
 & & & & 0
\end{pmatrix}, \tag{38.1}
$$

with independent nonzero entries drawn from a normal distribution. To obtain clean behavior in the limit $N \to \infty$, we follow the example of §35 and take this to be the normal distribution with mean zero and variance $1/N$, denoted by $N(0, N^{-1})$. We shall also be interested in variations of $\mathbf{A}$ that have random entries on the main diagonal instead of zeros, or complex entries instead of real.[1]

A hint that such upper triangular matrices may exhibit significant nonnormality comes from the fact that, aside from the trivial exception $\mathbf{A} = \mathbf{0}$, they are all nondiagonalizable. Since $\mathbf{A}$ is already in upper triangular form, it is simple to gauge its *departure from normality*, $\mathrm{dep_F}(\mathbf{A})$ (defined in §48). We expect each $N(0, N^{-1})$ entry to have magnitude on the order of $N^{-1/2}$. As $\mathbf{A}$ has $\mathcal{O}(N^2)$ such entries, we expect that $\mathrm{dep_F}(\mathbf{A}) = \|\mathbf{A}\|_F$ will grow at a rate $\mathcal{O}(N^{1/2})$ as $N \to \infty$. Though statistics like these hint that typical matrices $\mathbf{A}$ will be far from normal, they shed little light on their behavior. Figure 14.2 indicates that nonnormality is significant here.

More precise information was obtained by Viswanath and Trefethen, who considered random triangular matrices with nonzero entries on the main diagonal as well as above it. (The variance does not matter, as it cancels out in the formula $\kappa(\mathbf{A}) = \|\mathbf{A}\| \|\mathbf{A}^{-1}\|$.) They found that the norms of the inverses $\|\mathbf{A}^{-1}\|$ of such matrices grow exponentially with N [810]. As we shall see, the entries of $\mathbf{A}^{-1}$ satisfy a random recurrence relation, and by deriving the Lyapunov constant for this recurrence, Viswanath and Trefethen established almost sure (a.s.) exponential growth of the condition numbers:

$$
\text{real, random diagonal} : \quad \|\mathbf{A}^{-1}\|^{1/N} \to 2 \ \text{ a.s.,}
$$
$$
\text{complex, random diagonal} : \quad \|\mathbf{A}^{-1}\|^{1/N} \to \sqrt{e} \approx 1.6487 \ \text{ a.s.}
$$

---

[1]In this case $a_{jk} = x + iy$ for independent variables x and y, each taken from the $N(0, (2N)^{-1})$ distribution.

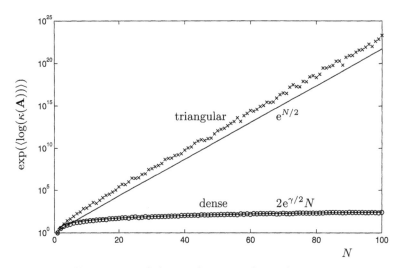

Figure 38.1: Comparison of the condition numbers of complex random upper triangular ($\times$) and dense ($\circ$) matrices. Each data point is a mean $\exp(\langle\log(\kappa(\cdot))\rangle)$ of 100 trials. The solid lines illustrate the expected behavior: the condition numbers grow exponentially for triangular matrices, but only linearly for dense matrices. Compare Figure 35.4.

Since $\|\mathbf{A}\|$ grows only algebraically with N, $\|\mathbf{A}\|^{1/N} \to 1$ a.s., and the above results imply exponential growth of the condition number $\kappa(\mathbf{A})$ at the same rates. Figure 38.1 contrasts this rapid growth in N with the linear growth of the condition numbers of dense random matrices shown in Figure 35.4.

One might think the growth of $\|\mathbf{A}^{-1}\|$ as $N \to \infty$ is caused by the small entries of the diagonal of $\mathbf{A}$, which may be arbitrarily close to zero. However, this is not so, for the diagonal entries are only algebraically close to zero, whereas the growth is exponential. To make this point precisely, Viswanath and Trefethen analyzed a modified set of matrices in which the diagonal entries are all exactly equal to $N^{-1/2}$, i.e., in the language of numerical analysis, matrices that are 'unit triangular' apart from the normalization factor $N^{-1/2}$. They found that here, too, the norms of the inverse grow exponentially, though at slightly lower rates:

$$\text{real, unit diagonal:} \qquad \|\mathbf{A}^{-1}\|^{1/N} \to 1.305683410\ldots \text{ a.s.,}$$
$$\text{complex, unit diagonal:} \quad \|\mathbf{A}^{-1}\|^{1/N} \to 1.347395784\ldots \text{ a.s.}$$

These growth rates for matrices with the fixed diagonal $a_{jj} = N^{-1/2}$ have a pseudospectral interpretation: They describe the norm of the resolvent of strictly upper triangular random matrices (i.e., zero on the main diagonal) at the point $z = N^{-1/2}$ as $N \to \infty$.

We wish to characterize the resolvent norm at finite values of z as

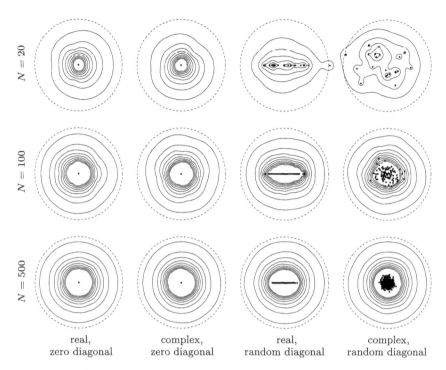

Figure 38.2: Spectra and ε-pseudospectra of various random triangular matrices for $\varepsilon = 10^{-2}, 10^{-3}, \ldots, 10^{-10}$. The dashed line is the circle of radius $1/2$.

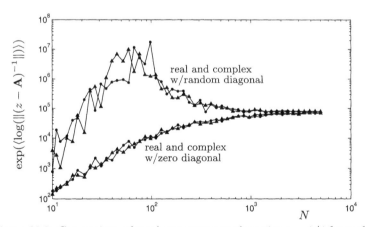

Figure 38.3: Comparison of resolvent norms at the point $z = 1/4$ for real ($\bullet$) and complex ($\blacktriangle$) random triangular matrices with either zeros or random entries on the main diagonal. While the resolvent norms differ for small N, they all appear to converge to the same limit. Each data point is the average of ten trials.

$N \to \infty$. Figure 38.2 illustrates pseudospectra for twelve matrices with real and complex random entries and with zero and random diagonals. Despite these differences, the pseudospectra appear to be converging to a common limit as N increases. Further evidence of this convergence is given in Figure 38.3, which shows $\|(z - \mathbf{A})^{-1}\|$ at $z = 1/4$ as a function of N for matrices of dimensions as large as $N = 5000$. The matrices with random diagonals have larger resolvent norms when N is sufficiently small that the diagonal entries fall near $z = 1/4$, but these differences recede as N grows. There is no consistent difference between the real and complex versions, with radial symmetry appearing as $N \to \infty$ for the real entries just as for the complex. For the remainder of this section, we shall concentrate on real, strictly upper triangular random matrices.

To gain a quantitative understanding of these plots, we follow the example of the last two sections and write the entries of $\mathbf{A}^{-1}$ as terms of a random recurrence. For example, when $N = 3$ we have

$$\mathbf{A} = \begin{pmatrix} 0 & a_{12} & a_{13} \\ 0 & 0 & a_{23} \\ 0 & 0 & 0 \end{pmatrix}$$

with resolvent

$$(z - \mathbf{A})^{-1} = \begin{pmatrix} z^{-1} & z^{-2}a_{12} & z^{-2}a_{13} + z^{-3}a_{12}a_{23} \\ 0 & z^{-1} & z^{-2}a_{23} \\ 0 & 0 & z^{-1} \end{pmatrix}.$$

For an arbitrary fixed N, let r_k denote the $(1, k)$ entry of the resolvent of $\mathbf{A}$. Once again, these entries obey a recurrence relation with random coefficients:

$$\begin{aligned} r_1 &= z^{-1}, \\ r_2 &= z^{-1}a_{21}r_1, \\ r_3 &= z^{-1}a_{31}r_1 + z^{-1}a_{32}r_2, \\ &\vdots \\ r_N &= z^{-1}(a_{N1}r_1 + a_{N2}r_2 + \cdots + a_{N,N-1}r_{N-1}). \end{aligned}$$

To analyze these numbers, we write them out explicitly, using the change of variables $\alpha_{jk} = z^{-1}a_{jk}$ to simplify the formulas:

$$\begin{aligned} r_1 &= z^{-1}, \\ r_2 &= z^{-1}\alpha_{21}, \\ r_3 &= z^{-1}(\alpha_{31} + \alpha_{32}\alpha_{21}), \\ r_4 &= z^{-1}(\alpha_{41} + \alpha_{42}\alpha_{21} + \alpha_{43}\alpha_{31} + \alpha_{43}\alpha_{32}\alpha_{21}), \\ &\vdots \end{aligned}$$

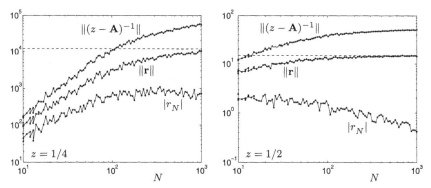

Figure 38.4: The corner entries of the resolvent of random triangular matrices decay as $N \to \infty$, while the norms of their first rows $\mathbf{r}^{\mathrm{T}}$ converge toward the predicted value (38.3) (dashed line), which is somewhat smaller than the resolvent norm. Each data point is the average, in the same sense as in Figures 38.1 and 38.3, of fifty trials.

Suppose that z is real and positive. Then α_{jk} is a real normal random variable with variance $z^{-2}N^{-1}$. Hence r_k is the scaled sum of products of independent, identically distributed random variables. The sum for r_k contains $\binom{k-2}{j-1}$ terms that are the product of j distinct variables.

With this characterization of r_k, we can readily obtain formulas for the mean $\langle r_k \rangle$ and variance $\mathrm{Var}[r_k]$ of r_k. Since each α_{jk} has mean zero, $\langle r_k \rangle = 0$. The variance is more interesting, for it indicates the magnitude of r_k: $\mathrm{Var}[r_k] = \langle r_k^2 \rangle - \langle r_k \rangle^2 = \langle r_k^2 \rangle$. Since the variance of a sum of independent variables equals the sum of the variances, we have

$$\mathrm{Var}[r_k] = z^{-2} \sum_{j=1}^{k-1} \binom{k-2}{j-1} v_j,$$

where v_j denotes the variance of the product of j independent $N(0, z^{-2}N^{-1})$ variables. If $x_1, \ldots, x_j$ denote such variables, then

$$
\begin{aligned}
v_j &= \mathrm{Var}[\textstyle\prod x_\ell] \\
&= \langle \textstyle\prod x_\ell^2 \rangle - \langle \textstyle\prod x_\ell \rangle^2 \\
&= \langle \textstyle\prod x_\ell^2 \rangle = \textstyle\prod \langle x_\ell^2 \rangle = \textstyle\prod \mathrm{Var}[x_\ell].
\end{aligned}
$$

It follows that $v_j = z^{-2j} N^{-j}$, and hence

$$
\begin{aligned}
\mathrm{Var}[r_k] &= z^{-2} \sum_{j=1}^{k-1} \binom{k-2}{j-1} z^{-2j} N^{-j} \\
&= z^{-4} N^{-1} \left(1 + z^{-2} N^{-1}\right)^{k-2}.
\end{aligned}
$$

For the corner entry, this formula gives $\text{Var}[r_N] = z^{-4}N^{-1}(1+z^{-2}N^{-1})^{N-2}$, and so

$$\text{Var}[r_N] = \langle r_N^2 \rangle \sim \frac{\exp(z^{-2})}{z^4 N}, \qquad N \to \infty.$$

Thus we expect the corner entry of the resolvent to *decay* as $N \to \infty$, as confirmed in Figure 38.4. (This is a consequence of the scaling of our matrix entries by $N^{-1/2}$; without this normalization, there would be strong exponential growth.) By the same reasoning, all the entries of the resolvent decay as $N \to \infty$. However, for small values of z the cumulative effect of these individually decaying entries is substantial, as shown in Figure 38.4. The resolvent norm is large, and is well estimated by the norm of the first row. Denoting this first row of $(z - \mathbf{A})^{-1}$ by $\mathbf{r}^{\mathrm{T}}$, we find that

$$\langle \|\mathbf{r}\|^2 \rangle = \sum_{k=1}^{N} \langle r_k^2 \rangle = z^{-2} + \sum_{k=2}^{N} |z|^{-4} N^{-1} (1 + z^{-2} N^{-1})^{k-2}$$

$$= z^{-2}(1 + z^{-2} N^{-1})^{N-1},$$

which implies

$$\lim_{N \to \infty} \langle \|\mathbf{r}\|^2 \rangle = \frac{\exp(z^{-2})}{z^2}. \tag{38.2}$$

One can prove more detailed results about the convergence of $\|\mathbf{r}\|^2$ to $\exp(z^{-2})/z^2$ as $N \to \infty$ in various probabalistic senses, and similar results follow (for all $z \in \mathbb{C}$) for matrices drawn from the complex normal distribution.[2]

We can interpret this characterization of $\|\mathbf{r}\|^2$ as yielding a lower bound on the norm of the resolvent in the sense that $\langle \|(z-\mathbf{A})^{-1}\|^2 \rangle \geq \langle \|\mathbf{r}\|^2 \rangle$. Does the norm of the first column capture the large-N behavior of the actual resolvent norm? In principle the resolvent norm might grow more rapidly, by as much as a factor of $N^{1/2}$, but computational evidence suggests that there is only mild, if any, N dependence in the limit $N \to \infty$. Thus, presuming that the size of the first row captures most of the nonnormality in $\mathbf{A}$, we propose the following model:

$$\|(z - \mathbf{A})^{-1}\| \approx \frac{\exp(|z|^{-2}/2)}{|z|}. \tag{38.3}$$

For values $100 \leq N \leq 1000$ and $0.15 < |z| < 0.5$, numerical experiments show that the left-hand side of (38.3) exceeds the right-hand side by roughly a factor of about 5. We do not have a theorem to make such agreement precise, but as a minimum it is presumably true that

$$\log \|(z - \mathbf{A})^{-1}\| \sim |z|^{-2}/2$$

---

[2] While we have spoken about the entries of $\mathbf{A}$ being $N(0, N^{-1})$ random variables, the above analysis holds without change for any triangular $\mathbf{A}$ with real random variables having mean 0 and variance $1/N$.

almost surely as $N \to \infty$ and $|z| \to 0$. Figure 38.5 compares the pseudo-spectra of real triangular random matrices of various dimensions with the model (38.3). Though the model overestimates the resolvent norm for small N in most of the displayed contours, as N grows, the resolvent norm eventually surpasses the prediction, which is consistent with Figure 38.4.

We mentioned in §35 that a number of results concerning dense random matrices can be derived by applying standard matrix decompositions from numerical linear algebra to these matrices; for one such example, see [691]. Suppose we attempt a similar approach here. Since unitary similarity transformations do not change the 2-norm pseudospectra, we can imagine applying such operations to reduce $\mathbf{A}$ to some deterministic form, with statistically small entries elsewhere. In particular, we shall consider applying the standard technique for upper Hessenberg reduction [327, 776] to the matrix $\mathbf{A}^*$. (This algorithm would do nothing to $\mathbf{A}$, since it is already upper triangular.) Unlike the clean situations that arise in the dense matrix case, this operation introduces dependencies among the combined entries as it progresses. Still, it provides a useful heuristic. At the first step, a unitary similarity $\mathbf{Q}_1 \mathbf{A}^* \mathbf{Q}_1^*$ zeros the first column below the $(2,1)$ entry and sets the magnitude of the $(2,1)$ entry equal to the 2-norm of the first column of $\mathbf{A}^*$. Since this column consists of $N-1$ entries of variance N^{-1}, the $(2,1)$ entry will be approximately 1; the remaining 'mass' is now distributed throughout the lower $(N-1) \times (N-1)$ submatrix. The next step applies a unitary similarity to $\mathbf{Q}_1 \mathbf{A}^* \mathbf{Q}_1^*$ to zero the second column below the $(3,2)$ entry. As a consequence of the previous transformation, the norm of this eliminated column approximately equals $1/\sqrt{2}$. Continuing in this fashion suggests that the pseudospectra of $\mathbf{A}$ will resemble those of

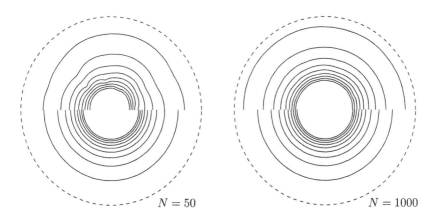

Figure 38.5: ε-pseudospectra (top) and corresponding predictions from the model (38.3) (bottom) for two random real strictly upper triangular matrices, with $\varepsilon = 10^{-2}, 10^{-3}, \ldots, 10^{-10}$. The dashed curve is the circle of radius $1/2$.

$$\mathbf{H} = \begin{pmatrix} 0 & & & & & & \\ 1 & 0 & & & & & \\ & \frac{1}{\sqrt{2}} & 0 & & & & \\ & & \frac{1}{\sqrt{3}} & \ddots & & & \\ & & & \ddots & 0 & & \\ & & & & \frac{1}{\sqrt{N-1}} & 0 & \end{pmatrix}. \qquad (38.4)$$

Let $\mathbf{h}^{\mathrm{T}}$ denote the first row of the resolvent $(z - \mathbf{H})^{-1}$. With a short calculation, one finds

$$\|\mathbf{h}^{\mathrm{T}}\|^2 \;=\; \frac{1}{|z|^2} \sum_{j=0}^{N-1} \frac{1}{j!|z|^{2j}} \;\geq\; \frac{\exp(|z|^{-2})}{|z|^2}\left(1 - \frac{1}{N!|z|^{2N}}\right),$$

which is consistent, to leading order, with the expected value of the square of the first row sum of $\mathbf{A}$ computed above. Figure 38.6 shows the excellent agreement between the pseudospectra of $\mathbf{H}$ and the model (38.3) and also illustrates the subdiagonal entries of an upper Hessenberg reduction of a sample $\mathbf{A}$ of dimension $N = 1000$.

We turn to the question of what these random triangular matrices reveal about more general matrices. Every matrix can be reduced by a unitary similarity transformation to Schur upper triangular form, and as a consequence, the pseudospectra associated with a general $\mathbf{A} \in \mathbb{C}^{N \times N}$ must

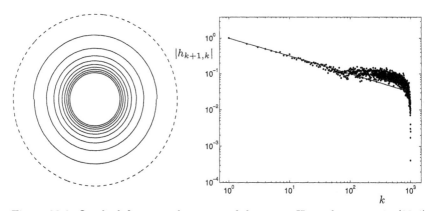

Figure 38.6: On the left, ε-pseudospectra of the upper Hessenberg matrix (38.4) (top half) of dimension $N = 1000$ and those predicted by the model (38.3) (bottom half) for $\varepsilon = 10^{-2}, 10^{-3}, \ldots, 10^{-10}$. One can hardly see the difference between the two halves. On the right, a comparison of the subdiagonal entries of an upper Hessenberg reduction of a random triangular matrix of dimension $N = 1000$ with those of (38.4) (solid line).

match those of some upper triangular matrix. Similarly, any possible matrix behavior (§47) or GMRES convergence curve (§26) must occur for some triangular matrix. Yet throughout this section, we have seen that triangular matrices drawn at random from a canonical distribution have very special behavior. Understanding, from a probabilistic point of view, how algorithms behave for this class of matrices will not say much about performance for general matrices.

In particular, studies of the numerical stability of Gaussian elimination with pivoting (i.e., row interchanges) have drawn attention to questions involving random triangular matrices [776, 779]. A longstanding mystery in numerical analysis is, Why is this algorithm so effective in practice, given its potential for exponential instability? Gaussian elimination factors a matrix $\mathbf{A} \in \mathbb{C}^{N \times N}$ into $\mathbf{PA} = \mathbf{LU}$, where $\mathbf{P}$ is a permutation matrix, $\mathbf{L}$ is unit lower triangular, and $\mathbf{U}$ is upper triangular. Wilkinson showed that this process is stable provided that $\mathbf{L}$ is well-conditioned [825]. Examples exist where $\mathbf{L}$ is exponentially ill-conditioned, but years of experience have shown them to be vanishingly rare in practice. Now to explain this behavior, one might be tempted to guess that the triangular matrices Gaussian elimination produces, to a reasonable approximation, should be random triangular matrices along the lines considered in this section. If random triangular matrices had well-behaved condition numbers, this would then give some explanation of the good behavior of Gaussian elimination. In fact, we have seen just the opposite: The condition numbers of random triangular matrices could hardly be worse, and if Gaussian elimination behaved like this it would be useless as a numerical method. The triangular matrices that arise in Gaussian elimination are actually nothing like random. There is an exponentially pronounced difference between random triangular matrices and the triangular factors of random dense matrices.

Similar effects are familiar in connection with other matrix factorizations, which are generally simpler than $\mathbf{PA} = \mathbf{LU}$. If the entries of a square random matrix are governed by one joint probability distribution, factorization of that matrix will change that distribution entirely in a precisely analyzable fashion. One considers the Jacobian of the transformation corresponding to the matrix factorization, and for most of the standard matrix factorizations of numerical linear algebra, these Jacobians are known explicitly. A classic early paper on this subject is by Olkin and Sampson [584], and an extended treatment is given in the book by Mathai [536].

IX. Computation of Pseudospectra

39 · Computation of matrix pseudospectra _____

Nearly every section of this book includes plots of pseudospectra. In this section and the next, which are built largely on [774] and [837], we explain the algorithms behind the pictures. We first concentrate on methods most appropriate for dense matrices. At the time of this writing, we regard the 'core EigTool algorithm' presented on page 375 as the procedure of choice for matrices of dimension $N \leq 1000$; when properly implemented, as in the EigTool system itself [838], it provides black-box functionality for such problems. Larger matrices require either patience or the specialized approaches detailed in §§40 and 41.

Basic SVD algorithm. All algorithms for computing pseudospectra start from the definitions

$$\sigma_\varepsilon(\mathbf{A}) = \{z \in \mathbb{C} : \|(z - \mathbf{A})^{-1}\| > \varepsilon^{-1}\},$$

or, if $\| \cdot \| = \| \cdot \|_2$,

$$\sigma_\varepsilon(\mathbf{A}) = \{z \in \mathbb{C} : s_{\min}(z - \mathbf{A}) < \varepsilon\},$$

where $s_{\min}(\cdot)$ denotes the minimum singular value. What we call the 'basic algorithm', which is far from optimal, computes $s_{\min}(z - \mathbf{A})$ on a regular grid of points in the complex plane, then visualizes the data, typically via a contour plot. Since we assume $\mathbf{A}$ is of modest dimension, it is possible to obtain $s_{\min}(z - \mathbf{A})$ by computing the entire set of singular values of $z - \mathbf{A}$ using library software such as LAPACK [6]. If the computational grid consists of m points in both the real and imaginary directions, with coordinates indexed by the vectors x and y, this algorithm can be implemented in four lines of MATLAB:

```
for k=1:m, for j=1:m
    sigmin(j,k) = min(svd((x(k)+y(j)*1i)*eye(N)-A));
end, end
contour(x,y,log10(sigmin))
```

This algorithm was the method used in 1991 to compute pseudospectra of thirteen matrices of dimension 32 on a Cray I supercomputer for the paper [772]. It is also at the heart of Higham's MATLAB routine pscont, a popular early code for pseudospectral computation [390]. The remainder of this section describes improvements that typically increase the efficiency of this method by factors of 10 to 100. We shall compare the various methods we describe in this section on a single test problem arising in laser theory, one of the first applications of pseudospectral ideas [479, 480].

Specifically, we take $\mathbf{A}$ to be a spectral discretization (described at the end of §43) of the compact integral operator presented in §60 as equation (60.3),

$$(\mathcal{A}u)(x) = \sqrt{\mathrm{i}F/\pi} \int_{-1}^{1} \mathrm{e}^{-\mathrm{i}F(x-y)^2} u(y)\, \mathrm{d}y, \qquad (39.1)$$

for $u \in L_2[-1, 1]$. Here F is a large constant called the *Fresnel number*. Since $\mathcal{A}$ is compact, it is bounded; its eigenvalues spiral toward the origin, as shown in Figure 60.2. The algorithm described above required about 249 minutes[1] to compute the pseudospectra shown in Figure 39.1 on a grid of 100×100 points for a discretization $\mathbf{A}$ of dimension $N = 400$.

Inverse iteration. The basic algorithm just described is simple but inefficient: It computes all the singular values of $z - \mathbf{A}$. The natural improvement, computing only the smallest singular value, can be accomplished through a variety of iterative algorithms. Since

$$s_{\min}(z - \mathbf{A}) = \sqrt{\text{smallest eigenvalue of } (z - \mathbf{A})^*(z - \mathbf{A})}$$

$$= \text{smallest positive eigenvalue of } \begin{pmatrix} \mathbf{0} & z - \mathbf{A} \\ \bar{z} - \mathbf{A}^* & \mathbf{0} \end{pmatrix},$$

the minimal singular value can be computed using iterative methods for Hermitian eigenvalue problems. Many such algorithms are available (for a survey, see [23]), and some of these have been advocated when $\mathbf{A}$ is large and sparse, as discussed in §41. But even when $\mathbf{A}$ is dense, simple iterative methods still can result in a significant speedup.

Let $\mathbf{B} = z - \mathbf{A}$. The simplest way to compute the smallest eigenvalue of $\mathbf{B}^*\mathbf{B}$ is to use *inverse iteration*, i.e., apply the power method to $(\mathbf{B}^*\mathbf{B})^{-1}$. The smallest eigenvalue of $\mathbf{B}^*\mathbf{B}$ corresponds to the largest eigenvalue of $(\mathbf{B}^*\mathbf{B})^{-1}$, and inverse iteration will generally converge at a rate dictated by the separation between this and the next eigenvalue, i.e., between the smallest two singular values of $\mathbf{B}$ (see the beginning of §28 for analysis of the power method). Inverse iteration requires multiplications by $\mathbf{B}^{-*}$ and $\mathbf{B}^{-1}$, which is implemented by computing an LU factorization of $\mathbf{B}$ [327, 776]. The factorization of $z - \mathbf{A}$ at each grid point is a major bottleneck: this procedure requires $\mathcal{O}(N^3)$ operations per grid point, which is the same asymptotic complexity as computing the full singular value decomposition. On a grid of m^2 points, one has the total complexity of $\mathcal{O}(m^2 N^3)$.

In MATLAB, the algorithm takes the following form.

---

[1] The timed computations in this chapter were performed on a 1.7GHz Pentium III Xeon using MATLAB version 6.1, which incorporates LAPACK for dense linear algebra, including the svd command.

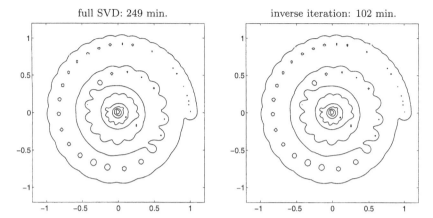

Figure 39.1: ε-pseudospectra of a matrix discretization of (39.1) for $F = 40\pi$ and $N = 400$ with $\varepsilon = 10^{-1}, 10^{-2}, \ldots, 10^{-8}$, computed using different algorithms with differing degrees of accuracy on a grid of 100×100 points. On the left, $\sigma_{\min}(z-\mathbf{A})$ is computed using the full SVD (code on p. 371); on the right, inverse iteration is used with an LU decomposition at each grid point (code below). The plots are indistinguishable.

```
for k=1:m, for j=1:m
    B = (x(k)+y(j)*1i)*eye(N)-A;
    u = randn(N,1)+1i*randn(N,1);
    [L,U] = lu(B); Ls = L'; Us = U';
    for p=1:maxit
        u = Ls\(Us\(U\(L\u))); sig = 1/norm(u);
        if abs(sigold/sig-1) < 1e-2, break, end
        u = sig*u; sigold = sig;
    end
    sigmin(j,k) = sqrt(sig);
end, end
contour(x,y,log10(sigmin))
```

Preliminary triangularization. In a 1997 paper, Lui [520] made a key observation. Recall that every square matrix $\mathbf{A}$ has a Schur decomposition; that is, $\mathbf{A}$ can be reduced to triangular form via a unitary similarity transformation,

$$\mathbf{A} = \mathbf{U}\mathbf{T}\mathbf{U}^*,$$

with upper triangular $\mathbf{T}$. Since unitary similarity transformations do not alter the 2-norm pseudospectra, $\sigma_\varepsilon(\mathbf{T}) = \sigma_\varepsilon(\mathbf{A})$, and we have a much more efficient way to perform inverse iterations: At each grid point $z - \mathbf{T}$ is already in triangular form, so no LU factorization is required. The Schur decomposition is performed only once, before any grid points are processed, imposing a one-time $\mathcal{O}(N^3)$ cost. At each grid point, solving the triangular systems requires only $\mathcal{O}(N^2)$ operations, and as inverse iteration typically converges in only a few iterations, one has the overall complexity $\mathcal{O}(N^3 + m^2 N^2)$, an improvement that makes an enormous difference in practice.

Lui's reduction of $\mathbf{A}$ to triangular form is the cornerstone of dense pseudospectra computations:

1. Reduce $\mathbf{A}$ to triangular form (Schur decomposition).
2. Iteratively compute $s_{\min}(z - \mathbf{A})$ on a grid of z values.

Lanczos iteration. We can improve performance over inverse iteration by using a more sophisticated iterative method for computing the minimal singular value. A natural choice is the Lanczos method, which approximates the minimal singular value of $\mathbf{B} = z_j - \mathbf{A}$ by taking linear combinations of the iterates generated by inverse iteration. Essentially, it applies

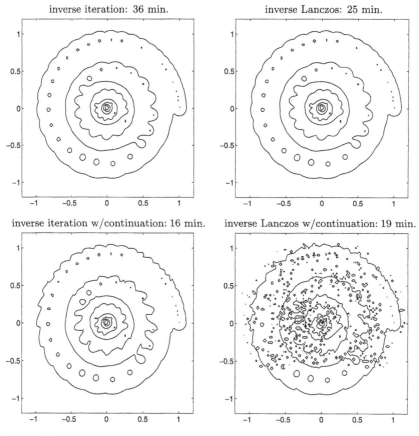

Figure 39.2: ε-pseudospectra of the same matrix as in Figure 39.1 with the same values of ε, but computed with four different algorithms, each of which performs a preliminary Schur factorization.

```
┌─────────────────────────────────────────────────────────────────────┐
│                      Core EigTool algorithm                          │
├─────────────────────────────────────────────────────────────────────┤
│  T = schur(A,'complex');                                             │
│  for k=1:m, for j=1:m                                                │
│      T1 = (x(k)+y(j)*1i)*eye(N)-T; T2 = T1';                         │
│      sigold = 0; qold = zeros(n,1); beta = 0; H = [];               │
│      q = randn(N,1)+1i*randn(N,1); q = q/norm(q);                    │
│      for p=1:maxit                                                    │
│          v = T1\(T2\q) - beta*qold;                                  │
│          alpha = real(q'*v); v = v - alpha*q;                        │
│          beta = norm(v); qold = q; q = v/beta;                       │
│          H(p+1,p) = beta; H(p,p+1) = beta; H(p,p) = alpha;           │
│          sig = max(eig(H(1:p, 1:p)));                                │
│          if abs(sigold/sig-1)<1e-3, break, end                      │
│          sigold = sig;                                                │
│      end                                                              │
│      sigmin(j,k) = sqrt(sig);                                         │
│  end, end                                                             │
│  contour(x,y,log10(sigmin))                                          │
└─────────────────────────────────────────────────────────────────────┘
```

Figure 39.3: Core EigTool algorithm for computing pseudospectra using inverse Lanczos iteration with preliminary triangularization, adapted from psa.m in [774] and efficiently implemented in the EigTool system [838].

the Arnoldi method (§28) to $(\mathbf{B}^*\mathbf{B})^{-1}$, and since $\mathbf{B}^*\mathbf{B}$ is Hermitian, this can be accomplished using only three-term recurrences. At typical grid points inverse Lanczos iteration quickly converges to acceptable accuracy (say, five or fewer iterations). The overall algorithm does not improve the asymptotic complexity of the inverse iteration algorithm (both require $\mathcal{O}(N^3 + m^2N^2)$ operations), but the speedup is still worthwhile: See the results reported in Figure 39.2.

The MATLAB algorithm in Figure 39.3 computes pseudospectra using Schur decomposition and inverse Lanczos iteration. It is adapted from the program psa.m from [774], and forms the basis of the core algorithm in EigTool [838].

Continuation. One of Lui's motivations for using an iterative method was to apply *continuation*, that is, to use the singular vector computed at one grid point as the starting vector for an adjacent grid point. While this idea has intuitive appeal, it proves problematic near points where the minimal singular value is not unique, as occurs almost invariably along certain curves in the complex plane in a pseudospectral plot, typically in regions between two eigenvalues. This is illustrated in Figure 39.4, which shows pseudospectra for the matrix discretization of (39.1) on a smaller axis than in the previous figures. Both plots were computed using inverse iteration. For the top one, the algorithm used continuation as the iteration progressed

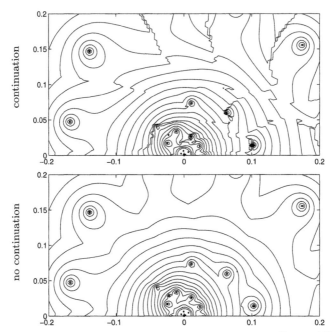

Figure 39.4: Boundaries of ε-pseudospectra for $\varepsilon = 10^{-10}$, $10^{-9.75}, \ldots, 10^{-4}$. The top plot was computed using inverse iteration with continuation, the bottom one without continuation, both with 100 grid points in the horizontal direction. The plots should be identical; discrepancies in the top one result from errors introduced by the continuation procedure.

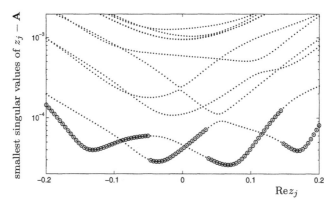

Figure 39.5: Lower singular values of $z_j - \mathbf{A}$. The smallest singular values form a slice of data used in the contour plots in Figure 39.4 for $\mathrm{Im}\, z_j \approx 0.17778$. The solid dots are true singular values of $z_j - \mathbf{A}$; the inverse iteration procedure will virtually always find the smallest one. The circles represent the results of inverse iteration with continuation, which tracks the wrong singular value for several points after two minimal singular values cross.

across horizontal lines on the computational grid; no continuation was used for the bottom plot. The same stringent stopping criterion was used for each calculation. Why the great discrepancy? As one traverses the computational grid along the real axis, the singular value that was once smallest drifts larger in magnitude, crossing with another singular value, which becomes the new smallest. Since the starting vector is continued from one iteration to the next, at some points the starting vector is almost orthogonal to the desired singular vector, and as a result, the wrong singular value is tracked. This is highlighted in Figure 39.5. Finer grids—where there is less difference between consecutive grid points—can potentially exacerbate this problem, even though they appear to improve the execution time of inverse iteration with continuation, as there is a smaller change in the minimal singular vector between adjacent grid points.[2]

The lower right pane of Figure 39.2 illustrates that crudely combining continuation with the inverse Lanczos algorithm can yield disastrous results, much worse than the errors introduced by continued inverse iteration for the same example.

In summary, although Figure 39.2 indicates that continuation methods can yield a tempting performance improvement, the errors they introduce are troublesome. Thus, we recommend inverse Lanczos iteration with no continuation for most computations.[3]

Grid selection. Some implementation details remain. In the absence of information from the user, how should an algorithm automatically determine a grid on which to compute $\sigma_\varepsilon(\mathbf{A})$? Braconnier et al. suggest using the numerical range $W(\mathbf{A})$ to establish the outer limits of the grid [104]. Theorem 17.2 offers justification for this: The ε-pseudospectrum can never exceed $W(\mathbf{A})$ by a distance greater than ε. However, the numerical range is often much larger than the pseudospectral region of interest. (For example, consider the stabilized Boeing 767 matrix whose pseudospectra are illustrated in Figure 15.3. The numerical range is approximately a disk centered at the origin with radius 8.45×10^6, an area far too large to be interesting for most purposes.) Thus, while no method is perfect, we rec-

---

[2]Close examination of Figure 39.5 reveals eleven points where singular value curves cross, and several other points (notably one near Re $z_j = 0$) where rather than crossing, two curves come close together but then turn aside to avoid each other. This phenomenon of avoidance is familiar in parameterized eigenvalue problems, where indeed it is generic, essentially because matrices with degenerate eigenvalues form a space of codimension at least 2 rather than 1, since one free parameter is absorbed by having two eigenvalues equal and another is absorbed by two invariant subspaces fusing into one [60], [487], §9.5], [747], [812]. Avoidance of singular values in parameterized matrix problems is not generic, however.

[3]Clever programming can lead to even better performance. For the pseudospectra shown in Figures 39.1 and 39.2, the EigTool system (which uses the same Lanczos algorithm as shown here, but with key components implemented in compiled C code) requires less than five minutes, making it more than five times quicker than the standard MATLAB implementation and fifty times quicker than the basic SVD algorithm.

ommend deriving the grid limits from the eigenvalues obtained from the
Schur decomposition. EigTool does this, roughly doubling the span of the
spectrum to obtain its default grid.

When one knows the largest value of ε of interest before computing, various strategies exist for excluding points from the computational domain by
quickly identifying those that fall outside the outermost pseudospectrum.
For small matrices, the ability to adjust ε after the computation (as facili-
tated by the graphical interface of EigTool) usually outweighs the benefit
of such exclusion methods. For very large problems, where every singular
value computation is costly, they can prove more important. Details are
given in §41.

Parallel implementation. The grid algorithm is 'embarassingly parallel': it can easily take advantage of multiple processors. Once the Schur
decomposition has been performed, the computations at each grid point
are independent, and thus the labor can be divided among all available
processors with minimal need for interprocessor communication. (Perhaps
the first parallel computations of pseudospectra, though without the Schur
decomposition, were reported in [780].) To minimize communication, one
might allocate an entire line of the computational domain at a time to each
processor. For very large matrices, it may be necessary to split the matrix
itself over numerous processors and have all processors compute for one
grid point at a time in unison; see §41.

Poor man's pseudospectrum. To this point, our algorithms have been
based on the definition of pseudospectra based on resolvent norms. The
equivalent definition based on perturbed eigenvalues,

$$\sigma_\varepsilon(\mathbf{A}) = \bigcup_{\|\mathbf{E}\| < \varepsilon} \sigma(\mathbf{A} + \mathbf{E}),$$

also suggests an algorithm: Select random matrices $\mathbf{E}$ with $\|\mathbf{E}\| = \varepsilon$, and
superimpose plots of $\sigma(\mathbf{A} + \mathbf{E})$, computed using standard dense matrix
eigenvalue algorithms. The result is a 'poor man's pseudospectrum', a
cloud of eigenvalues surrounding the spectrum whose density depends upon
the number of perturbations taken and the probability distribution of the
random perturbations. As can be seen from the proof of Theorem 2.1,
one can obtain any point in the pseudospectrum by restricting $\mathbf{E}$ to be of
rank 1. If $\mathbf{E}$ is generated as a full-rank random matrix, then the cost of
normalizing this matrix (so that $\|\mathbf{E}\| \leq \varepsilon$) is $\mathcal{O}(N^3)$ operations, whereas
rank-1 matrices can be constructed and normalized in $\mathcal{O}(N^2)$ operations.
This observation was perhaps first made by Riedel [640]. (Naturally, these
rank-1 matrices will have different statistical properties from matrices with
independent random entries.)

The above exercise must be repeated if ε-pseudospectra for other values
of ε are required. Such perturbation plots have intuitive appeal, but they

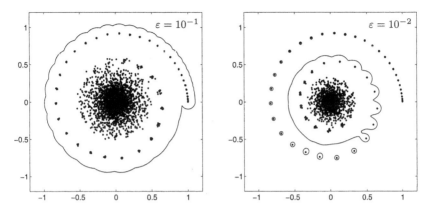

Figure 39.6: 'Poor man's ε-pseudospectra' for a discretization of (39.1) with $N = 400$ and $\varepsilon = 10^{-1}, 10^{-2}$. Each plot shows the spectrum of $\mathbf{A} + \mathbf{E}$ for ten different (full-rank) random perturbations $\mathbf{E}$ of norm ε. The boundary of $\sigma_\varepsilon(\mathbf{A})$ is also shown for comparison. Ideally, one would hope for the perturbed eigenvalues to fill the entire region $\sigma_\varepsilon(\mathbf{A})$ or trace out its boundary. A good rule, appropriate especially for matrices that are not so far from normal, is that these random perturbations instead tend to fill the region $\sigma_{\varepsilon/\sqrt{N}}(\mathbf{A})$.

necessarily provide only lower bounds on $\sigma_\varepsilon(\mathbf{A})$, and from the plots in Figure 39.6, one observes that these bounds need not be sharp. Their utility is in providing an inexpensive approximation, though with developments in fast grid-based pseudospectral computations, even this attraction has diminished.

Pseudospectra in other norms. The algorithms described above assume we seek pseudospectra in the standard Euclidean norm. It is a simple matter to adapt these techniques to different norms based upon an inner product; cf. §§44, 45, and 51. Any finite-dimensional inner product can be written as $(\mathbf{x}, \mathbf{y})_{\mathbf{L}} \equiv \mathbf{y}^* \mathbf{L}^* \mathbf{L} \mathbf{x}$ for some invertible matrix $\mathbf{L}$, giving

$$
\begin{aligned}
\|(z - \mathbf{A})^{-1}\|_{\mathbf{L}} &= \max_{\mathbf{x} \in \mathbb{C}^N} \left(\frac{\mathbf{x}^*(z - \mathbf{A})^{-*}\mathbf{L}^*\mathbf{L}(z - \mathbf{A})^{-1}\mathbf{x}}{\mathbf{x}^*\mathbf{L}^*\mathbf{L}\mathbf{x}} \right)^{1/2} \\
&= \max_{\mathbf{y} \in \mathbb{C}^N} \left(\frac{\mathbf{y}^*(\mathbf{L}^{-*}(z - \mathbf{A})^{-*}\mathbf{L}^*)(\mathbf{L}(z - \mathbf{A})^{-1}\mathbf{L}^{-1})\mathbf{y}}{\mathbf{y}^*\mathbf{y}} \right)^{1/2} \\
&= \| \mathbf{L}(z - \mathbf{A})^{-1}\mathbf{L}^{-1} \| = \|(z - \mathbf{L}\mathbf{A}\mathbf{L}^{-1})^{-1}\|.
\end{aligned}
$$

Thus the $\mathbf{L}$-norm pseudospectra of $\mathbf{A}$ can be computed as the standard 2-norm pseudospectra of $\mathbf{L}\mathbf{A}\mathbf{L}^{-1}$. (The same result appears with different notation in equation (45.9).) This norm flexibility is especially helpful when considering discretizations of infinite-dimensional operators, where the physically relevant inner product (e.g., describing energy) may differ from the Euclidean inner product, as discussed in §43. Further algorithms for computation of pseudospectra in weighted norms for large, sparse matrices are outlined in §44 and described in detail in [837, Chap. 5].

For norms associated with Banach rather than Hilbert spaces, the situation at present is more difficult. The basic procedure requires explicit computation of the resolvent (an $\mathcal{O}(N^3)$ process) at each grid point, followed by computation of the appropriate norm. Higham and Tisseur [393] have considered computation of 1-norm and ∞-norm pseudospectra, which are relevant for examples in applications in heat flow (§12) and probability theory (§§56, 57). They advocate use of a norm estimation scheme that can provide excellent estimates of the pseudospectra at an efficiency similar to that of the 2-norm algorithm described above.

40 · Projection for large-scale matrices _____

Many applications require the computation of pseudospectra of matrices of dimension $N \gg 1000$, for which the algorithms detailed in the last section are too expensive. Preliminary reduction of $\mathbf{A}$ to Schur triangular form, with its $\mathcal{O}(N^3)$ complexity, becomes intractable.

In this section, we review an important technique that reduces such matrices to a fraction of their original size by orthogonally projecting $\mathbf{A}$ onto an appropriate low-dimensional subspace. While, strictly speaking, this procedure only approximates the pseudospectra of $\mathbf{A}$, in many applications the results are in excellent agreement with the exact pseudospectra in a region of the complex plane of interest, and, if desired, descriptive error bounds can be explicitly computed. Furthermore, such projections are immediate byproducts of large, sparse eigenvalue calculations. The satisfying outcome is that, whereas fifteen years ago a supercomputer was used to calculate the pseudospectra of matrix of dimension $N = 32$ [772], one can now handle matrices with N on the order of a million on a mass-market machine.

That it is a challenge to compute pseudospectra of large matrices comes as no surprise, since computing the entire spectrum of such a matrix is itself an $\mathcal{O}(N^3)$ operation. Instead, eigenvalues of large matrices are routinely calculated by identifying those of interest for an application (e.g., rightmost eigenvalues for stability analysis of continuous time dynamical systems) and then projecting the matrix onto a carefully constructed low-dimensional subspace, such that the eigenvalues of the projected matrix are accurate approximations to the true eigenvalues of interest; for details, see §28 and the surveys [23, 653]. This methodology is at the core of all major large-scale non-Hermitian eigenvalue algorithms. The Arnoldi, bi-orthogonal Lanczos, rational Krylov, Jacobi–Davidson, and subspace iteration algorithms differ mainly in the way they construct and refine the approximating subspace.

The idea of computing only a small subset of the spectrum is not simply a practical expedient; it makes good sense for applications. For example, when $\mathbf{A}$ results from the discretization of a differential operator, some of the eigenvalues may have no physical significance. Furthermore, discretizations often induce spurious eigenvalues that pollute the spectrum. With projection, we only compute the physically significant eigenvalues.

Similarly, one often seeks accurate pseudospectra in a particular region of the complex plane. By restricting $\mathbf{A}$ to an appropriate low-dimensional subspace, one can approximate the pseudospectra of $\mathbf{A}$ by those of a much smaller matrix, which can be efficiently computed using the dense matrix techniques described in §39. How should we select the approximating sub-space? Here we consider two choices: projection onto invariant subspaces, and projection onto Krylov subspaces. A combination of these two ideas naturally emerges from large-scale eigenvalue software.

Projection onto an invariant subspace. It is natural to take the projection subspace $\mathcal{U} \subset \mathbb{C}^N$ to be the span of the eigenvectors or, more generally, the invariant subspace, associated with the physically interesting eigenvalues of $\mathbf{A}$. If $\mathbf{U} \in \mathbb{C}^{N \times p}$ is a matrix whose columns form an orthonormal basis for $\mathcal{U}$, then the eigenvalues of the projected matrix $\mathbf{U}^* \mathbf{A} \mathbf{U}$ are simply the eigenvalues of $\mathbf{A}$ associated with $\mathcal{U}$,

$$\sigma(\mathbf{U}^* \mathbf{A} \mathbf{U}) \subseteq \sigma(\mathbf{A}).$$

The pseudospectra obey the analogous containment, since, in the 2-norm,

$$
\begin{aligned}
\|(z - \mathbf{A})^{-1}\| &= \max_{\substack{\mathbf{x} \in \mathbb{C}^N \\ \|\mathbf{x}\|=1}} \|(z - \mathbf{A})^{-1}\mathbf{x}\| \\
&\geq \max_{\substack{\mathbf{y} \in \mathbb{C}^p \\ \|\mathbf{y}\|=1}} \|(z - \mathbf{A})^{-1}\mathbf{U}\mathbf{y}\| \\
&= \max_{\substack{\mathbf{y} \in \mathbb{C}^p \\ \|\mathbf{y}\|=1}} \|(z - \mathbf{U}^* \mathbf{A} \mathbf{U})^{-1}\mathbf{y}\| = \|(z - \mathbf{U}^* \mathbf{A} \mathbf{U})^{-1}\|.
\end{aligned}
$$

(The third line holds because range($\mathbf{U}$) is an invariant subspace of $\mathbf{A}$; it is not true for general matrices $\mathbf{U}$ with orthonormal columns.) We state this bound on the resolvent norm in terms of pseudospectra.

Proposition 40.1 *If* range($\mathbf{U}$) *is an invariant subspace of* $\mathbf{A}$, *then*

$$\sigma_\varepsilon(\mathbf{U}^* \mathbf{A} \mathbf{U}) \subseteq \sigma_\varepsilon(\mathbf{A}).$$

This invariant subspace projection technique was proposed by Reddy, Schmid, and Henningson [624], who used it in their analysis of the Orr–Sommerfeld operator (§22). We illustrate the method with two examples; a third study involving the differential operator (11.21) can be found in the survey [774].

The first example takes $\mathbf{A}$ to be a spectral discretization with $N = 1000$ of the Orr–Sommerfeld operator for plane Poiseuille flow with Reynolds number equal to the critical value 5772. To analyze the continuous time stability of this operator (see §15), we require the rightmost part of the spectrum and pseudospectra, and thus we project $\mathbf{A}$ onto invariant sub-spaces associated with the rightmost eigenvalues. Figure 40.1 shows the approximate pseudospectra resulting from projection onto such subspaces of dimension $p = 20$, 40, and 60. In the latter case, we obtain an excellent

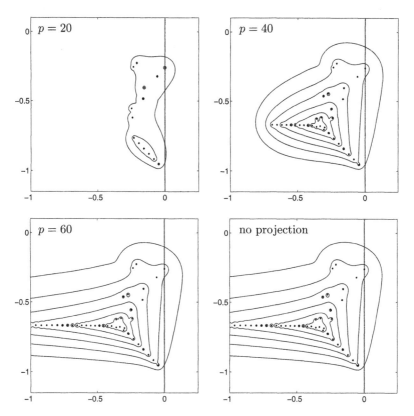

Figure 40.1: Approximation of ε-pseudospectra for a discretization of the Orr–Sommerfeld operator with $N = 1000$ by projection onto invariant subspaces of dimensions $p = 20$, 40, and 60 associated with the rightmost eigenvalues; $\varepsilon = 10^{-2}, 10^{-3}, \ldots, 10^{-7}$. Note that the eigenvalues of the projected matrix are always a subset of the unprojected spectrum, and the pseudospectra are nested in agreement with Proposition 40.1. It is difficult to find any difference between the two bottom plots, but the unprojected computation on the right took nearly four hours, whereas the $p = 60$ case took under five minutes, the bulk of it spent constructing the basis vectors for $\mathbf{U}$. Compare Figure 22.4.

approximation of the true pseudospectra in the region of interest, while reducing the dimension of the problem from $N = 1000$ to $p = 60$ and accelerating the computation by a factor of 50.

The second example is the laser integral operator (39.1), which formed the central test problem in the last section. Here we take a discretization of dimension $N = 1000$ with Fresnel number $F = 100\pi$. For this application one is concerned with the behavior of powers of the operator, and thus we seek the largest magnitude components of the pseudospectra. This compact operator has infinitely many eigenvalues that spiral toward the origin, and we project onto the invariant subspaces associated with the eigenvalues of largest modulus. Figure 40.2 shows the results.

How should one obtain the subspace $\mathcal{U}$ and its representation $\mathbf{U}$? When the dimension of $\mathbf{A}$ is not prohibitive (say, $N \leq 1000$), one can compute the Schur factorization of $\mathbf{A}$, and it is then a simple matter to reorder this decomposition to put the p desired eigenvalues in the first p positions on the diagonal of the Schur factor $\mathbf{T}$; for details, see [327, §7.6.2], [774]. This reordering also transforms the unitary factor in the Schur decomposition into the matrix $[\mathbf{U} \ \widehat{\mathbf{U}}]$, where $\mathbf{U}$ has orthonormal columns spanning $\mathcal{U}$ as required. The Schur factorization takes the form

$$\mathbf{T} = [\mathbf{U} \ \widehat{\mathbf{U}}]^* \mathbf{A} [\mathbf{U} \ \widehat{\mathbf{U}}] = \begin{bmatrix} \mathbf{U}^* \mathbf{A} \mathbf{U} & \mathbf{U}^* \mathbf{A} \widehat{\mathbf{U}} \\ \widehat{\mathbf{U}}^* \mathbf{A} \mathbf{U} & \widehat{\mathbf{U}}^* \mathbf{A} \widehat{\mathbf{U}} \end{bmatrix} = \begin{bmatrix} \mathbf{T}_{11} & \mathbf{T}_{12} \\ \mathbf{0} & \mathbf{T}_{22} \end{bmatrix}, \quad (40.1)$$

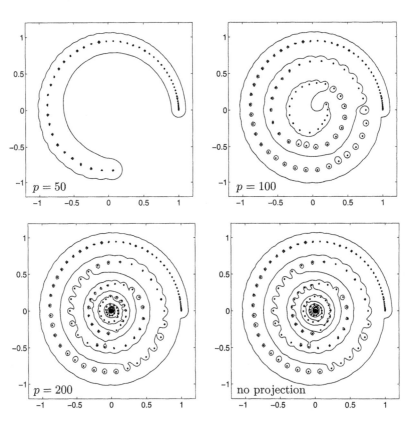

Figure 40.2: Spectra and ε-pseudospectra of the laser operator (60.3) and its projections onto invariant subspaces for $\varepsilon = 10^{-1}, 10^{-2}, \ldots, 10^{-10}$. The original matrix has dimension $N = 1000$ with Fresnel number $F = 100\pi$. Projecting onto an invariant subspace of dimension 200 gives an accurate approximation of the outermost contours.

where $\mathbf{T}_{11}$ and $\mathbf{T}_{22}$ are upper triangular matrices, with $\mathbf{T}_{11}$ equal to the desired compression $\mathbf{U}^*\mathbf{A}\mathbf{U}$. When the dimension of $\mathbf{A}$ is too large for Schur factorization, $\mathbf{U}$ can be obtained using algorithms for computing eigenvalues and invariant subspaces of large matrices. We discuss this option at greater length later in this section.

Can we obtain a better estimate of the agreement between $\sigma_\varepsilon(\mathbf{U}^*\mathbf{A}\mathbf{U})$ and $\sigma_\varepsilon(\mathbf{A})$? How sharp is the inequality in the proof of Proposition 40.1? Grammont and Largillier have developed bounds on the pseudospectra of block matrices of the form (40.1) [335]. In particular, they demonstrate that

$$\sigma_\varepsilon(\mathbf{A}) \subseteq \sigma_\eta(\mathbf{T}_{11}) \cup \sigma_\eta(\mathbf{T}_{22}),$$

for $\eta = \varepsilon\sqrt{1 + \|\mathbf{T}_{12}\|/\varepsilon} \geq \varepsilon$. A slightly different approach gives a bound on the resolvent norm,

$$\|(z - \mathbf{A})^{-1}\| \leq \|\mathbf{P}\| \left(\|(z - \mathbf{T}_{11})^{-1}\| + \|(z - \mathbf{T}_{22})^{-1}\| \right), \qquad (40.2)$$

where $\mathbf{P}$ is the spectral projector onto the invariant subspace $\mathcal{U}$. If $\mathbf{T}_{12} = \mathbf{0}$, then range$(\widehat{\mathbf{U}})$ is itself an invariant subspace, orthogonal to $\mathcal{U}$, and $\|\mathbf{P}\| = 1$. Thus $\mathcal{U}$ is perfectly conditioned, and from Theorem 2.4 we have

$$\sigma_\varepsilon(\mathbf{A}) = \sigma_\varepsilon(\mathbf{T}_{11}) \cup \sigma_\varepsilon(\mathbf{T}_{22}).$$

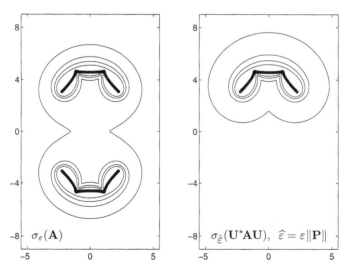

Figure 40.3: Example of the projection bound (40.2). The matrix $\mathbf{A}$ is block upper triangular with shifted and rotated versions of the Grcar matrix (7.14) of dimension 50 on the main diagonal and an off-diagonal block giving $\|\mathbf{P}\| \approx 168.8$. The left plot shows $\sigma_\varepsilon(\mathbf{A})$ for $\varepsilon = 10^{-2}, 10^{-3}, 10^{-4}, 10^{-5}$, while the right plot shows $\sigma_{\widehat{\varepsilon}}(\mathbf{U}^*\mathbf{A}\mathbf{U})$, where $\mathcal{U}$ is the invariant subspace associated with the top eigenvalues and $\widehat{\varepsilon} = \varepsilon\|\mathbf{P}\|$ for the same values of ε used in the left plot.

Near the eigenvalues of $\mathbf{T}_{11}$, the approximate pseudospectra must match the true pseudospectra. When $\mathbf{T}_{12} \neq \mathbf{0}$, Figure 40.3 illustrates that the bound (40.2) can be quite descriptive for smaller values of ε, provided that the spectrum of the projected matrix is not too near the rest of the spectrum of $\mathbf{A}$.

The bound (40.2) suggests that projection can be misleading when the invariant subspace being projected upon is ill-conditioned, i.e., when $\|\mathbf{P}\|$ is large and $\mathcal{U}$ forms a small angle with its complementary invariant subspace.[1] In applications where the resolvent norm grows exponentially with the matrix dimension N, such as examples in §§7, 8, and 36, projection will typically produce misleading results.

Projection onto Krylov subspaces. Projection onto invariant subspaces can be an effective tool for approximating pseudospectra in a particular region of the complex plane. However, for applications that require approximation of the pseudospectra over the whole outer part of the spectrum, other projections can prove useful. One powerful choice for the projection space $\mathcal{U}$, proposed in [761], is the *Krylov subspace*

$$\mathcal{K}_p(\mathbf{A}, \mathbf{x}) = \mathrm{span}\{\mathbf{x}, \mathbf{A}\mathbf{x}, \dots, \mathbf{A}^{p-1}\mathbf{x}\}$$

generated by $\mathbf{A}$ with starting vector $\mathbf{x} \in \mathbb{C}^N$. Let the columns of $\mathbf{U}$ form an orthonormal basis for $\mathcal{K}_p(\mathbf{A}, \mathbf{x})$. Typically the eigenvalues of $\mathbf{U}^*\mathbf{A}\mathbf{U}$, known as *Ritz values*, converge quickly as p increases to well-separated eigenvalues of $\mathbf{A}$ on the periphery of the spectrum.

Krylov subspaces form the core of many sparse matrix eigenvalue algorithms. As detailed in §26, the Arnoldi algorithm constructs an orthonormal basis for subspaces $\mathcal{K}_p(\mathbf{A}, \mathbf{x})$ of increasing dimension $p < N$, building matrices $\mathbf{U}_{p+1} = [\mathbf{U}_p \; \mathbf{u}_{p+1}]$ and $\widetilde{\mathbf{H}}_p$ that satisfy

$$\mathbf{A}\mathbf{U}_p = \mathbf{U}_{p+1}\widetilde{\mathbf{H}}_p, \tag{40.3}$$

where $\mathrm{range}(\mathbf{U})_j = \mathcal{K}_j(\mathbf{A}, \mathbf{x})$ with $\mathbf{U}_j^*\mathbf{U}_j = \mathbf{I}$ for all $1 \leq j \leq p$ and $\widetilde{\mathbf{H}}_p$ is a $(p+1) \times p$ upper Hessenberg matrix. The matrix $\mathbf{U}_p$ can be augmented by a matrix $\widehat{\mathbf{U}}_p$ such that $[\mathbf{U}_p \; \widehat{\mathbf{U}}_p]$ is unitary and

$$[\mathbf{U}_p \; \widehat{\mathbf{U}}_p]^* \mathbf{A} [\mathbf{U}_p \; \widehat{\mathbf{U}}_p] = \begin{bmatrix} \mathbf{H}_p & \mathbf{U}_p^*\mathbf{A}\widehat{\mathbf{U}}_p \\ h_{p+1,p}\mathbf{e}_1\mathbf{e}_p^* & \widehat{\mathbf{U}}_p^*\mathbf{A}\widehat{\mathbf{U}}_p \end{bmatrix}. \tag{40.4}$$

Here, $\mathbf{H}_p = \mathbf{U}_p^*\mathbf{A}\mathbf{U}_p \in \mathbb{C}^{p \times p}$ is a square upper Hessenberg matrix, the upper $p \times p$ part of $\widetilde{\mathbf{H}}_p$. In general, $h_{p+1,p} \neq 0$, and because of the nonzero $(2,1)$ block in (40.4),

$$\sigma_\varepsilon(\mathbf{H}_k) \not\subseteq \sigma_\varepsilon(\mathbf{A})$$

---

[1] Note that an invariant subspace can be well-conditioned even when the eigenvalues it contains are ill-conditioned [729]; see Figure 28.5.

in typical circumstances. This must be so whenever the Ritz values have not exactly converged, i.e., $\sigma(\mathbf{H}_k) \not\subseteq \sigma(\mathbf{A})$. As we shall see in §46, augmenting a matrix with additional columns only enlarges its pseudospectra, and so

$$\sigma_\varepsilon(\widetilde{\mathbf{H}}_p) = \sigma_\varepsilon\left(\left[\begin{array}{c} \mathbf{H}_p \\ h_{p+1,p}\mathbf{e}_1\mathbf{e}_p^* \end{array} \right] \right) \subseteq \sigma_\varepsilon(\mathbf{A}).$$

Because of this containment, $\sigma_\varepsilon(\widetilde{\mathbf{H}}_p)$ is often preferred over $\sigma_\varepsilon(\mathbf{H}_p)$ in practice, despite the possibility that $\sigma_\varepsilon(\widetilde{\mathbf{H}}_p) = \emptyset$ for some values of $\varepsilon > 0$; see §46 for details.

How accurate are such approximations? The results are mixed [761], and two extremes can be seen in the Orr–Sommerfeld and laser examples shown earlier. For the former case, shown in Figure 40.4, projection onto a Krylov subspace of dimension $p = 400$ yields only a crude approximation of the pseudospectra of physical interest. Projection onto a degree 40 invariant subspace performs much better; cf. Figure 40.1. For the latter case, shown in Figure 40.5, Krylov subspace projection actually does better than invariant subspace projection when $p = 50$; cf. Figure 40.2. An explanation is that there are more than fifty eigenvalues on the outer ring of the spiral. The Krylov subspace projection cannot give accurate estimates for them all, but it does have some rough estimate to which they all contribute, and this is good enough to provide a decent approximation to the 10^{-1}-pseudospectrum.

While the fact that the Krylov subspace is influenced by all the eigenvalues of $\mathbf{A}$ was an advantage for the approximation of the pseudospectra

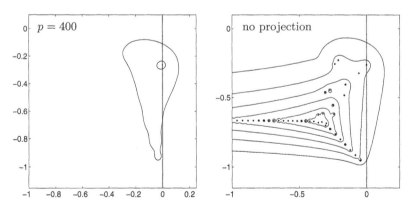

Figure 40.4: Analogue of Figure 40.1, but with projection onto a 400-dimensional Krylov subspace rather than invariant subspaces. The left plot shows pseudospectra of the rectangular matrix $\widetilde{\mathbf{H}}_{400}$; only contours for $\varepsilon = 10^{-2}$ and 10^{-3} are clearly visible. (The $\varepsilon = 10^{-4}$ contour appears as dot on the imaginary axis; there is no contour for $\varepsilon \leq 10^{-5}$.) For this unbounded operator, the basic Krylov subspace does not estimate many of the relevant eigenvalues well even when $p = 400$. None of the eigenvalues of $\mathbf{H}_{400}$ appear on these axes.

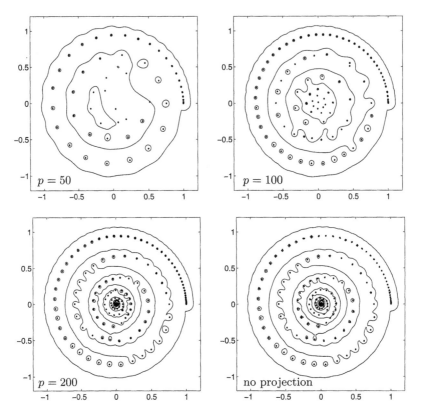

Figure 40.5: Analogue of Figure 40.2, but with projection onto Krylov subspaces rather than invariant subspaces. The first three plots show pseudospectra of the $(p + 1) \times p$ upper Hessenberg matrix $\widetilde{\mathbf{H}}_p$; the black dots show the Ritz values, eigenvalues of the square upper Hessenberg matrix $\mathbf{H}_p$. Krylov subspaces of modest dimension provide good estimates of the outermost eigenvalues and surrounding pseudospectra.

of the integral operator (39.1), it is often a significant limitation, especially for discretizations of unbounded operators like the Orr–Sommerfeld example. One might expect the approximate pseudospectra to be accurate to some relative tolerance, which could be quite slack when $\|\mathbf{A}\|$ is large due to physically irrelevant parts of the spectrum.

The Arnoldi eigenvalue algorithm, which uses the Ritz values to approximate $\sigma(\mathbf{A})$, suffers from the same difficulty. In order to obtain accurate eigenvalue estimates with a modest subspace dimension p, one often must *restart* the iteration with an improved starting vector, thus enhancing the components of $\mathbf{x}$ in the direction of eigenvectors associated with the desired eigenvalues [648]. Sorensen designed a numerically stable implementation and proposed a very effective technique for improving the starting vector [709]. This variant, the *implicitly restarted Arnoldi algorithm*, is implemented in the ARPACK software package [494], which can be accessed in MATLAB via the `eigs` command. To execute ARPACK, the user specifies the number of desired eigenvalues k, the maximal Krylov subspace dimension $p > k$, and the part of the spectrum to be computed (e.g., rightmost eigenvalues). At the end of a successful run, the algorithm produces an Arnoldi factorization (40.3), where the first k columns of $\mathbf{U}_p$ span, up to a prescribed tolerance, a k-dimensional invariant subspace associated with the desired eigenvalues. However, range($\mathbf{U}_p$) also contains a $(p-k)$-dimensional Krylov subspace orthogonal to the invariant subspace. By projecting onto range($\mathbf{U}_p$), one combines the advantages

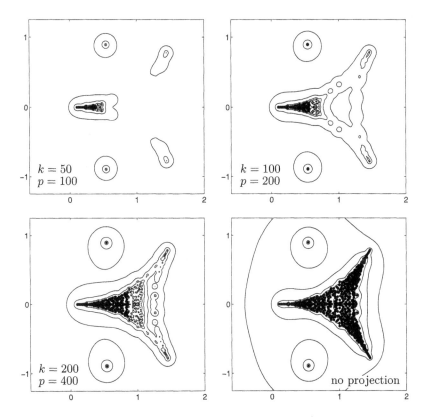

Figure 40.6: Spectra and ε-pseudospectra approximated via projection onto Krylov subspaces, as computed by EigTool for a matrix of dimension $N = 1500$ arising from the discretization of an invariant measure associated with a random Fibonacci recurrence (§37); $\varepsilon = 10^{-1}, 10^{-1.5}, \ldots, 10^{-4}$. In the first three plots, $k = 50$, 100, and 200 eigenvalues are computed; $\sigma_\varepsilon(\mathbf{A})$ is approximated by an $(p+1) \times p$ upper Hessenberg matrix, where $p = 2k$. The final plot shows the exact pseudospectra computed using the dense techniques of §39.

of invariant subspace and Krylov subspace projection. This approach was proposed in [840] and is implemented in EigTool [838]. For typical large-scale applications, computing $\sigma_\varepsilon(\widetilde{\mathbf{H}}_p)$ takes a fraction of the time required by ARPACK to find the k eigenvalues of $\mathbf{A}$; see [837, Fig. 2.14] for sample timings. Thus we strongly recommend that any sparse eigenvalue computation for a non-Hermitian matrix be followed by an approximation of pseudospectra. By doing this, one can estimate the accuracy of the computed spectrum, just as one would check a condition number estimate after solving a dense system of linear equations.

The quality of the approximations one obtains from this ARPACK projection technique will vary with the matrix, the choices for k and p, and the tolerance to which ARPACK computes the k eigenvalues. Wright's thesis provides a comprehensive study of such issues with many more illustrations than we present here. We content ourselves with just one example, shown in Figure 40.6. The matrix, of dimension $N = 1500$, arises from calculation of the Lyapunov constant for the random Fibonacci recurrence $x_{n+1} = x_n \pm x_{n-1}$ discussed in §37; see [246].

While we have focused on Krylov subspaces and the Arnoldi algorithm, we emphasize that any algorithm that generates an orthogonal basis approximating an invariant subspace of $\mathbf{A}$ could similarly be used for computation of pseudospectra. For example, to approximate $\sigma_\varepsilon(\mathbf{A})$ near the point $\mu \in \mathbb{C}$, one could apply the shift-and-invert Arnoldi method (i.e., Arnoldi iteration on $(\mu - \mathbf{A})^{-1}$ instead of $\mathbf{A}$) to generate the projection subspace. Similarly, Ruhe has investigated using the rational Krylov algorithm for pseudospectra approximation [646], and Wright has performed experiments with the Jacobi–Davidson QR (JDQR) method [837].

41 · Other computational techniques _____

When $\mathbf{A}$ is too large for the core EigTool algorithm of page 375 to be practical, projection onto key subspaces (as performed by the implicitly restarted Arnoldi approach that is implemented in EigTool and described in the last section) is usually effective. Sometimes, however, neither of these approaches is workable, and one looks for further alternatives to reduce the computational cost. This section focuses on three of these:

- Exploit sparse matrix methods when computing $s_{\min}(z - \mathbf{A})$,

- Reduce the number of grid points at which $s_{\min}(z - \mathbf{A})$ is computed,

- Use curve-tracing methods to find a single pseudospectral boundary.

These techniques can also be useful for matrices of more modest dimension in particular applications.

Sparse methods for computation of $s_{\min}(z - \mathbf{A})$. The core EigTool algorithm determines the 2-norm of the resolvent by computing $s_{\min}(z - \mathbf{A})$ on a regular grid in the complex plane using an inverse Lanczos iteration, as described in §39. At most grid points, convergence is attained in just a few steps. The dominant cost arises from the solution of linear systems involving $z - \mathbf{A}$ and its adjoint, the complexity of which can be reduced from $\mathcal{O}(N^3)$ to $\mathcal{O}(N^2)$ operations by transforming $\mathbf{A}$ to Schur triangular form before the singular value computations commence. When $\mathbf{A}$ is too large to triangularize, one can proceed without the Schur factor, instead expediting the computation of $s_{\min}(z - \mathbf{A})$ by exploiting sparsity or other structure in $\mathbf{A}$ itself.

Several options are available. First, one can still apply the inverse Lanczos iteration, either to the matrix

$$(z - \mathbf{A})^*(z - \mathbf{A}), \qquad (41.1)$$

whose eigenvalues are the squares of the singular values of $z - \mathbf{A}$, or to the augmented matrix

$$\begin{pmatrix} \mathbf{0} & z - \mathbf{A} \\ \overline{z} - \mathbf{A}^* & \mathbf{0} \end{pmatrix}, \qquad (41.2)$$

whose eigenvalues are the singular values of $z - \mathbf{A}$ and their negatives. The former approach is the same one implemented in EigTool, but now $\mathbf{A}$ does not have triangular form. Both approaches require the solution of linear systems involving $z - \mathbf{A}$ and $\overline{z} - \mathbf{A}^*$. If $\mathbf{A}$ is sparse, one can compute a sparse direct LU factorization of $z - \mathbf{A}$ at each grid point [218], or solve the systems with a (possibly preconditioned) iterative method [541]. A variety of fast algorithms are also available when $z - \mathbf{A}$ has special structure (e.g., Toeplitz). With such methods one can sometimes recover the $\mathcal{O}(N^2)$ complexity obtained with triangular systems, or do even better. Alternatives to inverse Lanczos iteration, including preconditioned eigensolvers like the Jacobi–Davidson algorithm [405, 697], can also be employed. Braconnier and Higham use an inverse Lanczos method with explicit Chebyshev restarting and reorthogonalization [103]; for other approaches based on variations of the Lanczos theme, see [530, 531, 842]. A Davidson method is used in [129].

As an alternative technique for computing $s_{\min}(z - \mathbf{A})$, one can apply the standard Lanczos algorithm to $(z - \mathbf{A})^*(z - \mathbf{A})$. Avoiding inverse iteration, this method relies on matrix-vector multiplications with $z - \mathbf{A}$ and $\bar{z} - \mathbf{A}^*$ rather than the solution of linear systems. Algorithms such as the implicitly restarted Lanczos method [126, 494] can be applied here, though convergence may be slow if $z - \mathbf{A}$ has many small singular values. (The augmented form (41.2) is not well suited for this direct iteration, as the desired eigenvalue of smallest modulus is located in the interior of the spectrum, which causes eigenvalue iterations to converge slowly.)

Any grid-based algorithm for computing pseudospectra of dense matrices is 'embarassingly parallel', since the individual singular value computations can be distributed over multiple processors, either in a single machine or over a network. For very large matrices the situation may change, for the entire matrix may be stored across the available processors.[1] Then the processors can collaborate on the computation of $s_{\min}(z - \mathbf{A})$, processing one grid point at a time with either of the Lanczos approaches described above. The Qualitative Computing research group at CERFACS made numerous early contributions concerning the computation of large-scale pseudospectra on high-performance computers. Among the fruits of their research are the numerous 'spectral portraits' of large matrices included in the MatrixMarket collection [68].

To compute the pseudospectra of the Hatano–Nelson matrix of dimension 10^6 shown in Figure 36.6, we applied the inverse Lanczos algorithm to $(z - \mathbf{A})^*(z - \mathbf{A})$, exploiting the fact that $\mathbf{A}$ was already in bidiagonal form and stored in a sparse data structure. In order to get a high-resolution image, a fine grid was required, and sparse matrix methods alone were insufficient. Using the asymptotics of the problem, described in §36, to predict where the desired pseudospectral boundaries should fall, we eliminated the need for singular value computations at most grid points.

Reducing the number of singular value computations. In many cases, like the one just mentioned, the pseudospectra of interest fill only a portion

---

[1]In this case, one can still use the ideas of the last section by running ARPACK in parallel to compute the projection subspace.

of the rectangular domain on which a grid-based algorithm like EigTool computes singular values. If one seeks only a single ε-pseudospectrum, there are a variety of techniques that improve performance by eliminating singular value computations in regions that can be identified as far from the boundary of $\sigma_\varepsilon(\mathbf{A})$. (EigTool does not incorporate such ideas, in order to allow the user to adjust the desired ε levels without requiring further computation.)

The most primitive approach along these lines is to simply stop the iterative computation of $s_{\min}(z-\mathbf{A})$ in Figure 39.3 after a few steps, judging when the eigenvalue estimate `sig` appears to be converging far from ε^2. Since the minimal singular value estimates generated by the inverse Lanczos method must decrease monotonically, as soon as `sig` $< \varepsilon^2$, one can conclude that $z \in \sigma_\varepsilon(\mathbf{A})$ without the need for further computation. Heuristics can be designed to rule out cases where `sig` $\gg \varepsilon^2$.

A more elegant approach, due to Koutis and Gallopoulos [462], has the power to eliminate many grid points from $\sigma_\varepsilon(\mathbf{A})$ with just a single resolvent norm computation. Suppose one knows some $\widetilde{z} \notin \sigma_\varepsilon(\mathbf{A})$. Then by elementary arguments (essentially the same as will be used for the proof of Theorem 52.4 below), one can show

$$\operatorname{dist}(\widetilde{z}, \sigma_\varepsilon(\mathbf{A})) \geq \frac{1}{\|(\widetilde{z}-\mathbf{A})^{-1}\|} - \varepsilon, \qquad (41.3)$$

which implies

$$\widetilde{z} + \Delta_r \ \cap \ \sigma_\varepsilon(\mathbf{A}) \ = \ \emptyset,$$

where Δ_r is the open disk of radius $r = 1/\|(\widetilde{z}-\mathbf{A})^{-1}\| - \varepsilon$. The quality of the bound (41.3) is not difficult to deduce: we expect it to be accurate if $\mathbf{A}$ is nearly normal, while it will likely be poor in areas of the complex plane where the resolvent norm is rapidly changing. Figure 41.1 illustrates the use of exclusion regions to eliminate points from the 10^{-2}-pseudospectrum of the Orr–Sommerfeld operator described in §22 and used as an example in the last section. The exclusion regions roughly describe the bottom edge of the pseudospectrum, but are less helpful near the upper boundary. This can be explained by looking at Figure 40.1, where one observes larger separation of the pseudospectral boundaries above the eigenvalues than below them.

One can imagine how this tool could be implemented in coordination with a grid-based algorithm, where the resolvent norm would be first computed on a coarse grid of points, followed by calculations on a finer grid at points not yet excluded from $\sigma_\varepsilon(\mathbf{A})$ by the coarse-grid calculations. Such an approach resembles Gallestey's 'SH algorithm', which uses the fact that the function $\|(z-\mathbf{A})^{-1}\|$ is subharmonic away from the eigenvalues of $\mathbf{A}$ (Theorem 4.2) to exclude regions from $\sigma_\varepsilon(\mathbf{A})$ [297]. Gallestey's algorithm first divides the computational domain into large rectangles, some of which

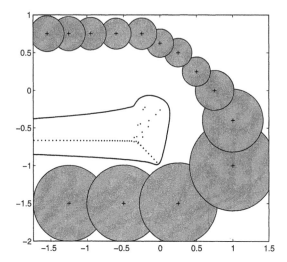

Figure 41.1: Exclusion disks eliminate areas of the complex plane from the 10^{-2}-pseudospectrum of a degree-500 spectral discretization of the Orr–Sommerfeld operator with Reynolds number equal to the critical value 5772. The radius of each disk is derived from a resolvent norm computation at its center (+), based on a technique of Koutis and Gallopoulos [462].

are entirely disjoint from $\sigma_\varepsilon(\mathbf{A})$; the remaining rectangles are subdivided until a satisfactory approximation to $\sigma_\varepsilon(\mathbf{A})$ is obtained.

Curve-tracing to compute a pseudospectral boundary. For a more dramatic reduction in singular value computations, one can depart from grid-based computations altogether and instead attempt to trace the boundary $\partial\sigma_\varepsilon(\mathbf{A})$. That is, given some starting point $z_0 \in \partial\sigma_\varepsilon(\mathbf{A})$ and some fixed $\varepsilon > 0$, compute the level curve

$$\{z \in \mathbb{C} : \|(z - \mathbf{A})^{-1}\| = \varepsilon^{-1}\}.$$

Kostin proposed this approach in an early pseudospectra paper [460, §3], and Brühl developed a full implementation for his Diploma thesis; see [111]. The method exploits the fact that $\|(z-\mathbf{A})^{-1}\|_2 = 1/s_{\min}(z-\mathbf{A})$. It consists of two main components:

- *Find a starting point $z_0 \in \partial\sigma_\varepsilon(\mathbf{A})$.*
  Brühl advocates using Newton's method to find z_0, though one could also apply techniques used by Burke, Lewis, Overton, and Mengi [117, 545], described in the next section, for the computation of the pseudospectral abscissa and radius.

- *Given $z_{k-1} \in \partial\sigma_\varepsilon(\mathbf{A})$, find a nearby point z_k also in $\partial\sigma_\varepsilon(\mathbf{A})$.*
  To follow a level curve of a function, one searches orthogonally to

the direction of steepest descent. Breaking z into real and imaginary components, define

$$g(x,y) = \sigma_{\min}((x+iy) - \mathbf{A}) = \frac{1}{\|((x+iy) - \mathbf{A})^{-1}\|_2}.$$

Then, provided the minimal singular value of $(x+iy) - \mathbf{A}$ is simple, one can show that

$$\nabla g(x,y) = \begin{pmatrix} \mathrm{Re}\,(\mathbf{v}^*\mathbf{u}) \\ \mathrm{Im}\,(\mathbf{v}^*\mathbf{u}) \end{pmatrix},$$

where $\mathbf{u}$ and $\mathbf{v}$ are the left and right singular vectors corresponding to the minimal singular value of $(x+iy) - \mathbf{A}$; see [111]. In the complex plane, the gradient direction is $\mathbf{v}^*\mathbf{u}$, so the direction orthogonal is $i\mathbf{v}^*\mathbf{u}$. After updating z_{k-1} by a step of length τ orthogonal to the gradient, Brühl advocates one step of Newton's method to correct the approximation. In pseudocode, we have:

$(\sigma, \mathbf{u}, \mathbf{v}) =$ minimum singular triplet of $z_0 - \mathbf{A}$
repeat
    $\widehat{z}_k = z_{k-1} + \tau i(\mathbf{v}^*\mathbf{u})/|\mathbf{v}^*\mathbf{u}|$
    $(\sigma, \mathbf{u}, \mathbf{v}) =$ minimum singular triplet of $\widehat{z}_k - \mathbf{A}$
    $z_k = \widehat{z}_k - (\sigma - \varepsilon)/(\mathbf{u}^*\mathbf{v})$
until the contour is completed.

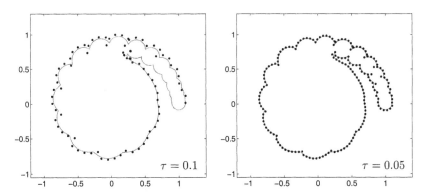

Figure 41.2: Examples of the curve-tracing algorithm for the laser integral operator (60.3) with $N = 100$ and $F = 10\pi$. In each plot, points generated by a curve-tracing routine are superimposed upon the boundary of the 10^{-1}-pseudospectrum, computed using the dense matrix techniques described in §39. The left plot shows that when the algorithm is executed with the relatively large step size $\tau = 1/10$, it generates an inaccurate curve, and eventually fails completely. Reducing the step size to $\tau = 1/20$ leads to improved performance, as seen in the right plot.

To accelerate his algorithm, Brühl uses gradient information from the predicted point $\widehat{z}_k$ above, rather than the corrected point z_k, thus requiring only one singular value computation for each point of the contour. One could alternatively determine the minimal singular triplet of z_k and use these values in the formula for $\widehat{z}_k$. Brühl also suggests the possibility of adjusting τ at each step.

Several complications may arise in a curve-tracing algorithm. If $\sigma_\varepsilon(\mathbf{A})$ comprises several disjoint components, one must repeat the procedure for each component. Also, it is common for $\partial\sigma_\varepsilon(\mathbf{A})$ to have cusps, where the curve-tracing algorithm may break down. Finally, if other level curves are desired (e.g., as required to compute an envelope of ε-dependent bounds like those described in §14), the curve-tracing procedure must be repeated. Improvements to the basic procedure that address these issues have been proposed by Bekas and Gallopoulos [42, 43] and Mezher and Philippe [550]; these works also address parallel implementation. Mezher, Najem, and Philippe have implemented a path-following algorithm in freely available software [549] designed to run on parallel computers or networks of workstations. For recent work on curve-tracing techniques for large-scale problems, see [44].

Figure 41.2 shows results of the curve-tracing algorithm applied to the laser integral operator (60.3) with discretization size $N = 100$ and Fresnel number $F = 10\pi$. The step-length parameter τ controls the number of points that describe the contour, as well as the accuracy.

42 · Pseudospectral abscissae and radii _____

The last three sections described efficient ways to compute the pseudospectra of dense and sparse matrices. However, in a variety of circumstances one does not need to know entire pseudospectra, but only certain scalar quantities derived from them. For example, to apply many of the bounds presented in §§15 and 16, one only needs to know the resolvent norm at a single point. In this section we describe algorithms for determining two quantities that are fundamental to such bounds, the ε-*pseudospectral abscissa* and the ε-*pseudospectral radius*:

$$\alpha_\varepsilon(\mathbf{A}) = \sup_{z \in \sigma_\varepsilon(\mathbf{A})} \mathrm{Re}\, z, \qquad \rho_\varepsilon(\mathbf{A}) = \sup_{z \in \sigma_\varepsilon(\mathbf{A})} |z|.$$

In words, we seek algorithms that find the points with largest real part and largest magnitude on the boundary of the ε-pseudospectrum of $\mathbf{A}$, denoted by $\partial \sigma_\varepsilon(\mathbf{A})$. This task is closely related to the problem of computing the *distance to instability*, the smallest value of ε for which $\partial \sigma_\varepsilon(\mathbf{A})$ intersects the imaginary axis (continuous time) or the unit circle (discrete time); see §49. Indeed, methods for computing the distance to instability [100, 123] are at the root of the algorithms we discuss for $\alpha_\varepsilon(\mathbf{A})$ and $\rho_\varepsilon(\mathbf{A})$.

An obvious algorithm for computing pseudospectral abscissae and radii springs to mind. One could approximate $\partial \sigma_\varepsilon(\mathbf{A})$ with a curve-tracing algorithm, as described in the last section, then find the points on this boundary with largest real part and largest magnitude. This approach requires considerable computation, which multiplies when $\sigma_\varepsilon(\mathbf{A})$ contains disconnected components. Refinement strategies would be necessary to obtain high-accuracy estimates of $\alpha_\varepsilon(\mathbf{A})$ and $\rho_\varepsilon(\mathbf{A})$.

Fortunately, there is a startlingly efficient alternative. Motivated by spectral optimization problems for nonnormal matrices (see, e.g., the Boeing 767 example in §15), Burke, Lewis, Overton, and Mengi have developed robust, globally convergent 'criss-cross' algorithms for computing the pseudospectral abscissa [117] and radius [545]. These closely related procedures are included in the EigTool system [838].

As the computation of the pseudospectral radius introduces additional subtleties, we start with the pseudospectral abscissa. The algorithm begins from the rightmost eigenvalue of $\mathbf{A}$ and then proceeds with a series of horizontal and vertical searches to find the boundary $\partial \sigma_\varepsilon(\mathbf{A})$. (Convergence at the end is guaranteed.) Each horizontal search gives an estimate for $\alpha_\varepsilon(\mathbf{A})$, while each vertical search identifies favorable locations for the next round of horizontal searches. The details are described below, with each step illustrated in Figure 42.1.

Criss-cross algorithm for the pseudospectral abscissa

---

1. Find a rightmost eigenvalue $\lambda \in \sigma(\mathbf{A})$, i.e., $\mathrm{Re}\lambda = \alpha(\mathbf{A})$.

2. Find the rightmost point on $\partial\sigma_\varepsilon(\mathbf{A})$ that intersects the horizontal line through λ, $\{z \in \mathbb{C} : \mathrm{Im}z = \mathrm{Im}\lambda\}$. Let z_1 denote this point; note that $\mathrm{Re}z_1$ is a lower bound for $\alpha_\varepsilon(\mathbf{A})$.

For $k = 1, 2, \ldots$ until convergence:

3. Find all the points at which $\partial\sigma_\varepsilon(\mathbf{A})$ intersects the vertical line through z_k, $\{z \in \mathbb{C} : \mathrm{Re}z = \mathrm{Re}z_k\}$. From these intersections, determine the intervals along this vertical line that intersect $\overline{\sigma_\varepsilon(\mathbf{A})}$.

4. Compute the midpoints of these intervals, each of which is contained in $\overline{\sigma_\varepsilon(\mathbf{A})}$. From each midpoint, search horizontally for the rightmost intersection with $\partial\sigma_\varepsilon(\mathbf{A})$. Call the rightmost point of all these intersections z_{k+1}. Check convergence; if necessary, return to step 3.

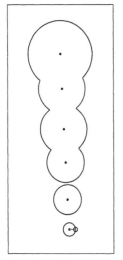 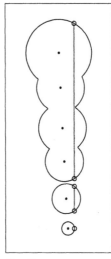 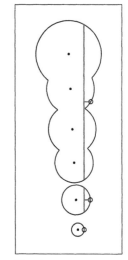

Steps 1 and 2
Find λ: $\mathrm{Re}\lambda = \alpha(\mathbf{A})$
Horizontal search

Step 3
Vertical search to find
intervals in $\sigma_\varepsilon(\mathbf{A})$

Step 4
Horizontal search
from midpoints

Figure 42.1: The initial iteration of the criss-cross algorithm for computing the pseudospectral abscissa. Steps 1 and 2 find an initial point on $\partial\sigma_\varepsilon(\mathbf{A})$. Step 3 identifies intervals where the vertical line intersects $\partial\sigma_\varepsilon(\mathbf{A})$; here the lowest interval is degenerate, i.e., a single point. Step 4 searches horizontally for $\partial\sigma_\varepsilon(\mathbf{A})$ from the midpoint of each interval; the rightmost of the resulting points gives the new estimate of $\alpha_\varepsilon(\mathbf{A})$.

The Burke–Lewis–Overton algorithm relies on the ability to repeatedly find the intersection of a line in the complex plane with the boundary $\partial\sigma_\varepsilon(\mathbf{A})$. One might expect that this would require the solution of a non-linear optimization problem, but there is a far better approach. First we consider the problem in step 3 of finding the intersection of a vertical line with $\partial\sigma_\varepsilon(\mathbf{A})$. (The horizontal search can be derived from the same technique.) Suppose that ε is a singular value of $(x+iy) - \mathbf{A}$, which is the case if and only if ε is an eigenvalue of the augmented matrix

$$\begin{pmatrix} \mathbf{0} & (x+iy) - \mathbf{A} \\ (x-iy) - \mathbf{A}^* & \mathbf{0} \end{pmatrix}.$$

Equivalently, the matrix

$$\begin{pmatrix} -\varepsilon\mathbf{I} & (x+iy) - \mathbf{A} \\ (x-iy) - \mathbf{A}^* & -\varepsilon\mathbf{I} \end{pmatrix}$$

is singular, as is any matrix that results if we scale and swap its rows. In particular, multiply the first block-row by -1 and exchange it with the second to obtain the singular matrix

$$\begin{pmatrix} (x - \mathbf{A}^*) - iy & -\varepsilon\mathbf{I} \\ \varepsilon\mathbf{I} & (\mathbf{A} - x) - iy \end{pmatrix}.$$

We have proved the following fundamental lemma [117]. It is essentially due to Byers, who proved a similar result and applied it in his algorithm for the distance to instability [123].

Lemma 42.1 *The matrix* $(x+iy) - \mathbf{A}$ *has a singular value* ε *if and only if* iy *is an eigenvalue of*

$$\begin{pmatrix} x - \mathbf{A}^* & -\varepsilon\mathbf{I} \\ \varepsilon\mathbf{I} & \mathbf{A} - x \end{pmatrix}. \qquad (42.1)$$

Suppose for a moment that $\varepsilon \ll 1$ and $\mathbf{A}$ is far from normal. We would then expect the eigenvalues of (42.1) to be highly sensitive to perturbations, and it would be risky to build an algorithm upon such a foundation. But Lemma 42.1 is special, for though the matrix (42.1) may be far from normal, it is *Hamiltonian* (see p. 491 for a definition and discussion). Algorithms exist to compute the eigenvalues of matrices with such structure accurately, even when $\mathbf{A}$ is large and sparse; see, e.g., [51, 52].

We now turn to the problem of determining the intervals of the vertical line $\{z \in \mathbb{C} : \mathrm{Re}\,z = x\}$ that intersect $\overline{\sigma_\varepsilon(\mathbf{A})}$. Suppose that $i\hat{y}$ is a purely imaginary eigenvalue of the matrix (42.1). Note that $x + i\widehat{y} \in \partial\sigma_\varepsilon(\mathbf{A})$ if and only if ε is a *minimal* singular value of $(x + i\widehat{y}) - \mathbf{A}$. Lemma 42.1 ensures that $(x + i\widehat{y}) - \mathbf{A}$ has a singular value ε, but we do not yet know if ε is minimal. Thus, for each purely imaginary eigenvalue of (42.1), we must compute the minimal singular value (or, equivalently, the norm of the inverse) of an $N \times N$ matrix. Suppose that ε is the minimal singular value of $(x + iy_j) - \mathbf{A}$ for m of the purely imaginary eigenvalues, $iy_1, \ldots, iy_m$, labeled by increasing imaginary part. Then the points $\{x + iy_j\}$ are the ends of the intervals where $\overline{\sigma_\varepsilon(\mathbf{A})}$ intersects the line $\{z \in \mathbb{C} : \mathrm{Re}\,z = x\}$. As the matrix (42.1) has dimension $2N$, it follows that $m \leq 2N$; i.e., a vertical line cannot intersect the boundary of a pseudospectrum at more than $2N$ points.[1]

We now have the endpoints of the intervals at which $\overline{\sigma_\varepsilon(\mathbf{A})}$ intersects the vertical line $\{z \in \mathbb{C} : \mathrm{Re}\,z = x\}$, but still must determine whether each of these points falls at the top or bottom of an interval. Since $\sigma_\varepsilon(\mathbf{A})$ is bounded, it must be that y_1 corresponds to the bottom of an interval, while y_m marks the top. This would be enough to determine the nature of $y_2, \ldots, y_{m-1}$, except for the possibility that some intervals could be degenerate; i.e., y_j could be both the bottom and top of an interval. This 'noncrossing' situation occurs when the vertical line intersects $\overline{\sigma_\varepsilon(\mathbf{A})}$ at $x + iy_j$ but no neighboring points. Conveniently, there is an easy test for this degenerate case that usually requires no further computation [117, Lemma 2.4]: Provided the minimal singular value ε of $(x+i\widehat{y})-\mathbf{A}$ is simple, y_j is a noncrossing point if and only if iy_j is an eigenvalue of (42.1) with *even algebraic multiplicity*. Such noncrossing cases may seem like rarities, but they do arise in practice; the 'step 3' plot in Figure 42.1 is such an example.

Once the top and bottom endpoints of the intervals where the vertical line $\{z \in \mathbb{C} : \mathrm{Re}\,z = x\}$ intersects $\sigma_\varepsilon(\mathbf{A})$ have been found, step 4 of the criss-cross algorithm computes the midpoint of each interval, then searches horizontally from each midpoint for the rightmost point on $\partial\sigma_\varepsilon(\mathbf{A})$. Again, this is a problem of finding where $\partial\sigma_\varepsilon(\mathbf{A})$ intersects a line. We can solve it by transforming the horizontal search into a vertical search and then applying Lemma 42.1: $(x + iy) - \mathbf{A}$ has a singular value ε if and only if $i(x + iy - \mathbf{A})$ does as well. Thus from Lemma 42.1, we seek the purely imaginary eigenvalues ix of the matrix

$$\begin{pmatrix} -y + i\mathbf{A}^* & -\varepsilon\mathbf{I} \\ \varepsilon\mathbf{I} & i\mathbf{A} + y \end{pmatrix}, \tag{42.2}$$

which again is Hamiltonian.

---

[1] As we shall see, the fact that circles centered at the origin can intersect $\partial\sigma_\varepsilon(\mathbf{A})$ at infinitely many points introduces a technical complication for the analogous pseudospectral radius algorithm.

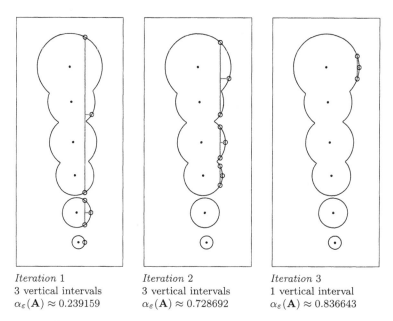

Iteration 1	*Iteration* 2	*Iteration* 3
3 vertical intervals	3 vertical intervals	1 vertical interval
$\alpha_\varepsilon(\mathbf{A}) \approx 0.239159$	$\alpha_\varepsilon(\mathbf{A}) \approx 0.728692$	$\alpha_\varepsilon(\mathbf{A}) \approx 0.836643$

Figure 42.2: Steps 3 and 4 for three iterations of the criss-cross pseudospectral abscissa algorithm for the same example as in Figure 42.1. In this case the algorithm terminates exactly at the third iteration.

Since the goal of our computation is the rightmost point in $\partial\sigma_\varepsilon(\mathbf{A})$, we are interested only in the rightmost point at which the horizontal line $\{z \in \mathbb{C} : \mathrm{Re}\, z = x\}$ intersects $\partial\sigma_\varepsilon(\mathbf{A})$. One might expect this would require us to check whether each purely imaginary eigenvalue $i\widehat{x}$ corresponds to a matrix $(\widehat{x}+iy) - \mathbf{A}$ with minimal singular value ε, but in fact we are spared this labor. Burke, Lewis, and Overton show that the largest of the purely imaginary eigenvalues of (42.2) always corresponds to a case where ε is the minimal singular value [117].

With the vertical and horizontal searches in place, all that remains is to repeat this procedure until satisfactory accuracy is attained. Figure 42.2 shows further iterations of the computation begun in Figure 42.1. The estimates for the pseudospectral abscissa are monotonically increasing and bounded above by $\alpha_\varepsilon(\mathbf{A})$. Burke, Lewis, and Overton show that, in exact arithmetic, the algorithm will always converge, and locally, the convergence rate is quadratic [117]. (If the estimates are not strictly increasing, then $\alpha_\varepsilon(\mathbf{A})$ has been computed exactly.) In floating-point arithmetic, the algorithm is backward stable: it computes the exact ε-pseudospectral abscissa of a matrix that differs from $\mathbf{A}$ by entries on the order of machine precision times $\|\mathbf{A}\|$. (To achieve this, the algorithm requires a backward stable Hamiltonian eigensolver, not used in the current EigTool implementation.)

The Mengi–Overton algorithm for computing the pseudospectral radius follows a similar pattern [545]. The horizontal searches are replaced by searches along radial lines through the origin, each of which provides an estimate for $\rho_\varepsilon(\mathbf{A})$. The vertical intervals are replaced by arc-intervals of a circle centered at the origin. Several steps of this algorithm are illustrated in Figure 42.3.

Criss-cross algorithm for the pseudospectral radius

1. Find an eigenvalue $\lambda \in \sigma(\mathbf{A})$ of largest magnitude, i.e., $|\lambda| = \rho(\mathbf{A})$.

2. Find the largest magnitude point on $\partial\sigma_\varepsilon(\mathbf{A})$ that intersects the line through the origin and λ, $\{z \in \mathbb{C} : \arg z = \arg\lambda\}$. Let z_1 denote this point; note that $|z_1|$ is a lower bound for $\rho_\varepsilon(\mathbf{A})$.

For $k = 1, 2, \ldots$ until convergence:

3. Find all the points at which $\partial\sigma_\varepsilon(\mathbf{A})$ intersects the circle of radius $|z_k|$. From these intersections, determine the arcs of this circle that intersect $\overline{\sigma_\varepsilon(\mathbf{A})}$.

4. Compute the midpoint of each of these arcs. From each midpoint, search radially for the largest magnitude intersection with $\partial\sigma_\varepsilon(\mathbf{A})$. Call the largest magnitude point of all these intersections z_{k+1}. Check convergence; if necessary, return to step 3.

The algorithm begins by finding an eigenvalue $\lambda = \varrho e^{i\theta} \in \sigma_\varepsilon(\mathbf{A})$ of largest magnitude, then searching for the largest magnitude point along the ray $\{re^{i\theta} : r \geq 0\}$. Finding the points at which a radial line intersects $\partial\sigma_\varepsilon(\mathbf{A})$ is essentially the same as the computation of horizontal and vertical intersections in the pseudospectral abscissa algorithm. Just as we rotated the horizontal search to a vertical search by multiplying $z - \mathbf{A}$ by i, now we multiply $re^{i\theta} - \mathbf{A}$ by $ie^{-i\theta}$ and apply Lemma 42.1 to see that ε is a singular value of $re^{i\theta} - \mathbf{A}$ if and only if ir is an eigenvalue of

$$\begin{pmatrix} ie^{i\theta}\mathbf{A}^* & -\varepsilon\mathbf{I} \\ \varepsilon\mathbf{I} & ie^{-i\theta}\mathbf{A} \end{pmatrix}.$$

As before, this is a Hamiltonian matrix, and if $i\hat{r}$ is the largest purely imaginary eigenvalue with $\hat{r} \geq 0$, then ε is the *minimum* singular value of $\hat{r}e^{i\theta} - \mathbf{A}$. Thus $\hat{r}e^{i\theta}$ is the largest magnitude intersection of $\partial\sigma_\varepsilon(\mathbf{A})$ with the ray $\{z = re^{i\theta} : r \geq 0\}$.

Now one must compute all intersections of $\partial\sigma_\varepsilon(\mathbf{A})$ with the circle of radius r. Unfortunately, this is significantly more difficult than the search for intersections of $\partial\sigma_\varepsilon(\mathbf{A})$ with a line. In fact, for some $\mathbf{A}$ the boundary $\partial\sigma_\varepsilon(\mathbf{A})$ intersects the circle of radius r at infinitely many points. (For

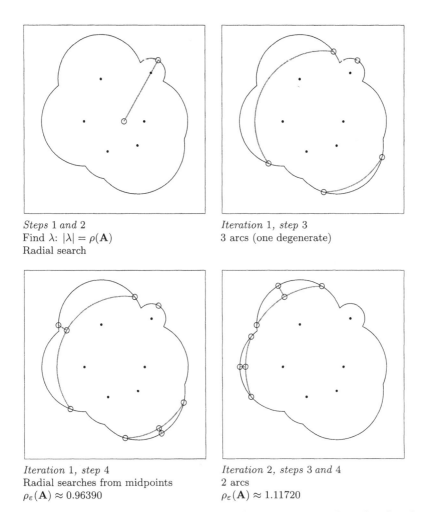

Steps 1 *and* 2
Find λ: $|\lambda| = \rho(\mathbf{A})$
Radial search

Iteration 1, *step* 3
3 arcs (one degenerate)

Iteration 1, *step* 4
Radial searches from midpoints
$\rho_\varepsilon(\mathbf{A}) \approx 0.96390$

Iteration 2, *steps* 3 *and* 4
2 arcs
$\rho_\varepsilon(\mathbf{A}) \approx 1.11720$

Figure 42.3: Illustration of two iterations of the criss-cross algorithm for the pseudospectral radius. In this case, the exact solution is determined at the end of the second iteration.

example, let $\mathbf{A}$ be a Jordan block with eigenvalue zero; then $re^{i\theta} - \mathbf{A}$ has the same minimal singular value for all $\theta \in [0, 2\pi)$.) Mengi and Overton [545], again building on the work of Byers [123], show that the intersections of $\partial\sigma_\varepsilon(\mathbf{A})$ with the circle of radius r occur for values of θ for which $e^{i\theta}$ solves the generalized eigenvalue problem

$$\begin{pmatrix} -\varepsilon\mathbf{I} & \mathbf{A} \\ r\mathbf{I} & 0 \end{pmatrix} \mathbf{v} = e^{i\theta} \begin{pmatrix} 0 & r\mathbf{I} \\ \mathbf{A}^* & -\varepsilon\mathbf{I} \end{pmatrix} \mathbf{v} \qquad (42.3)$$

for some nonzero $\mathbf{v}$. On occasion this will hold for all values of $e^{i\theta}$, in which case one can only conclude that there are infinitely many points of intersection of the circle and $\partial\sigma_\varepsilon(\mathbf{A})$. Mengi and Overton show that this situation can arise only during the first iteration of the algorithm. The generalized eigenvalue problem (42.3) has structure, but there are not yet backward stable algorithms for its solution. Given such a stable method, this pseudospectral radius algorithm would also be backward stable. In exact arithmetic, like the pseudospectral abscissa method, this algorithm converges globally with a local quadratic rate.

Here is a numerical example with the flavor of [80]. Let $\mathbf{A}$ be the 100×100 Grcar matrix shown in Figure 7.5, except multiplied by 0.4. This matrix is power-bounded, with spectral radius $\rho(\mathbf{A}) \approx 0.9052$, but its powers grow as large as $\|\mathbf{A}^{104}\| \approx 60{,}060{,}433$ before eventually decaying. Suppose we set $\varepsilon = 10^{-8}$ and use the criss-cross algorithm to compute the ε-pseudospectral radius. The result is $\rho_\varepsilon(\mathbf{A}) \approx 1.0321$, attained for $z \approx 0.1576 \pm 1.0200\mathrm{i}$. From equation (16.15) of Theorem 16.4 we can infer from this value alone that the norms $\|\mathbf{A}^k\|$ must rise at least as high as $3.2 \cdot 10^6$ on a time scale roughly on the order of $k = 100$.

43 · Discretization of continuous operators

To compute pseudospectra of differential or integral operators, the usual procedure is to approximate them by matrices and then apply the algorithms for matrices described in the last few sections. This is not the only approach to such problems, nor always the most powerful. But it is certainly the most flexible, and it is the method we have used to generate about fifty of the figures in this book. We discretize by spectral methods rather than finite differences or finite elements, because high accuracy is needed if one wants plots of ε-pseudospectra for small ε.

In this section we outline the techniques we have found so useful, employing a 'how to' style illustrated by MATLAB code segments. An earlier discussion of some of these ideas appears in [774], and in §30 we have already mentioned the pseudospectra of some of the matrices that arise in spectral discretizations. We make no attempt to survey systematically the vast and highly developed field of numerical discretization of operators. Our methods are essentially those described in Trefethen's textbook on spectral methods [775]. A software suite for such computations has been developed by Weideman and Reddy [817], and the books by Boyd and Fornberg offer a wealth of practical information [99, 282]. Our computations of pseudospectra and other norm-dependent quantities rely on discretization of integrals of smooth functions and are thus mainly restricted to Hilbert spaces, e.g., the L^2-norm. (The L^1-norm, for example, is also defined by an integral, but contains an absolute value that introduces derivative discontinuities.)

If the domain is periodic, we use a *Fourier spectral collocation method*.[1] To treat a periodic function u defined on $[0, 2\pi]$, say, we pick an integer $N > 0$ and consider vectors of sample values

$$v_j \approx u(x_j), \quad x_j = \frac{2\pi j}{N}, \quad j = 1, 2, \ldots, N.$$

We associate each such vector $\mathbf{v} \in \mathbb{C}^N$ with its *trigonometric interpolant* $v(x)$ defined on $[0, 2\pi]$, a trigonometric polynomial satisfying $v(x_j) = v_j$ for each j, which takes one of the forms

$$v(x) = \sum_{k=-N/2}^{N/2} a_k e^{ikx}, \qquad v(x) = \sum_{k=(1-N)/2}^{(N-1)/2} a_k e^{ikx}$$

depending on whether N is even or odd, respectively. The interpolant is unique if we impose the condition $a_{-N/2} = a_{N/2}$ in the former case.

---

[1] See the footnote on p. 289.

The coefficients a_k in these expansions are not computed explicitly; all computations work instead with the data vector $\mathbf{v}$. Thus the derivative of $v(x)$ satisfies the identity $v'(x) = \sum ika_k e^{ikx}$, but we do not work with it in this form. Instead we consider the vector $\mathbf{w}$ of its values on the grid and write

$$w_j = v'(x_j) = (\mathbf{Dv})_j$$

for an $N \times N$ circulant matrix $\mathbf{D}$ known as a *differentiation matrix*. The (i,j) entry of $\mathbf{D}$ is equal to the value at x_i of the derivative of the trigonometric polynomial that interpolates the vector that is 1 at x_j and zero elsewhere. The entries of $\mathbf{D}$ are known explicitly,

$$(\mathbf{D})_{ij} = \tfrac{1}{2}(-1)^{j-i+1} \cot \frac{(j-i)\pi}{N}, \tag{43.1}$$

and similarly, the entries of the second derivative matrix $\mathbf{D}^{(2)}$ are

$$(\mathbf{D}^{(2)})_{ij} = \begin{cases} -\tfrac{1}{12}N^2 - \tfrac{1}{6} & (i=j), \\ \tfrac{1}{2}(-1)^{j-i+1} \csc^2 \dfrac{(j-i)\pi}{N} & (i \neq j), \end{cases} \tag{43.2}$$

where csc denotes the cosecant.[2] (These expressions are valid for even N; for odd N, the details change.) Alternatively, rather than using formulas like these, one can construct $\mathbf{D}$ or $\mathbf{D}^{(2)}$ on the fly by taking a Discrete Fourier Transform (DFT), multiplying by ik or $(ik)^2$, and inverse transforming. The latter approach is very flexible and can be generalized to pseudodifferential operators.

For example, here is a slightly stripped-down version of the MATLAB code that was used to generate Figure 12.8, concerning a linear operator of Benilov et al. [49, 50]. The operator is

$$\mathcal{L}u = h^2 \sin(x)u_{xx} + hu_x \tag{43.3}$$

with $h = 1/10$ and periodic boundary conditions on $[-\pi, \pi]$, and we approximate it by the $N \times N$ matrix

$$\mathbf{L} = h^2 \mathbf{S}\mathbf{D}^{(2)} + h\mathbf{D}, \tag{43.4}$$

where $\mathbf{S}$ is the diagonal matrix with entries $\sin(x_1), \ldots, \sin(x_N)$. This code uses the DFT method of discretization and takes N odd in order to exclude spurious sawtoothed pseudoeigenvectors that appear if N is even.

```
% Set up Fourier grid:
  N = 271;                           % N must be odd
  dx = 2*pi/N; x = -pi+dx*(1:N)';    % step size and grid
```

---

[2]$\mathbf{D}^{(2)}$ is not quite the same as $\mathbf{D}^2$, because of differing treatments of the maximal wave number $k = \pm N/2$, but the difference would be unimportant for most applications.

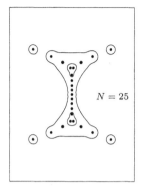 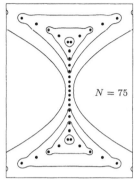 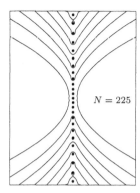

Figure 43.1: Eigenvalues and ε-pseudospectra of Fourier spectral discretizations of various dimensions of the operator (43.3) of Benilov et al. Each image shows the rectangle $-1.5 \leq \mathrm{Re}\lambda \leq 1.5$, $-2 \leq \mathrm{Im}\lambda \leq 2$, and the contours correspond to $\varepsilon = 10^{-1}, 10^{-2}, \ldots$. Compare Figure 12.8.

```
% Fourier spectral differentiation matrices:
  omega = exp(pi*2i/N);                      % Nth root of unity
  A = omega.^((0:N-1)'*(0:N-1))/sqrt(N);     % DFT matrix
  ik = 1i*([0:(N-1)/2 (1-N)/2:-1]);          % Fourier multipliers
  D = real(A*diag(ik)*A');                   % 1st-order diff. matrix
  D2 = real(A*diag(ik.^2)*A');               % 2nd-order diff. matrix
% The Benilov-O'Brian-Sazonov operator:
  h = 0.1;
  A = h^2*diag(sin(x))*D2 + h*D;
% Call EigTool to compute eigenvalues and pseudospectra:
  opts.ax = [-3 3 -2 2];
  opts.levels = -8:-1;
  eigtool(A,opts)
```

Figure 43.1 shows eigenvalues and pseudospectra generated by this code for three values of N. The images hint at a common property of spectral discretizations: A computed number, such as a particular eigenvalue of $\mathcal{L}$, tends to 'snap in' to the correct value when N becomes sufficiently large, so that, roughly speaking, any particular computed quantity will be either highly accurate or entirely wrong. The snapping-in generally occurs once the grid is fine enough that the function of interest is resolved by at least two points per wavelength. As N increases beyond this point, one observes 'spectral accuracy', which means convergence at a rate $\mathcal{O}(C^N)$ for some $C < 1$ if u is analytic or $\mathcal{O}(N^{-M})$ for all M if u is C^∞. For example, Figure 43.2 shows the error in the resolvent norm $\|(z - \mathbf{L})^{-1}\|$ as a function of N for the particular choice $z = 1 + 1.5\mathrm{i}$, the value marked by the cross in Figure 12.8. The convergence is geometric, satisfying

$$\|(z - \mathbf{L}^{(N)})^{-1}\| - \|(z - \mathcal{L})^{-1}\| = \mathcal{O}(C^N).$$

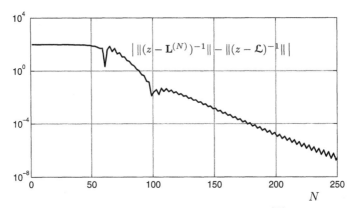

Figure 43.2: Convergence of the resolvent norm $\|(z - \mathbf{L}^{(N)})^{-1}\|$ to the correct value $98.403046\ldots$ as the matrix dimension is increased for the example (43.3) of Benilov et al. with $z = 1 + 1.5\mathrm{i}$. The anomalously small errors near $N = 60$ and $N = 100$ are caused by sign changes.

Note that the first norm on the left-hand side is a matrix 2-norm, while the second is an operator L^2-norm.

Theorems establishing the spectral accuracy of spectral discretizations have been published in many papers and books, including [56, 127, 294, 775], and a small literature exists on approximation of pseudospectra per se [373, 374, 375, 836]. Three related effects combine to make the computation of a resolvent norm by a Fourier spectral method successful. First are phenomena of *approximation* that ensure that a smooth function can be approximated by trigonometric polynomials with rapidly improving accuracy as $N \to \infty$. Second are effects of *collocation*, ensuring that the trigonometric polynomial implicitly utilized by the spectral method is sufficiently close to optimal for this rapid convergence rate to be realized. Third is a matter of *quadrature*. The definition of the L^2-norm involves an integral, and when we approximate the resolvent norm of an operator by that of a matrix derived by Fourier discretization, we are implicitly taking advantage of the fact that for smooth functions on a periodic domain, the trapezoid rule is a spectrally accurate quadrature formula. See Chapters 4 and 12 of [775].

As a second illustration of a Fourier spectral method, the next code was used to plot the pseudospectra of Davies' complex harmonic oscillator in Figure 11.3. The operator is

$$\mathcal{L}u = -h^2 u_{xx} + \mathrm{i}x^2 u \qquad (43.5)$$

on the whole real line, and therefore, this might not seem a candidate for a Fourier spectral method. However, the eigenfunctions and pseudoeigenfunctions decay exponentially, so one can cut off the domain to a finite interval $[-L, L]$ with negligible error. On that interval, zero boundary con-

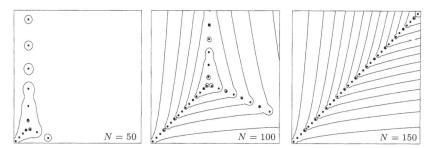

Figure 43.3: Eigenvalues and ε-pseudospectra of Fourier spectral discretizations of Davies' operator (43.5). Each image shows the square $0 \leq \mathrm{Re}\,\lambda, \mathrm{Im}\,\lambda \leq 5$, and the contours correspond to $\varepsilon = 10^{-1}, 10^{-2}, \ldots$. Compare Figures 5.1 and 11.3.

ditions would be appropriate, which could be imposed by a Chebyshev spectral method. Alternatively, it is just as effective to take the boundary conditions to be periodic, and this enables one to get the same accuracy with smaller values of N, because fewer grid points are wasted near the boundary. Our matrix approximation is

$$\mathbf{L} = -h^2 \mathbf{D}^{(2)} + \mathrm{i}\mathbf{S}, \qquad (43.6)$$

where $\mathbf{S}$ is the diagonal matrix with entries $x_1^2, \ldots, x_N^2$. This time the code uses the explicit formula (43.2) via MATLAB's 'toeplitz' command; results for three values of N are shown in Figure 43.3. Plots (not shown) like those of Figure 43.2 again reveal geometric convergence of eigenvalues and resolvent norms.

```
% Fourier grid:
  L = 6;                    % real line is approximated by [-L,L]
  N = 150;                  % N must be even
  x = 2*L*(1-N/2:N/2)/N;    % regular grid in [-L,L]
% 2nd-order differentiation matrix:
  column = [-N^2/12-1/6 -.5*(-1).^(1:N-1)./sin(pi*(1:N-1)/N).^2];
  D2 = (pi/L)^2*toeplitz(column);
% The Davies operator:
  h = 0.1;
  A = -h^2*D2 + 1i*diag(x.^2);
% Call EigTool to compute eigenvalues and pseudospectra:
  opts.ax = [0 5 0 5];
  opts.levels = -13:-1;
  eigtool(A,opts)
```

Now we turn to problems on bounded nonperiodic domains, where we use *Chebyshev spectral collocation methods*. To treat a function u defined on $[-1, 1]$, we pick an integer $N > 0$ and consider vectors

$$v_j \approx u(x_j), \quad x_j = \cos(j\pi/N), \quad j = 0, 1, \ldots, N.$$

These *Chebyshev points* $\{x_j\}$ (given earlier in (30.1)) are not uniformly distributed, but clustered near the endpoints, for unavoidable reasons explained in Chapter 5 of [775]. We associate a vector $\mathbf{v}$ with its unique *polynomial interpolant* $v(x)$ defined on $[-1, 1]$, an algebraic polynomial satisfying $v(x_j) = v_j$ for each j, taking the form

$$v(x) = \sum_{k=0}^{N} a_k T_k(x),$$

where T_k is the kth Chebyshev polynomial. As before, the coefficients a_k in these expansions are not computed explicitly, nor do we make explicit use of Chebyshev polynomials. Again we work with derivatives via their values on the grid, writing, for example,

$$w_j = v'(x_j) = (\mathbf{D}\mathbf{v})_j$$

for a differentiation matrix $\mathbf{D}$ (no longer circulant) of dimension $(N+1) \times (N+1)$. The (i, j) entry of $\mathbf{D}$ is equal to the value at x_i of the derivative of the polynomial that interpolates the vector that is 1 at x_j and zero elsewhere, and these numbers are known explicitly; they were apparently first published in [330]. If the rows and columns are indexed from 0 to N, the off-diagonal entries are

$$\mathbf{D}_{ij} = \frac{c_i}{c_j} \frac{(-1)^{i+j}}{(x_i - x_j)}, \qquad i \neq j, \quad i, j = 0, \ldots, N,$$

where $c_0 = c_N = 2$ and $c_i = 1$ otherwise, and the diagonal entries are defined by the condition that each row of $\mathbf{D}$ sums to zero. Alternatively, as in the Fourier case, it is possible to compute $\mathbf{D}$ on the fly by use of the DFT [775, Chap. 8].

The following code segment, building upon code listed on page 290, illustrates the use of such methods for the advection-diffusion operator (12.3). The operator is

$$\mathcal{L}u = \eta u_{xx} + u_x, \qquad x \in (0, 1) \tag{43.7}$$

with $\eta = 0.015$, and the eigenvalues and pseudospectra were plotted in Figure 12.4. In Figure 43.4 we repeat this computation for several values of N.

```
% Chebyshev differentiation matrix:
  N = 60;
  x = cos(pi*(0:N)/N)';                    % Chebyshev points
  c = [2; ones(N-1,1); 2].*(-1).^(0:N)';
  X = repmat(x,1,N+1); dX = X-X';
  D = (c*(1./c)')./(dX+(eye(N+1)));        % off-diagonal entries
  D = D - diag(sum(D'));                    % diagonal entries
  D = 2*D; x = (x+1)/2;                     % rescale [-1,1] -> [0,1]
```

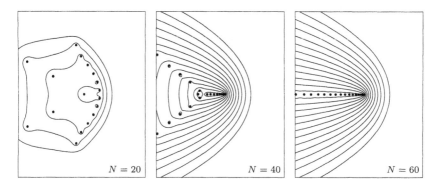

Figure 43.4: Eigenvalues and ε-pseudospectra of Chebyshev spectral discretizations of the advection-diffusion operator (43.7). Each image shows the rectangle $-60 \leq \text{Re}\lambda \leq 20$, $-50 \leq \text{Im}\lambda \leq 50$, and the contours correspond to $\varepsilon = 10^{-1}, 10^{-2}, \ldots$. Compare Figure 12.4.

```
% Advection-diffusion operator:
  eta = 0.015;
  L = eta*D^2 + D;
  L = L(2:N,2:N);                        % impose BCs u(0)=u(1)=0
  W = diag(sqrt(pi*sqrt(x-x.^2)/(2*N))); % Gauss-Chebyshev weights
  L = W*L/W;                             % similarity transformation
% Call EigTool to compute eigenvalues and pseudospectra:
  opts.ax = [-60 20 -50 50];
  opts.levels = -12:0;
  eigtool(L,opts)
```

Two new issues arise with this computation that did not appear for periodic problems. The first is the enforcement of boundary conditions. For this problem, with $u(0) = u(1) = 0$, this is simply a matter of stripping off the first and last rows and columns of the differentiation matrix by the command L = L(2:N,2:N). More complicated boundary conditions can be treated by methods discussed, for example, in [775, Chap. 13].

The more interesting issue pertains to the next two lines of the program:

```
  W = diag(sqrt(pi*sqrt(x-x.^2)/(2*N)));
  L = W*L/W;
```

For many purposes in the numerical solution of differential equations, there is no need to calculate a norm. Pseudospectra, however, are norm-dependent, as are associated quantities such as $\|e^{t\mathcal{L}}\|$. Thus we are faced with the third of the three numerical effects mentioned earlier, that of *quadrature*. For periodic problems, one generally does not need to think about quadrature, because the matrix norm amounts to a spectrally accurate approximation of the appropriate operator norm (due to the accuracy of the trapezoid rule in this setting). On an irregular grid such as a Chebyshev grid, however, simply taking the matrix norm will lead to the wrong answer. For example, if we delete the two lines just listed from the code above, we find convincing convergence as $N \to \infty$ to the following three numbers:

$$\|(-10 - \mathcal{L})^{-1}\| \approx 6992.73, \quad \|e^{0.1\mathcal{L}}\| \approx 1.0446, \quad \alpha(\mathcal{L}) \approx 0.87637.$$

(Here α denotes the numerical abscissa, i.e., the maximal real part of all points in the numerical range; see §17.) All three of these numbers are wrong. Indeed, the latter two can be seen to be wrong by inspection, since $\mathcal{L}$ is a dissipative operator in $L^2[0,1]$. The correct results are

$$\|(-10 - \mathcal{L})^{-1}\| \approx 6618.37, \quad \|e^{0.1\mathcal{L}}\| \approx 0.98392, \quad \alpha(\mathcal{L}) \approx -0.148044.$$

To compute quantities like these correctly, we must introduce a weighting to compensate for the irregularity of the grid. Suppose that $\|\cdot\|_2$ is the usual vector 2-norm and $\|\cdot\|$ denotes the weighted vector norm intended to approximate the continuous L^2-norm on functions. Then we have

$$\|\mathbf{u}\|^2 = \sum w_k |u_k|^2$$

for some set of quadrature weights $\{w_k\}$. If $\mathbf{W} = \text{diag}\,(\{(w_k)^{1/2}\})$, then this implies

$$\|\mathbf{u}\| = \|\mathbf{W}\mathbf{u}\|_2,$$

and at the matrix level (see (45.9)),

$$\sigma_\varepsilon(\mathbf{A}) = \sigma_\varepsilon^{\text{2-norm}}(\mathbf{W}\mathbf{A}\mathbf{W}^{-1}).$$

If we cared only about convergence, not efficiency, then any set of weights with asymptotic density proportional to $(x - x^2)^{1/2}$ would do, since this would compensate for the asymptotic density proportional to $(x - x^2)^{-1/2}$ of the Chebyshev points scaled to $[0,1]$. For efficiency, however, one would like to choose the weights more carefully so as to achieve rapid convergence. As it happens, weights proportional to $(x - x^2)^{1/2}$ are an accurate choice as well as an adequate one, differing from our two lines of code only by a constant factor, which cancels in the similarity transformation. The weights in the code are the *Gauss–Chebyshev–Lobatto* weights for the interval $[0,1]$, and they will compute the L^2-norm $\|u\|$ exactly for any function u whose square is equal to $\sqrt{x - x^2}$ times a polynomial of degree $\leq 2N - 1$.

An alternative set of weights that are more appropriate in principle and moderately better in practice are *Clenshaw–Curtis weights* [153, 189]. With this discretization, the result will be exact for any u whose square is a polynomial of degree $\leq 2N - 1$. The Clenshaw–Curtis weights are not given by quite as simple a formula, but they can be computed by means of the DFT or by expressions encoded, for example, in the program `clencurt` of [775]; it is then enough to replace the two lines above by

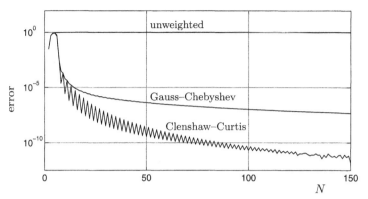

Figure 43.5: Convergence of the computed numerical abscissa $\alpha(\mathbf{L}^{(N)})$ of the advection-diffusion operator (43.7) to the correct value $-\eta\pi^2$ as $N \to \infty$. With unweighted Chebyshev matrix approximations, there is no convergence. Both Gauss–Chebyshev and Clenshaw–Curtis formulas are effective, the latter more so. Nevertheless, the convergence for this discretization falls short of spectral accuracy.

```
[s,w] = clencurt(N);
W = diag(sqrt(w(2:N))/2);
L = W*L/W;
```

Figure 43.5 shows the results of using these various methods to compute the numerical abscissa of $\mathcal{L}$. We see that Clenshaw–Curtis quadrature outperforms Gauss–Chebyshev. However, the difference is not too important in practice, since the errors in the Gauss–Chebyshev approach are relative to the quantity being computed, not to the scale of the matrix; thus they are invisible to the eye in a typical plot of ε-pseudospectra, even for very small ε. We also note that the convergence in Figure 43.5 is not spectral, but only algebraic: The error is $\mathcal{O}(N^{-2})$ for the Gauss–Chebyshev weights and $\mathcal{O}(N^{-4})$ for Clenshaw–Curtis. No doubt one could develop better discretizations of this problem.

Our fourth and final example concerns an integral rather than differential operator: the complex symmetric operator (60.3) that arises in the theory of lasers,

$$\mathcal{A}u(x) = \sqrt{\frac{\mathrm{i}F}{\pi}} \int_{-1}^{1} \mathrm{e}^{-\mathrm{i}F(x-s)^2} u(s)\,\mathrm{d}s, \qquad (43.8)$$

where $F > 0$ is the Fresnel number. Since the spatial domain is $[-1,1]$, the operator (43.8) could be discretized by Chebyshev methods. However, for maximal accuracy and to show a variety of methods, the following code uses Gauss quadrature instead. (The appearance of W*A*W rather than W*A/W in this code is not an error; the extra factor of W*W on the right

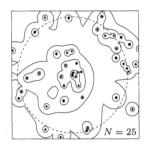 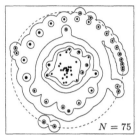 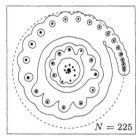

Figure 43.6: Eigenvalues and ε-pseudospectra of Gauss quadrature discretizations of the laser integral operator (43.8) with $F = 16\pi$. Each image shows the rectangle $-1.1 \leq \mathrm{Re}\lambda \leq 1.2$, $-1.1 \leq \mathrm{Im}\lambda \leq 1.1$, and the contours correspond to $\varepsilon = 10^{-1}, 10^{-1.5}, 10^{-2}, \ldots$. Compare Figure 60.2.

arises from the fact that in addition to the matter of appropriate weighting of vectors, the operator itself is defined by an integral that is approximated by Gauss quadrature.) Pseudospectra for three values of N are shown in Figure 43.6, and a look at the numbers reveals satisfying spectral accuracy. For example, Figure 43.7 shows rapid convergence to the numbers

$$\|\mathcal{A}\| \approx 1.000000, \quad \|\mathcal{A}^{-1}\| \approx 88.6952, \quad \alpha(\mathcal{A}) \approx 0.999714.$$

```
% Nodes and weights for Gauss quadrature:
  N = 150;
  beta = 0.5*(1-(2*(1:N-1)).^(-2)).^(-1/2);
  T = diag(beta,1) + diag(beta,-1);    % tridiagonal Jacobi matrix
  [V D] = eig(T);                      % eigenvalues and vectors
  [x ii] = sort(diag(D)');             % nodes
  w = 2*V(1,ii).^2;                    % weights

% Integral operator A:
  F = 16*pi;
  A = zeros(N);
  for k=1:N
    A(k,:) = sqrt(1i*F/pi)*exp(-1i*F*(x(k)-x).^2);
  end
  W = diag(sqrt(w));
  A = W*A*W;

% Call EigTool to compute eigenvalues and pseudospectra:
  opts.ax = [-1.1 1.2 -1.1 1.1];
  opts.levels = -3:.5:-1;
  eigtool(A,opts)
```

All of our discussion has concerned operators acting in one space dimension, and indeed, because the one-dimensional case suffices to illustrate most effects of nonnormality, multivariate operators hardly appear in

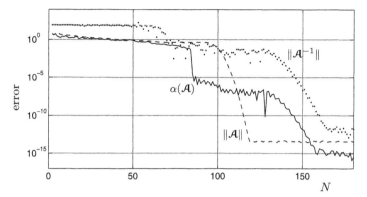

Figure 43.7: Errors as a function of N in $\|\mathcal{A}\|$, $\|\mathcal{A}^{-1}\|$, and the numerical abscissa $\alpha(\mathcal{A})$, computed via Gauss quadrature for the laser operator (43.8). In each case, once N is large enough, there is rapid convergence down to the level of rounding errors.

this book at all. Such operators can be discretized by spectral methods in which the dimensions are treated independently. Often one utilizes a mix of Fourier discretization in some directions and Chebyshev in others, e.g., if the domain is a disk or a sphere or a cylinder. In such problems the contributions along a periodic or unbounded coordinate with constant coefficients may be orthogonal, in which case there may be no need to discretize at all: the pseudospectra problem decouples into a set of problems of lower dimension. For example, the pseudospectra of hydrodynamic stability operators in §20 correspond to three-dimensional problems, conceptually speaking, but in each case two of the dimensions are eliminated by Fourier transformation, leaving a two-parameter family of spectral discretizations in one dimension.

A more challenging matter is the question of what one can do to compute operator norms that are not defined by integrals of smooth functions, that is, norms in Banach rather than Hilbert spaces. In such cases there is usually no quadrature formula, and the 1- or ∞-norms of a spectral discretization matrix, say, cannot be expected to match those of the corresponding operator to high accuracy. There must be effective methods for such computations, but we do not know them.

44 · A flow chart of pseudospectra algorithms _____

The appearance in 2002 of Thomas Wright's Oxford D.Phil. thesis *Algorithms and Software for Pseudospectra* [837] was a landmark in the development of methods for the computation of pseudospectra. This thesis introduced the EigTool software system [838] and the method of computing pseudospectra of large matrices via implicitly restarted Arnoldi iteration (§40). It also put forward a wider vision of how pseudospectra might be calculated for a diverse range of problems. Though some of his ideas were not implemented in EigTool, Wright also proposed algorithms for dense and sparse pseudospectra computations associated with generalized eigenvalue problems (§45), rectangular matrices (§46), and weighted norms (§51). He summarized this vision in four pages of flow charts.

With Wright's permission, we have reproduced his flow charts in this section. They are reprinted directly, without being re-typeset to match the rest of the book. Our view is that Wright's summary is so comprehensive, and such an interesting snapshot of the state of the art in 2002, that it is most appropriate to present it exactly as he did. As the years go by, some of the methods he proposed will undoubtedly be improved upon, but as we write, just two years later, there is little we would change.

Given a matrix pseudospectra problem, one begins in the first flow chart, Figure 44.1, by checking if the problem concerns a weighted norm as opposed to the standard matrix 2-norm. After appropriate preliminary calculations, one then moves to one of the next three flow charts, Figures 44.2–44.4, depending on whether the matrix is dense and square (the standard case discussed in §39), sparse and square (typically for dimensions in the thousands or higher), or dense and rectangular, respectively. The last case is of great importance in practice because a calculation by the implicitly restarted Arnoldi method, as described in §40, reduces a large sparse matrix to a dense rectangular one, with one more row than column, as indicated in the bottom right portion of the flow chart of Figure 44.3. Cases with many more rows than columns have received much less attention as yet.

Wright's schema covers algorithms for a broad range of 2-norm and weighted 2-norm (i.e., Hilbert space) computations, but it does not incorporate 1-, ∞-, or other Banach space norms. Details of each of these computational steps can be found in [837].

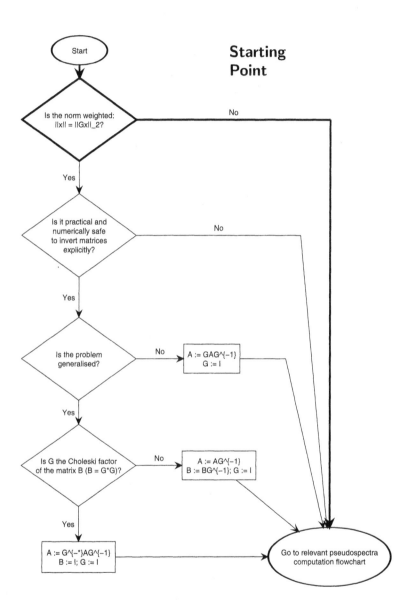

Figure 44.1: Starting point of a pseudospectra computation; the operations shown set up data for the other three flow charts. Boldface marks the path explicitly implemented in EigTool. Reproduced with permission from [837].

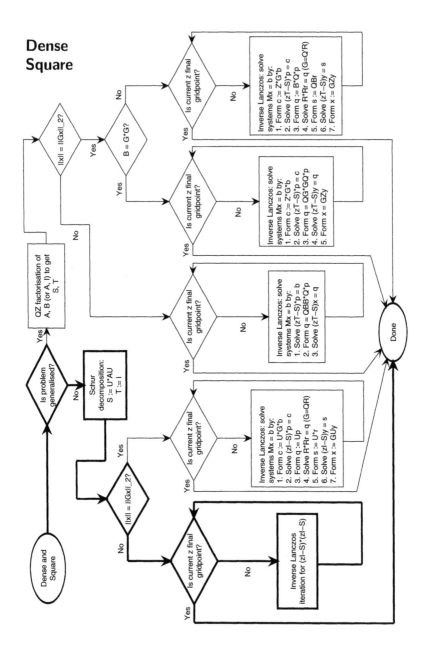

Figure 44.2: Algorithms for computing pseudospectra of dense square matrices, from [837]. Boldface marks the path implemented in EigTool.

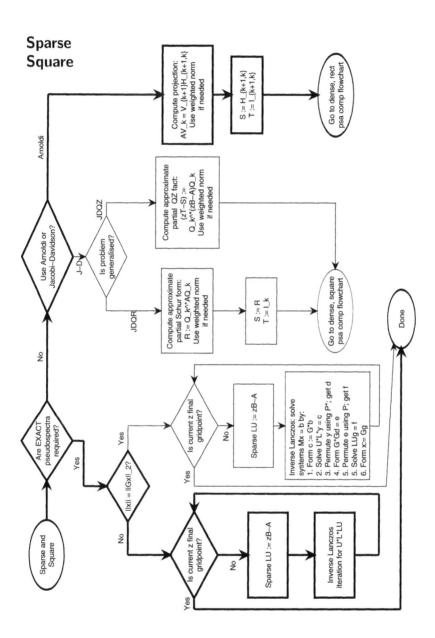

Figure 44.3: Algorithms for computing pseudospectra of sparse square matrices, from [837]. Boldface marks the paths implemented in EigTool.

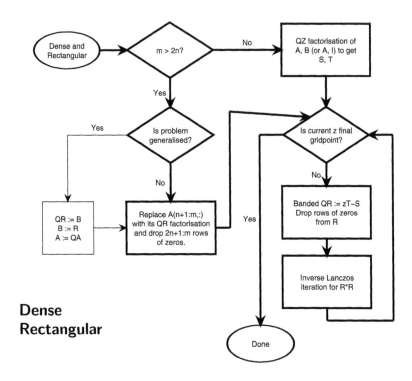

Figure 44.4: Algorithms for computing pseudospectra of dense rectangular matrices, from [837]. In the common situation where this flow chart is entered with an $(n+1) \times n$ upper Hessenberg matrix from Arnoldi projection as in Figure 44.3, there is no need to carry out the QZ factorization indicated in the upper right box: One proceeds directly to banded QR factorization.

X. Further Mathematical Issues

45 · Generalized eigenvalue problems _____

If $\mathbf{A}$ is a square matrix, then the evolution equation

$$\frac{d\mathbf{u}}{dt} = \mathbf{A}\mathbf{u}$$

has solutions

$$\mathbf{u}(t) = e^{\lambda t}\mathbf{v} \tag{45.1}$$

whenever λ and $\mathbf{v}$ solve the eigenvalue problem

$$\mathbf{A}\mathbf{v} = \lambda\mathbf{v}. \tag{45.2}$$

In some applications, however, the equation takes the more general form

$$\mathbf{B}\frac{d\mathbf{u}}{dt} = \mathbf{A}\mathbf{u}, \tag{45.3}$$

where $\mathbf{B}$ is another square matrix of the same dimension. Now there will be solutions of the form (45.1) provided λ and $\mathbf{v}$ solve a *generalized eigenvalue problem*,

$$\mathbf{A}\mathbf{v} = \lambda\mathbf{B}\mathbf{v}. \tag{45.4}$$

It is also common to write this as $(\mathbf{A}-\lambda\mathbf{B})\mathbf{v} = \mathbf{0}$; the parameter-dependent matrix $\mathbf{A} - \lambda\mathbf{B}$ is known as a *matrix pencil*. In finite element analysis, $\mathbf{B}$ is called the 'mass matrix' and $\mathbf{A}$ is the 'stiffness matrix', since (45.3) is like Newton's first law for a mass $\mathbf{B}$ attached to a spring of stiffness $\mathbf{A}$. (To be precise, in that application there would be a minus sign and the time derivative would be of second order.) In many applications $\mathbf{B}$ will be Hermitian positive definite, but in others it may be non-Hermitian, indefinite, and even singular. In the singular case, (45.1) is known as a system of *differential-algebraic equations* (DAEs) [14, 365, 610].

These remarks apply to continuous time dynamics, but of course, there are analogous statements for discrete time. The evolution equation

$$\mathbf{u}^{(k+1)} = \mathbf{A}\mathbf{u}^{(k)}$$

has solutions

$$\mathbf{u}^{(k)} = \lambda^k\mathbf{v} \tag{45.5}$$

whenever λ and $\mathbf{v}$ satisfy (45.2), and the more general evolution equation

$$\mathbf{B}\mathbf{u}^{(k+1)} = \mathbf{A}\mathbf{u}^{(k)} \tag{45.6}$$

has solutions (45.5) provided λ and $\mathbf{v}$ satisfy (45.4). These notions can be generalized to linear operators, but for simplicity we shall confine our attention to matrices.

In principle, many evolution problems originate in generalized form. Nevertheless, one sees standard eigenvalue problems more often than generalized ones, and there are various reasons for this. In many applications, $\mathbf{B}$ may be a constant (i.e., a multiple of the identity matrix, typically with physical dimensions attached), in which case it is trivially eliminated from the problem. More generally, if $\mathbf{B}$ is nonsingular, then it can again be eliminated by reducing (45.4) to the standard eigenvalue problem

$$(\mathbf{B}^{-1}\mathbf{A})\mathbf{v} = \lambda\mathbf{v}. \tag{45.7}$$

Conventionally, there would be just two main reasons to work with the generalized eigenvalue problem (45.4) rather than its equivalent standard form (45.7). First, one might be concerned with a case where $\mathbf{B}$ was singular. Second, the generalized form might be preferable for reasons of insight, scaling, or computation. For example, in an application in structural mechanics, the mass matrix may be invertible mathematically, but of too large a dimension to be inverted explicitly on a computer.

When we turn to nonnormal dynamics and pseudospectra, the matter of standard vs. generalized formulations requires more careful thought, because now, norms matter. Multiplying a vector by $\mathbf{B}^{-1}$ changes its norm, and thus from a quantitative point of view, (45.4) and (45.7) begin to look very different.

What is the 'right' definition of pseudospectra for a matrix pencil? Should we think of perturbations of $\mathbf{A}$ alone? Should we consider independent perturbations of both $\mathbf{A}$ and $\mathbf{B}$? Should we reduce the problem to that of the pseudospectra of $\mathbf{B}^{-1}\mathbf{A}$? There is a small literature on such questions dating to 1994, and the definitions considered by the dozen or so authors involved can be summarized as follows.

Definition 1. If $\mathbf{B}$ is nonsingular, one can define $\sigma_\varepsilon(\mathbf{A}, \mathbf{B}) = \sigma_\varepsilon(\mathbf{B}^{-1}\mathbf{A})$. This amounts to saying that the boundary of $\sigma_\varepsilon(\mathbf{A}, \mathbf{B})$ is the ε^{-1} level curve of $\|(\lambda - \mathbf{B}^{-1}\mathbf{A})^{-1}\|$, and this is the definition used by Ruhe [646].

Definition 2. Another approach is to say that $\sigma_\varepsilon(\mathbf{A}, \mathbf{B})$ is bounded by the ε^{-1} level curve of $\|(\lambda\mathbf{B} - \mathbf{A})^{-1}\|$. This is equivalent to considering perturbations of $\mathbf{A}$ only, not $\mathbf{B}$, and is the course followed by van Dorsselaer [794, 795]. It is also mentioned by Riedel [640], though he prefers Definition 4, below.

Definition 3. More generally, one could define $\sigma_\varepsilon(\mathbf{A}, \mathbf{B})$ by perturbing both $\mathbf{A}$ and $\mathbf{B}$ by independently controllable amounts. This approach requires the introduction of parameters to control the relative perturbations of the two matrices. It is favored by Fraysse et al., Higham and Tisseur, Lavallée, and Toumazou [290, 485, 754, 764].

Definition 4. Suppose $\mathbf{B}$ is Hermitian positive definite with Cholesky factorization $\mathbf{B} = \mathbf{F}^*\mathbf{F}$. Riedel proposes defining $\sigma_\varepsilon(\mathbf{A}, \mathbf{B})$ to be the set bounded by the ε^{-1} level curve of $\|(\lambda - \mathbf{F}^{-*}\mathbf{A}\mathbf{F}^{-1})^{-1}\|$ [640], where $\mathbf{F}^{-*} = (\mathbf{F}^{-1})^*$ as usual.

Rather than treat all of these ideas as equal we shall take a stand on which one is 'right'. Throughout this book, we take the view that the most important applications of pseudospectra are to the *behavior* of nonnormal matrices and operators, and that applications to eigenvalue perturbations are secondary. This judgment is a matter of taste, and of course, we recognize that perturbation of eigenvalues is an important subject too and that in some applications it is the key issue. Nevertheless, we shall follow our emphasis on behavior here and note that it suggests the following principle: *The pseudospectra associated with a pencil* $\mathbf{A} - \lambda\mathbf{B}$ *should not change if both* $\mathbf{A}$ *and* $\mathbf{B}$ *are multiplied by a nonsingular matrix* $\mathbf{C}$. If pseudospectra are to shed light on dynamical behavior such as transients in the dependence of $\|\mathbf{u}(t)\|$ on t or $\|\mathbf{u}^{(k)}\|$ on k, for example, then since premultiplying the evolution equation (45.3) or (45.6) by $\mathbf{C}$ does not change this dependence, it should not change the pseudospectra.

This principle immediately suggests that Definition 1 above is the right one, but with an important caveat. The caveat helps to explain why there have been a multiplicity of definitions in use, and specifically, it explains the rationale behind Definition 4.

Definition of pseudospectra for generalized eigenvalue problems

For any $\varepsilon > 0$, if $\mathbf{B}$ is nonsingular, the ε-*pseudospectrum* of the pencil $\mathbf{A} - \lambda\mathbf{B}$ is

$$\sigma_\varepsilon(\mathbf{A}, \mathbf{B}) = \sigma_\varepsilon(\mathbf{B}^{-1}\mathbf{A}). \tag{45.8}$$

However, one must note that as always, $\sigma_\varepsilon(\mathbf{B}^{-1}\mathbf{A})$ is defined with respect to a particular choice of norm $\|\cdot\|$ in the space of vectors $\mathbf{u}$ to which $\mathbf{B}^{-1}\mathbf{A}$ is applied. In many applications, this norm is given by $\|\mathbf{u}\| = \|\mathbf{F}\mathbf{u}\|_2$ for some nonsingular matrix $\mathbf{F}$, in which case (45.8) reduces to

$$\sigma_\varepsilon(\mathbf{A}, \mathbf{B}) = \sigma_\varepsilon^{2\text{-norm}}(\mathbf{F}\mathbf{B}^{-1}\mathbf{A}\mathbf{F}^{-1}). \tag{45.9}$$

In some of these applications, the matrix $\mathbf{F}$ satisfies $\mathbf{B} = \mathbf{F}^*\mathbf{F}$, in which case we have the further simplification

$$\sigma_\varepsilon(\mathbf{A}, \mathbf{B}) = \sigma_\varepsilon^{2\text{-norm}}(\mathbf{F}^{-*}\mathbf{A}\mathbf{F}^{-1}). \tag{45.10}$$

Equation (45.9) can be derived as follows. By (45.8), $\sigma_\varepsilon(\mathbf{A}, \mathbf{B})$ is bounded by the ε^{-1} level curve of $\|(z - \mathbf{B}^{-1}\mathbf{A})^{-1}\|$, for which we compute

$$\|(z - \mathbf{B}^{-1}\mathbf{A})^{-1}\| = \sup_{\mathbf{u} \neq 0} \frac{\|\mathbf{F}(z - \mathbf{B}^{-1}\mathbf{A})^{-1}\mathbf{u}\|_2}{\|\mathbf{F}\mathbf{u}\|_2}$$

$$= \sup_{\mathbf{u} \neq 0} \frac{\|\mathbf{F}(z - \mathbf{B}^{-1}\mathbf{A})^{-1}\mathbf{F}^{-1}\mathbf{u}\|_2}{\|\mathbf{F}\mathbf{F}^{-1}\mathbf{u}\|_2}$$

$$= \|\mathbf{F}(z - \mathbf{B}^{-1}\mathbf{A})^{-1}\mathbf{F}^{-1}\|_2,$$

which establishes (45.9). As for (45.10), if $\mathbf{B} = \mathbf{F}^*\mathbf{F}$, then this last expression is the same as $\|(z - \mathbf{F}\mathbf{F}^{-1}\mathbf{F}^{-*}\mathbf{A}\mathbf{F}^{-1})^{-1}\|_2$, that is, $\|(z - \mathbf{F}^{-*}\mathbf{A}\mathbf{F}^{-1})^{-1}\|_2$. (Essentially the same calculation was presented in different notation on p. 379.)

We have just defined $\sigma_\varepsilon(\mathbf{A}, \mathbf{B})$ by reducing it to $\sigma_\varepsilon(\mathbf{B}^{-1}\mathbf{A})$. This is a mathematical definition, and it is not intended to imply anything about how $\sigma_\varepsilon(\mathbf{A}, \mathbf{B})$ should be computed in practice. Sometimes the explicit reduction to $\mathbf{B}^{-1}\mathbf{A}$ or $\mathbf{F}\mathbf{B}^{-1}\mathbf{A}\mathbf{F}^{-1}$ or $\mathbf{F}^{-*}\mathbf{A}\mathbf{F}^{-1}$ is a good idea; it is certainly simple to carry out when the matrices are of small enough dimensions (e.g., hundreds rather than thousands) and $\mathbf{B}$ and $\mathbf{F}$ are well-conditioned. On the other hand, there may be advantages of speed, feasibility, or numerical accuracy in working with a formulation that does not require inverting $\mathbf{B}$ or $\mathbf{F}$. Methods of this kind are advocated by Riedel [640], who relates $\sigma_\varepsilon(\mathbf{A}, \mathbf{B})$ to the generalized singular value decomposition of the pair $\mathbf{A}, \mathbf{B}$ [327, 800]. Such computational methods were outlined in the last section. For a survey of related issues concerning large-scale generalized eigenvalue calculations, see [542].

In a number of cases in the literature, authors have transformed their matrices to achieve the effect of (45.9) or (45.10) without referring explicitly to generalized eigenvalue problems. An early example is the 1993 paper of Reddy, Schmid, and Henningson on the Orr–Sommerfeld operator, considered in §22 [624]. Equation (1.5) of [624] is $\mathbf{A}\mathbf{u} = \lambda\mathbf{B}\mathbf{u}$, and in equation (4.3) and at the bottom of page 27 of that paper it is explained that the appropriate energy norm for the problem is $\|\mathbf{u}\| = \|\mathbf{F}\mathbf{u}\|_2$ with $\mathbf{F}^*\mathbf{F} = \mathbf{B}$. This leads to $\|(z - \mathbf{B}^{-1}\mathbf{A})^{-1}\| = \|\mathbf{F}(z - \mathbf{B}^{-1}\mathbf{A})^{-1}\mathbf{F}^{-1}\|_2$ just as we have described, and Reddy et al. present the details in their Appendix A.

We have seen that investigation of dynamical problems leads naturally to generalized eigenvalue problems. In fact, it very often leads to *second-order* generalized problems, and indeed, virtually all of the problems that arise in the analysis of structures and vibrations are of second order. We shall now consider a general equation of this kind.

Suppose we have the evolution equation

$$\mathbf{M}\mathbf{x}'' + \mathbf{C}\mathbf{x}' + \mathbf{K}\mathbf{x} = 0, \tag{45.11}$$

where $\mathbf{x} = \mathbf{x}(t)$ is a vector, $'$ denotes the time derivative, and $\mathbf{M}$, $\mathbf{C}$, and $\mathbf{K}$ are square matrices. In an application, $\mathbf{M}$ is typically associated with mass and $\mathbf{K}$ with stiffness, and we shall assume that both of these matrices are Hermitian positive definite. We also assume that the energy of the system at any time t is given by the sum

$$\text{energy} = (\mathbf{x}')^* \mathbf{M} \mathbf{x}' + \mathbf{x}^* \mathbf{K} \mathbf{x}, \tag{45.12}$$

with the two terms corresponding to kinetic and potential energy. If $\mathbf{C}$ is zero or very small in norm, the dynamics will be close to normal. If $\mathbf{C}$ is large and positive definite, the system is strongly damped. We do not assume that $\mathbf{C}$ is positive definite, however, and some interesting applications involve matrices that are not.

All the questions we have discussed for first-order generalized eigenvalue problems arise again for the second-order problem (45.11). In particular, how should pseudospectra be defined? Of the many possibilities, we shall again select the one most closely tied to dynamical behavior. We rewrite (45.11) as the first-order equation

$$\begin{pmatrix} \mathbf{M} & 0 \\ 0 & \mathbf{M} \end{pmatrix} \begin{pmatrix} \mathbf{x} \\ \mathbf{x}' \end{pmatrix}' = \begin{pmatrix} 0 & \mathbf{M} \\ -\mathbf{K} & -\mathbf{C} \end{pmatrix} \begin{pmatrix} \mathbf{x} \\ \mathbf{x}' \end{pmatrix}, \tag{45.13}$$

which has the form (45.3) with

$$\mathbf{A} = \begin{pmatrix} 0 & \mathbf{M} \\ -\mathbf{K} & -\mathbf{C} \end{pmatrix}, \quad \mathbf{B} = \begin{pmatrix} \mathbf{M} & 0 \\ 0 & \mathbf{M} \end{pmatrix}, \quad \mathbf{u} = \begin{pmatrix} \mathbf{x} \\ \mathbf{x}' \end{pmatrix}. \tag{45.14}$$

At this point, following the ideas of (45.8)–(45.10), let us suppose that $\mathbf{K}$ and $\mathbf{M}$ have factorizations

$$\mathbf{K} = \mathbf{G}^* \mathbf{G}, \quad \mathbf{M} = \mathbf{H}^* \mathbf{H}; \tag{45.15}$$

in practice these would typically be Cholesky factorizations with $\mathbf{G}$ and $\mathbf{H}$ upper triangular. Then according to (45.12), the appropriate energy norm is defined by

$$\|\mathbf{u}\| = \|\mathbf{F}\mathbf{u}\|_2, \quad \mathbf{F} = \begin{pmatrix} \mathbf{G} & \\ & \mathbf{H} \end{pmatrix}, \tag{45.16}$$

with the blank entries representing zero matrices. From here, the definition (45.9) can be applied, revealing that the matrix describing the dynamics in first-order, 2-norm form is $\mathbf{F}\mathbf{B}^{-1}\mathbf{A}\mathbf{F}^{-1}$. To derive the same result explicitly, we may rewrite (45.13) as

$$\begin{pmatrix} \mathbf{x} \\ \mathbf{x}' \end{pmatrix}' = \begin{pmatrix} 0 & \mathbf{I} \\ -\mathbf{M}^{-1}\mathbf{K} & -\mathbf{M}^{-1}\mathbf{C} \end{pmatrix} \begin{pmatrix} \mathbf{x} \\ \mathbf{x}' \end{pmatrix}, \tag{45.17}$$

or equivalently,

$$\begin{pmatrix} \mathbf{G} & \\ & \mathbf{H} \end{pmatrix} \begin{pmatrix} \mathbf{x} \\ \mathbf{x}' \end{pmatrix}'$$
$$= \begin{pmatrix} \mathbf{G} & \\ & \mathbf{H} \end{pmatrix} \begin{pmatrix} 0 & \mathbf{I} \\ -\mathbf{M}^{-1}\mathbf{K} & -\mathbf{M}^{-1}\mathbf{C} \end{pmatrix} \begin{pmatrix} \mathbf{G}^{-1} & \\ & \mathbf{H}^{-1} \end{pmatrix} \begin{pmatrix} \mathbf{G} & \\ & \mathbf{H} \end{pmatrix} \begin{pmatrix} \mathbf{x} \\ \mathbf{x}' \end{pmatrix}. \tag{45.18}$$

This formulation is simply

$$\mathbf{Fu}' = \mathbf{FB}^{-1}\mathbf{AF}^{-1}\mathbf{Fu}.$$

Hence by multiplying the first three matrices on the right of (45.18) we find

$$\mathbf{FB}^{-1}\mathbf{AF}^{-1} = \begin{pmatrix} 0 & \mathbf{GH}^{-1} \\ -\mathbf{HM}^{-1}\mathbf{KG}^{-1} & -\mathbf{HM}^{-1}\mathbf{CH}^{-1} \end{pmatrix},$$

or upon using the identities $\mathbf{KG}^{-1} = \mathbf{G}^*$ and $\mathbf{HM}^{-1} = \mathbf{H}^{-*}$,

$$\mathbf{FB}^{-1}\mathbf{AF}^{-1} = \begin{pmatrix} 0 & \mathbf{GH}^{-1} \\ -\mathbf{H}^{-*}\mathbf{G}^* & -\mathbf{H}^{-*}\mathbf{CH}^{-1} \end{pmatrix}. \tag{45.19}$$

This is the matrix one could use to investigate the dynamics of (45.11) via pseudospectra of a standard (not generalized) problem. Note that in the undamped case, $\mathbf{C} = \mathbf{0}$, this matrix is skew-Hermitian and hence normal, with eigenvalues on the imaginary axis, but in general, (45.19) is nonnormal.

All of this discussion, including the definition (45.8), has assumed that $\mathbf{B}$ is nonsingular. The theory of pseudospectra for cases in which $\mathbf{B}$ is singular has begun to receive some attention, for example in [394], but such studies are not far advanced as yet. Among the complications that arise are the fact that when $\mathbf{B}$ is singular and $\mathbf{A}$ is not, the pencil must have an infinite eigenvalue, $\lambda = \infty$. When $\mathbf{A}$ is also singular and its null space has a nontrivial intersection with that of $\mathbf{B}$, then there exist $\mathbf{x} \neq \mathbf{0}$ such that $\mathbf{Ax} = \lambda\mathbf{Bx}$ for *all* $\lambda \in \mathbb{C}$, and the pencil is said to be *singular*; see, e.g., [299, 729]. Rather than exploring such situations systematically, we shall simply give an example to illustrate some of the issues that may arise if one wishes to apply the principle of dynamical behavior followed in this section to problems with singular $\mathbf{B}$.

Consider the linear differential-algebraic system

$$u = v, \qquad v' = v,$$

which we may write in matrix form as the system

$$\begin{pmatrix} 0 & 0 \\ 0 & 1 \end{pmatrix} \begin{pmatrix} u' \\ v' \end{pmatrix} = \begin{pmatrix} 1 & -1 \\ 0 & 1 \end{pmatrix} \begin{pmatrix} u \\ v \end{pmatrix}, \tag{45.20}$$

where the matrix on the left is singular. What should we say are the pseudospectra of the associated pencil? One answer comes by interpreting this system as the $\delta \to 0$ limit of

$$\begin{pmatrix} -\delta & \delta \\ 0 & 1 \end{pmatrix} \begin{pmatrix} u' \\ v' \end{pmatrix} = \begin{pmatrix} 1 & -1 \\ 0 & 1 \end{pmatrix} \begin{pmatrix} u \\ v \end{pmatrix},$$

that is,

$$\begin{pmatrix} u' \\ v' \end{pmatrix} = \begin{pmatrix} -1/\delta & 1+1/\delta \\ 0 & 1 \end{pmatrix} \begin{pmatrix} u \\ v \end{pmatrix}.$$

As $\delta \to 0$, one eigenvalue of this last matrix diverges to $-\infty$ while the other stays fixed at 1. Meanwhile, one component of the ε-pseudospectrum moves off to $-\infty$ and the other component converges to the open disk of radius $\sqrt{2}\varepsilon$ around the point $z = 1$. This latter set would be a possible choice for the ε-pseudospectrum of the pencil associated with (45.20).

In closing, we reiterate that we have focused in this section on a single definition of pseudospectra of matrix pencils, which is not the only reasonable one. Theorems, algorithms, and applications appropriate to other definitions can be found in the references. In particular, many authors consider independent perturbations of **A** and **B**, and such investigations are related to perturbation theory for generalized eigenvalue problems [200, 291, 725, 729]. Related notions have been proposed for polynomial eigenvalue problems [394, 474, 754].

For rectangular, rather than square, pencils, see the remarks at the end of the next section.

46 · Pseudospectra of rectangular matrices

Eigenvalues are ordinarily defined for square matrices, but it is possible to extend the idea to the rectangular case. If $\mathbf{A}$ is an $M \times N$ matrix with $M > N$, we define an *eigenvalue* λ and *eigenvector* $\mathbf{v} \neq \mathbf{0}$ of $\mathbf{A}$ by the equation

$$(\mathbf{A} - \lambda \widetilde{\mathbf{I}})\mathbf{v} = \mathbf{0}, \tag{46.1}$$

where $\widetilde{\mathbf{I}}$ denotes the $M \times N$ 'rectangular identity' with 1 on the main diagonal and 0 elsewhere. This idea has appeared, for example, in [70, 725, 750]. However, the consideration of eigenvalues of rectangular matrices is not common. One indication of its limitations follows from (46.1): Most rectangular matrices have no eigenvalues at all, and for those that do, an infinitesimal perturbation will in general remove them.

Pseudospectra are better behaved. The pseudospectra of a rectangular matrix are stable under perturbations and can be used in various applications. In particular, they are a crucial tool in the algorithm for computing pseudospectra of large-scale square matrices presented in §40, based on implicitly restarted Arnoldi iterations. Every time EigTool computes the pseudospectra of a large sparse matrix by iterative methods, the final phase of the process is the computation of the pseudospectra of a smaller $(N{+}1) \times N$ rectangular matrix [837]. Even for Hermitian matrices, pseudospectra of rectangular projected submatrices give interesting insight into the convergence of Lanczos iterations.

We collect the definitions in the following theorem, which closely follows the developments of §2 for square matrices. These ideas have a rather short history: we only know of the references [97, 118, 394, 761, 841]. The first two conditions of the theorem apply to any matrix norm subordinate to a pair of vector norms (one in the domain $\mathbb{C}^N$, the other in the range $\mathbb{C}^M$). The second two are restricted to the case where $\|\cdot\| = \|\cdot\|_2$ in both spaces. In this circumstance we define the *pseudoinverse* of an $M \times N$ matrix $\mathbf{A}$ ($M \geq N$) with singular value decomposition

$$\mathbf{A} = \mathbf{U}\mathbf{\Sigma}\mathbf{V}^* = \sum_{j=1}^{N} s_j \mathbf{u}_j \mathbf{v}_j^*$$

by

$$\mathbf{A}^+ = (\mathbf{A}^*\mathbf{A})^{-1}\mathbf{A}^* = \sum_{j=1}^{N} \frac{1}{s_j} \mathbf{v}_j \mathbf{u}_j^*, \tag{46.2}$$

when s_N is nonzero; i.e., $\mathbf{A}$ has full rank [327, 776]. If $\mathbf{A}$ is rank-deficient, we take $\mathbf{A}^+$ to be undefined and interpret its norm to be ∞ in condition (iv).

Equivalent definitions of pseudospectra of a rectangular matrix

Theorem 46.1 *Let* $\mathbf{A}$ *be an* $M \times N$ *matrix and* $\widetilde{\mathbf{I}}$ *the* $M \times N$ *'identity', and let* $\varepsilon > 0$ *be arbitrary. The following two definitions of the* ε-*pseudospectrum* $\sigma_\varepsilon(\mathbf{A})$ *are equivalent: It is the set of* $z \in \mathbb{C}$ *such that*

  (i) *there exists* $\mathbf{v} \in \mathbb{C}^N$ *with* $\|\mathbf{v}\| = 1$ *such that* $\|(z - \mathbf{A})\mathbf{v}\| < \varepsilon$;

  (ii) z *is an eigenvalue of* $\mathbf{A} + \mathbf{E}$ *for some* $\mathbf{E} \in \mathbb{C}^{M \times N}$ *with* $\|\mathbf{E}\| < \varepsilon$.

If $\| \cdot \| = \| \cdot \|_2$, *two further equivalent conditions are*

  (iii) $s_N(z - \mathbf{A}) < \varepsilon$;

  (iv) $\|(z - \mathbf{A})^+\| > \varepsilon^{-1}$.

Proof (compare Theorem 2.1). If $z \in \sigma(\mathbf{A})$, the equivalence of (i)–(iv) is trivial, so assume $z \notin \sigma(\mathbf{A})$. To prove (ii)⇒(i), suppose $(\mathbf{A} + \mathbf{E})\mathbf{v} = z\mathbf{v}$ for some $\mathbf{E} \in \mathbb{C}^{M \times N}$ with $\|\mathbf{E}\| < \varepsilon$ and $\mathbf{v} \in \mathbb{C}^N$ with $\|\mathbf{v}\| = 1$. Then $\|(z - \mathbf{A})\mathbf{v}\| = \|\mathbf{E}\mathbf{v}\| < \varepsilon$, as required. Conversely, if $\|(z - \mathbf{A})\mathbf{v}\| < \varepsilon$ for some $\mathbf{v} \in \mathbb{C}^N$ with $\|\mathbf{v}\| = 1$, a rank-1 matrix $\mathbf{E} = c\mathbf{v}\mathbf{w}^*$ with $\|\mathbf{E}\| < \varepsilon$ and $(\mathbf{A} + \mathbf{E})\mathbf{v} = z\mathbf{v}$ can be constructed as in the proof of Theorem 2.1.

Turning to (iii) and (iv), suppose $\| \cdot \| = \| \cdot \|_2$. Since $\|(z - \mathbf{A})^+\| = 1/s_N(z - \mathbf{A})$, the equivalence of these two conditions is immediate, and their equivalence to (ii) follows from the Schmidt–Mirsky theorem, which states that the 2-norm distance of a matrix to the space of rank-deficient matrices is equal to its smallest singular value [327, 729]. ∎

We now come to a theorem that is important for applications. In words, condition (i) below states that if $\mathbf{A}$ is a matrix and $\widetilde{\mathbf{A}}$ is a submatrix of $\mathbf{A}$ obtained by selecting certain columns, then the pseudospectra of $\widetilde{\mathbf{A}}$ are contained in those of $\mathbf{A}$. For this to be true we assume that the norms involved are consistent in the following sense: If $\widetilde{\mathbf{v}}$ is a vector obtained from a vector $\mathbf{v}$ by selecting certain entries, then $\|\widetilde{\mathbf{v}}\| \leq \|\mathbf{v}\|$. This property holds, for example, if $\| \cdot \| = \| \cdot \|_p$ for some p with $1 \leq p \leq \infty$. (This theorem uses MATLAB notation for submatrices.)

Monotonicity of pseudospectra

Theorem 46.2 *Let* $\mathbf{A}$ *be an* $M \times N$ *matrix with* $M \geq N$. *Then for any* $\varepsilon > 0$,

  (i) $\sigma_\varepsilon(\mathbf{A}(:, 1{:}k)) \subseteq \sigma_\varepsilon(\mathbf{A}(:, 1{:}k{+}1))$, $1 \leq k < N$,

  (ii) $\sigma_\varepsilon(\mathbf{A}(1{:}k{+}1, :)) \subseteq \sigma_\varepsilon(\mathbf{A}(1{:}k, :))$, $1 \leq k < M$,

assuming that the norms are consistent in the sense defined above.

Proof. These assertions follow from condition (i) of Theorem 46.1, together with the fact that the quantity $\min_{\|\mathbf{v}\|=1} \|(z - \mathbf{A})\mathbf{v}\|$ can only increase (or remain constant) if a column is removed from $\mathbf{A}$ or if a new row is added to $\mathbf{A}$. ∎

It is time for some examples, based as usual on the norm $\|\cdot\| = \|\cdot\|_2$. Our first, following [841], is the 4×3 matrix

$$\mathbf{A} = \begin{pmatrix} 1 & 10 & 10 \\ 0 & 2.1 & 4.2 \\ 0 & 0.1 & 0.2 \\ 0 & 0.1 & 0.2 \end{pmatrix}. \tag{46.3}$$

Unlike most rectangular matrices, this one has some eigenvalues: $\lambda_1 = 0$ and $\lambda_2 = 1$, with eigenvectors $(10, -2, 1)^{\mathrm{T}}$ and $(1, 0, 0)^{\mathrm{T}}$. The first panel of Figure 46.1 shows these points together with some of the pseudospectra. We see immediately that in addition to the two eigenvalues, where $\|(z - \mathbf{A})^{-1}\|$ goes to infinity, there is a third area in the complex plane where $\|(z - \mathbf{A})^{-1}\|$ has a finite local maximum. This could not happen for a square matrix, for which $\|(z - \mathbf{A})^{-1}\|$ would satisfy a maximum principle. The local maximum is further illustrated in the surface plot of Figure 46.2.

This lack of a maximum principle reflects some profound differences between the square and rectangular eigenvalue problems. For a square matrix $\mathbf{A}$, the resolvent $(z - \mathbf{A})^{-1}$ is an analytic function of z for $z \notin \sigma(\mathbf{A})$. If $\|\cdot\| = \|\cdot\|_2$, this implies that $\log \|(z - \mathbf{A})^{-1}\|$ is a subharmonic function, and this is one way to derive the maximum principle for $\|(z - \mathbf{A})^{-1}\|$ (cf. Theorem 4.2). It also implies that projectors and other matrix functions can be computed by Cauchy integrals. All these convenient properties vanish in the rectangular case. For example, because the pseudoinverse of a matrix $\mathbf{B}$ is defined by a formula $\mathbf{B}^+ = (\mathbf{B}^*\mathbf{B})^{-1}\mathbf{B}^*$ that contains a complex conjugate, it is evident that $(z - \mathbf{A})^+$ does not depend analytically on z. To analyze it one has to use more cumbersome techniques, and

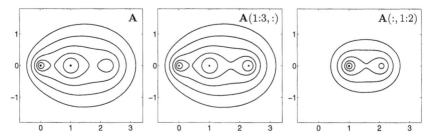

Figure 46.1: ε-pseudospectra of the matrix $\mathbf{A}$ of (46.3) and two of its submatrices, for $\varepsilon = 10^{-1}, 10^{-1.25}, \ldots, 10^{-2}$. The pseudospectra are nested, as established by Theorem 46.2.

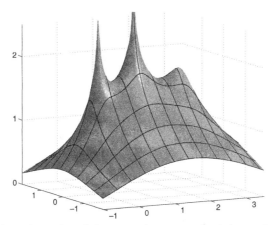

Figure 46.2: A surface plot of the same data as in the left panel of Figure 46.1, that is, $\|(z - \mathbf{A})^{-1}\|$ as a function of z for the 4×3 matrix (46.3). The first two peaks are eigenvalues where the height goes to infinity, but the third is finite. This could not happen for a square matrix.

in fact, it is not known precisely how many local maxima $\|(z - \mathbf{A})^{-1}\|$ can have for an $M \times N$ matrix $\mathbf{A}$ with $M > N$, or perhaps equivalently, how many connected components the pseudospectrum $\sigma_\varepsilon(\mathbf{A})$ can have. Byers [124] and Gu [352] emphasize that the answer must be $\mathcal{O}(N)$ or even greater, and Burke, Lewis, and Overton have established an upper bound of $2N(4N - 1)$ [118].

The second and third panels of Figure 46.1 illustrate Theorem 46.2. In the second panel we have deleted a row to obtain a square matrix, and as established in condition (ii) of the theorem, the pseudospectra grow bigger and there is now a full set of three eigenvalues. In the right panel we have deleted a column, and as established in condition (i), the pseudospectra shrink.

For a quite different set of examples, Figure 46.3 shows pseudospectra of three rectangular matrices adapted from examples appearing elsewhere in this book. In each case we start from an $N \times N$ matrix and then delete its final column, thereby obtaining a rectangular matrix of dimensions $N \times (N - 1)$. The first panel comes from the Grcar matrix of Figure 7.5 with $N = 100$, the second from the Scottish flag matrix (9.7) of Figure 9.4 with $N = 101$, and the third from the complex random matrix of Figure 35.3 with $N = 256$. In all three cases, there are no eigenvalues. Nevertheless, the pseudospectra are quite similar to those displayed in the figures indicated. (This does not happen for all matrices.) We do not claim that these particular $(N - 1) \times N$ matrix sections have much significance for any applications. Instead, our point is a more general one. Throughout this book it is illustrated that for square matrices, pseudospectra may reveal structure that is not shown by eigenvalues. We now see that for rectangular matrices, pseudospectra may reveal structure even when there

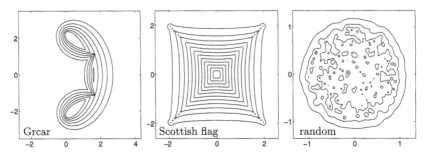

Figure 46.3: Pseudospectra of rectangular matrices obtained from the matrices of Figures 7.5, 9.4, and 35.3 by deleting the final columns. In each case there are no eigenvalues, but the pseudospectra look much the same as before.

are no eigenvalues at all.

The problem of how best to compute pseudospectra of rectangular matrices is complex. We shall not give details here, but refer the reader to §44 and [841] for a description of available algorithms and to EigTool [838] for their implementation. The problem is made challenging by the fact that in contrast to the square case (§39) [520], no method appears to be known for reducing a rectangular matrix $\mathbf{A}$ to triangular or other simple form in a way that is preserved under shifts $z - \mathbf{A}$. As a result, EigTool may compute the pseudospectra of an $N \times N$ matrix twice or more times as fast as those of an $N \times (N-1)$ submatrix. Fortunately, even with slowdowns like these, the algorithms available for the rectangular case are still eminently practical. Assuming $\| \cdot \| = \| \cdot \|_2$, an important step in the case of an $M \times N$ matrix with $M \geq 2N$ is to reduce the problem to a trapezoidal matrix of size $2N \times N$ by a preliminary QR factorization. For $N < M < 2N$, and in particular when M is just slightly bigger than N, one can benefit from a different trapezoidal reduction involving the QZ factorization; see Figure 44.4.

Let us briefly mention some applications of pseudospectra of rectangular matrices. More details can be found in [841].

Arnoldi and related iterations. As mentioned at the beginning, pseudospectra of rectangular matrices have found a major application as part of a computational algorithm for estimating pseudospectra of large, typically sparse, square matrices. This idea originated with a suggestion of Toh and Trefethen [761] that one might estimate pseudospectra by taking some steps of an Arnoldi iteration to project a large $N \times N$ matrix $\mathbf{A}$ to an upper Hessenberg matrix $\mathbf{H}_k$ of dimensions $N \times k$ for some $k \ll N$. By Theorem 46.2(i), the pseudospectra of $\mathbf{H}_k$ are lower bounds for those of $\mathbf{A}$, and because of the Hessenberg structure, we may delete rows of zeros and treat $\mathbf{H}_k$ as a matrix of dimension $(k + 1) \times k$. In its original form this idea proved not very powerful, but when the pure Arnoldi iteration is replaced by the implicitly restarted Arnoldi process [494, 709, 710], it becomes powerful indeed. This is a basic part of the method used by EigTool for computing pseudospectra of large matrices; see §40. The idea can be generalized to block iterations and other variants.

Bounds for Lanczos iterations. The Lanczos iteration is the special case of the Arnoldi iteration for $\mathbf{A} = \mathbf{A}^*$, and again there is an implicitly restarted variant. As in the non-Hermitian case, these iterations construct rectangular Hessenberg matrices $\mathbf{H}_k$ that represent the compression of $\mathbf{A}$ to a k-dimensional subspace. As before, we have $\sigma_\varepsilon(\mathbf{H}_k) \subseteq \sigma_\varepsilon(\mathbf{A})$, and since $\mathbf{A}$ is normal, this means that any $z \in \sigma_\varepsilon(\mathbf{H}_k)$ is within a distance ε of an eigenvalue of $\mathbf{A}$. This suggests a view of convergence of Lanczos processes that focuses not on the usual Ritz values (the eigenvalues of the upper square $k \times k$ block of $\mathbf{H}_k$) but on pseudospectra that shrink down to the eigenvalues of $\mathbf{A}$ as k increases. As pointed out in [841], this approach is related to work of Lehmann [493] (Hermitian case), Beattie and Ipsen [41] (non-Hermitian), and Jia [431].

Control theory. A fundamental problem in control theory is that of estimating the distance of a controllable linear control system

$$\mathbf{x}' = \mathbf{A}\mathbf{x} + \mathbf{B}\mathbf{u} \tag{46.4}$$

to the nearest uncontrollable system (a system that cannot be driven from an arbitrary starting state to an arbitrary finishing state) [443]. In the notation of our discussion (which differs from that of control theory), $\mathbf{A}$ is $N \times N$, $\mathbf{B}$ is $N \times (M - N)$, $\mathbf{x}$ is an N-vector of states, and $\mathbf{u}$ is an $(M - N)$-vector of control inputs. The distance to uncontrollability was defined by Paige [594] and shown by Eising [237] to be equal to the smallest singular value, minimized over all values of λ, of the $M \times N$ matrix

$$\begin{pmatrix} \lambda - \mathbf{A}^* \\ \mathbf{B}^* \end{pmatrix}.$$

As pointed out by Higham and Tisseur [394], this is equal to the infimum of all ε values for which

$$\sigma_\varepsilon\begin{pmatrix} \mathbf{A}^* \\ -\mathbf{B}^* \end{pmatrix} \neq \emptyset;$$

see also [332]. In other words, to find the minimum distance to an uncontrollable system, we need to find the maximum of a resolvent norm surface like the one illustrated in Figure 46.2. Algorithms for this problem have been proposed by Boley, Byers, Gu, and Burke, Lewis, and Overton [69, 118, 124, 352].

Game theory. Applications of eigenvalues of rectangular matrices to two-player games in which one player has M options and the other has

$N < M$ options have been considered by Thompson and Weil [750]. As described in [841], the pseudospectra of the same matrices provide bounds on the maximum and minimum winnings of each player.

This section has concerned rectangular matrices, but there is an important generalization to rectangular *pencils* $\mathbf{A} - z\mathbf{B}$, where $\mathbf{A}$ and $\mathbf{B}$ are matrices with $M > N$. Most of the articles we have cited in this section actually deal with this general situation rather than with the very specific case $\mathbf{B} = \widetilde{\mathbf{I}}$ that we have emphasized. However, what is the right definition of the pseudospectra $\sigma_\varepsilon(\mathbf{A}, \mathbf{B})$ of such a pencil? The answer is not obvious. One would like a definition that reduces to what we have discussed in the case $\mathbf{B} = \widetilde{\mathbf{I}}$ and also reduces to the pseudospectra defined in §45 in the case of a square pencil with $m = n$. In §45, we discussed various possibilities for pseudospectra of square pencils and took the view that the 'right' definition should have the property that $\sigma_\varepsilon(\mathbf{A}, \mathbf{B})$ is the same as the ordinary matrix pseudospectrum $\sigma_\varepsilon(\mathbf{B}^{-1}\mathbf{A})$. Is there a natural way to define pseudospectra for a rectangular pencil $\mathbf{A} - z\mathbf{B}$ in such a way that they uphold this principle in the square case? Such problems are beginning to get some attention; see, e.g., [97, 394].

47 · Do pseudospectra determine behavior? _____

A central theme of this book is the question, What connections can be made between the location of a matrix or operator in the complex plane and its behavior?[1] In this section we consider a precise formulation of this question.

Both ends of the question need to be pinned down. Concerning 'location in the complex plane', among the obvious sets that one might consider are the spectrum, the numerical range, and the pseudospectra. As explained in §17, the first two of these are determined by the ε-pseudospectra in the limits $\varepsilon \to 0$ and $\varepsilon \to \infty$, respectively, and thus we shall concentrate just on pseudospectra.

Our notion of 'behavior' will be based upon *norms of functions of matrices and operators*. Let $\mathbf{A}$ be a square matrix or bounded operator acting in a Banach space, and let f be a function that is analytic in a neighborhood of the spectrum $\sigma(\mathbf{A})$. Then the operator $f(\mathbf{A})$ can be defined by a variety of techniques, such as the Dunford–Taylor integral; see, e.g., (14.9) and the accompanying discussion on page 139.

Here we shall take the view that the behavior of $\mathbf{A}$ consists of all the 'measurements' that might be made of it, where by a measurement, we mean the quantity $\|f(\mathbf{A})\|$ for some function f. This point of view is made precise in the following definition.

Norm behavior
Two operators $\mathbf{A}$ and $\mathbf{B}$ have the same *norm behavior* if $\sigma(\mathbf{A}) = \sigma(\mathbf{B})$ and, for every function f analytic in a neighborhood of this set, $$\|f(\mathbf{A})\| = \|f(\mathbf{B})\|.$$

As always, $\|\cdot\|$ denotes the norm of the Banach space under study. $\mathbf{A}$ and $\mathbf{B}$ may operate on different Banach spaces, however—even when they are matrices, since their dimensions might differ. In constructing matrix examples, we shall assume that $\|\cdot\|$ is simply the 2-norm of the appropriate dimensions. Also, if $\mathbf{A}$ and $\mathbf{B}$ are matrices, it is enough to consider polynomials rather than general functions f; see, e.g., [415, Thm. 6.2.9].

The notion of norm behavior may seem artificial at first, but it is a natural one, and it is implicit in most of the sections of this book. For

---

[1]This section is adapted from [343].

example, the following four quantities $\|f(\mathbf{A})\|$ arise frequently:

$$\|e^{t\mathbf{A}}\| \qquad \text{(continuous time evolution processes)};$$

$$\|\mathbf{A}^k\| \qquad \text{(discrete time evolution processes)};$$

$$\|p_k(\mathbf{A})\| \qquad \text{(iterative methods in linear algebra)};$$

$$\|(z - \mathbf{A})^{-1}\| \qquad \text{(response of forced systems)}.$$

(In the third item, p_k denotes a polynomial of degree k.) Typical questions of concern in applications are *stability*, which is related to the boundedness of $\|e^{t\mathbf{A}}\|$ or $\|\mathbf{A}^k\|$ for all $t \geq 0$ or $n \geq 0$ (e.g., §§20, 31, 58); *convergence*, which is related to the rate of decrease of $\|\mathbf{A}^k\|$ or $\|p_k(\mathbf{A})\|$ to zero as $n \to \infty$ (§§24–29, 56); and *resonance* or *pseudoresonance*, which is quantified by $\|(z - \mathbf{A})^{-1}\|$ [780]. A fifth example of a quantity $\|f(\mathbf{A})\|$ of recurring interest is

$$\|g(\mathbf{A}) - p_k(\mathbf{A})\| \qquad \text{(polynomial approximation of a function)},$$

where $g(\mathbf{A})$ is a given function such as $e^{t\mathbf{A}}$ or $\mathbf{A}^{-1}$. Here, again, one is typically concerned with the rate of convergence to zero as $k \to \infty$.

If $\mathbf{A}$ is normal, its norm behavior is fully determined by its spectrum. For any set $S \subseteq \mathbb{C}$ and function f defined on S, let us define

$$\|f\|_S = \sup_{z \in S} |f(z)|. \tag{47.1}$$

The following theorem appeared already as equation (2.14) for the special case $f(z) = (\lambda - z)^{-1}$.

Norm behavior of normal matrices and operators

Theorem 47.1 *Let $\mathbf{A}$ be a matrix or bounded operator in a Hilbert space. If $\mathbf{A}$ is normal, then*

$$\|f(\mathbf{A})\| = \|f\|_{\sigma(\mathbf{A})} \tag{47.2}$$

for every function f analytic in a neighborhood of $\sigma(\mathbf{A})$. Consequently, if $\mathbf{A}$ and $\mathbf{B}$ are normal, then $\mathbf{A}$ and $\mathbf{B}$ have the same norm behavior if and only if $\sigma(\mathbf{A}) = \sigma(\mathbf{B})$.

Proof. If $\mathbf{A}$ is a normal matrix, it can be unitarily diagonalized in the form $\mathbf{A} = \mathbf{V}\mathbf{\Lambda}\mathbf{V}^*$. Since any function f applied to a matrix of dimension N is equivalent to some polynomial p of degree $\leq N$, this implies $f(\mathbf{A}) = \mathbf{V}f(\mathbf{\Lambda})\mathbf{V}^*$, from which (47.2) follows readily. If $\mathbf{A}$ is a bounded operator, we can carry out the proof as follows. If $\mathbf{A}$ is normal, then $f(\mathbf{A})$ is normal also, implying that $\|f(\mathbf{A})\|$ is equal to the spectral radius of $f(\mathbf{A})$. But by the spectral mapping theorem [448, 641], $\sigma(f(\mathbf{A})) = f(\sigma(\mathbf{A}))$, so this spectral radius is equal to $\|f\|_{\sigma(\mathbf{A})}$. ∎

Of course, our main concern is the study of matrices and operators that are not normal. Any reader of this book knows that in this case, the spectrum alone cannot determine norm behavior. For the rest of this section, we shall focus on finite-dimensional matrices with $\|\cdot\| = \|\cdot\|_2$. The matrices

$$\begin{pmatrix} 2 & 0 \\ 0 & 0 \end{pmatrix}, \quad \begin{pmatrix} 2 & 1 \\ 0 & 0 \end{pmatrix}, \quad \begin{pmatrix} 2 & 2 \\ 0 & 0 \end{pmatrix}$$

have the same spectra but different norms $\|\mathbf{A}^k\|$ and $\|e^{t\mathbf{A}}\|$. In fact, even their norms $\|\mathbf{A}\|$ are different. For an example where more general functions f must be brought into play to reveal distinct norm behavior, consider the block diagonal matrices

$$\begin{pmatrix} 2 & 0 & \\ 0 & 0 & \\ & & 3 \end{pmatrix}, \quad \begin{pmatrix} 2 & 1 & \\ 0 & 0 & \\ & & 3 \end{pmatrix}.$$

(The omitted entries are zero.) These matrices have identical norms $\|\mathbf{A}^k\|$ for all $n \geq 0$ and $\|e^{t\mathbf{A}}\|$ for all $t \geq 0$, but they differ when it comes to other functions such as $\|(\mathbf{I} - \mathbf{A})^{-1}\|$ or $\|\mathbf{A} - 3\mathbf{I}\|$.

When do two matrices have identical norm behavior? A sufficient condition is unitary equivalence, since $\|\cdot\|$ is unitarily invariant. Thus, for example, the matrices

$$\begin{pmatrix} 2 & 0 \\ 0 & 0 \end{pmatrix}, \quad \begin{pmatrix} 0 & 0 \\ 0 & 2 \end{pmatrix}, \quad \begin{pmatrix} 1 & 1 \\ 1 & 1 \end{pmatrix}$$

have the same norm behavior. Unitary equivalence is not necessary, however, as one can see from the matrices

$$\begin{pmatrix} 2 & 0 & 0 \\ 0 & 0 & 0 \\ 0 & 0 & 0 \end{pmatrix}, \quad \begin{pmatrix} 2 & 0 & 0 \\ 0 & 2 & 0 \\ 0 & 0 & 0 \end{pmatrix},$$

which are behaviorally equivalent to each other as well as to the preceding 2×2 examples (by Theorem 47.1, since all five of these matrices are normal and have spectrum $\{0, 2\}$). Evidently changes of eigenvalue multiplicities do not affect norm behavior. Another example of a transformation that does not change norm behavior is transposition. For any matrix $\mathbf{A}$, $\|f(\mathbf{A})\| = \|f(\mathbf{A}^{\mathrm{T}})\|$, and thus $\mathbf{A}$ and $\mathbf{A}^{\mathrm{T}}$ have the same norm behavior, although they are not in general unitarily equivalent if $N \geq 3$.[2]

---

[2]For example, the 3×3 matrices $\mathbf{A} = [1\ 1\ 0;\ 0\ 1\ 1;\ 0\ 0\ 2]$ and $\mathbf{B} = [1\ 2\ 0;\ 0\ 3\ 4;\ 0\ 0\ 5]$ (MATLAB notation) are not unitarily equivalent to their transposes. We are indebted to Roger Horn for providing these examples (private communication, 1991), which can be verified by application of the theorem of Pearcy given as Theorem 2.2.8 in [414].

It is noteworthy that the three operations just considered—unitary similarity, change of multiplicity, and transposition—leave invariant not only the norm behavior of a matrix but also all of its pseudospectra. This observation suggests the conjecture, Might the pseudospectra of a matrix or operator determine its norm behavior? If so, this would be a very satisfying state of affairs, establishing that in a certain sense at least, pseudospectra contain all the behavioral information one could ask for. And the converse is certainly true: Norm behavior determines pseudospectra.

Unfortunately, the conjecture is false [343].

Pseudospectra $\not\Rightarrow$ norm behavior

Theorem 47.2 *There exist matrices* $\mathbf{A}$ *and* $\mathbf{B}$ *for which* $\|(z-\mathbf{A})^{-1}\| = \|(z-\mathbf{B})^{-1}\|$ *for all* $z \in \mathbb{C}$ *but* $\|p(\mathbf{A})\| \neq \|p(\mathbf{B})\|$ *for some polynomial* p.

Proof. Consider the Jordan blocks

$$\mathbf{J}_1 = \begin{pmatrix} 0 & 1 & \\ & 0 & 1 \\ & & 0 \end{pmatrix}, \qquad \mathbf{J}_2 = \begin{pmatrix} 0 & \sqrt{2} \\ & 0 \end{pmatrix}.$$

(The entry $\sqrt{2}$ can be replaced by any number in the interval $(1, \sqrt{2}\,]$.) We have $\|\mathbf{J}_1\| = 1$ and $\|\mathbf{J}_2\| = \sqrt{2}$, hence $\|\mathbf{J}_2\| > \|\mathbf{J}_1\|$. On the other hand, $\|(z - \mathbf{J}_2)^{-1}\| \leq \|(z - \mathbf{J}_1)^{-1}\|$ for all $z \in \mathbb{C}$, with equality achieved at $z = 0$ and also in the limit $|z| \to \infty$. To prove this, we note first that for any Jordan block $\mathbf{J}$, $\|(z - \mathbf{J})^{-1}\|$ depends only on $|z|$; i.e., the pseudospectra are disks about the origin, as can be proved by a diagonal unitary similarity transformation. Thus it is enough to establish $\|(z - \mathbf{J}_2)^{-1}\| \leq \|(z - \mathbf{J}_1)^{-1}\|$ for z real and positive, which can be easily checked numerically or proved by algebraic methods.

The proof of the theorem is completed by taking

$$\mathbf{A} = \mathbf{J}_1, \qquad \mathbf{B} = \begin{pmatrix} \mathbf{J}_1 & \\ & \mathbf{J}_2 \end{pmatrix},$$

for which we have $\|\mathbf{B}\| = \sqrt{2} > \|\mathbf{A}\| = 1$ but $\|(z - \mathbf{A})^{-1}\| = \|(z - \mathbf{B})^{-1}\| = \|(z - \mathbf{J}_1)^{-1}\|$ for all $z \in \mathbb{C}$. (Recall from Theorem 2.4 that $\sigma_\varepsilon(\mathbf{B}) = \sigma_\varepsilon(\mathbf{J}_1) \cup \sigma_\varepsilon(\mathbf{J}_2)$.) If one one prefers a counterexample in which $\mathbf{A}$ and $\mathbf{B}$ have the same dimension, this can be arranged by padding $\mathbf{A}$ with two additional rows and columns of zeros. ∎

Although Theorem 47.2 places a limit on the uses of pseudospectra, it leaves some doors open. In the Frobenius (Hilbert–Schmidt) norm $\|\mathbf{A}\|_{\mathrm{F}}^2 = \mathrm{tr}(\mathbf{A}^*\mathbf{A})$, it can be shown that the pseudospectra of a matrix *do* determine its behavior [343]. Also, even in the standard 2-norm that is the subject of the theorem, it may be that the pseudospectra–behavior link can be made solid by adding an additional hypothesis. For example, it is an open question whether or not pseudospectra determine behavior for matrices that are nonderogatory, that is, if each eigenvalue is associated with just a single Jordan block.

For the moment, however, the state of our knowledge is that pseudo-spectra do not determine norm behavior exactly. Of course, a central theme of this book is that they determine norm behavior approximately, providing much closer bounds than spectra alone. Such bounds are the subject of Part IV of this book.

In closing this section we mention another approach that has been taken to problems of this kind. Given a matrix or operator $\mathbf{A}$, a set S for which the identity $\|f(\mathbf{A})\| \leq \|f\|_S$ holds for all f is called a *spectral set* for $\mathbf{A}$ [605], [641, §§153–155]. It is known that $\sigma(\mathbf{A})$ is a spectral set for $\mathbf{A}$ if and only if $\mathbf{A}$ is normal. On the other hand, according to a result of von Neumann, the closed disk $\overline{\Delta}_{\|\mathbf{A}\|}$ is a spectral set for any $\mathbf{A}$ [641]. These observations sound like promising steps toward settling the question of where a matrix 'lives' in $\mathbb{C}$; isn't the answer simply the smallest spectral set? Unfortunately, this notion is not well-defined. The intersection of two spectral sets is in general not a spectral set, and in fact, the intersection of all of the spectral sets of a matrix $\mathbf{A}$ is always just $\sigma(\mathbf{A})$, regardless of normality.

48 · Scalar measures of nonnormality _____

There are an abundance of ways to describe a normal matrix; together, the papers of Grone et al. [351] and Elsner and Ikramov [242] enumerate eighty-nine equivalent characterizations. When a matrix fails to satisfy these conditions, one might naturally ask, 'How nonnormal is it?' Compact answers to this question are provided by various scalar measures of the *departure from normality*.

One appealing starting point is the distance of $\mathbf{A} \in \mathbb{C}^{N \times N}$ from the set $\mathcal{N}$ of normal matrices,

$$\operatorname{dist}(\mathbf{A}, \mathcal{N}) = \min_{\mathbf{N} \in \mathcal{N}} \|\mathbf{A} - \mathbf{N}\|, \tag{48.1}$$

a quantity that has proved more difficult to calculate than one might expect. In fact, no satisfactory characterization of (48.1) is yet known in the 2-norm. However, for the Frobenius norm $\|\mathbf{A}\|_F \equiv (\sum_{j,k=1}^{N} |a_{jk}|^2)^{1/2} = \operatorname{trace}(\mathbf{A}^* \mathbf{A})^{1/2}$, a solution has been found. The key observation is that for this norm, the unitary matrix $\mathbf{Q}$ that diagonalizes the minimizing $\mathbf{N}$ in (48.1) also solves the optimization problem

$$\max_{\mathbf{Q} \text{ unitary}} \|\operatorname{diag}(\mathbf{Q}^* \mathbf{A} \mathbf{Q})\|_F;$$

see [296, 645]. This suggests that one might approximate the nearest normal matrix by making a series of elementary unitary similarity transformations to $\mathbf{A}$, transferring the magnitude of entries off the diagonal toward the main diagonal. This is accomplished by the Jacobi algorithm, which was originally designed for computing eigenvalues [223]. The Frobenius norm is induced by the matrix inner product $(\mathbf{X}, \mathbf{Y})_F \equiv \operatorname{trace}(\mathbf{Y}^* \mathbf{X})$, and thus approximation theory for inner product spaces can be used to characterize the (nonunique) nearest normal matrix [296, 645]. Unfortunately, the Jacobi algorithm often converges slowly, making such a computation challenging even for small matrices.

Here are two examples of nearest normal matrices in the Frobenius norm. First, consider the Jordan block of dimension $N = 32$, with ones on the first superdiagonal and zeros everywhere else. Figure 48.1 compares the pseudospectra of this Toeplitz matrix with the spectrum of a nearest normal matrix

$$\mathbf{N} = \tfrac{31}{32} \begin{pmatrix} 0 & 1 & & \\ & 0 & \ddots & \\ & & \ddots & 1 \\ 1 & & & 0 \end{pmatrix} \in \mathbb{C}^{32 \times 32};$$

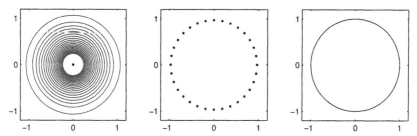

Figure 48.1: Spectrum and ε-pseudospectra of a Jordan block of dimension $N = 32$ for $\varepsilon = 10^{-1}, \ldots, 10^{-20}$ (left), together with the spectrum of a nearest normal matrix (middle), and the range of the symbol (right), which in this case is the unit circle. Here, $\mathrm{dist}_F(\mathbf{A}, \mathcal{N}) = \sqrt{31/32}$.

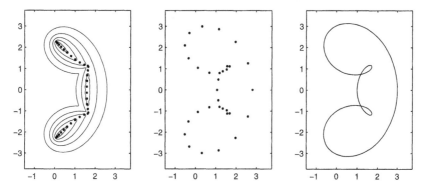

Figure 48.2: Spectrum and ε-pseudospectra of the Grcar matrix of dimension $N = 32$ for $\varepsilon = 10^{-1}, \ldots, 10^{-5}$ (left), together with the spectrum of a nearest normal matrix (middle), and the range of the symbol (right). Here $\mathrm{dist}_F(\mathbf{A}, \mathcal{N}) \approx 2.1997$.

see [645, p. 596]. (This $\mathbf{N}$ is not unique: its lower left entry can be replaced by $\frac{31}{32}e^{i\theta}$ for any $\theta \in [0, 2\pi)$.) Figure 48.2 makes the same comparison for the 32×32 Grcar matrix introduced on page 58.

Though the distance (48.1) has intrinsic appeal, other scalar measures of nonnormality are simpler to compute and arise more naturally in analysis. Perhaps such metrics were first considered by Wielandt in 1953, who proposed a measure of nonnormality and applied it to describe eigenvalue inclusion regions [823]. For a diagonalizable matrix $\mathbf{A} = \mathbf{V}\mathbf{\Lambda}\mathbf{V}^{-1}$, Wielandt measured nonnormality via the 'deformation angle'

$$\theta(\mathbf{A}) = \min_{\substack{\mathbf{x}^*\mathbf{y}=0 \\ \mathbf{x}, \mathbf{y} \neq 0}} \angle(\mathbf{V}\mathbf{x}, \mathbf{V}\mathbf{y}).$$

This angle effectively measures how far an eigenvector matrix is from orthogonal, with normal matrices having $\theta(\mathbf{A}) = \pi/2$, while nearly defective matrices must have small values of $\theta(\mathbf{A})$. A drawback typical of metrics based on diagonalization is that the departure from normality can vary when $\mathbf{A}$ is fixed but the columns of $\mathbf{V}$ are scaled.

Related to Wielandt's metric is the condition number of a matrix whose columns are eigenvectors of $\mathbf{A}$,

$$\kappa(\mathbf{V}) = \|\mathbf{V}\|_2 \, \|\mathbf{V}^{-1}\|_2.$$

If $\mathbf{A}$ is normal, we can take $\mathbf{V}$ so that this condition number is 1, while $\kappa(\mathbf{V})$ grows without bound as $\mathbf{A}$ approaches a defective matrix. Householder [416, §3.4] showed that

$$\kappa(\mathbf{V}) = \cot(\tfrac{1}{2}\theta(\mathbf{A})).$$

As seen throughout this book, $\kappa(\mathbf{V})$ arises frequently in analysis, especially when bounding the norm of a function of a matrix:

$$\|f(\mathbf{A})\|_2 = \|\mathbf{V}f(\mathbf{\Lambda})\mathbf{V}^{-1}\|_2 \le \kappa(\mathbf{V}) \max_{\lambda \in \sigma(\mathbf{A})} |f(\lambda)|, \qquad (48.2)$$

where f is any function analytic on $\sigma(\mathbf{A})$. The Bauer–Fike inclusion (2.19) of Theorem 2.3 uses $\kappa(\mathbf{V})$ to bound eigenvalues of perturbations of $\mathbf{A}$. Notice that $\kappa(\mathbf{V})$ varies with the scaling of the columns of $\mathbf{V}$. The search for the optimal scaling is a classical problem in the study of conditioning for linear systems of equations. When $\mathbf{A}$ has simple eigenvalues, a theorem of van der Sluis [788] guarantees that if each column of $\mathbf{V}$ has unit 2-norm, then $\kappa(\mathbf{V})$ is within $\sqrt{N}$ of its optimal value.

A considerable drawback of both $\theta(\mathbf{A})$ and $\kappa(\mathbf{V})$ becomes apparent from studying the perturbed identity matrix

$$\begin{pmatrix} 1 & 0 \\ 0 & 1 \end{pmatrix} + \begin{pmatrix} 0 & \xi \\ 0 & 0 \end{pmatrix}. \qquad (48.3)$$

When $\xi = 0$, $\theta(\mathbf{A}) = 1/2$ and $\kappa(\mathbf{V}) = 1$, while for all nonzero ξ, the matrix is no longer diagonalizable, yielding $\theta(\mathbf{A}) = 0$ and $\kappa(\mathbf{V}) = \infty$. Thus, both these measures of nonnormality are discontinuous functions of the matrix entries. While these quantities can be descriptive in certain cases, they tend to exaggerate the importance of nondiagonalizability. The matrix (48.3) with $|\xi| \ll 1$ is an example of a nondiagonalizable matrix whose nonnormality is inconsequential for many purposes.

As an alternative, Henrici proposed a departure from normality that is more broadly applicable [380]. For any square matrix $\mathbf{A}$, there is a Schur decomposition $\mathbf{A} = \mathbf{U}\mathbf{T}\mathbf{U}^* = \mathbf{U}(\mathbf{\Lambda} + \mathbf{R})\mathbf{U}^*$, where $\mathbf{U}$ is unitary, $\mathbf{\Lambda}$ is diagonal, and $\mathbf{R}$ is strictly upper triangular; see, e.g., [327, 414]. When $\mathbf{R}$ is zero, this is a unitary diagonalization, and hence $\mathbf{A}$ must be normal. Thus $\|\mathbf{R}\|$ provides a measure of nonnormality. Since the Schur decomposition is generally not unique, Henrici took the minimum value over all possible decompositions,

$$\text{dep}(\mathbf{A}) = \min_{\substack{\mathbf{A}=\mathbf{U}(\Lambda+\mathbf{R})\mathbf{U}^* \\ (\text{Schur decomposition})}} \|\mathbf{R}\|.$$

This dependence on the particular Schur decomposition is not an issue when working in the Frobenius norm, in which case $\|\mathbf{A}\|_F^2 = \|\Lambda + \mathbf{R}\|_F^2 = \|\Lambda\|_F^2 + \|\mathbf{R}\|_F^2$, and thus

$$\text{dep}_F(\mathbf{A}) = \sqrt{\|\mathbf{A}\|_F^2 - \|\Lambda\|_F^2} = \Big(\sum_{j=1}^N \sigma_j^2 - \sum_{j=1}^N |\lambda_j|^2\Big)^{1/2},$$

where $\{\sigma_j\}$ and $\{\lambda_j\}$ are the singular values and eigenvalues of $\mathbf{A}$, respectively. Since $\mathbf{U}\Lambda\mathbf{U}^*$ is a normal matrix approximating $\mathbf{A}$, in any unitarily invariant norm (such as the 2-norm and Frobenius norm),

$$\text{dist}(\mathbf{A}, \mathcal{N}) \le \|\mathbf{A} - \mathbf{U}\Lambda\mathbf{U}^*\| = \|\mathbf{U}^*\mathbf{R}\mathbf{U}\| = \text{dep}(\mathbf{A}).$$

The normal approximation $\mathbf{U}\Lambda\mathbf{U}^*$ has the same eigenvalues as $\mathbf{A}$, but we can see from Figures 48.1 and 48.2 that the spectrum of a *nearest* normal matrix can differ significantly from that of $\mathbf{A}$. In the Frobenius norm, László has shown that $\text{dist}_F(\mathbf{A}, \mathcal{N}) \ge \text{dep}_F(\mathbf{A})/\sqrt{N}$ [482].

The Frobenius norm departure from normality, $\text{dep}_F(\mathbf{A})$, has been employed in a variety of applications. Henrici used it to bound the norms of matrix powers [380], and Descloux generalized these bounds to a wider class of functions [203]. Eigenvalue bounds involving $\text{dep}_F(\mathbf{A})$ have been described by van der Sluis [789], among others. Lee has suggested bounds for $\text{dep}_F(\mathbf{A})$ based directly upon the entries of $\mathbf{A}$ [491, 492]. The simple inclusion result below, akin to the Bauer–Fike theorem, gives a flavor of how the departure from normality arises.

Theorem 48.1 *In the 2-norm,*

$$\sigma_\varepsilon(\mathbf{A}) \subseteq \sigma(\mathbf{A}) + \Delta_{\varepsilon + \text{dep}_2(\mathbf{A})}. \tag{48.4}$$

Proof. Let $z \in \sigma_\varepsilon(\mathbf{A})$. Then there exists some unit vector $\mathbf{v} \in \mathbb{C}^N$ such that $\|\mathbf{A}\mathbf{v} - z\mathbf{v}\|_2 < \varepsilon$. Taking the Schur decomposition that minimizes the 2-norm departure from normality, $\mathbf{A} = \mathbf{U}(\Lambda + \mathbf{R})\mathbf{U}^*$, we have

$$\varepsilon > \|\mathbf{A}\mathbf{v} - z\mathbf{v}\|_2 = \|\mathbf{U}(\Lambda + \mathbf{R})\mathbf{U}^*\mathbf{v} - z\mathbf{U}\mathbf{U}^*\mathbf{v}\|_2 = \|(\Lambda + \mathbf{R})\mathbf{y} - z\mathbf{y}\|_2,$$

where $\mathbf{y} = \mathbf{U}^*\mathbf{v}$ has unit norm. Defining $\mathbf{x} = (\Lambda + \mathbf{R})\mathbf{y} - z\mathbf{y}$, we have $\|\mathbf{x}\|_2 < \varepsilon$ and $\Lambda\mathbf{y} - z\mathbf{y} = \mathbf{x} - \mathbf{R}\mathbf{y}$. Thus,

$$\|\Lambda\mathbf{y} - z\mathbf{y}\|_2 \le \|\mathbf{x}\|_2 + \|\mathbf{R}\|_2\|\mathbf{y}\|_2 < \varepsilon + \text{dep}_2(\mathbf{A}),$$

implying (48.4). ∎

Many other possible metrics for nonnormality can be derived by measuring how much each of the various equivalent characterizations of normality is violated for a particular $\mathbf{A}$. For example, since $\mathbf{A}\mathbf{A}^* = \mathbf{A}^*\mathbf{A}$ when $\mathbf{A}$ is normal, one can quantify the distance from normality as $\|\mathbf{A}\mathbf{A}^* - \mathbf{A}^*\mathbf{A}\|$. Elsner and Paardekooper present a family of bounds relating eleven different departures from normality to one another [243]. Chaitin-Chatelin and Fraysse compare several measures on a descriptive 2×2 example, along with a set of matrices from the Harwell–Boeing test matrix collection [133, §10.1.2].

In the end, all scalar measures of nonnormality suffer from a basic limitation: Nonnormality is too complex to be summarized in a single number. Even when a matrix is 'highly nonnormal', one must know the geometry of that nonnormality in order to judge whether it will have important consequences for a given application. For example, when considering the size of matrix powers, nonnormality associated with eigenvalues of small magnitude may lead to a large value of $\mathrm{dep}(\mathbf{A})$ but have little effect on $\|\mathbf{A}^k\|$. In this setting, a more descriptive quantity may be

$$\|\mathbf{A}\| - \rho(\mathbf{A}),$$

where $\rho(\mathbf{A})$ denotes the spectral radius. Strictly speaking, this quantity does not measure distance from normality, as it can be zero for nonnormal $\mathbf{A}$.

To obtain more descriptive information than is available from any single scalar measure of nonnormality, one could use the condition numbers of the individual eigenvalues (§52) or the numerical range (§17). The ε-pseudospectra, for a range of ε values, provide even more information by interpolating between these two extremes.

49 · Distance to singularity and instability _____

Given a matrix $\mathbf{A}$, what is

$$\min\{\|\mathbf{E}\| : \mathbf{A} + \mathbf{E} \text{ is singular}\}\,? \qquad (49.1)$$

This is the question of *distance to singularity*, and it is equivalent to seeking the minimum norm of $\mathbf{E}$ that gives $\mathbf{A} + \mathbf{E}$ a zero eigenvalue. The answer is well-known to be $\|\mathbf{E}\| = \|\mathbf{A}^{-1}\|^{-1}$, as can be seen from Theorem 4.1.[1] In the 2-norm, the formula reduces to $s_{\min}(\mathbf{A})$, the smallest singular value of $\mathbf{A}$ [327, 388]. The numerical examples in this section use $\|\cdot\| = \|\cdot\|_2$.

It is straightforward to translate (49.1) into a question of pseudospectra. What is the largest value of ε for which $0 \notin \sigma_\varepsilon(\mathbf{A})$? Equivalently, for which ε does the boundary of $\sigma_\varepsilon(\mathbf{A})$ pass through the point 0? Thus the distance to singularity has a geometric interpretation, as sketched in Figure 49.1.

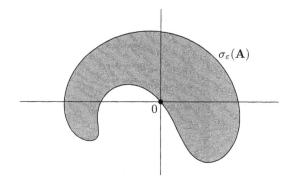

Figure 49.1: The distance of $\mathbf{A}$ to singularity is equal to the value of ε for which the boundary of $\sigma_\varepsilon(\mathbf{A})$ passes through the origin.

For example, consider the matrix whose pseudospectra are plotted in Figure 49.2. The dot shows that the origin lies inside the contour corresponding to 10^{-16}. It follows that this matrix can be made singular by a perturbation $\mathbf{E}$ with $\|\mathbf{E}\| < 10^{-16}$. The actual minimum is roughly 1.38×10^{-18}.

One's motivation for asking how close a matrix is to singular is not the possibility that a perturbation might make it exactly singular. After all, even if the distance to singularity is 10^{-10}, there is zero probability that a random perturbation will lead to exact singularity. One asks how

---

[1] In the numerical linear algebra literature, this result is attributed to Gastinel; see [441], [391, Thm. 6.5], [729, p. 133].

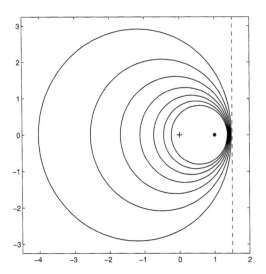

Figure 49.2: ε-pseudospectra of the triangular matrix (49.2) for $N = 64$, $\varepsilon = 10^{-4}, 10^{-6}, 10^{-8}, \ldots, 10^{-16}$; the origin is marked by a cross. In standard double precision arithmetic with $\epsilon_{\text{machine}} \approx 10^{-16}$, this matrix is 'numerically singular'. In the limit $N \to \infty$, the pseudospectra fill the half-plane $\text{Re}\,z < 3/2$ (dashed).

close $\mathbf{A}$ is to singular in order to learn something about $\mathbf{A}$ itself. If $\mathbf{A}$ is nearly singular, then $\mathbf{A}^{-1}$ is large, and any process that depends upon $\mathbf{A}^{-1}$ can be expected to amplify the data it is applied to, as well as any errors in that data. This association of ill-conditioning with nearness to singularity arises in many contexts and has been investigated by Kahan and Demmel [197, 442].

In other words, the problem of distance to singularity is related to the problem of *robust rank determination*. The most robust way to determine the rank of a matrix, at least for problems where the 2-norm is appropriate, is to compute its singular value decomposition and then count the number of singular values that are significantly nonzero according to some suitable measure [327]. This process is expensive, however, and besides that, it is an infinite algorithm (in principle) for a problem that is mathematically finite. Thus it is common to look for *rank-revealing factorizations* of matrices that are simpler than the SVD. We shall take a moment to discuss briefly two instances of such factorizations that lead to interesting examples of pseudospectra. For further details about rank-revealing factorizations, see [63, 372, 408, 726].

The simplest rank-revealing factorization is an LU decomposition computed by Gaussian elimination with partial or complete pivoting, $\mathbf{P}_1\mathbf{A}\mathbf{P}_2 = \mathbf{LU}$. To determine the rank of $\mathbf{A}$, one counts the number of significantly nonzero elements along the diagonal of $\mathbf{U}$. However, this algorithm cannot be trusted always to detect near rank-deficiency correctly. A well-known example is the $N \times N$ triangular matrix of the form

$$\mathbf{A} = \mathbf{U} = \begin{pmatrix} 1 & -1 & -1 & -1 & -1 \\ & 1 & -1 & -1 & -1 \\ & & 1 & -1 & -1 \\ & & & 1 & -1 \\ & & & & 1 \end{pmatrix}. \qquad (49.2)$$

All the diagonal elements of $\mathbf{A}$ are far from zero, so $\mathrm{rank}(\mathbf{A}) = N$, and this is the answer that would be delivered by Gaussian elimination with either kind of pivoting. However, a perturbation of norm $2^{1-N}\sqrt{N}$ is enough to make $\mathbf{A}$ singular (subtract 2^{1-N} from each element a_{j1}). With $N > 60$, for example, the smallest singular value of $\mathbf{A}$ is smaller than machine precision in IEEE double precision arithmetic. The matrix is *numerically singular*, and for general purposes, one would want a numerical computation of the rank to deliver the result $N - 1$.

Figure 49.2 shows a pseudospectral interpretation of the near-singularity of (49.2) for $N = 64$. The plot shows that the pseudospectra are approximately disks in the left half-plane with $\mathrm{Re}\, z < 3/2$. To explain this configuration we can make use of the ideas of §7, since $\mathbf{A}$ is triangular and Toeplitz. The symbol of $\mathbf{A}$ is

$$\begin{aligned} f_N(z) &= 1 - z^{-1} - z^{-2} - \cdots - z^{1-N} \\ &= 2 - \frac{1 - z^{-N}}{1 - z^{-1}} = \frac{1 - 2z^{-1} + z^{-N}}{1 - z^{-1}}, \end{aligned} \qquad (49.3)$$

or (formally) in the limit $N \to \infty$,

$$f_\infty(z) = \frac{1 - 2z^{-1}}{1 - z^{-1}}.$$

This function is a Möbius transformation that maps the unit disk in the z^{-1}-plane onto the half-plane $\mathrm{Re}\, z < 3/2$ and smaller disks D_r about the origin onto disks contained in that half-plane. For a numerical estimate we can apply the ideas that led to the proof of Theorem 7.2. Equation (49.3) implies that $f_N(z) = 0$ for a value $z^{-1} = 1/2 + \mathcal{O}(2^{-N})$ as $N \to \infty$. We accordingly estimate that $0 \in \sigma_\varepsilon(\mathbf{A})$ for a value of ε on the order of 2^{-N}, which matches beautifully the actual value $2^{1-N}\sqrt{N}$.

Since Gaussian elimination may give misleading results, another rank-revealing factorization was proposed years ago: QR factorization with column pivoting. In this algorithm, $\mathbf{A}$ is factored according to $\mathbf{AP} = \mathbf{QR}$, where $\mathbf{Q}$ is unitary, $\mathbf{R}$ is upper triangular, and $\mathbf{P}$ is a permutation matrix constructed to enforce the condition that for each j, $|r_{jj}|$ is at least as large as the 2-norm of each vector $(r_{jk}, \ldots, r_{kk})^{\mathrm{T}}$, $j < k$. The number of significantly nonzero elements on the diagonal of $\mathbf{R}$ is then counted. With this algorithm, the matrix of Figure 49.2 could not cause problems, as the later columns have greater norm than the earlier ones. Indeed, QR factorization with column pivoting is highly reliable in practice. Nevertheless, in a classic paper in 1966, Kahan presented an example showing that this algorithm, too, is not fail-safe [327, 441]. Kahan's example is

$$\mathbf{A} = \mathbf{R} = \begin{pmatrix} 1 & -c & -c & -c & -c \\ & s & -sc & -sc & -sc \\ & & s^2 & -s^2c & -s^2c \\ & & & \ddots & \vdots \\ & & & & s^{N-1} \end{pmatrix} \in \mathbb{C}^{N \times N}, \qquad (49.4)$$

where s is a number slightly less than 1 and $s^2 + c^2 = 1$. One can verify that this matrix is already in column-pivoted form. If $s^{N-1} = \gamma$ is significantly larger than zero, then the rank determination algorithm will conclude that $\mathbf{A}$ has rank N. However, Kahan showed that a perturbation of norm

$$\|\mathbf{E}\| \approx \exp(-\sqrt{2N|\log\gamma|}) \qquad (49.5)$$

is enough to make $\mathbf{A}$ singular. In other words, for large N, this matrix too should properly be detected as 'numerically singular'.

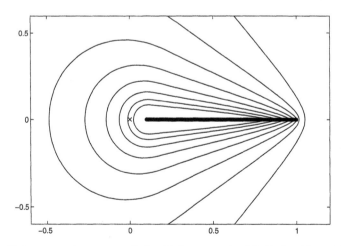

Figure 49.3: ε-pseudospectra of the Kahan matrix (49.4) for $N = 256$, $\varepsilon = 10^{-2}, 10^{-4}, 10^{-6}, \ldots, 10^{-16}$; again the origin is marked by a cross. Evidently this matrix is at the edge of numerical singularity in double precision arithmetic, and numerically singular in single precision.

Pseudospectra of (49.4) are shown in Figure 49.3 [772]. This example takes $\gamma = s^{N-1} = 1/10$, so the spectrum lies in the interval $[1/10, 1]$. The pseudospectra, however, lie lopsidedly to the left of this interval. Kahan's ingenuity lay in constructing a triangular matrix with this property that also satisfies the column pivoting condition. The cross shows that the origin lies inside the 10^{-14}-pseudospectrum, so $s_{\min}(\mathbf{A})$ must be less than 10^{-14}. The actual number is $s_{\min}(\mathbf{A}) \approx 2.41 \times 10^{-15}$, and the estimate (49.5) has the value $\approx 1.23 \times 10^{-15}$.

Now let us move on to the second, related topic of this section: distance to instability. A matrix $\mathbf{A}$ is said to be *stable* if $\sigma(\mathbf{A})$ is a subset of the open left half-plane. Otherwise it is *unstable*. If the eigenvalues lie far to the left of the imaginary axis, however, this is no guarantee that $\mathbf{A}$ is far from unstable, unless $\mathbf{A}$ is close to normal. Thus it is natural to investigate the *distance to instability* or *stability radius* of $\mathbf{A}$,

$$\min\{\|\mathbf{E}\| : \mathbf{A} + \mathbf{E} \text{ is unstable}\}. \tag{49.6}$$

The interpretation of (49.6) in terms of pseudospectra is straightforward: What is the smallest ε for which the boundary of $\sigma_\varepsilon(\mathbf{A})$ touches the imaginary axis? See Figure 49.4.

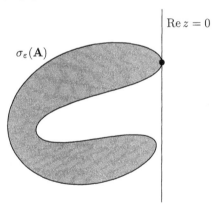

Figure 49.4: The distance of $\mathbf{A}$ to instability is equal to the smallest ε for which the boundary of $\sigma_\varepsilon(\mathbf{A})$ touches the imaginary axis.

The distance to instability problem has received a great deal of attention in the control theory literature since the mid-1980s; it belongs to the subfield known as *robust control* [220, 848]. The term 'stability radius' was introduced by Hinrichsen and Pritchard [398]; other publications on this subject include [119, 123, 388, 399, 613]. The usual motivation behind such studies is to understand the behavior of the dynamical system $\mathbf{x}' = \mathbf{A}\mathbf{x}$ and the associated continuous evolution process $e^{t\mathbf{A}}$.[2] If $\mathbf{A}$ is a stable matrix,

---

[2]For the discrete process $\mathbf{A}^k$, the stable set is the unit disk (§§16, 18), and of course,

so the usual reasoning goes, one would like to know whether its stability is robust in the sense that it will be unaffected by small perturbations. Since the right half-plane is a continuum rather than a single point, a random perturbation large enough to make a matrix unstable actually has a finite probability of doing so, in contrast to the situation with distance to singularity. Thus this usual motivation of (49.6) in terms of perturbations is, to a degree, justifiable. However, we shall argue in the last two pages of this section that even here, perturbations are usually not as significant as is commonly thought.

To illustrate what the pseudospectra of a nearly unstable matrix may look like, here is an example with a bit of history. In 1985 Van Loan suggested as a heuristic that for any matrix $\mathbf{A}$, at least one of the points at which $\sigma_\varepsilon(\mathbf{A})$ first touches the imaginary axis (the solid dot in Figure 49.4) should have the same imaginary part as one of the eigenvalues of $\mathbf{A}$ [799]. To one familiar with pseudospectra, this heuristic is implausible, for as numerous examples in this book have shown, there may be little connection between the shapes of pseudospectral boundaries and the positions of the eigenvalues. By reasoning in just this way, Demmel in 1987 derived a counterexample to Van Loan's heuristic [196]. First he observed that the validity of the heuristic would imply that for any matrix $\mathbf{A}$ with just a single eigenvalue, all of the pseudospectra $\sigma_\varepsilon(\mathbf{A})$ must be convex, since, otherwise, a scaling and rotation yields a counterexample. Then he showed this convexity corollary is not valid. Figures 7.5 (the top part involving a limaçon), 14.4, and 30.2 represent suitable counterexamples. (Indeed, Van Loan's heuristic would require the pseudospectra of any matrix with a single eigenvalue to be circular.)

Demmel's original counterexample is also a Toeplitz matrix, and though it is not as simple as Figure 7.5, it is more dramatic. The matrix is

$$
\mathbf{A} = \begin{pmatrix}
-1 & -M & -M^2 & \cdots & -M^{N-1} \\
 & -1 & -M & \ddots & \vdots \\
 & & -1 & -M & -M^2 \\
 & & & -1 & -M \\
 & & & & -1
\end{pmatrix}, \tag{49.7}
$$

where M^{N-1} is a large number. (This is the inverse of the bidiagonal Toeplitz matrix with entries -1, M.) The case $N = 3$, $M^{N-1} = 10^4$ is enough to yield nonconvex pseudospectra, and Demmel showed this with the plot of pseudospectral contours reproduced as Figure 6.4. Following [772], the case $N = 32$, $M^{N-1} = 10^8$ is illustrated in Figures 49.5 and 49.6.

---

an analogous discussion of distance to instability would be possible for this problem, or for problems involving other special sets in the complex plane.

Consider first Figure 49.5. On a large scale, the pseudospectra of $\mathbf{A}$ are dominated by the large entries in its upper right-hand corner and look like disks about the origin; we see this with $\varepsilon = 10^{-4}$. Smaller ε-values reveal more structure near the origin, apparent in Figure 49.6. For very small ε, $\sigma_\varepsilon(\mathbf{A})$ approximates a disk around the single eigenvalue -1. As ε increases, these disks pinch near the origin, eventually folding over into the right half-plane and giving the crescent-shaped configuration seen for $\varepsilon = 10^{-5}$ in Figure 49.5. When ε reaches a value of about 1.158×10^{-5}, the two horns of the crescent touch each other and $\sigma_\varepsilon(\mathbf{A})$ acquires the topology of an annulus instead of a disk. The pseudospectra are no longer simply connected, let alone convex. The point $z \approx 14.5$ at which the horns first touch is a saddle point of $\|(z - \mathbf{A})^{-1}\|$ and is marked by the cross in Figure 49.6. Figure 49.7 shows a surface plot of resolvent norms for this example, illustrating the local minimum.

What makes this behavior possible? At a certain point $z \approx 0.44$ on the positive real axis, the resolvent norm has a local minimum $\|(z - \mathbf{A})^{-1}\| \approx 0.79$. (Local maxima of $\|(z - \mathbf{A})^{-1}\|$ are impossible, but not local minima; see Theorem 4.2.) The minimal point is marked by a diamond in Figure 49.6. One can explain this behavior qualitatively by noting that the symbol of $\mathbf{A}$ is

$$f(z^{-1}) \;=\; -1 - (Mz^{-1}) - \cdots - (Mz^{-1})^{N-1} \;=\; \frac{(Mz^{-1})^N - 1}{1 - Mz^{-1}}, \quad (49.8)$$

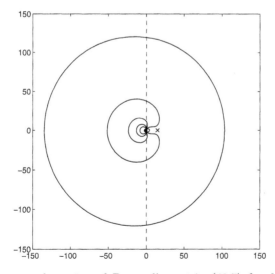

Figure 49.5: ε-pseudospectra of Demmel's matrix (49.7) for $N = 32$, $\varepsilon = 10^{-4}, 10^{-5}, \ldots, 10^{-8}$. The dashed line is the imaginary axis, and the $\times$ marks a saddle point in $\|(z - \mathbf{A})^{-1}\|$.

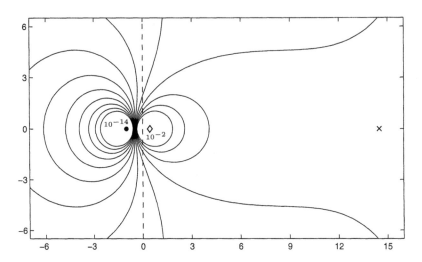

Figure 49.6: Closeup of Figure 49.5 with curves for $\varepsilon = 10^{-2}, 10^{-3}, \ldots, 10^{-14}$. The $\diamond$ marks a local minimum of $\|(z - \mathbf{A})^{-1}\|$.

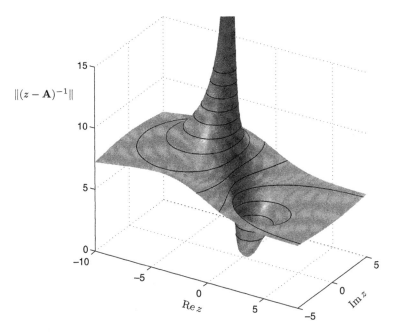

Figure 49.7: Resolvent norm surface of Demmel's matrix, with contours corresponding to ε-pseudospectral boundaries for $\varepsilon = 10^{-2}, 10^{-3}, \ldots, 10^{-15}$. The spike at the eigenvalue -1 is infinite, but the local minimum near $z = 0.44$ is finite.

and for $|Mz^{-1}| < 1$, this approximates the Möbius transformation $-1/(1 - Mz^{-1})$. The inner loops to the right of the origin in Figure 49.6 look approximately like images under this Möbius transformation of circles about the origin in the z^{-1}-plane. Quantitative estimates can be obtained with the aid of Theorem 7.2.

From Figure 49.6 it is evident that a perturbation of norm between 10^{-7} and 10^{-6} is enough to make this matrix unstable. The true value is about 2.84×10^{-7}. By contrast, it would require a perturbation of norm $s_{\min}(\mathbf{A}) \approx 0.356$ to make the matrix singular.

We now turn to a specific problem in this area that has attracted a great deal of attention. Many researchers in control theory, motivated in part by the behavior of linear systems under explicit perturbations, have been concerned with the problem of distance to instability when only *real* perturbations $\mathbf{E}$ are allowed, i.e., the problem of the *real stability radius* of $\mathbf{A}$, as opposed to the (complex) stability radius we have discussed thus far. The real case is harder than the complex one, for the answer is no longer given by the norm of the resolvent. Indeed, the problem of real stability radii has been a major motivation for the study of 'real pseudospectra', called by Hinrichsen, Kelb, and Pritchard *spectral value sets* [397, 398, 399]; see §§6 and 50. This problem of computing the real stability radius was solved for the 2-norm in a 1995 paper of Qiu, Bernhardsson, Rantzer, Davison, Young, and Doyle [616], confirming a 1992 conjecture by Qiu and Davison. The following theorem records this solution together with the simpler result for the complex case implicit in our discussion of the last few pages, expressed in a form to highlight the analogy between the real and complex cases.

Complex and real stability radii

Theorem 49.1 *The 2-norm complex stability radius of a matrix* $\mathbf{A}$ *is equal to*

$$\left(\sup_{\mathrm{Re}\,z=0} s_1(\mathbf{R}(z)) \right)^{-1},$$

where s_1 *denotes the largest singular value and* $\mathbf{R}(z) = (z - \mathbf{A})^{-1}$. *The 2-norm real stability radius is equal to*

$$\left(\sup_{\mathrm{Re}\,z=0} \inf_{\gamma \in (0,1]} s_2\left(\begin{bmatrix} \mathrm{Re}\,\mathbf{R}(z) & -\gamma\,\mathrm{Im}\,\mathbf{R}(z) \\ \gamma^{-1}\,\mathrm{Im}\,\mathbf{R}(z) & \mathrm{Re}\,\mathbf{R}(z) \end{bmatrix} \right) \right)^{-1},$$

where s_2 *denotes the second-largest singular value. The function inside the 'inf' is unimodal, so its local minimum is its global minimum.*

To find stability radii defined by stability regions other than the left half-plane, it is enough to modify the condition $\mathrm{Re}\,z = 0$ in this theorem appropriately.

We would like to finish this section by presenting an example that is mathematically elementary yet has surprising implications.

Given a real dynamical system, $\mathbf{x}' = \mathbf{A}\mathbf{x}$ for some $\mathbf{A} \in \mathbb{R}^{N \times N}$, should one expect the real stability radius or the complex stability radius to give more insight? Many would assume it is the former, since the real stability radius determines the robustness of asymptotic stability if $\mathbf{A}$ is subject to real perturbations. However, even for real dynamical systems, it is the complex stability radius that has implications for the behavior of $\mathbf{x}(t)$ at short time scales. Consider this 2×2 example:

$$\mathbf{A} = \begin{pmatrix} -1 & M^2 \\ -1 & -1 \end{pmatrix}, \quad M \gg 1. \tag{49.9}$$

This matrix has eigenvalues $-1 \pm \mathrm{i}M$, so it is stable, but for large values of M it is far from normal. Real perturbations smear the eigenvalues along the line $\mathrm{Re}\,z = -1$, but large real perturbations are required to move the eigenvalues significantly toward the imaginary axis. Since $\mathbf{A}+\mathbf{E}$ is real, any complex eigenvalues must come as conjugate pairs; the most efficient way to shift a pair to the right is to perturb $\mathbf{A}$ by a multiple of the identity. An $\mathcal{O}(\varepsilon)$ real perturbation can only shift eigenvalues a distance $\mathcal{O}(\varepsilon)$ to the right. Complex perturbations have far more potency: a perturbation of size ε can move eigenvalues $\mathcal{O}(M\varepsilon)$ to the right. For $M = 100$, the complex stability radius is $200/10001 \approx 0.019998$ (attained at $z \approx 99.995\mathrm{i}$, not at $z = 100\mathrm{i}$ as one might expect). The real stability radius, on the

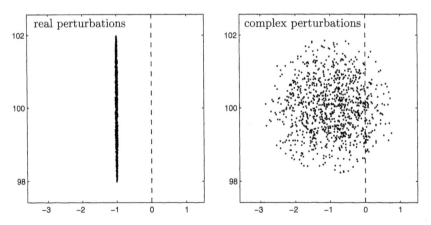

Figure 49.8: Eigenvalues of 1000 random real and complex perturbations of the matrix (49.9) with $M - 100$. The perturbations are of size $\varepsilon = 400/10001$, twice the complex stability radius. (Only the top half of the spectrum is shown; similar perturbed eigenvalues appear near $-1 - 100\mathrm{i}$.) Compare Figure 50.3.

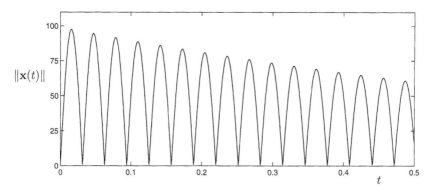

Figure 49.9: The norm of the solution of the real dynamical system $\mathbf{x}' = \mathbf{A}\mathbf{x}$ for $\mathbf{A}$ of (49.9) with $M = 100$, where $\mathbf{x}(0)$ is a random real unit vector. The 100-fold amplification of the initial vector can be explained by the complex eigenvalue perturbations of Figure 49.8, but not the real ones.

other hand, is exactly 1. Taking $M \to \infty$ will magnify the difference between these two numbers. Figure 49.8 confirms these claims; it shows eigenvalues of $\mathbf{A}$ perturbed by random real and complex matrices of norm $\varepsilon = 400/10001$. But is it the real or complex stability radius that gives insight into the behavior of the solution of the dynamical system $\mathbf{x}' = \mathbf{A}\mathbf{x}$? Figure 49.9 shows the norm of a solution $\mathbf{x}(t)$ for a random real starting vector $\mathbf{x}(0)$. While the asymptotic decay rate is governed by the spectral abscissa $\alpha(\mathbf{A}) = -1$, there is immediate growth in $\|\mathbf{x}(t)\|$ on the order of $M = 100$. Analysis based on the real stability radius will miss this.

Once this effect has been noted, it is easy to justify it mathematically. The theorems of §15 relate the norm of the resolvent—not the more complicated quantity that is inverted in the second half of Theorem 49.1—to the size of $\|e^{t\mathbf{A}}\|$. For example, applying (15.11) of Theorem 15.4 with $z = 200 + 100i$ shows that $\|e^{t\mathbf{A}}\| \geq 32.3$ for this 2×2 example for some $t \leq 0.03$. Since $\mathbf{A}$ is real, the norm $\|e^{t\mathbf{A}}\|$ must be attained for some real starting vector; so complex pseudospectra really do shed light on real dynamics.

50 · Structured pseudospectra _____

Throughout this book, at least until the last three pages, we have focused on the effect of general complex perturbations $\mathbf{E} \in \mathbb{C}^{N \times N}$ on the spectrum of a matrix $\mathbf{A}$. In many cases, it might seem more natural to tailor $\mathbf{E}$ to match specific properties of $\mathbf{A}$. For example, perhaps one should perturb a real matrix $\mathbf{A}$ with a real $\mathbf{E}$, or restrict the nonzero pattern of $\mathbf{E}$ to match that of a sparse matrix $\mathbf{A}$? Does a Toeplitz $\mathbf{A}$ call for a Toeplitz $\mathbf{E}$? In situations where one is specifically interested in eigenvalues of perturbed matrices, such structured analysis is often appropriate; applications include floating-point error analysis and studies where matrix entries suffer from experimental uncertainty. In this section, we survey a variety of structured perturbation problems, but before delving too deeply, we must reiterate a crucial point, just illustrated by the example of Figures 49.8 and 49.9. We have used perturbed eigenvalues primarily as a means to infer matrix *behavior*. Our analysis of powers, polynomials and exponentials of $\mathbf{A}$ relies on contour integrals of the resolvent (see (14.9)), whose norm is directly related, via Theorem 2.1, to the eigenvalues that arise from general complex perturbations. By contrast, the eigenvalues that arise from structured perturbations do not bear as close a relation to the resolvent norm and may not provide much information about matrix behavior.

We begin with a general definition, then focus on the familiar case of real perturbations of real matrices.

Structured pseudospectra

Sets of the form

$$\bigcup_{\substack{\mathbf{E} \text{ structured} \\ \|\mathbf{E}\| < \varepsilon}} \sigma(\mathbf{A} + \mathbf{E})$$

are called *structured ε-pseudospectra* of $\mathbf{A}$. In particular, the *real structured ε-pseudospectrum* of $\mathbf{A} \in \mathbb{R}^{N \times N}$ is defined as

$$\sigma_\varepsilon^{\mathbb{R}}(\mathbf{A}) = \bigcup_{\substack{\mathbf{E} \in \mathbb{R}^{N \times N} \\ \|\mathbf{E}\| < \varepsilon}} \sigma(\mathbf{A} + \mathbf{E}). \tag{50.1}$$

The study of real perturbations of real matrices has its roots in the theory of backward error analysis for numerical linear algebra algorithms, the subject of §53. Indeed, Wilkinson includes a sketch of a real structured pseudospectrum in *The Algebraic Eigenvalue Problem* [827, p. 454]. In recent years, real structured pseudospectra have been studied by the control theory community. Motivated by stability theory for dynamical systems under real versus complex perturbations, Hinrichsen and Pritchard were among the first to study $\sigma_\varepsilon^{\mathbb{R}}(\mathbf{A})$, which they called a *spectral value set* [400]. Karow's recent dissertation contains Theorem 50.1 below, along with many additional contributions to the theory of structured pseudospectra [447].

To illustrate how real structured pseudospectra can differ from conventional pseudospectra, we begin by considering perturbations of the upper triangular matrix

$$\mathbf{A} = \begin{pmatrix} -1 & 20 & 0 & 0 \\ & -1/3 & 20 & 0 \\ & & 1/3 & 20 \\ & & & 1 \end{pmatrix}. \tag{50.2}$$

Figure 50.1 shows the eigenvalues of random complex and real perturbations to $\mathbf{A}$. The complex perturbations produce a cloud of perturbed eigenvalues, familiar from similar plots throughout this book. The eigenvalues of the real perturbations, on the other hand, form just a skeleton of the complex set, covering a portion of the real axis with narrow curved regions extending into the complex plane. (Of course, $\sigma_\varepsilon^{\mathbb{R}}(\mathbf{A}) \subseteq \sigma_\varepsilon(\mathbf{A})$.)

As a rule, it is more difficult to compute structured pseudospectra than unstructured ones. While it is simple to compute plots of eigenvalues of random real perturbations of $\mathbf{A}$ to obtain a 'poor man's structured pseudospectrum', precise determination of the boundaries of $\sigma_\varepsilon^{\mathbb{R}}(\mathbf{A})$ is a greater challenge. Hinrichsen and Pritchard derived lower bounds on $\sigma_\varepsilon^{\mathbb{R}}(\mathbf{A})$ to complement the obvious upper bound of $\sigma_\varepsilon(\mathbf{A})$ [400], but these need not be sharp. Fortunately, the formula derived by Qiu et al. [616] for the real stability radius, presented in Theorem 49.1, can be readily adapted into an algorithm for computing the boundaries of $\sigma_\varepsilon^{\mathbb{R}}(\mathbf{A})$ (in the 2-norm). For any $z \notin \sigma(\mathbf{A})$, denote the resolvent by $\mathbf{R}(z) = (z - \mathbf{A})^{-1}$. The 2-norm of the smallest real matrix $\mathbf{E}$ that makes z an eigenvalue of $\mathbf{A} + \mathbf{E}$ is given by

$$\wp(\mathbf{A}, z) \equiv \left(\inf_{\gamma \in (0,1]} s_2 \left(\begin{bmatrix} \operatorname{Re} \mathbf{R}(z) & -\gamma \operatorname{Im} \mathbf{R}(z) \\ \gamma^{-1} \operatorname{Im} \mathbf{R}(z) & \operatorname{Re} \mathbf{R}(z) \end{bmatrix} \right) \right)^{-1}, \tag{50.3}$$

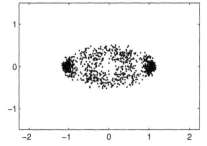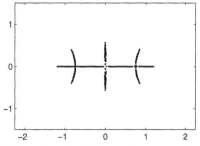

Figure 50.1: Eigenvalues of 250 complex (left) and real (right) random perturbations $\mathbf{E}$ of the matrix $\mathbf{A}$ in (50.2); in each case $\|\mathbf{E}\| = 10^{-4}$.

where $s_2(\cdot)$ denotes the second-largest singular value. (Qiu et al. also provide a construction for $\mathbf{E}$.) From this formula follows an equivalent characterization of the real structured pseudospectra.

Formula for real structured pseudospectra

Theorem 50.1 *The 2-norm real structured ε-pseudospectrum of $\mathbf{A} \in \mathbb{R}^{N \times N}$ is given by*

$$\sigma_\varepsilon^{\mathbb{R}}(\mathbf{A}) = \{z \in \mathbb{C} : \wp(\mathbf{A}, z) < \varepsilon\}. \tag{50.4}$$

With this formulation in hand, we can write a MATLAB algorithm for computing real pseudospectra, an analogue of the the 'basic SVD algorithm' presented on page 371 for standard pseudospectra. (Undoubtedly this algorithm, like the basic SVD algorithm, could be significantly improved.)

```
for k=1:m, for j=1:m
   [ignore,dist] = fminbnd('objfun',1e-9,1,[],A,x(k)+1i*y(j));
   pertdist(j,k) = 1/dist;
end, end
contour(x,y,log10(pertdist))
```

Here the function `objfun` is defined as

```
function obj = objfun(gam,A,z)
Rz  = inv(z*eye(size(A))-A);
sig = svd([real(Rz) -gam*imag(Rz); (1/gam)*imag(Rz) real(Rz)]);
obj = sig(2);
```

As an example of the use of this algorithm, Figure 50.2 illustrates the real structured pseudospectra of the matrix (50.2). Note that the figure confirms that the 'ribs' extending into the complex plane for the real perturbations in Figure 50.1 do not simply fall on curves, but rather form thin sets of finite width.

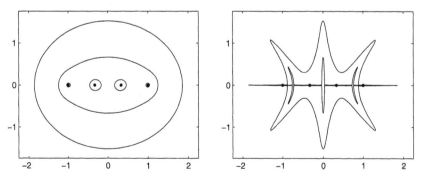

Figure 50.2: Eigenvalues and pseudospectra of the matrix (50.2), with $\sigma_\varepsilon(\mathbf{A})$ (left) and $\sigma_\varepsilon^{\mathbb{R}}(\mathbf{A})$ (right) for $\varepsilon = 10^{-3}, 10^{-4}, 10^{-5}$.

Our second example will give an even greater discrepancy between $\sigma_\varepsilon(\mathbf{A})$ and $\sigma_\varepsilon^{\mathbb{R}}(\mathbf{A})$. The following matrix is a special case of (49.7), designed by Demmel to address a question about stability radii [196] described in the last section:

$$\mathbf{A} = \begin{pmatrix} -1 & -10 & -100 & -1000 & -10000 \\ & -1 & -10 & -100 & -1000 \\ & & -1 & -10 & -100 \\ & & & -1 & -10 \\ & & & & -1 \end{pmatrix}. \qquad (50.5)$$

Figure 50.3 illustrates both $\sigma_\varepsilon(\mathbf{A})$ and $\sigma_\varepsilon^{\mathbb{R}}(\mathbf{A})$ for this matrix, together with eigenvalues of both complex and real perturbations of $\mathbf{A}$. As in the previous example, the real structured pseudospectra look 'spikier' than their smooth complex counterparts.

We do not wish to give the impression that discrepancies as great as those seen as seen in Figure 50.3 are typical. Thus we offer a third example, taking $\mathbf{A}$ to be a Jordan block of dimension 16 with ones on the first superdiagonal and all other entries zero. Figure 50.4 compares $\sigma_\varepsilon(\mathbf{A})$ and $\sigma_\varepsilon^{\mathbb{R}}(\mathbf{A})$; in this case these sets are quite similar for a wide range of ε values, though the real structured pseudospectra have characteristic rippled boundaries.

Many applications give rise to structured matrices, where certain entries must have prescribed values (e.g., zero or one) and one wishes to investigate the effect of perturbations of the other entries. For example, consider the companion matrix

$$\mathbf{A} = \begin{pmatrix} 0 & 1 & & & \\ & 0 & 1 & & \\ & & \ddots & \ddots & \\ & & & 0 & 1 \\ -c_0 & -c_1 & \cdots & \cdots & -c_{N-1} \end{pmatrix},$$

whose eigenvalues are zeros of the characteristic polynomial $p(z) = c_0 + c_1 z + \cdots + c_{N-1} z^{N-1} + z^N$. Perturbations of the polynomial coefficients correspond to perturbations only of the last row of $\mathbf{A}$. The eigenvalues that result from all such structured perturbations form the *pseudozero sets*, details and illustrations of which are included in our discussion of polynomial zerofinding in §55. Perturbations of magnitude ε to only the $-c_0$ entry yield the lemniscate $|p(z)| = \varepsilon$.

Matrices with similar companion-like structure arise in population dynamics. Simple models divide a population into N age brackets, each y years long. Assume that a member of the jth bracket will produce $f_j \geq 0$ offspring over the next y years, and let $r_j \in [0, 1]$ denote the probability that a member of bracket j will survive the next y years. Then the

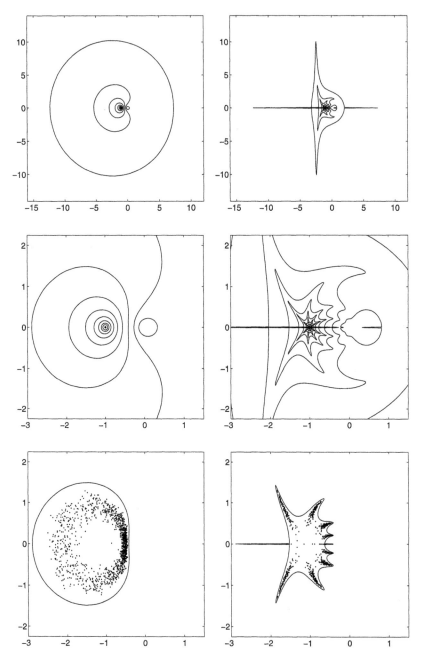

Figure 50.3: In the top row, pseudospectra $\sigma_\varepsilon(\mathbf{A})$ (left) and real structured pseudospectra $\sigma_\varepsilon^{\mathbb{R}}(\mathbf{A})$ (right) for the Demmel matrix (50.5) with $\varepsilon = 10^{-2}$, $10^{-3}, \ldots, 10^{-10}$. The second row is a zoom of the one above it, and the bottom row shows eigenvalues of 250 perturbations of norm 10^{-4}, together with the boundary of the 10^{-4}-pseudospectrum.

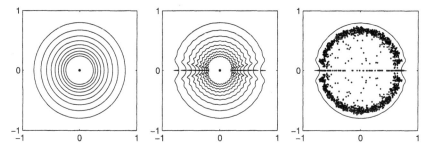

Figure 50.4: Pseudospectra $\sigma_\varepsilon(\mathbf{A})$ (left) and real structured pseudospectra $\sigma_\varepsilon^{\mathbb{R}}(\mathbf{A})$ (middle) for $\varepsilon = 10^{-2}, 10^{-3}, \ldots, 10^{-10}$ for a Jordan block of dimension 16. The rightmost plot shows eigenvalues of 100 real random perturbations $\mathbf{A} + \mathbf{E}$, where $\|\mathbf{E}\| = 10^{-2}$.

population after yk years is given by $\|\mathbf{A}^k \mathbf{p}\|_1$, where $\mathbf{p}$ denotes the initial population distribution, and $\mathbf{A}$ is the *Leslie matrix*

$$\mathbf{A} = \begin{pmatrix} f_1 & f_2 & f_3 & \cdots & f_N \\ r_1 & & & & \\ & r_2 & & & \\ & & \ddots & & \\ & & & r_{N-1} & \end{pmatrix} ;$$

see, e.g., [130]. The population $\mathbf{A}^k \mathbf{p}$ will decay as $k \to \infty$ provided the spectral radius $\rho(\mathbf{A})$ is less than one. Given the obvious uncertainties and simplifications in this model, one wonders how $\sigma(\mathbf{A})$ and $\rho(\mathbf{A})$, in particular, change with perturbations to $f_j \geq 0$ and $r_j \in [0,1]$; we shall investigate this question using structured pseudospectra. If one wishes to study transient dynamics, an issue of matrix behavior, the standard pseudospectra $\sigma_\varepsilon(\mathbf{A})$ should be used.

For example, consider a Leslie matrix $\mathbf{A}$ with $N = 4$, fertility parameters $f_1 = f_3 = f_4 = 1/10$, $f_2 = 1/2$, and survivability probabilities $r_1 = r_2 = r_3 = 9/10$. The spectral radius of this matrix is less than one $(\rho(\mathbf{A}) \approx 0.85445)$, so the population $\mathbf{A}^k \mathbf{p}$ will decay as $k \to \infty$. But suppose that both the fertility and survivability parameters could have errors as large as $\pm 1/10$. Is it possible, given this level of uncertainty, that the population could instead grow exponentially? We examine eigenvalues of structured random perturbations $\mathbf{A} + \mathbf{E}$, where $\mathbf{A} + \mathbf{E}$ is also a valid Leslie matrix. Figure 50.5 shows the result: among our random perturbations was a matrix $\mathbf{E}$ giving $\rho(\mathbf{A} + \mathbf{E}) \approx 1.0381$, which implies that exponential growth is possible. With this level of uncertainty, we cannot conclude that the population will decay.

For our final example of structured pseudospectra, we take $\mathbf{P}$ to be the transition matrix for a Markov chain, as described in §56. The entry

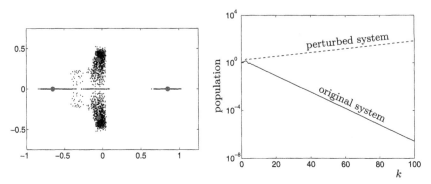

Figure 50.5: Eigenvalues (gray dots) and perturbed eigenvalues of a stable Leslie matrix, where we take 1000 random matrices $\mathbf{E}$ such that $\|\mathbf{E}\| \leq 10^{-1}$ and $\mathbf{A} + \mathbf{E}$ is also of Leslie form (left). The plot on the right shows population evolution for the original system $\mathbf{A}$ and for a matrix $\mathbf{A} + \mathbf{E}$ with $\|\mathbf{E}\| \leq 10^{-1}$.

$p_{jk} \in [0, 1]$ denotes the probability of moving from state j to state k in one step of the Markov chain; each row of $\mathbf{P}$ must sum to 1 to conserve probability. This conservation condition guarantees that $\mathbf{P}$ has an eigenvalue $\lambda = 1$, and the rate of convergence of the Markov chain to its steady state distribution is given by the magnitude of the second largest eigenvalue. If the probabilities that determine $\mathbf{P}$ are estimated from some physical, biological, or economic model, one might wish to quantify how errors in the system affect the convergence behavior of the Markov chain. In this case, the perturbed matrix $\mathbf{P} + \mathbf{E}$ must satisfy the basic properties of a Markov chain: All its entries must be in $[0, 1]$, and each row must sum to one.

Perturbation theory for Markov chains addresses precisely this situation; see, e.g., [148, 149]. Rather than considering these results in detail, we shall simply look at perturbed eigenvalues for two small transition matrices:

$$\mathbf{P}_1 = \begin{pmatrix} 1/2 & 1/2 & 0 \\ 1/8 & 3/4 & 1/8 \\ 0 & 1/2 & 1/2 \end{pmatrix}, \qquad \mathbf{P}_2 = \begin{pmatrix} 1/4 & 3/4 & 0 \\ 1/4 & 1/4 & 1/2 \\ 1/2 & 1/4 & 1/4 \end{pmatrix},$$

with the ∞-norm pseudospectra. (In §56 we explain why $\| \cdot \|_\infty$ is the appropriate norm for this problem.) Figure 50.6 illustrates eigenvalues of $\mathbf{P}_j + \mathbf{E}$, where the perturbations $\mathbf{E}$ are constructed so that $\mathbf{P}_j + \mathbf{E}$ is a valid Markov chain transition matrix and $\|\mathbf{E}\|_\infty \leq 1/4$. Compare these perturbed eigenvalues to the boundaries of the ∞-norm $1/4$-pseudospectrum, also shown in Figure 50.6. Note that for both $\mathbf{P}_1$ and $\mathbf{P}_2$, none of the perturbation matrices is sufficient to cause the eigenvalues of smaller modulus to merge with the eigenvalue $\lambda = 1$. Since $\mathbf{P}_j + \mathbf{E}$ is a transition matrix, it must also have an eigenvalue $\lambda = 1$.

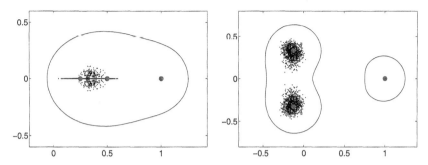

Figure 50.6: Boundary of the ∞-norm ε-pseudospectrum for the Markov chain transition matrices $\mathbf{P}_1$ (left) and $\mathbf{P}_2$ (right) for $\varepsilon = 1/4$, together with eigenvalues of 1000 random perturbations $\mathbf{P}_j + \mathbf{E}$ such that $\|\mathbf{E}\|_\infty \leq 1/4$ and $\mathbf{P}_j + \mathbf{E}$ remains a valid transition matrix.

In addition to the applications illustrated here, there are many more ways in which one can imagine introducing structured perturbations. For example, Hinrichsen, Kelb, and Pritchard consider the spectrum of $\mathbf{A} + \mathbf{D}_1 \mathbf{E} \mathbf{D}_2$ as $\mathbf{E}$ varies and $\mathbf{D}_1$ and $\mathbf{D}_2$ remain fixed; see, e.g., [397, 399]. Related is the study of $\sigma(\mathbf{A} + t\mathbf{E})$ for some *fixed* $\mathbf{E} \in \mathbb{C}^{N \times N}$ and $t \in \mathbb{C}$ [5, 133, 692]; cf. §52. In other cases, one might study how the spectrum of $\mathbf{A}$ evolves as changes are made to one particular entry at a time. As these changes need not be small in magnitude, they may be called *modifications* rather than perturbations [510]. (Wilkinson liked to call them 'blunders'.) For results concerning such perturbations of banded Toeplitz matrices, see [37, 82, 81, 510].

51 · Similarity transformations and canonical forms ⎯

Similarity transformations are one of the basic tools of linear algebra, and they also have generalizations to linear operators in Hilbert and Banach spaces. This section mainly reviews the matrix case, with a few comments at the end about linear operators.

From the perspective of theoretical linear algebra, a matrix is only one of many possible representations of a linear transformation resulting from a particular choice of a coordinate system for a vector space. This view affords the theoretician the luxury of choosing the most convenient coordinate system. Given a matrix $\mathbf{A} \in \mathbb{C}^{N \times N}$, a change of coordinates can be performed via a *similarity transformation*,

$$\widehat{\mathbf{A}} = \mathbf{S}^{-1} \mathbf{A} \mathbf{S},$$

for any invertible $\mathbf{S} \in \mathbb{C}^{N \times N}$. If $(\lambda, \mathbf{v})$ is an eigenpair of $\mathbf{A}$, then

$$\widehat{\mathbf{A}} \mathbf{S}^{-1} \mathbf{v} = \lambda \mathbf{S}^{-1} \mathbf{v}, \tag{51.1}$$

so $(\lambda, \mathbf{S}^{-1} \mathbf{v})$ is an eigenpair of $\widehat{\mathbf{A}}$. Thus similarity transformations leave the spectrum of $\mathbf{A}$ unchanged, revealing one of the great attractions of eigenvalues: they are properties of the linear transformation, independent of the choice of basis. The eigenvectors, however, can be significantly altered by such a transformation.

If $\mathbf{A}$ is normal, it has an orthogonal basis of eigenvectors and one can write $\mathbf{A} \mathbf{U} = \mathbf{U} \mathbf{\Lambda}$, or

$$\mathbf{\Lambda} = \mathbf{U}^* \mathbf{A} \mathbf{U}, \tag{51.2}$$

for a unitary matrix $\mathbf{U}$ and diagonal matrix of eigenvalues, $\mathbf{\Lambda}$. This diagonalizing similarity performs an orthogonal change of basis. In this section, we are particularly interested in two generalizations of this similarity for nonnormal matrices: the Jordan canonical form and the Schur factorization. To begin with, suppose $\mathbf{A}$ is nonnormal but can still be diagonalized,

$$\mathbf{\Lambda} = \mathbf{V}^{-1} \mathbf{A} \mathbf{V}; \tag{51.3}$$

since $\mathbf{A}$ is nonnormal, $\mathbf{V}$ cannot be unitary. The eigenvectors of $\mathbf{A}$, which potentially form an ill-conditioned basis for $\mathbb{C}^N$, are transformed via (51.3) into the ideal elementary basis vectors. The condition number of $\mathbf{V}$, $\kappa(\mathbf{V}) = \|\mathbf{V}\| \|\mathbf{V}^{-1}\|$, provides a measure of the distortion involved in this transformation (see §18). For example, the tridiagonal Toeplitz matrix (3.1) has distinct eigenvalues, which implies it is diagonalizable. However, the condition number of the diagonalizing transformation grows exponentially with the matrix dimension [632], as seen below for three values of N (accurate to the digits listed).

N	$\kappa(\mathbf{V})$
25	1.813×10^7
50	6.120×10^{14}
100	6.912×10^{29}

The choice of $\mathbf{V}$ is not unique, though as noted on page 19, by taking the columns of $\mathbf{V}$ to have unit length we obtain a value for $\kappa(\mathbf{V})$ within $\sqrt{N}$ of the optimal value, since the eigenvalues of $\mathbf{A}$ are distinct.

Figure 51.1 shows how the diagonalizing similarity transformation alters the eigenvectors of this nonnormal matrix when $N = 25$. The top plot shows that the eigenvectors form a poor basis for $\mathbb{C}^N$, consistent with the large value of $\kappa(\mathbf{V})$ given above. The bottom plot trivially shows that the jth eigenvector of $\mathbf{\Lambda} = \mathbf{V}^{-1}\mathbf{A}\mathbf{V}$ is equal to the elementary basis vector $\mathbf{e}_j$, which has one in the jth component and zero elsewhere. These bottom basis vectors are orthonormal, so if one expands an arbitrary vector with unit 2-norm as a linear combination of them, none of the expansion coefficients can exceed 1. Expanding the same vector in the top basis will likely lead to much larger coefficients, a consequence of cancellation among the various nearly-aligned basis vectors. In applications of nonnormal matrices, such cancellation leads to interesting transient effects that would be missed if

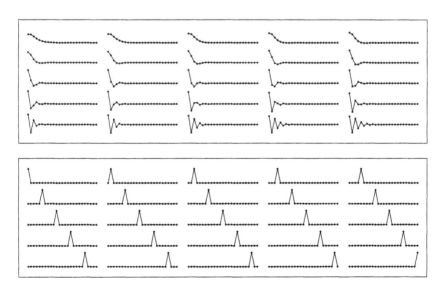

Figure 51.1: Ill-conditioned eigenvectors $\{\mathbf{v}_j\}_{j=1}^N$ of the tridiagonal Toeplitz matrix (3.1) for $N = 25$ (top) compared to the eigenvectors $\{\mathbf{V}^{-1}\mathbf{v}_j\}_{j=1}^N$ of the diagonalized matrix (bottom). Compare Figure 12.3.

one analyzed the problem in the convenient orthonormal basis.

When the eigenvectors of $\mathbf{A}$ fail to span $\mathbb{C}^N$, the matrix is said to be *defective* or *nondiagonalizable*, and no similarity transformation will reduce $\mathbf{A}$ to diagonal form. In this instance, (51.3) generalizes to the *Jordan canonical form*,

$$\mathbf{J} = \mathbf{V}^{-1}\mathbf{A}\mathbf{V}, \tag{51.4}$$

where $\mathbf{J}$ is a structured bidiagonal matrix. The development of the Jordan form is algebraically elegant but intricate; here we describe just those properties needed for the perturbation theory described in the next section. For a complete account, see, e.g., [414, 548, 475].

Suppose λ_j is an eigenvalue of $\mathbf{A}$ with algebraic multiplicity a_j (multiplicity of λ_j as a root of the characteristic polynomial) and geometric multiplicity g_j (number of linearly independent eigenvectors associated with λ_j). There exist matrices $\mathbf{V}_j \in \mathbb{C}^{N \times a_j}$ and $\mathbf{J}_j \in \mathbb{C}^{a_j \times a_j}$ such that

$$\mathbf{A}\mathbf{V}_j = \mathbf{V}_j\mathbf{J}_j, \tag{51.5}$$

where $\mathbf{V}_j$ has linearly independent columns and $\mathbf{J}_j = \mathrm{diag}(\mathbf{J}_{j,1}, \ldots, \mathbf{J}_{j,g_j})$ with

$$\mathbf{J}_{j,\ell} = \begin{pmatrix} \lambda_j & 1 & & \\ & \lambda_j & \ddots & \\ & & \ddots & 1 \\ & & & \lambda_j \end{pmatrix}, \qquad \ell = 1, \ldots, g_j. \tag{51.6}$$

The dimensions of the $\mathbf{J}_{j,\ell}$ depend upon properties of $\mathbf{A}$; they must sum to a_j. Overall, $\mathbf{J}_j$ has exactly $a_j - g_j$ superdiagonal ones. The columns of $\mathbf{V}_j$ form a basis for

$$\mathcal{V}_j = \{\mathbf{v} \in \mathbb{C}^N : (\lambda_j - \mathbf{A})^k \mathbf{v} = \mathbf{0} \text{ for some } k\},$$

known as an *invariant subspace* of $\mathbf{A}$ because $\mathbf{A}\mathcal{V}_j \subseteq \mathcal{V}_j$. As $\mathcal{V}_j$ is finite-dimensional, there must exist some smallest integer k_j such that $(\mathbf{A} - \lambda_j)^{k_j} \mathbf{v} = \mathbf{0}$ for all $\mathbf{v} \in \mathcal{V}_j$. This k_j is called the *index* of λ_j; it equals the largest dimension of all Jordan blocks $\mathbf{J}_{j,\ell}$ associated with λ_j, and hence $k_j \leq a_j$. The Jordan form (51.4) is obtained by assembling $\mathbf{V} = [\mathbf{V}_1 \, \mathbf{V}_2 \cdots \mathbf{V}_m]$ and $\mathbf{J} = \mathrm{diag}(\mathbf{J}_1, \ldots, \mathbf{J}_m)$, where m is the number of distinct eigenvalues of $\mathbf{A}$.

The reduction to Jordan form is numerically problematic [328]. Difficulties arise in simply trying to determine a_j and g_j. How does one discern if two computed eigenvalues are identical in the presence of rounding errors? Given a multiple eigenvalue, are the associated eigenvectors linearly dependent, or just nearly so? Many authors have suggested more numerically-suitable alternatives [34, 195, 440]. In place of (51.4), one can seek a well-conditioned similarity transformation that *block-diagonalizes*

A, imposing less restrictive conditions than those required for the Jordan form. Eigenvalues should be grouped together based on the angles formed between eigenvectors: Small angles are collected in the same block. Since the eigenvectors associated with each group form a poor basis for their invariant subspace, they are replaced by an orthogonal basis for that subspace. This yields the equation

$$\mathbf{AS}_j = \mathbf{S}_j\mathbf{B}_j$$

of the form (51.5), but now the columns of $\mathbf{S}_j$ are orthogonal. Constructing $\mathbf{S}_j$ via the Gram–Schmidt process leads to an upper triangular $\mathbf{B}_j$. The matrix $\mathbf{S}$ is assembled from the rectangular $\mathbf{S}_j$ matrices.

Taking this approach to its logical extreme, all eigenvalues with nonorthogonal eigenvectors could be grouped together. The resulting similarity transformation would then be unitary, but the cost of this perfectly conditioned transformation is that **A** is only reduced to general triangular form, rather than the bidiagonal Jordan form. This is the *Schur factorization*,

$$\mathbf{T} = \mathbf{U}^*\mathbf{AU}, \tag{51.7}$$

which exists for any matrix $\mathbf{A} \in \mathbb{C}^{N \times N}$. Like (51.4), this form reduces to the unitary diagonalization (51.2) when **A** is normal.

We have only addressed the two best-known canonical forms, but other options are possible. One of them is the *rational canonical form* [827], which reduces **A** to a block diagonal matrix with companion matrix blocks (see §55). For a survey of canonical forms based on unitary similarity transformations, see [674].

How does a similarity transformation affect pseudospectra? When the transformation matrix is ill-conditioned, as is possible with the Jordan form, the change can be significant. On the other hand, 2-norm pseudospectra are invariant under unitary similarities.

Pseudospectra of similarity transformations

Theorem 51.1 *For any invertible* $\mathbf{S} \in \mathbb{C}^{N \times N}$ *and any* $\varepsilon > 0$,

$$\sigma_{\varepsilon/\kappa(\mathbf{S})}(\mathbf{S}^{-1}\mathbf{AS}) \subseteq \sigma_\varepsilon(\mathbf{A}) \subseteq \sigma_{\varepsilon\kappa(\mathbf{S})}(\mathbf{S}^{-1}\mathbf{AS}),$$

where $\kappa(\mathbf{S}) = \|\mathbf{S}\|\|\mathbf{S}^{-1}\|$. *In particular, when* $\mathbf{S}$ *is unitary and* $\|\cdot\| = \|\cdot\|_2$, $\sigma_\varepsilon(\mathbf{A}) = \sigma_\varepsilon(\mathbf{S}^{-1}\mathbf{AS})$.

Proof. Suppose $z \in \sigma_\varepsilon(\mathbf{A})$. Then

$$\varepsilon^{-1} < \|(z - \mathbf{A})^{-1}\| = \|(z - \mathbf{SS}^{-1}\mathbf{ASS}^{-1})^{-1}\|$$
$$= \|\mathbf{S}(z - \mathbf{S}^{-1}\mathbf{AS})^{-1}\mathbf{S}^{-1}\| \leq \kappa(\mathbf{S})\|(z - \mathbf{S}^{-1}\mathbf{AS})^{-1}\|,$$

implying that $z \in \sigma_{\varepsilon\kappa(\mathbf{S})}(\mathbf{S}^{-1}\mathbf{A}\mathbf{S})$, and thus giving the second inclusion. The first follows similarly. When $\mathbf{S}$ is a unitary matrix, the inequality becomes an equality due to the unitary invariance of the 2-norm. ∎

Since unitary transformations do not affect the 2-norm pseudospectra, (51.7) implies one could learn all there is to know about those pseudospectra by considering only triangular matrices.[1] Thus, the range of 2-norm pseudospectral behavior is controlled by no more than $\frac{1}{2}N(N+1)$ parameters. Jordan matrices, with their $2N - 1$ parameters, form only a small subset of triangular matrices. In particular, the pseudospectra of the Jordan factor $\mathbf{J} = \mathbf{V}^{-1}\mathbf{A}\mathbf{V}$ consist of the union of disks. To see this, note that $\mathbf{J}$ is the direct sum of matrices of the form (51.6). If $\mathbf{J}_0 \in \mathbb{C}^{k \times k}$ is of the form (51.6) with zero on the main diagonal, then for any $\theta \in [0, 2\pi)$, the diagonal unitary matrix

$$\mathbf{S} = \mathrm{diag}(\mathrm{e}^{\mathrm{i}\theta}, \mathrm{e}^{2\mathrm{i}\theta}, \ldots, \mathrm{e}^{k\mathrm{i}\theta})$$

rotates $\mathbf{J}_0$ without affecting the pseudospectra, $\mathbf{S}\mathbf{J}_0\mathbf{S}^* = \mathrm{e}^{\mathrm{i}\theta}\mathbf{J}_0$. Thus, the pseudospectra of a Jordan factor must equal the union of disks, by Theorem 2.4(iii). The radii of these circles depend upon the dimension of the blocks. (The Zabczyk operator discussed in §19 is built from shifted and scaled Jordan blocks. As can be seen in Figure 19.1, the pseudospectra of this operator consist of the union of the circular pseudospectra associated with each Jordan block.)

Lavalée, Malyshev, and Sadkane propose a method for computing approximate pseudospectra based on the principle in Theorem 51.1 [486]. They suggest that one should block-diagonalize $\mathbf{A}$ by a well-conditioned (but not necessarily unitary) similarity transformation, and then use the union of the pseudospectra of the individual diagonal blocks as an approximation to $\sigma_\varepsilon(\mathbf{A})$. The quality of the approximation is controlled by the condition of the similarity transformation. To illustrate how this procedure can be useful for certain nonnormal matrices, we consider an N-dimensional matrix derived from a stabilized finite element discretization of a two-dimensional advection-diffusion equation on a square grid, as described in [276] and used as a model problem in §26. For this particular example, the eigenvalues of the discretization matrix fall on $\sqrt{N}$ vertical lines in the complex plane with $\sqrt{N}$ eigenvalues per line. The eigenvectors associated with each line form small angles with one another but are orthogonal to the eigenvectors associated with every other line. Thus, there is a unitary similarity that reduces $\mathbf{A}$ to a block diagonal matrix with $\sqrt{N}$ blocks each of size $\sqrt{N}$. The ε-pseudospectrum of the entire matrix is simply the union of the ε-pseudospectra of each individual block. Figure 51.2

---

[1]This does not seem an efficient program, since so much of the structure that facilitates analysis, e.g., of Toeplitz matrices, would be lost in the reduction to Schur form. See also the comments on p. 367. For computation of pseudospectra, though, the Schur form is indispensable, as described in §39.

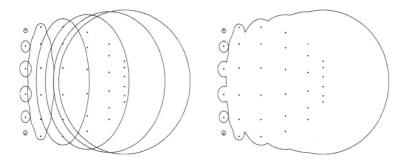

Figure 51.2: Spectrum and 10^{-4}-pseudospectrum for an advection-diffusion discretization matrix of dimension $N = 36$ [276]. Six eigenvalues fall on each of six vertical lines. The figure on the left superimposes the 10^{-4}-pseudospectra associated with each of the six different lines (6×6 matrices). The 10^{-4}-pseudospectrum of the entire matrix is the union of these pseudospectra, shown on the right.

illustrates this for $N = 36$ and $\varepsilon = 10^{-4}$. Liesen and Strakoš use this fact to analyze the convergence of the GMRES algorithm for solving linear systems involving these matrices [504].

Similarity transformations provide a convenient means for changing from the standard Euclidean norm to a weighted norm induced by the inner product $(\mathbf{x}, \mathbf{y})_{\mathbf{L}} \equiv (\mathbf{Lx}, \mathbf{Ly}) = \mathbf{y}^* \mathbf{L}^* \mathbf{Lx}$ for an invertible matrix $\mathbf{L}$: from page 379 or (45.10),

$$\|\mathbf{A}\|_{\mathbf{S}} = \|\widehat{\mathbf{A}}\|_2,$$

where $\widehat{\mathbf{A}} = \mathbf{LAL}^{-1}$. Similarly,

$$\|(z - \mathbf{A})^{-1}\|_{\mathbf{S}} = \|\mathbf{L}(z - \mathbf{A})^{-1}\mathbf{L}^{-1}\|_2 \leq \kappa(\mathbf{L}) \|(z - \mathbf{A})^{-1}\|,$$

suggesting the following analogue to Theorem 51.1 that bounds how much a change of inner product can alter the pseudospectra.

Theorem 51.2 *For any invertible* $\mathbf{L} \in \mathbb{C}^{N \times N}$ *and any* $\varepsilon > 0$,

$$\sigma_\varepsilon^{\mathbf{L}\text{-norm}}(\mathbf{A}) \subseteq \sigma_{\varepsilon \kappa(\mathbf{L})}^{2\text{-norm}}(\mathbf{A}).$$

Suppose $\mathbf{A}$ is diagonalizable, $\mathbf{\Lambda} = \mathbf{V}^{-1}\mathbf{AV}$, where $\mathbf{\Lambda}$ is diagonal and the columns $\mathbf{v}_1, \dots, \mathbf{v}_N$ of $\mathbf{V}$ are eigenvectors of $\mathbf{A}$. Even when these columns form an ill-conditioned basis, $\mathbf{V}$ induces an inner product in which $\mathbf{A}$ is *normal*. To see this, note that

$$(\mathbf{v}_j, \mathbf{v}_k)_{\mathbf{V}^{-1}} = \mathbf{v}_k^* \mathbf{V}^{-*} \mathbf{V}^{-1} \mathbf{v}_j = \mathbf{e}_k^* \mathbf{e}_j;$$

i.e., the eigenvectors of $\mathbf{A}$ are orthogonal in the $\mathbf{V}^{-1}$ inner product. Thus we have the following elementary observations: Any diagonalizable matrix is normal in some inner product, and in this inner product, the norm of $\mathbf{A}$ equals its spectral radius, $\|\mathbf{A}\|_{\mathbf{V}^{-1}} = \rho(\mathbf{A})$. (When $\mathbf{A}$ is nondiagonalizable, the fact that for any $\varepsilon > 0$ there exists some norm for which $\|\mathbf{A}\| < \rho(\mathbf{A}) + \varepsilon$ can be derived from the fact that the superdiagonal ones in the Jordan form $\mathbf{J}$ can be replaced by an arbitrary nonzero value by adjusting the similarity transformation.)

We close by outlining how some of the ideas of similarity transformations for matrices can be generalized to operators in infinite-dimensional spaces [184, 323]. If X is a Banach space with norm $\|\cdot\|$, a *basis* for X is a sequence of vectors $\mathbf{x}_j \in X$, $j = 1, 2, \ldots$, such that any vector in X has a unique representation as a series $\sum_{j=1}^{\infty} c_j \mathbf{x}_j$ that converges in the norm $\|\cdot\|$; a *normalized basis* is a basis with $\|\mathbf{x}_j\| = 1$ for each j. An *unconditional basis* is a basis with the property that every permutation of the sequence is also a basis. If X is a Hilbert space, a sequence $\{\mathbf{x}_j\}$ is an unconditional basis if and only if it has 'finite condition number' in the sense that there exists a bounded invertible operator $\mathbf{S}$ on X such that $\{\mathbf{e}_j \equiv \mathbf{S}^{-1}\mathbf{x}_j\}$ is a complete orthonormal set in X. A basis with this property is also known as a *Riesz basis*. The main results of this section carry over to operators in this context. For example, if an operator $\mathbf{A}$ in a Hilbert space X has eigenvalues λ_j and a complete set of associated normalized eigenvectors $\{\mathbf{v}_j\}$ that form a Riesz basis for X, then with $\mathbf{V}$ defined by $\mathbf{v}_j = \mathbf{V}\mathbf{e}_j$ for some complete orthonormal set $\{\mathbf{e}_j\}$ in X, we find that $\mathbf{A}$ has the 'diagonalization' $\mathbf{A} = \mathbf{V}\mathbf{\Lambda}\mathbf{V}^{-1}$, where $\mathbf{\Lambda}$ denotes the operator that maps $\mathbf{v}_j$ to $\lambda_j \mathbf{v}_j$ for each j. We define the *condition number* of $\mathbf{V}$, or of the set of eigenvectors $\{\mathbf{v}_j\}$, by $\kappa(\mathbf{V}) = \kappa(\{\mathbf{v}_j\}) = \|\mathbf{V}\|\|\mathbf{V}^{-1}\|$. A part of the *spectral theorem* asserts that if $\mathbf{A}$ is a bounded operator on a Hilbert space with a complete set of eigenvectors that form a Riesz basis, then $\mathbf{A}$ is normal (i.e., $\mathbf{A}\mathbf{A}^* = \mathbf{A}^*\mathbf{A}$) if and only if these eigenvectors can be chosen to be orthonormal, i.e., $\kappa(\mathbf{V}) = 1$. The results in this book involving the condition number of a matrix of eigenvectors, such as Theorems 2.2 and 2.3 and equations (14.5), (14.22), and (48.2), can be extended to this generalized context.

52 · Eigenvalue perturbation theory _____

Since the eigenvalues of a matrix are continuous functions of the matrix entries, it is natural to ask how the spectrum moves when matrix coefficients are altered by small amounts. Classical perturbation theory addresses this question. Early in the twentieth century, it was found that when applied to the Hermitian and nearly Hermitian operators of quantum mechanics, this theory could explain all kinds of phenomena of atomic and molecular physics, notably the splitting of spectral lines by second-order physical effects. Since this time, spectral perturbation theory has been one of the central topics of mathematical physics. See, for example, [562, 629].

Our interest here is in more strongly nonnormal finite-dimensional matrices. We shall find that perturbation theory provides a description of the ε-pseudospectra in the limit $\varepsilon \to 0$. When ε is very small, the pseudospectra look like the union of disks whose radii depend upon the conditioning and Jordan structure of the individual eigenvalues.

Given a matrix $\mathbf{A} \in \mathbb{C}^{N \times N}$, we are interested in the evolution of the eigenvalues of

$$\mathbf{A}(t) = \mathbf{A} + t\mathbf{E},$$

where $\|\mathbf{E}\| = 1$ and t is a complex variable, usually assumed to be small in magnitude. Comprehensive sources for such analysis include the books by Baumgärtel [33] and Kato [448].

To begin with, consider the most straightforward case: Suppose all the eigenvalues $\lambda_1, \ldots, \lambda_N$ of $\mathbf{A}$ are distinct, which implies the existence of a full set of left and right eigenvectors determined up to scaling,

$$\mathbf{u}_j^* \mathbf{A} = \lambda_j \mathbf{u}_j^*, \qquad \mathbf{A} \mathbf{v}_j = \lambda_j \mathbf{v}_j,$$

for $j = 1, \ldots, N$. We focus on the eigenpair $(\lambda_1, \mathbf{v}_1)$ of $\mathbf{A}$, which is perturbed to become the eigenpair $(\lambda_1(t), \mathbf{v}_1(t))$ of $\mathbf{A}(t)$:

$$\mathbf{A}(t)\mathbf{v}_1(t) = \lambda_1(t)\mathbf{v}_1(t). \tag{52.1}$$

Since the eigenvalue $\lambda_1 \in \sigma(\mathbf{A})$ is a simple root of the characteristic equation $\det(\lambda - \mathbf{A}) = 0$, the eigenvalue $\lambda_1(t)$ of $\mathbf{A}(t)$ is given by the convergent Taylor series

$$\lambda_1(t) = \lambda_1 + \alpha_1 t + \alpha_2 t^2 + \cdots$$

for sufficiently small $|t|$ and appropriate constants $\alpha_1, \alpha_2, \ldots$. Similarly, we can always scale the eigenvector $\mathbf{v}_1(t)$ to take the form

$$\mathbf{v}_1(t) = \mathbf{v}_1 + \sum_{j=2}^{n} \left(t\beta_{1j} + t^2\beta_{2j} + t^3\beta_{3j} + \cdots \right)\mathbf{v}_j.$$

Note that $\mathbf{u}_j^* \mathbf{v}_k = 0$ when $j \neq k$, and so premultiplying (52.1) by $\mathbf{u}_1^*$ yields

$$
\begin{aligned}
\lambda_1(t)\mathbf{u}_1^*\mathbf{v}_1 &= \mathbf{u}_1^* \mathbf{A}(t)\mathbf{v}_1(t) \\
&= \mathbf{u}_1^* \mathbf{A}\mathbf{v}_1(t) + t\mathbf{u}_1^* \mathbf{E}\mathbf{v}_1(t) \\
&= \lambda_1 \mathbf{u}_1^*\mathbf{v}_1 + t\mathbf{u}_1^* \mathbf{E}\mathbf{v}_1(t).
\end{aligned}
$$

Thus the perturbation $t\mathbf{E}$ moves the eigenvalue λ_1 by the distance

$$
|\lambda_1 - \lambda_1(t)| = |t| \frac{|\mathbf{u}_1^* \mathbf{E}\mathbf{v}_1(t)|}{|\mathbf{u}_1^* \mathbf{v}_1|} \tag{52.2}
$$

$$
\leq |t| \frac{\|\mathbf{u}_1\|\|\mathbf{v}_1(t)\|}{|\mathbf{u}_1^* \mathbf{v}_1|} = |t| \frac{\|\mathbf{u}_1\|\|\mathbf{v}_1\|}{|\mathbf{u}_1^* \mathbf{v}_1|} + \mathcal{O}(|t|^2). \tag{52.3}
$$

The coefficient of $|t|$ in this final bound is the *condition number* of λ_1, written $\kappa(\lambda_1)$. For any simple eigenvalue λ_j,

$$
\kappa(\lambda_j) = \frac{\|\mathbf{u}_j\|\|\mathbf{v}_j\|}{|\mathbf{u}_j^* \mathbf{v}_j|}.
$$

The Cauchy–Schwarz inequality implies that $|\mathbf{u}_j^* \mathbf{v}_j| \leq \|\mathbf{u}_j\|\|\mathbf{v}_j\|$, so $\kappa(\lambda_j) \geq 1$. The condition number equals 1 when equality holds in the Cauchy–Schwarz inequality, i.e., when $\mathbf{u}_j$ and $\mathbf{v}_j$ are collinear. This is always the case when $\mathbf{A}$ is a normal matrix, since left and right eigenvectors can be taken to be the same; it can also occur for some, though never all, eigenvalues of a nonnormal matrix. An eigenvalue for which $\kappa(\lambda_j) = 1$ is called a *normal eigenvalue*. Eigenvalue condition numbers were popularized by Wilkinson, whose derivation we have followed above [827, Chap. 2].

If we normalize $\mathbf{u}_j$ and $\mathbf{v}_j$ so that $\mathbf{u}_j^* \mathbf{v}_j = 1$, then the matrix

$$
\mathbf{P}_j = \mathbf{v}_j \mathbf{u}_j^* \tag{52.4}
$$

is a *projector*, since $\mathbf{P}_j^2 = \mathbf{v}_j \mathbf{u}_j^* \mathbf{v}_j \mathbf{u}_j^* = \mathbf{v}_j \mathbf{u}_j^* = \mathbf{P}_j$. The range of $\mathbf{P}_j$ is simply the (right) eigenspace associated with λ_j, $\mathrm{Ran}\, \mathbf{P}_j = \mathrm{span}\{\mathbf{v}_j\}$. Since $\mathbf{u}_j$ is orthogonal to $\mathbf{v}_k$ for $k \neq j$, $\mathbf{P}_j \mathbf{v}_k = \mathbf{0}$ for $k \neq j$, and therefore the projector $\mathbf{P}_j$ is said to project *along* (or *parallel to*) $\mathrm{span}\{\mathbf{v}_k\}_{k \neq j}$. Because $\mathbf{P}_j$ projects onto an eigenspace and along the complementary eigenspaces, it is called the *spectral projector* for λ_j. Among its many useful properties, notice that $\mathbf{A}\mathbf{P}_j = \mathbf{P}_j \mathbf{A} = \lambda_j \mathbf{P}_j$. Furthermore,

$$
\|\mathbf{P}_j\| = \frac{\|\mathbf{v}_j \mathbf{u}_j^*\|}{|\mathbf{u}_j^* \mathbf{v}_j|} = \max_{\|\mathbf{x}\|=1} \frac{\|\mathbf{v}_j \mathbf{u}_j^* \mathbf{x}\|}{|\mathbf{u}_j^* \mathbf{v}_j|} = \frac{\|\mathbf{v}_j\|\|\mathbf{u}_j\|}{|\mathbf{u}_j^* \mathbf{v}_j|} = \kappa(\lambda_j),
$$

where we have taken $\mathbf{x} = \mathbf{u}_j/\|\mathbf{u}_j\|$ in the maximization. Thus the eigenvalue condition number is simply the norm of the spectral projector. The connection between spectral projectors and pseudospectra can be better appreciated through the equivalent integral

$$\mathbf{P}_j = \frac{1}{2\pi i} \int_\Gamma (z - \mathbf{A})^{-1} \, dz, \tag{52.5}$$

where Γ is a contour surrounding λ_j but none of the other eigenvalues.

Taking the union of the asymptotic bounds in (52.3) over all eigenvalues $\lambda_j \in \sigma(\mathbf{A})$ and all matrices $\mathbf{E}$ with $\|\mathbf{E}\| = 1$, we deduce the following result.

Asymptotic pseudospectra inclusion regions

Theorem 52.1 *Suppose* $\mathbf{A} \in \mathbb{C}^{N \times N}$ *has* N *distinct eigenvalues,* $\sigma(\mathbf{A}) = \{\lambda_j\}_{j=1}^N$. *Then as* $\varepsilon \to 0$,

$$\sigma_\varepsilon(\mathbf{A}) \subseteq \bigcup_{j=1}^N (\lambda_j + \Delta_{\varepsilon \kappa(\lambda_j) + \mathcal{O}(\varepsilon^2)}),$$

where $\Delta_\delta = \{z \in \mathbb{C} : |z| < \delta\}$.

As we shall see, the set obtained by neglecting $\mathcal{O}(\varepsilon^2)$ effects,

$$\Omega_\varepsilon = \bigcup_{j=1}^N (\lambda_j + \Delta_{\varepsilon \kappa(\lambda_j)}),$$

provides a good indication of the pseudospectra for small values of ε. Bauer and Fike described similar inclusion regions that are valid for all $\varepsilon > 0$: The cost of omitting the $\mathcal{O}(\varepsilon^2)$ term is an increase in the radius of the inclusion disks by a factor of N [32, Thm. 4], [199, §4.3].

Bauer–Fike theorem based on $\kappa(\lambda_j)$

Theorem 52.2 *Suppose* $\mathbf{A} \in \mathbb{C}^{N \times N}$ *has* N *distinct eigenvalues,* $\sigma(\mathbf{A}) = \{\lambda_j\}_{j=1}^N$. *Then for all* $\varepsilon > 0$,

$$\sigma_\varepsilon(\mathbf{A}) \subseteq \bigcup_{j=1}^N (\lambda_j + \Delta_{\varepsilon N \kappa(\lambda_j)}).$$

Proof. If $z \in \sigma_\varepsilon(\mathbf{A})$, then there exists a unit vector $\mathbf{x} \in \mathbb{C}^N$ such that

$$(\mathbf{A} + \mathbf{E})\mathbf{x} = z\mathbf{x} \tag{52.6}$$

for some $\mathbf{E}$ with $\|\mathbf{E}\| < \varepsilon$. Since $\mathbf{A}$ has N distinct eigenvalues, its spectral projectors $\mathbf{P}_j$ are defined by (52.4). For any fixed j, premultiply (52.6) by $\mathbf{P}_j$ and rearrange to obtain $\mathbf{P}_j \mathbf{E} \mathbf{x} = \mathbf{P}_j(z - \mathbf{A})\mathbf{x} = (z - \lambda_j)\mathbf{P}_j \mathbf{x}$. Provided $\mathbf{P}_j \mathbf{x} \neq \mathbf{0}$, this implies

$$|\lambda_j - z| = \frac{\|\mathbf{P}_j \mathbf{E} \mathbf{x}\|}{\|\mathbf{P}_j \mathbf{x}\|} < \varepsilon \frac{\|\mathbf{P}_j\|}{\|\mathbf{P}_j \mathbf{x}\|}.$$

Since $1 = \|\mathbf{x}\| = \|\sum_{j=1}^{N} \mathbf{P}_j \mathbf{x}\| \leq \sum_{j=1}^{N} \|\mathbf{P}_j \mathbf{x}\|$, there must be at least one j for which $\|\mathbf{P}_j \mathbf{x}\| \geq 1/N$. For this j, $|\lambda_j - z| < \varepsilon N \|\mathbf{P}_j\| = \varepsilon N \kappa(\lambda_j)$ as required. ∎

The more familiar version of the Bauer–Fike theorem (Theorem 2.3) follows this same pattern but with $N\kappa(\lambda_j)$ replaced by the condition number of the eigenvector matrix, $\kappa(\mathbf{V}) = \|\mathbf{V}\|\|\mathbf{V}^{-1}\|$, so that the inclusion disks for all the eigenvalues have the same radius. Theorem 52.2 can be more descriptive, for example, when there are well-conditioned eigenvalues separated from the rest of the spectrum.

Theorem 52.2 trades the $\mathcal{O}(\varepsilon^2)$ term of Theorem 52.1 for a potentially much larger constant. To see how these regions compare for a specific example, consider

$$\mathbf{A} = \begin{pmatrix} -1 & 6 & 0 \\ 0 & i & 8 \\ 0 & 0 & \frac{1}{2} \end{pmatrix}. \tag{52.7}$$

Figure 52.1 illustrates $\sigma_\varepsilon(\mathbf{A})$ along with the region Ω_ε and the inclusion region from Theorem 52.2. As expected, the sets Ω_ε become increasingly descriptive as ε gets smaller, though they need never be inclusion sets since they neglect $\mathcal{O}(\varepsilon^2)$ terms. For larger matrices the multiplicative factor n is more significant, and the bounds of Theorems 52.1 and 52.2 differ more acutely.

What if we relax the requirement that $\mathbf{A}$ have distinct eigenvalues? When $\mathbf{A}$ has repeated eigenvalues but remains diagonalizable (i.e., the algebraic multiplicity of each eigenvalue matches the geometric multiplicity), the theory outlined above carries through with only minor modifications. For example, in Theorem 52.2, $\kappa(\lambda_j)$ must be replaced by $\|\mathbf{P}_j\|$, where $\mathbf{P}_j$ is the spectral projector onto the entire eigenspace associated with λ_j, and the factor N can be replaced by the number of distinct eigenvalues.

On the other hand, if the algebraic multiplicity a_j of λ_j exceeds the geo-

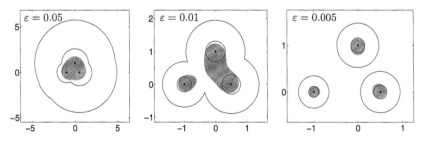

Figure 52.1: Comparison of the regions described by Theorems 52.1 (excluding $\mathcal{O}(\varepsilon^2)$ term) and 52.2 for the matrix (52.7). The gray sets are $\sigma_\varepsilon(\mathbf{A})$, the inner black lines are the boundaries of Ω_ε, and the outer black lines are the boundaries of the Bauer–Fike inclusion regions of Theorem 52.2.

metric multiplicity g_j, then λ_j is *defective* and $\mathbf{A}$ is not diagonalizable. This significantly changes the asymptotic picture, for now, in contrast to (52.3), λ_j can respond to a perturbation in a nonlinear fashion. To describe this behavior, we shall use the Jordan structure of the matrix $\mathbf{A}$ outlined in the last section. There exists some $\mathbf{V}_j \in \mathbb{C}^{N \times a_j}$ such that $\mathbf{A}\mathbf{V}_j = \mathbf{V}_j\mathbf{J}_j$ as in (51.5), where $\mathbf{J}_j$ is a direct sum of Jordan blocks $\mathbf{J}_{j,\ell}$. Similarly one can find a $\mathbf{U}_j \in \mathbb{C}^{N \times a_j}$ such that $\mathbf{U}_j^*\mathbf{A} = \mathbf{J}_j\mathbf{U}_j^*$ and $\mathbf{U}_j^*\mathbf{V}_j = \mathbf{I} \in \mathbb{C}^{a_j \times a_j}$. The spectral projector $\mathbf{P}_j$ for λ_j is then

$$\mathbf{P}_j = \mathbf{V}_j\mathbf{U}_j^*.$$

The contour integral formula (52.5) also holds in this case. Suppose λ_j is a defective eigenvalue associated with the Jordan blocks $\{\mathbf{J}_{j,\ell}\}_{\ell=1}^{g_j}$. The characteristic polynomial $\phi(\mathbf{A}) = \det(\lambda - \mathbf{A})$ has λ_j as a root of multiplicity a_j. If there are several Jordan blocks associated with λ_j (i.e., $g_j > 1$), each block will potentially split into a separate set of distinct eigenvalues, and the manner in which this occurs depends upon the dimension of each block. The eigenvalues associated with a block of dimension d can split into d distinct but related eigenvalues according to the formula

$$\lambda_{j,h}(t) = \lambda_j + \alpha_1\omega^h t^{1/d} + \alpha_2\omega^{2h}t^{2/d} + \cdots, \qquad h = 0, \ldots, d-1, \quad (52.8)$$

called a *Puiseux series*, where $\omega^h = e^{2h\pi i/d}$ is a principal dth root of unity; see Baumgärtel [33, §A1] or Kato [448, p. 65]. The size of the largest Jordan block associated with λ_j is called the *index* of λ_j, denoted by k_j. The formula (52.8) shows that $\mathbf{A} + t\mathbf{E}$ can have a cluster of eigenvalues at a distance $\mathcal{O}(|t|^{1/k_j})$ from λ_j as $t \to 0$; compare Theorem 16.6. The role of the Jordan block dimension is illustrated in Figures 52.2 and 52.3.

Elaborate formulas specify the values of the coefficients α_q in (52.8), depending on $\mathbf{E}$, the Jordan structure of $\mathbf{A}$ as characterized by $\mathbf{J}_j$, and the matrices $\mathbf{V}_j$ and $\mathbf{U}_j$ that reveal that structure. This perturbation series was studied in the 1960s by Lidskii [502], whose analysis is elegantly described

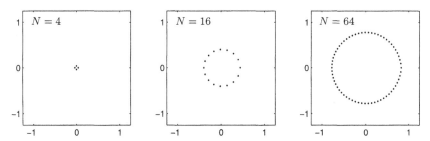

Figure 52.2: Eigenvalues of $\mathbf{A} + t\mathbf{E}$ for three dimensions N, with $\varepsilon = 10^{-5}$ and $\mathbf{E}$ having all its components equal to $1/N$ (so that $\|\mathbf{E}\| = 1$). $\mathbf{A}$ is a degree-N Jordan block, zero everywhere except for ones on the first superdiagonal.

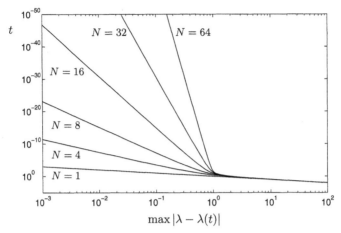

Figure 52.3: Maximum movement of the eigenvalues of $\mathbf{A} + t\mathbf{E}$ as t varies for several values of N. The matrices $\mathbf{A}$ and $\mathbf{E}$ are the same as in Figure 52.2. The unusual orientation of the axes will facilitate comparison with Figure 52.5.

by Moro, Burke, and Overton [561]. When $\mathbf{A}$ has only one Jordan block associated with λ_j, the first coefficient is

$$\alpha_1 = \mathbf{u}_{j,k_j}^* \mathbf{E} \mathbf{v}_{j,1}, \tag{52.9}$$

where $\mathbf{v}_{j,1}$ and $\mathbf{u}_{j,k_j}$ are the first and last columns of $\mathbf{V}_j$ and $\mathbf{U}_j$, respectively; i.e., $\mathbf{v}_{j,1}$ and $\mathbf{u}_{j,k_j}$ are the right and left eigenvectors of $\mathbf{A}$ associated with λ_j. Compare this expression for α_1 to the coefficient of $|t|$ (52.2), which equals $\mathbf{u}_1^* \mathbf{E} \mathbf{v}_1$ with the analogous normalization $\mathbf{u}_1^* \mathbf{v}_1 = 1$. (For the examples in Figures 52.2 and 52.3, the formula (52.9) implies that $\alpha_1 = 1/N$, since $\mathbf{V}_j = \mathbf{U}_j = \mathbf{I}$.)

Random perturbations will normally induce perturbations of $\mathcal{O}(|t|^{k_j})$, but structured perturbations can yield other effects. For example, if

$$\mathbf{A} = \begin{pmatrix} 0 & 1 & 0 \\ 0 & 0 & 1 \\ 0 & 0 & 0 \end{pmatrix}, \quad \mathbf{E}_1 = \begin{pmatrix} 0 & 0 & 1 \\ 0 & 0 & 0 \\ 0 & 0 & 0 \end{pmatrix}, \quad \mathbf{E}_2 = \begin{pmatrix} 0 & 0 & 0 \\ 1 & 0 & 0 \\ 0 & 0 & 0 \end{pmatrix}, \quad \mathbf{E}_3 = \begin{pmatrix} 0 & 0 & 0 \\ 0 & 0 & 0 \\ 1 & 0 & 0 \end{pmatrix},$$

then $\mathbf{A} + t\mathbf{E}_1$ has the same eigenvalues as $\mathbf{A}$, $\mathbf{A} + t\mathbf{E}_2$ has two eigenvalues $\mathcal{O}(|t|^2)$ from $\lambda = 0$, and $\mathbf{A} + t\mathbf{E}_3$ has all three eigenvalues $\mathcal{O}(|t|^3)$ from $\lambda = 0$ as $t \to 0$.

The series (52.8) indicates that small perturbations can cause Jordan blocks to split into clusters, as observed in Figure 52.2. Figure 52.4 illustrates similar splitting, but for a matrix consisting of three different Jordan blocks, each associated with a different eigenvalue. The block diagonal matrix $\mathbf{A} \in \mathbb{C}^{14 \times 14}$ is defined as the direct sum of three Jordan blocks: one of dimension 3 with $\lambda_1 = -1/2$, one of dimension 7 with $\lambda_2 = 0$, and one of

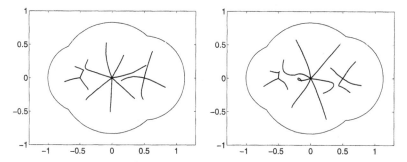

Figure 52.4: Eigenvalue trajectories of $\mathbf{A} + t\mathbf{E}$ for two complex random perturbations $\mathbf{E}$ to a matrix $\mathbf{A}$ with three Jordan blocks of differing dimension. The perturbation matrix is scaled so that $\|\mathbf{E}\| = 1$; t takes real values in $[0, 10^{-1}]$. The exterior curve denotes the boundary of the 10^{-1}-pseudospectrum of $\mathbf{A}$, a boundary for all these perturbed eigenvalues.

dimension 4 with $\lambda_3 = 1/2$. The figure shows the eigenvalue trajectories for $\mathbf{A} + t\mathbf{E}$ as t varies for two random complex perturbations $\mathbf{E}$. As t grows in magnitude, the series (52.8) becomes less descriptive, and the interactions between the individual eigenvalues become apparent. For similar examples, see [142, 692].

Since infinitesimal perturbations can alter the Jordan structure of a matrix discontinuously, it is not typically feasible to determine this canonical form numerically [328]. One approach to finding this structure, suggested by Chaitin-Chatelin and her coauthors (see, e.g., [142]), involves estimating the dimension of the Jordan blocks by examining the eigenvalues of small random perturbations in light of the growth dictated by the series (52.8). In applications, the actual structure of the Jordan form may be less relevant than the behavior of a cluster of eigenvalues, which could consist of any combination of simple or repeated eigenvalues. Kato shows that a perturbation of norm ε moves the *arithmetic mean* $\widehat{\lambda}$ of any given cluster according to

$$|\widehat{\lambda} - \widehat{\lambda}(t)| = t\|\mathbf{P}\| + \mathcal{O}(|t|^2),$$

where $\mathbf{P}$ is the spectral projector onto the invariant subspace associated with the cluster of eigenvalues [448, §2.2]. (This projector $\mathbf{P}$ is simply the sum of the spectral projectors associated with the individual eigenvalues that constitute the cluster.) Since $\mathbf{P}$ is a projector, $\|\mathbf{P}\| \geq 1$, with equality implying that $\mathbf{P}$ is an orthogonal projector, i.e., that taken together, the cluster is perfectly conditioned, though individual eigenvalues within it may be highly ill-conditioned. Thus $\|\mathbf{P}\|$ is said to be the condition number of the cluster of eigenvalues; see [24, §4.2], [442, §II]. (The conditioning of the invariant subspace associated with a cluster depends additionally on the separation of those eigenvalues from the rest of the spectrum, a separation conventionally measured by a positive real number known as 'sep'. This quantity, introduced by Stewart [723, 724] and studied by Varah [802], takes account of nonnormality.)

The formulas described by the expressions (52.3) and (52.8) naturally have an analogue in terms of the resolvent norm.

Asymptotic formula for the resolvent norm

Theorem 52.3 *Let* $\lambda_j \in \sigma(\mathbf{A})$ *be an eigenvalue of index* k_j. *Then there exist constants* $d_j > 0$ *and* $C_j > 0$ *such that*

$$\|(z - \mathbf{A})^{-1}\| \le C_j |z - \lambda_j|^{-k_j}$$

for all z *satisfying* $|z - \lambda_j| \le d_j$. *If* λ_j *is a simple eigenvalue, then the infimum of possible values for* C_j *is* $\kappa(\lambda_j)$.

Proof. Suppose $\mathbf{A} \in \mathbb{C}^{n \times n}$ is a matrix with $L \le N$ distinct eigenvalues, so that for each $j = 1, \ldots, L$, we have

$$\mathbf{A} \mathbf{V}_j = \mathbf{V}_j \mathbf{J}_j, \qquad \mathbf{U}_j^* \mathbf{A} = \mathbf{J}_j \mathbf{U}_j^*$$

for $\mathbf{U}_j^* \mathbf{V}_j = \mathbf{I}$ and $\mathbf{J}_j$ in Jordan form as described in §51. Let

$$\mathbf{V} = [\mathbf{V}_1 \ \mathbf{V}_2 \ \cdots \ \mathbf{V}_L], \ \mathbf{U} = [\mathbf{U}_1 \ \mathbf{U}_2 \ \cdots \ \mathbf{U}_L], \ \mathbf{J} = \mathrm{diag}(\mathbf{J}_1, \mathbf{J}_2, \ldots, \mathbf{J}_L).$$

Note that $\mathbf{V}^{-1} = \mathbf{U}^*$ and

$$\mathbf{A} = \mathbf{V} \mathbf{J} \mathbf{V}^{-1} = \sum_{j=1}^{L} \mathbf{V}_j \mathbf{J}_j \mathbf{U}_j^*.$$

Now consider the resolvent at $z \notin \sigma(\mathbf{A})$:

$$(z - \mathbf{A})^{-1} = (z - \mathbf{V} \mathbf{J} \mathbf{V}^{-1})^{-1}$$

$$= \mathbf{V}(z - \mathbf{J})^{-1} \mathbf{V}^{-1} = \sum_{j=1}^{L} \mathbf{V}_j (z - \mathbf{J}_j)^{-1} \mathbf{U}_j^*. \quad (52.10)$$

The matrix $\mathbf{J}_j$ is block diagonal with submatrices of the form (51.6) on the diagonal, the largest of which has dimension k_j. We can write $z - \mathbf{J}_j = (z - \lambda_j) - \mathbf{D}_j$, where $\mathbf{D}_j$ is a matrix that is zero everywhere except perhaps on the first superdiagonal. Expanding $(z - \mathbf{J}_j)^{-1}$ in a series, we obtain

$$(z - \mathbf{J}_j)^{-1} = \left((z - \lambda_j) - \mathbf{D}_j \right)^{-1}$$

$$= (z - \lambda_j)^{-1} \mathbf{I} + (z - \lambda_j)^{-2} \mathbf{D}_j + (z - \lambda_j)^{-3} \mathbf{D}_j^2$$

$$+ \cdots + (z - \lambda_j)^{-k_j} \mathbf{D}_j^{k_j - 1}.$$

The series terminates since $\mathbf{D}_j^\ell = \mathbf{0}$ for $\ell \geq k_j$. Using this formula for the resolvent of $\mathbf{J}_j$, we can expand (52.10) as

$$(z - \mathbf{A})^{-1} = \sum_{j=1}^{L} \left(\mathbf{V}_j \left[(z - \lambda_j)^{-1} \mathbf{I} + \sum_{\ell=1}^{k_j - 1} (z - \lambda_j)^{-(\ell+1)} \mathbf{D}_j^\ell \right] \mathbf{U}_j^* \right);$$

see Kato [448, p. 40]. As $z \to \lambda_j$,

$$\|(z - \mathbf{A})^{-1}\| = \|\mathbf{V}_j \mathbf{D}_j^{k_j - 1} \mathbf{U}_j^*\| |z - \lambda_j|^{-k_j} + \mathcal{O}(|z - \lambda_j|^{-k_j + 1}), \quad (52.11)$$

from which the main result follows. When λ_j is simple, the only associated term in the sum for $(z - \mathbf{A})^{-1}$ is $\mathbf{V}_j (z - \lambda_j)^{-1} \mathbf{U}_j^*$, and $\|\mathbf{V}_j (z - \lambda_j) \mathbf{U}_j^*\| = \|\mathbf{V}_j \mathbf{U}_j^*\| |z - \lambda_j| = \|\mathbf{P}_j\| |z - \lambda_j| = \kappa(\lambda_j) |z - \lambda_j|$. ∎

Notice that if $\mathbf{J}_j$ consists of a single Jordan block ($g_j = 1$), then $\mathbf{D}_j^{k_j - 1} \in \mathbb{C}^{k_j \times k_j}$ is zero everywhere except in the $(1, k_j)$ entry, so that

$$\|\mathbf{V}_j \mathbf{D}_j^{k_j - 1} \mathbf{U}_j^*\| = \|\mathbf{v}_{j,1} \mathbf{u}_{j,k_j}^*\| = \|\mathbf{v}_{j,1}\| \|\mathbf{u}_{j,k_j}\|,$$

where, as before, $\mathbf{v}_{j,1}$ and $\mathbf{u}_{j,k_j}$ are the first and last columns of $\mathbf{V}_j$ and $\mathbf{U}_j$. Compare this formula for the leading coefficient of series (52.11) with the corresponding coefficient $\alpha_1 = \mathbf{u}_{j,k_j}^* \mathbf{E} \mathbf{v}_{j,1}$ in the expansion (52.8) for $\lambda_{j,h}(t)$. Under the same $g_j = 1$ assumption,

$$|\alpha_1| = |\mathbf{u}_{j,k_j}^* \mathbf{E} \mathbf{v}_{j,1}| \leq \|\mathbf{v}_{j,1}\| \|\mathbf{u}_{j,k_j}\|,$$

with equality being attained for $\mathbf{E} = \mathbf{u}_{j,k_j} \mathbf{v}_{j,1}^* / \|\mathbf{u}_{j,k_j}\| \|\mathbf{v}_{j,1}\|$.

We now investigate several computational examples, analogues to the eigenvalue perturbation experiments earlier in this section. First, consider the N-dimensional Jordan block whose perturbed eigenvalues were examined in Figures 52.2 and 52.3. As $z \to \lambda = 0$, Theorem 52.3 asserts that the resolvent norm will grow like $|z|^{-N}$, which is confirmed in Figure 52.5. Figure 52.6 illustrates the pseudospectra of this Jordan block for three values of N. Near the eigenvalue, the pseudospectra differ significantly depending on the matrix dimension, though far from the eigenvalue they look very similar. This emphasizes that the results in this section are asymptotic and can potentially be misleading in applications where the behavior is not controlled by a single eigenvalue.

These examples illustrate a somewhat ideal situation, since the matrix is already in Jordan form, and thus the matrices $\mathbf{V}_j$ and $\mathbf{U}_j$ used in the analysis above are trivial. Now we complicate the situation by introducing

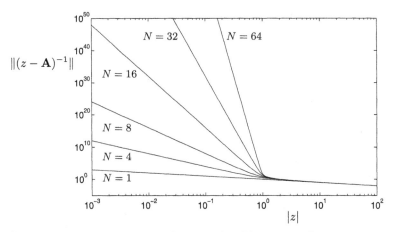

Figure 52.5: Resolvent norm for a Jordan block as in Figures 52.2–52.3 as a function of the distance from the eigenvalue $\lambda = 0$, for various dimensions N. The rate of growth of the resolvent norm depends on the block dimension as $|z| \to 0$. Note the agreement with Figure 52.3.

a nonzero constant on the second superdiagonal of the matrix,

$$\mathbf{A} = \begin{pmatrix} 0 & 1 & \gamma & & & \\ & 0 & 1 & \ddots & & \\ & & 0 & \ddots & \gamma & \\ & & & \ddots & 1 & \\ & & & & 0 \end{pmatrix} \in \mathbb{C}^{N \times N}; \qquad (52.12)$$

see Figure 7.5. Even for moderate values of γ, this matrix differs considerably from the Jordan block studied previously ($\gamma = 0$). Though the Jordan factor of $\mathbf{A}$ is independent of γ, the condition number of the similarity transformation that brings $\mathbf{A}$ into that Jordan form increases exponentially with γ. Since there is only a single Jordan block, it is easy to compute the constants for the leading terms in the expansions for the perturbed eigenvalues and the resolvent norm, which, as seen above, are necessarily the same. The right and left eigenvectors of $\mathbf{A}$ take the form $\mathbf{v}_{1,1} = (1, 0, \ldots, 0)^{\mathrm{T}}$ and $\mathbf{u}_{1,N} = (0, \ldots, 0, 1)^{\mathrm{T}}$, and thus $\|\mathbf{v}_{j,1}\| \|\mathbf{u}_{j,k_j}^*\| = 1$. As $|z| \to 0$, the resolvent norm must behave like that of the Jordan block of the same dimension, regardless of γ. This fact is observed in Figures 52.7 and 52.8, which also illustrate that asymptotic analysis does not reveal all properties of the resolvent norm, even for a matrix with a single eigenvalue. Near $\lambda = 0$, the pseudospectra are largely independent of γ, while at intermediate distances the influence of γ is significant. This distinction has important implications for the behavior of matrix functions, such as $\mathbf{A}^k$ for $0 < k < N$, as confirmed by the upper and lower bounds presented in §16.

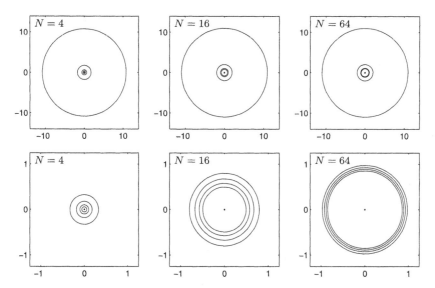

Figure 52.6: Spectrum and ε-pseudospectra for Jordan blocks as in the previous figure, for various N and $\varepsilon = 10^{-2}$, 10^{-1}, 10^0, 10^1 (top) and $\varepsilon = 10^{-5}$, 10^{-4}, 10^{-3}, 10^{-2} (bottom). For large ε, the pseudospectra depend only mildly on the block dimension, while for small ε, the block size is crucial.

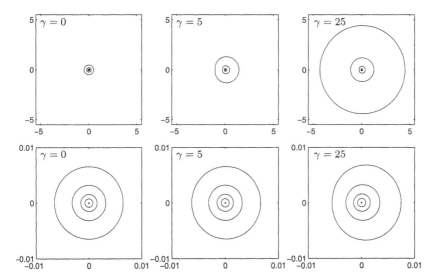

Figure 52.7: Spectrum and ε-pseudospectra for three matrices (52.12) of dimension $N = 16$ with $\varepsilon = 10^{-5}$, 10^{-10}, 10^{-15}, 10^{-20} (top) and $\varepsilon = 10^{-35}$, 10^{-40}, 10^{-45}, 10^{-50} (bottom). The pseudospectra on the top vary significantly with γ, whereas they agree more closely for the smaller values of ε shown on the bottom.

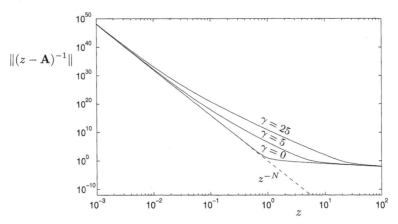

Figure 52.8: Resolvent norm for the matrix (52.12) as a function of the distance from the eigenvalue $\lambda = 0$, with $N = 16$ and various values of γ. As $|z| \to 0$, the Jordan structure alone determines the leading-order resolvent norm behavior: $\|(z - \mathbf{A})^{-1}\| \sim |z|^{-N}$ for all γ.

Perturbation theory tells us that as $\varepsilon \to 0$, the ε-pseudospectrum looks like the union of circular disks. The radii of these disks depend upon the conditioning of the eigenvalue problem and the size of the Jordan blocks associated with each eigenvalue. While this local information can be informative in some circumstances, in many applications one is concerned with nonnormality manifested at points in the complex plane far from the spectrum. In such situations, bounds derived from asymptotic perturbation theory can lack sufficient accuracy.

Throughout this book, we have seen examples of highly nonnormal matrices for which small perturbations to the matrix coefficients significantly move the spectrum. The utility of pseudospectra would be seriously compromised if they exhibited the same sensitivities. The following simple result shows they do not.

Perturbation of pseudospectra

Theorem 52.4 *For all* $\mathbf{A}, \mathbf{E} \in \mathbb{C}^{N \times N}$ *and* $\varepsilon > \|\mathbf{E}\|$,

$$\sigma_{\varepsilon - \|\mathbf{E}\|}(\mathbf{A}) \subseteq \sigma_\varepsilon(\mathbf{A} + \mathbf{E}) \subseteq \sigma_{\varepsilon + \|\mathbf{E}\|}(\mathbf{A}).$$

Proof. Suppose $z \in \sigma_{\varepsilon - \|\mathbf{E}\|}(\mathbf{A})$. Then there exists $\mathbf{F}$ with $\|\mathbf{F}\| < \varepsilon - \|\mathbf{E}\|$ such that $z \in \sigma(\mathbf{A} + \mathbf{F}) = \sigma((\mathbf{A} + \mathbf{E}) + (\mathbf{F} - \mathbf{E}))$. To obtain the first inclusion, note that $\|\mathbf{F} - \mathbf{E}\| \leq \|\mathbf{F}\| + \|\mathbf{E}\| < \varepsilon$. The second inclusion follows by swapping the roles of $\mathbf{A}$ and $\mathbf{A} + \mathbf{E}$ in the first. ∎

53 · Backward error analysis

In the last section we studied, from a theoretical perspective, the effect of a small perturbation on the eigenvalues of a matrix $\mathbf{A} \in \mathbb{C}^{N \times N}$. Here we address the practical problem of calculating eigenvalues on a computer. Such finite precision computations effectively perturb the entries of $\mathbf{A}$, leading to a natural connection to pseudospectra. We come to this topic rather late in this book, as our emphasis has been on matrix behavior instead of rounding errors. However, much of the early research on pseudospectra was driven by the goal of understanding the behavior of eigenvalue algorithms for non-Hermitian matrices, with contributions from Varah, Godunov's group at Novosibirsk, and Chaitin-Chatelin and her colleagues, as described in §6.

The study of the interaction between linear algebra algorithms and finite precision computers began in the 1940s and 1950s with the analysis of the stability of Gaussian elimination for solving $\mathbf{A}\mathbf{x} = \mathbf{b}$. The iterative methods for solving such systems discussed in §§24 and 25 are robust to errors in the sense that an arithmetic mistake made by a computer can be corrected by additional iterations. How does Gaussian elimination, a one-pass process, respond to similar errors? The answer to this question applies not only to a single procedure for solving linear systems, but throughout numerical analysis: there is a critical distinction between the *stability* of an algorithm and the *conditioning* of the problem the algorithm is asked to solve. Some problems are inherently sensitive to small changes in their data, and in such cases it is unrealistic to expect any finite precision algorithm, be it direct or iterative, to produce accurate answers. On the other hand, one can objectively evaluate algorithms based on how they perform when applied to insensitive problems.

One might hope that the computed solution $\widehat{\mathbf{x}}$ to the linear system $\mathbf{A}\mathbf{x} = \mathbf{b}$ would yield a small *forward error* $\|\mathbf{x} - \widehat{\mathbf{x}}\|$. For many problems, this goal is difficult to attain. A more realistic objective is that $\widehat{\mathbf{x}}$ be the exact solution to a nearby problem, e.g., $(\mathbf{A}+\mathbf{E})\widehat{\mathbf{x}} = \mathbf{b}$, where $\|\mathbf{E}\|$ is small relative to $\|\mathbf{A}\|$. The norm of the perturbation $\mathbf{E}$ is the *backward error*, and this style of *backward error analysis* was championed by Wilkinson and is a fundamental part of numerical linear algebra.

An algorithm is said to be *backward stable* if, for any problem (that fits in the computer's arithmetic) it produces solutions with a backward error that is small relative to the size of the data. The forward error, then, can be approximated by multiplying the backward error by the *condition number* of the problem, which measures the sensitivity of that problem to small changes in the data. Continuing with the linear system $\mathbf{A}\mathbf{x} = \mathbf{b}$, suppose that $(\mathbf{A} + \delta\mathbf{A})(\mathbf{x} + \delta\mathbf{x}) = \mathbf{b}$ for infinitesimal $\delta\mathbf{A}$ and $\delta\mathbf{x}$, i.e., that

$\mathbf{x} + \delta\mathbf{x}$ is the exact solution to an infinitesimally perturbed problem. By combining these equations and discarding second-order terms, one obtains

$$\frac{\|\delta\mathbf{x}\|}{\|\mathbf{x}\|} \leq \|\mathbf{A}\|\|\mathbf{A}^{-1}\|\frac{\|\delta\mathbf{A}\|}{\|\mathbf{A}\|}.$$

This formula shows that a change in $\mathbf{A}$ can alter $\mathbf{x}$ by a factor $\|\mathbf{A}\|\|\mathbf{A}^{-1}\|$ times larger. This magnification term is the condition number for the problem of solving $\mathbf{A}\mathbf{x} = \mathbf{b}$, which has appeared in various contexts throughout this book. There is an immediate connection to the distance to singularity addressed in §49: If $\mathbf{A}$ is normalized so that $\|\mathbf{A}\| = 1$, then the condition number is precisely the distance to singularity. The closer a matrix is to singular, the greater the forward error we can expect, if the solution is computed with a backward stable algorithm.

Backward error analysis has its roots in the work of Turing [785], and Wilkinson discovered that these ideas apply throughout numerical linear algebra. His work is summarized in the books *Rounding Errors in Algebraic Processes* [826] and *The Algebraic Eigenvalue Problem* [827]. For a survey and historical information, see the survey article [828] and Higham's treatise [391]. One fruit of these efforts of Wilkinson and his colleagues was a series of *Numerische Mathematik* articles that provided Algol codes and supporting documentation for a wide variety of matrix computations. These papers were collected in the *Handbook for Automatic Computation* volume on linear algebra [831], published in 1971, and led to the landmark EISPACK and LINPACK software libraries over the course of the next decade.

Pseudospectra can be used to gain insight into the behavior of numerous finite precision matrix computations. For example, Higham and Knight investigate matrix powers; see [392, 728]. Here we shall focus on the performance of methods for calculating eigenvalues. The method of choice for finding all the eigenvalues of a dense, non-Hermitian matrix is the QR algorithm, proposed independently in 1961 by Francis and Kublanovskaya [286, 287, 471]. This method applies a series of unitary similarity transformations to reduce $\mathbf{A}$ to upper triangular form, effectively computing a Schur factorization. The practical QR method does this in two steps. First, $\mathbf{A}$ is reduced to an upper Hessenberg matrix, which is accomplished in $\mathcal{O}(N^3)$ operations. The second phase of the algorithm uses further unitary similarity transformations to eliminate the remaining subdiagonal entries. (In principle, it is impossible to remove these entries entirely, as this would imply the ability to factor degree N polynomials exactly. However, they can be made arbitrarily small.) For a general overview, see, e.g., [199, 327, 776]; some interesting recent developments are described in [105, 106].

To quantify the behavior of this procedure in finite precision arithmetic, we must describe the floating-point number system in which the calculations are performed. The design of such systems and the accompanying basic arithmetic operations is an art. For our purposes we shall assume that fundamental properties hold, so that the quality of the floating-point system can be quantified by a single number, *machine epsilon*, written ε_{mach}, which equals the separation between 1 and the next larger number in the floating-point system [391, 592].

Suppose that $\mathbf{A}$ has real entries. In general, the entries in $\mathbf{A}$ will not be exactly represented in the floating-point system, so the matrix $\mathbf{A} + \mathbf{E}$ is actually stored, where $|e_{jk}| \leq \frac{1}{2}\varepsilon_{mach}|a_{jk}|$. Unitary similarity transformations do little to magnify these errors. Wilkinson demonstrated that the first phase of the QR algorithm computes an upper Hessenberg matrix $\mathbf{H}$ that is exactly unitarily similar to a matrix $\widehat{\mathbf{A}} = \mathbf{A} + \mathbf{E}$, where

$$\|\mathbf{E}\|_F \leq c_1 N^2 \varepsilon_{mach} \|\mathbf{A}\|_F,$$

where c_1 is a small constant; see [533], [327, §7.4], [827, §6.5]. Given an upper Hessenberg matrix $\mathbf{H}$, the second phase of the algorithm reduces the upper Hessenberg matrix $\mathbf{H}$ to a matrix $\mathbf{T}$ with 'real Schur form', meaning that $\mathbf{T}$ may have 2×2 blocks on the main diagonal corresponding to complex conjugate eigenvalues. This computed $\mathbf{T}$ is exactly unitarily similar to a matrix $\widehat{\mathbf{H}} = \mathbf{H} + \mathbf{E}$, where

$$\|\mathbf{E}\|_F \leq c_2 N p \varepsilon_{mach} \|\mathbf{H}\|_F,$$

where c_2 is a small constant and p is the number of similarity transformations the QR algorithm required to reach this upper triangular form [532, 831]; typically, $p \leq 2N$.

Combining these results, we expect the QR algorithm to return the exact eigenvalues of a matrix $\mathbf{A} + \mathbf{E}$ with $\|\mathbf{E}\|$ of order $\varepsilon_{mach}\|\mathbf{A}\|$; i.e., we roughly expect the computed eigenvalues to fall somewhere in $\sigma_\varepsilon(\mathbf{A})$ for $\varepsilon \approx \varepsilon_{mach}\|\mathbf{A}\|$. For this reason, our plots of pseudospectra later in this section show levels $\varepsilon/\|\mathbf{A}\|$. Some researchers go further, replacing ε in our definition of pseudospectra with $\varepsilon\|\mathbf{A}\|$ [317], [133, p. 164]; see the footnote on page 43.

We now turn to several computational examples. A classic test matrix for non-Hermitian eigenvalue algorithms was introduced by Werner Frank in 1958 [224, 288, 328, 803]. The Frank matrix is upper Hessenberg with subdiagonal entries $N - 1, N - 2, \ldots, 1$; on and above the main diagonal, $a_{jk} = N - k + 1$. For example, for $N = 5$,

$$\mathbf{A} = \begin{pmatrix} 5 & 4 & 3 & 2 & 1 \\ 4 & 4 & 3 & 2 & 1 \\ & 3 & 3 & 2 & 1 \\ & & 2 & 2 & 1 \\ & & & 1 & 1 \end{pmatrix}.$$

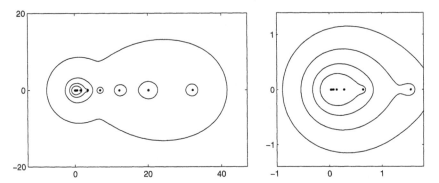

Figure 53.1: Spectra and ε-pseudospectra for the Frank matrix of dimension $N = 12$, with $\varepsilon/\|\mathbf{A}\| = 10^{-1}$, 10^{-2}, 10^{-3}, 10^{-4} (left) and a closeup with $\varepsilon/\|\mathbf{A}\| = 10^{-4}$, $10^{-5}, \ldots, 10^{-8}$ (right). In single precision arithmetic, the computed eigenvalues will fall near the innermost contour.

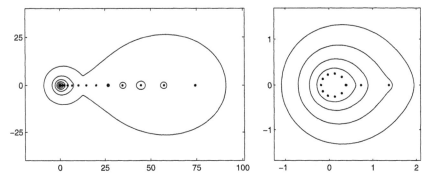

Figure 53.2: Numerically computed eigenvalues and ε-pseudospectra for the Frank matrix of dimension $N = 24$, with $\varepsilon/\|\mathbf{A}\| = 10^{-2}$, $10^{-4}, \ldots, 10^{-10}$ (left) and a closeup with $\varepsilon/\|\mathbf{A}\| = 10^{-10}$, 10^{-12}, 10^{-14}, 10^{-16} (right). The eigenvalues of smallest magnitude, which should all be real, are not computed accurately in double precision arithmetic. These computed values fall within the boundary of the ε-pseudospectrum for $\varepsilon = 10^{-16}\|\mathbf{A}\|$.

The eigenvalues of these matrices are always real and distinct and are related to the roots of Hermite polynomials [224, 803]. Hence $\mathbf{A}$ is similar to a real symmetric matrix, though the condition number of this similarity transformation grows exponentially with N. The rightmost eigenvalues are well-conditioned, while the leftmost eigenvalues are significantly ill-conditioned. The Frank matrix of dimension $N = 12$ became a popular test problem in the 1960s, because even for that dimension, the ill-conditioning of the smallest eigenvalues was enough to cause problems for eigenvalue algorithms in single precision arithmetic ($\varepsilon_{\text{mach}} \approx 10^{-7}$). This is explained by Figure 53.1, which shows $\sigma_\varepsilon(\mathbf{A})$ for $\varepsilon = 10^{-7}\|\mathbf{A}\|$. Though this figure, computed in double precision ($\varepsilon_{\text{mach}} \approx 2 \cdot 10^{-16}$), shows the eigenvalues correct to plotting accuracy, Figure 53.2 illustrates that by doubling the dimension to $N = 24$, we obtain a problem for which double precision computations in MATLAB produce eigenvalues with nonzero imaginary parts. Given the splitting of eigenvalues near the origin, one might suspect that $\mathbf{A}$ has a Jordan block of dimension 10 or 11 (cf. Figure 52.2), when, in fact, this matrix is diagonalizable. Examples like this highlight the difficulty of numerically determining the Jordan form [328].

Our second example is a 7×7 matrix with less apparent structure that was introduced by Godunov, again to test algorithms for computing eigenvalues; see, e.g., [315]:

$$\mathbf{A} = \begin{pmatrix} 289 & 2064 & 336 & 128 & 80 & 32 & 16 \\ 1152 & 30 & 1312 & 512 & 288 & 128 & 32 \\ -29 & -2000 & 756 & 384 & 1008 & 224 & 48 \\ 512 & 128 & 640 & 0 & 640 & 512 & 128 \\ 1053 & 2256 & -504 & -384 & -756 & 800 & 208 \\ -287 & -16 & 1712 & -128 & 1968 & -30 & 2032 \\ -2176 & -287 & -1565 & -512 & -541 & -1152 & -289 \end{pmatrix}. \quad (53.1)$$

A similarity transformation with the matrix

$$\mathbf{L} = \begin{pmatrix} 1 & 0 & 0 & 0 & 0 & 0 & 0 \\ 0 & 1 & 0 & 0 & 0 & 0 & 0 \\ 1 & 0 & 1 & 0 & 0 & 0 & 0 \\ 0 & 0 & 0 & 1 & 0 & 0 & 0 \\ 0 & 0 & 1 & 0 & 1 & 0 & 0 \\ 1 & 0 & 0 & 0 & 0 & 1 & 0 \\ 0 & 1 & 1 & 0 & 1 & 0 & 1 \end{pmatrix}$$

reduces $\mathbf{A}$ to upper triangular form:

$$\mathbf{LAL}^{-1} = \begin{pmatrix} 1 & 2048 & 256 & 128 & 64 & 32 & 16 \\ 0 & -2 & 1024 & 512 & 256 & 128 & 32 \\ 0 & 0 & 4 & 512 & 1024 & 256 & 64 \\ 0 & 0 & 0 & 0 & 512 & 512 & 128 \\ 0 & 0 & 0 & 0 & -4 & 1024 & 256 \\ 0 & 0 & 0 & 0 & 0 & 2 & 2048 \\ 0 & 0 & 0 & 0 & 0 & 0 & -1 \end{pmatrix},$$

and thus $\sigma(\mathbf{A}) = \{-4, -2, -1, 0, 1, 2, 4\}$. Clearly $\mathbf{LAL}^{-1}$ must be far from normal; since $\mathbf{L}$ has a small condition number, $\kappa(\mathbf{L}) \approx 5.9467$, Theorem 51.1 ensures that the pseudospectra of $\mathbf{A}$ will resemble those of $\mathbf{L}^{-1}\mathbf{AL}$. Figure 53.3 shows $\sigma_\varepsilon(\mathbf{A})$ and the computed eigenvalues. As with the $N = 24$ Frank matrix, in double precision arithmetic there are

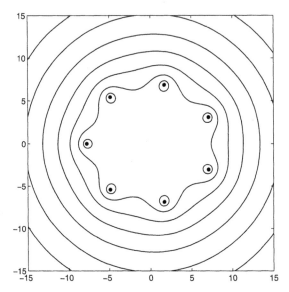

Figure 53.3: Numerically computed eigenvalues and ε-pseudospectra for the Godunov matrix (53.1) with $\varepsilon/\|\mathbf{A}\| = 10^{-16.5}, 10^{-16}, \ldots, 10^{-13.5}$. The exact eigenvalues are $\sigma(\mathbf{A}) = \{-4, -2, -1, 0, 1, 2, 4\}$.

computed eigenvalues that are far from the real line. The ε-pseudospectra computed by EigTool do not even include the true eigenvalues for the value $\varepsilon = 10^{-16.5}\|\mathbf{A}\| < \varepsilon_{\mathrm{mach}}\|\mathbf{A}\|$.[1]

Many large-scale applications require knowledge of only a small subset of the spectrum, and for such problems the iterative methods discussed in §28 are more suitable than the QR algorithm. Suppose that an iterative method like the power or Arnoldi algorithms returns an eigenpair $(\widehat{\lambda}, \widehat{\mathbf{x}})$. What can be said of its accuracy? Define the residual vector $\mathbf{r} = \mathbf{A}\widehat{\mathbf{x}} - \widehat{\lambda}\widehat{\mathbf{x}}$. Then $(\widehat{\lambda}, \widehat{\mathbf{x}})$ is an exact eigenpair of $\mathbf{A} + \mathbf{E}$ for

$$\mathbf{E} = -\frac{\mathbf{r}\widehat{\mathbf{x}}^*}{\|\widehat{\mathbf{x}}\|^2},$$

since

$$(\mathbf{A} + \mathbf{E})\widehat{\mathbf{x}} = \mathbf{A}\widehat{\mathbf{x}} - \mathbf{r} = \widehat{\lambda}\widehat{\mathbf{x}};$$

see [827, 727]. Since $\|\mathbf{E}\| = \|\mathbf{r}\|$, a small residual norm implies a small backward error in the eigenpair. For more sophisticated perturbation analysis

---

[1] As described in §39, before computing pseudospectra EigTool reduces $\mathbf{A}$ to upper triangular form, which amounts to performing the QR iteration on $\mathbf{A}$. This process introduces the rounding errors seen in Figure 53.3. The second stage of the Core EigTool Algorithm is applied not to $\mathbf{A}$, but to this computed upper triangular matrix, which is a modest perturbation of $\mathbf{A}$. From Theorem 52.4, we expect good agreement between the pseudospectra of these matrices, up to the level of the perturbation.

for invariant subspaces, see [724, 729].

Our brief discussion of backward error analysis has omitted many significant topics. For example, when $\mathbf{A}$ possesses special structure, one may wish that the perturbed matrix $\mathbf{A} + \mathbf{E}$ does, too. Similarly, if the entries of $\mathbf{A}$ differ significantly in size, it may be appropriate for the entries of $\mathbf{E}$ to be scaled accordingly. Such situations give rise to the related notions of *structured* and *componentwise backward error*, which are related to the structured pseudospectra discussed in §50. Chaitin-Chatelin and Frayssé propose a *componentwise ε-pseudospectrum* [133, p. 177].

In other cases, $\mathbf{A}$ is endowed with structure that can be exploited by specialized eigensolvers. *Hamiltonian matrices* are such an example; they arise in many control theory applications and play a critical role in the algorithms for computing pseudospectral abscissae and radii described in §42. Define

$$\mathbf{J} = \begin{pmatrix} \mathbf{0} & \mathbf{I} \\ -\mathbf{I} & \mathbf{0} \end{pmatrix}.$$

A matrix $\mathbf{H} \in \mathbb{C}^{2N \times 2N}$ is Hamiltonian provided that $\mathbf{JH}$ is Hermitian. Equivalently, $\mathbf{H}$ must have the form

$$\mathbf{H} = \begin{pmatrix} \mathbf{A} & \mathbf{B} \\ \mathbf{C} & -\mathbf{A}^* \end{pmatrix}, \qquad \mathbf{B} = \mathbf{B}^*, \ \mathbf{C} = \mathbf{C}^* \qquad (53.2)$$

for $\mathbf{A}, \mathbf{B}, \mathbf{C} \in \mathbb{C}^{N \times N}$. Hamiltonian matrices have beautiful spectral properties [483, 593]. If $\lambda \in \sigma(\mathbf{H})$, then $-\lambda \in \sigma(\mathbf{H})$, too. Thus the spectrum of real Hamiltonian matrices has a four-fold symmetry, as $\lambda \in \sigma(\mathbf{H})$ implies $\{\pm\lambda, \pm\bar{\lambda}\} \subseteq \sigma(\mathbf{H})$. When the standard QR algorithm is applied to a Hamiltonian matrix, the structure is generally destroyed in the transformation to upper Hessenberg form, and the eigenvalues computed by the iterative phase of the QR algorithm need not obey the symmetry property. There are specialized backward stable algorithms that compute the eigenvalues of $\mathbf{H}$ while respecting the Hamiltonian structure [52].

This section of the book is the only one devoted to rounding errors, and the reason for this was mentioned in the Preface: The main effects of nonnormality, and the main uses of pseudospectra, have nothing to do with rounding errors. Nevertheless, we have been distressed to find over the years that many people who hear us talk about pseudospectra get the wrong message. Perhaps because we are numerical analysts, and numerical analysts are known to be experts in computer arithmetic, people all too readily assume that if we are advocating a tool that has some connection with perturbations, then the essence of the matter must be rounding errors. This is a mistake. Most effects associated with nonnormality have nothing to do with computer arithmetic, and the useful values of ε are typically much larger than $\varepsilon_{\mathrm{mach}}$.

54 · Group velocity and pseudospectra _____

The purpose of this section is to describe an analogy between group velocity and pseudospectra. Group velocity is an old idea, while pseudospectra are more recent, but the two have much in common. Both are tools of linear analysis in which the key point is that the 'physics' of a system lies not in the individual modes, but in their superposition. In both cases this behavior is counterintuitive and requires some getting used to.

The notion of group velocity belongs to the theory of dispersive waves, that is, waves whose speed varies with the wave number [505, 820]. In one dimension, suppose we have a linear PDE or other linear system that admits solutions of the form

$$u(x,t) = e^{i(\omega t - kx)}, \tag{54.1}$$

for any $k \in \mathbb{R}$, provided that $\omega \in \mathbb{R}$ is related to k by a fixed *dispersion relation*,

$$\omega = \omega(k). \tag{54.2}$$

Here k is known as the *wave number* and ω is the *frequency*. From (54.1) it is easily seen that individual wave crests travel at the *phase velocity*

$$c = \frac{\omega}{k}. \tag{54.3}$$

However, this is not the velocity of a wave packet constructed from a superposition of wave numbers (Figure 54.1). If the packet is composed of wave numbers close to k, it travels at the *group velocity*

$$C = \frac{d\omega}{dk}. \tag{54.4}$$

That is, to find the velocity of propagation we must differentiate the dispersion relation with respect to the wave number (Figure 54.2).[1]

For example, one of the simplest of dispersive partial differential equations is the free-space time-dependent Schrödinger equation

$$u_t = -iu_{xx}. \tag{54.5}$$

Inserting (54.1) in (54.5) gives the dispersion relation $i\omega = ik^2$, that is,

$$\omega = k^2.$$

_____
[1] In multiple space dimensions, (54.1)–(54.4) generalize to $u(\mathbf{x}, t) = e^{i(\omega t - \mathbf{k} \cdot \mathbf{x})}$, $\omega = \omega(\mathbf{k})$, $\mathbf{c} = \omega \mathbf{k}/\|\mathbf{k}\|_2^2$, $\mathbf{C} = \nabla_{\mathbf{k}} \omega$.

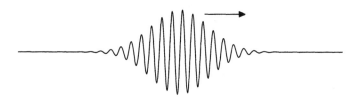

Figure 54.1: Individual wave crests travel at the phase velocity $c = \omega/k$, but a wave packet travels at the group velocity $C = d\omega/dk$.

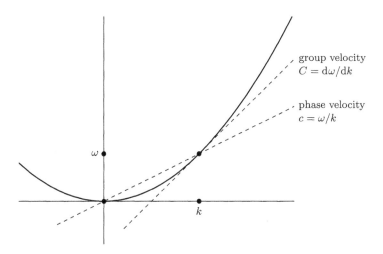

Figure 54.2: The group velocity for wave number k is obtained by differentiating the dispersion relation $\omega = \omega(k)$. This figure corresponds to the Schrödinger equation (54.5), with $|C| \geq |c|$. For water waves the dispersion relation curves downward and we have $|C| \leq |c|$.

From (54.3) and (54.4) we conclude that the phase and group velocities for the Schrödinger equation are

$$c = k, \qquad C = 2k.$$

There is a great deal of physics in these expressions, as was appreciated by Schrödinger early in 1926. We see that shorter waves (higher k) travel faster, and for any k, $C = 2c$. Thus an electron travels at twice the speed of the individual wave crests in its quantum state function [107].

Another example of a dispersive PDE is the *beam equation*,

$$u_{tt} = -u_{xxxx}, \tag{54.6}$$

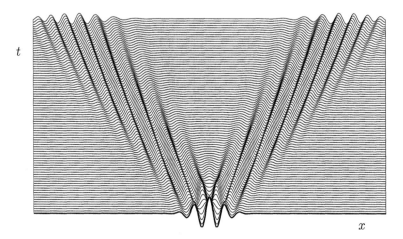

t

x

Figure 54.3: Left-going and right-going wave packets for the beam equation (54.6). Note that there is a well-defined wave number k at each point in space, that higher values of k (shorter wavelengths) move faster, and that each packet moves at twice the speed of the individual wave crests it is composed of. (Illustration adapted from a code by David Allwright.)

which governs vibrations in a solid bar. The dispersion relation is $\omega^2 = k^4$, or

$$\omega = \pm k^2,$$

suggesting that this equation is a kind of product of left-going and right-going Schrödinger equations; see Figure 54.3. This is indeed true, as is evident from the factorization $\partial_{tt}^2 + \partial_{xxxx}^4 = (\partial_t^2 + \mathrm{i}\partial_{xx}^2)(\partial_t^2 - \mathrm{i}\partial_{xx}^2)$.

The first hints of the idea of group velocity appeared in a paper of Hamilton in 1839, and the theory was initiated in earnest in the 1870s with the work of Stokes, Froude, Reynolds, and especially Rayleigh, whose *Theory of Sound* describes applications of group velocity in optics, fluids, and solids [617, 618]. Further contributions were made in the following decades by Kelvin, Lamb, Green, and others, and when the theory of special relativity appeared in 1905, with its absolute limit of the speed of light for all kinds of propagation of energy and information, group velocity became a subject of urgent interest. Sommerfeld, Brillouin, and others developed further necessary distinctions between group, energy, and signal velocities for dissipative media (i.e., media for which ω has a positive imaginary component when k is real and nonzero), and it was confirmed that there were no contradictions to the theory of relativity. Later, group velocity took on a new importance with the explosion of quantum mechanics in 1926 and the discovery that particles cannot be localized to points, only to wave packets. In the 1930s group velocity was also the focus of attention by experimentalists measuring the speed of light, for many such experiments depend on propagation of light in air, which is a slightly dispersive medium.

In this era, the measurements had begun to be accurate enough that it became important to make the corresponding group velocity correction, amounting to about one part in 10^5 or 3 km/sec.[2] By this time, group velocity was an established tool of physics and applied mathematics [75, 429, 472, 730], and when the Institute for Mathematics and Its Applications in Britain launched its *Journal* in 1965, the first paper in the first issue, by Lighthill, was a survey of group velocity [506].

Another famous example of dispersive wave propagation concerns waves in deep water, such as the waves that form when a stone is thrown into a pond. The dispersion relation for such waves turns out to be

$$\omega = \pm\sqrt{gk},$$

where g is the gravitational constant, which implies

$$c = \pm\sqrt{\frac{g}{k}}, \qquad C = \pm\frac{1}{2}\sqrt{\frac{g}{k}}.$$

Thus water waves are the opposite of Schrödinger wave packets: It is now the longer waves (lower k) that travel faster, with $C = c/2$ instead of $C = 2c$. The disturbance spreads outward from the rock at half the speed of the individual ripples, a fact that was known to Stokes, Rayleigh, and Reynolds.

The group velocity formula (54.4) can be derived in many ways, some with more mathematical power than others. Here are sketches of the four most common derivations. Stokes in 1876 noted that if two waves e^{ik_1x} and e^{ik_2x} are superimposed, the sum is a chain of 'beats' that moves with velocity $(\omega_2 - \omega_1)/(k_2 - k_1)$; in the limit $k_2 \to k_1$ we recover (54.4) (Figure 54.4). Rayleigh in 1877 considered perturbing k slightly to a value $k + \delta k$ with δk imaginary; the imaginary wave number provides shape in the envelope, revealing that the propagation speed is $\delta\omega/\delta k$, which again converges to (54.4) as $\delta k \to 0$ (Figure 54.5). Kelvin in 1887 introduced the method of stationary phase analysis of Fourier integrals: an initial wave packet $u(x, 0) = \int_{-\infty}^{\infty} \omega(k)\, e^{-ikx}\, dx$ evolves into

$$u(x, t) = \int_{-\infty}^{\infty} \omega(k)\, e^{it(\omega - ik(x/t))}\, dk,$$

and the rapidly oscillatory integral converges to zero as $t \to \infty$ along all lines $x/t = $ constant except if $x/t = d\omega/dk$. For dissipative media this stationary phase argument generalizes to a steepest descent analysis of a

---

[2]The speed of light in vacuo is 299792.458 km/sec. In air under typical conditions, the phase velocity is 88 km/sec less, whereas the group velocity is 91 km/sec less, and it is the latter that is measured in many experiments [55, 62]. Roughly speaking, the accuracy with which the speed of light is known has improved from 1 digit in the 18th century, to 3 digits by 1900, to 5 digits by 1930, to 8 digits today.

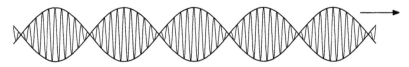

Figure 54.4: Stokes derived the formula for group velocity by considering the superposition of two sine waves. This wave train travels at velocity $(\omega_2 - \omega_1)/(k_2 - k_1)$, and taking the limit $k_2 \to k_1$ gives (54.4).

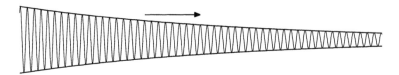

Figure 54.5: Rayleigh derived the same formula by perturbing k to a value $k + \delta k$ with δk small and imaginary. This decaying wave train travels at velocity $\delta\omega/\delta k$, and taking the limit $\delta k \to 0$ gives (54.4) again.

contour integral. Finally, Havelock in 1914 introduced the idea of viewing a dispersive system as a conservation law whose conserved quantity is the number of wave crests; if we define a local wave number $k = k(x, t)$ for a smoothly varying solution of a dispersive system, we find that k satisfies the hyperbolic PDE

$$\frac{\partial k}{\partial t} + \left(\frac{d\omega}{dk}\right)\frac{\partial k}{\partial x} = 0,$$

whose solutions are constant along characteristics propagating at speed $d\omega/dk$.

Now then, what does all of this have to do with pseudospectra?

As stated at the outset, the connection is one of analogy. We shall not belabor the many points of contact between group velocity and pseudospectra, but just offer a list of those that we have noticed and leave it to the reader to judge the significance of this analogy.

1. Both ideas are *linear*.

2. Both become important in cases where individual modes have little physical meaning, where *the physics is in the cancellation of modes*.

3. Both are *subtler and less obvious* than what they replace (namely, phase velocity on the one hand, eigenvalues and eigenvectors on the other).

4. Though both ideas in principle are generalizations of what they replace, in practice, both are *needed for just a minority of applications*.

5. Both were sometimes *overlooked in the early years*. In speed of light measurements, Michelson famously neglected the group velocity correction in 1927 [62]; in hydrodynamic stability theory, misconceptions about the

significance of eigenvalues were widespread until the 1990s [65, 669, 780]; see §20.

6. Both can be illustrated by *examples involving just two modes*. For group velocity, consider the beating of two sine waves; for pseudospectra, consider a non-Hermitian 2×2 matrix.

7. Both are equally *important in appropriate continuum limits*. For group velocity, most applications involve a continuum of wave numbers; for pseudospectra, matrices are often more properly replaced by differential or integral operators.

8. Rigorous analysis of both ideas makes use of *contour integrals in the complex plane*. For group velocity, we have stationary phase and steepest descent integrals; for pseudospectra, it is the matrix or operator Cauchy integral.

9. In both contexts, *small perturbations may reveal truth about the unperturbed system*. Thus we can discern the energy propagation velocity for a pure sine wave by measuring how the phase velocity changes when the wave number is perturbed, as in Figure 54.5; and we can learn about an evolution process governed by a highly nonnormal matrix by seeing how the eigenvalues change when the entries are perturbed.

10. Both notions are *more robust, more physical than what they replace*. Indeed, group velocity may be well-defined when phase velocity is not, for example on a discrete space-time finite difference lattice [767]; and eigenvalues may be nonexistent for an operator whose pseudospectra are entirely meaningful, such as the Hille–Phillips operator of Figure 19.3.

Of course, any analogy has its points of difference too, and here are some of them.

1. Group velocity is specific to the subject of dispersive wave propagation, whereas pseudospectra are applicable to all kinds of questions of behavior of matrices and operators.

2. When phase velocity fails, group velocity is indispensable and gives an exact answer. When eigenvalues fail, pseudospectra rarely give exact answers, and it is not clear that they are indispensable.

3. Quantum mechanics helped make the idea of group velocity famous. By contrast, quantum mechanics helped keep the idea of pseudospectra hidden, for quantum mechanical operators are Hermitian, and their Hermitian behavior shaped scientists' conceptions of eigenvalues and eigenvectors for generations.

XI. Further Examples and Applications

55 · Companion matrices and zeros of polynomials ____

A *companion matrix* has the special form

$$
\mathbf{A} = \begin{pmatrix} 0 & 1 & & & \\ & 0 & 1 & & \\ & & \ddots & \ddots & \\ & & & 0 & 1 \\ -c_0 & -c_1 & \cdots & & -c_{N-1} \end{pmatrix} \tag{55.1}
$$

or its transpose. The characteristic polynomial of $\mathbf{A}$ is

$$
p(z) = c_0 + c_1 z + \cdots + c_{N-1} z^{N-1} + z^N, \tag{55.2}
$$

as can be verified in many ways. For example, we may note that for any $\lambda \in \mathbb{C}$, the vector

$$
\mathbf{v} = (1, \lambda, \lambda^2, \ldots, \lambda^{N-1})^{\mathrm{T}}
$$

is mapped by $\mathbf{A}$ to the vector

$$
\mathbf{A}\mathbf{v} = (\lambda, \lambda^2, \lambda^3, \ldots, \lambda^N - p(\lambda))^{\mathrm{T}}.
$$

This implies that if $p(\lambda) = 0$, then $\mathbf{v}$ is an eigenvector of $\mathbf{A}$ with eigenvalue λ, which establishes that p is the characteristic polynomial of $\mathbf{A}$ in the case where p has distinct zeros. Another approach is to expand $\det(\lambda - \mathbf{A})$ by cofactors with respect to the bottom row.

It is also easily verified that $\mathbf{A}$ is *nonderogatory*, that is, there is a single Jordan block associated with each eigenvalue, or equivalently, the characteristic polynomial is the same as the minimal polynomial. To see this, note that for any λ, removing the first column and last row from $\lambda - \mathbf{A}$ leaves a triangular matrix with 1 on the diagonal, so that $\lambda - \mathbf{A}$ has rank at least $N - 1$. We summarize using the words of Horn and Johnson [414, Thm. 3.3.14]:

> **Theorem 55.1** *Every monic polynomial is both the minimal polynomial and the characteristic polynomial of its companion matrix.*

More generally, an arbitrary matrix $\mathbf{A}$ is similar to a block diagonal matrix with a companion matrix in each block; this is the *rational canonical form* mentioned on page 469 [414, 827].

Companion matrices are obviously non-Hermitian, so it is natural to investigate their pseudospectra. Let us consider as an example the companion matrix $\mathbf{A}$ corresponding to the monic polynomial with roots $1, 2, \ldots, 12$, namely,

$$
\begin{aligned}
p(z) = \ & 479001600 - 1486442880z + 1931559552z^2 - 1414014888z^3 \\
& + 657206836z^4 - 206070150z^5 + 44990231z^6 - 6926634z^7 \\
& + 749463z^8 - 55770z^9 + 2717z^{10} - 78z^{11} + z^{12}.
\end{aligned}
$$

One notes immediately that the coefficients are very large; for the polynomial with roots $1, 2, \ldots, N$, the largest will be somewhat bigger than $N!$. With entries of order 10^{10} but eigenvalues of order 10^1, it is clear that this matrix must be far from normal by any measure. For example, the 2-norm condition number of its matrix of normalized eigenvectors is $\kappa(\mathbf{V}) \approx 2.17 \times 10^{14}$. It is hardly surprising that in the left half of figure Figure 55.1, we see that the pseudospectra lie far from the eigenvalues.

The eigenvalues of $\mathbf{A}$ are distinct, and thus $\mathbf{A}$ is similar to any other matrix with the same distinct eigenvalues. For example, it is similar to $\mathrm{diag}(1, 2, \ldots, 12)$. But there is a more interesting similarity transformation for companion matrices. A standard tool in the field of the numerical solution of matrix eigenvalue problems is the procedure of *balancing* a matrix by a similarity transformation involving a permutation of a diagonal

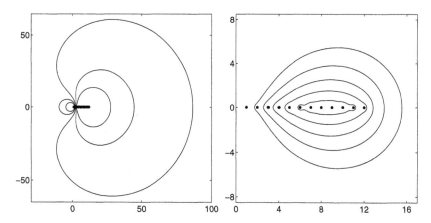

Figure 55.1: On the left, eigenvalues and ε-pseudospectra of the companion matrix $\mathbf{A}$ with eigenvalues $1, 2, \ldots, 12$, for $\varepsilon = 10^{-4}, 10^{-5}, \ldots, 10^{-8}$. Because of the large scale, the eigenvalues are nearly indistinguishable, appearing like the line segment $[1, 12]$. (The $\varepsilon = 10^{-4}$ and 10^{-5} pseudospectral boundaries are the small loops to the left of the origin, signaling a region where the resolvent norm achieves a local minimum; cf. Figure 49.7. The corresponding pseudospectra contain the whole square region shown except for the area enclosed by those loops.) On the right, the same, but for the matrix $\mathbf{B}$ obtained from $\mathbf{A}$ by Parlett–Reinsch balancing.

matrix, $\widehat{\mathbf{A}} = \mathbf{DAD}^{-1}$. This technique was introduced in 1969 by Parlett and Reinsch [602] and is implemented in the subroutines BALANC in EISPACK and xGEBAL in LAPACK [6, 700]. In MATLAB, it can be achieved by the command `balance`, and this is done by default in MATLAB eigenvalue computations unless the flag `'nobalance'` is specified. In Parlett–Reinsch balancing, the entries of $\mathbf{D}$ are chosen so that for each j, the jth row and the jth column of $\widehat{\mathbf{A}}$ have approximately equal sums of magnitudes. At the same time, to avoid introducing rounding errors, each nonzero d_{jk} is taken as a power of 2. Years of experience have shown that balancing a matrix before computing its eigenvalues usually leads to more accurate results.

When our 12×12 companion matrix is balanced, it changes greatly. Instead of -479001600, 1486442880, $\ldots$, the entries of the last row become approximately

$$-0.89,\ 1.4,\ -1.8,\ 2.6,\ -2.4,\ 3.1,\ -2.7,\ 3.3,\ -5.7,\ 13.6,\ -42.4,\ 78,$$

and instead of all ones, the superdiagonal entries become

$$1/2,\ 1,\ 2,\ 2,\ 4,\ 4,\ 8,\ 16,\ 32,\ 64,\ 64.$$

(A matrix like this with companion structure but nonconstant positive entries on the superdiagonal is sometimes known as a *Leslie matrix* [130].) The condition number of the matrix of normalized eigenvectors shrinks to $\kappa(\mathbf{V}) \approx 5.99 \times 10^8$, indicating that in this measure at least, the balanced matrix is six orders of magnitude 'closer to normal' than the unbalanced one. The right half of Figure 55.1 shows that there is a pronounced change in the pseudospectra.

What makes the difference between the two images of Figure 55.1 so interesting is their connection with the problem of computing zeros of polynomials. Suppose we are given the coefficients of a monic polynomial p. Determining the roots of p is a notoriously ill-conditioned problem in general, and it was Wilkinson who made this fact famous in the 1950s and 1960s [609, 826, 829]. Suppose that a polynomial zerofinding algorithm is implemented in floating-point arithmetic on a computer. Then according to Wilkinson's principle of backward error analysis (§53), the best one can normally hope for is that the zeros obtained computationally are equal to the exact zeros of a perturbed polynomial $\widetilde{p}$ with coefficients $\widetilde{c}_j = c_j(1+\varepsilon_j)$, with $|\varepsilon_j| = \mathcal{O}(\varepsilon_{\text{mach}})$, where $\varepsilon_{\text{mach}}$ is the machine precision [592]. Now in Figure 55.2, we see the results of perturbing the coefficients of p in just this fashion. Suppose we define a polynomial $\widetilde{p}$ by coefficients $\widetilde{c}_j = c_j(1+\varepsilon_j)$, where each ε_j is a normally distributed complex number with mean zero and standard deviation ε. The figure shows the superimposed results of fifty such perturbations with $\varepsilon = 10^{-5}$ and 10^{-8}. The approximate agreement with the $\varepsilon = 10^{-5}$ and 10^{-8} pseudospectral boundaries in the right plot of Figure 55.1 is striking.

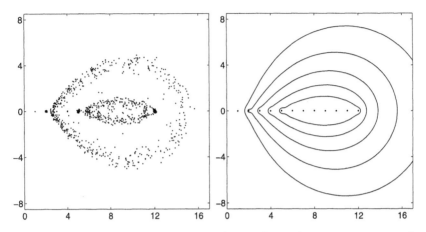

Figure 55.2: Pseudozeros of the monic polynomial p with roots $1, 2, \ldots, 12$. On the left, superimposed zeros of 50 polynomials $\widetilde{p}$ obtained by random coefficient perturbations of relative magnitude 10^{-5} and 10^{-8}. On the right, pseudozero sets $Z_\varepsilon(p)$ for relative coefficient perturbations as defined by Theorem 55.2 with $\varepsilon = 10^{-4}, 10^{-5}, \ldots, 10^{-8}$.

The left half of Figure 55.2 suggests that just as we speak of pseudo-eigenvalues of matrices, it may be useful to consider *pseudozeros of polynomials*. This idea was put forward by Mosier in 1986 [566], who wrote of 'root neighborhoods of a polynomial', plotted examples, and presented theorems. Concerning earlier work of a similar flavor, Mosier recommends the book by Ostrowski [591].

Given a polynomial p of degree N, Mosier defines $Z_\varepsilon(p)$ to be the set of all roots of all polynomials $\widetilde{p}$ with $d(p, \widetilde{p}) \leq \varepsilon$, where the metric d is defined by

$$d(p, \widetilde{p}) = \max \frac{|c_j - \widetilde{c}_j|}{m_j}.$$

Here $\{c_j\}$ and $\{\widetilde{c}_j\}$ are the coefficients of p and $\widetilde{p}$, and the numbers m_j are prescribed weights. By choosing $m_j = |c_j|$ we get the case of relative perturbations. Mosier shows that $Z_\varepsilon(p)$ can be characterized as follows.

Theorem 55.2 *The ε-pseudozero set $Z_\varepsilon(p)$ for the degree-N polynomial p with respect to weights m_j is the set of numbers $z \in \mathbb{C}$ satisfying*

$$\frac{|p(z)|}{m_0 + m_1|z| + \cdots + m_N|z|^N} \leq \varepsilon.$$

The second image of Figure 55.2 shows boundaries of $Z_\varepsilon(p)$ for our degree-12 example p with $m_j = |c_j|$ and, as in Figure 55.1, $\varepsilon = 10^{-4}, 10^{-5}, \ldots, 10^{-8}$.

Note how these curves resemble those in the second half of Figure 55.1.

We have just seen evidence of approximate agreement between the pseudospectra of a balanced companion matrix and the pseudozero sets for componentwise relative perturbations of the associated monic polynomial. Toh and Trefethen [760] showed that while this association need not hold for all polynomials (as further demonstrated by Edelman and Murakami [228]), it is quite common. Figure 55.3 illustrates the agreement for three polynomials chosen somewhat arbitrarily. An explanation of this phenomenon is as follows. To a reasonable approximation, the operation of balancing renders the nonzero entries of a companion matrix of similar magnitudes. If its eigenvalues could be computed in a manner that delivered exact results for a matrix slightly perturbed in only the nonzero positions (cf. §50), then this would be tantamount to computing pseudozeros. In fact, a matrix eigenvalue algorithm will effectively perturb other entries too; but evidently this does not usually change the result too much. For a discussion of the relationship between $Z_\varepsilon(p)$ and the ε-pseudospectrum of the associated companion matrix in the d-metric, see [760].

In concluding this section we comment on the curious situation of algorithms for computing zeros of polynomials, which must in principle be no more difficult than computing eigenvalues of general matrices since it is a special case. The polynomial zerofinding problem is one of the oldest problems of mathematics, associated with names stretching back half a millennium such as Khayyam, Tartaglia, Ferrari, Cardano, Newton, Bernoulli, Abel, and Galois. Since the invention of computers, algorithms for finding zeros of polynomials have been considered endlessly at all levels of sophistication. For example, McNamee has compiled a bibliography of several thousand papers in this area [540]. Virtually all of the vast number of algorithms that have been developed for these problems are designed to take advantage of the special structure of the zerofinding problem, aiming for solutions in close to $\mathcal{O}(N^2)$ time with $\mathcal{O}(N)$ storage. The idea of setting up a companion matrix and finding its eigenvalues, which requires $\mathcal{O}(N^3)$ time and $\mathcal{O}(N^2)$ storage at least by standard methods, has never seemed a glamorous alternative and does not have much of a published literature.

Yet companion matrix eigenvalues often seem to provide the best polynomial roots after all! This was the conclusion of a strongly worded article by Goedecker, whose tests suggested that in practice the companion matrix approach is more accurate and often faster than standard codes such as ZPORC in the IMSL Library and C02AGF in the NAG Library [320]. It was also the conclusion of Toh and Trefethen, who compared companion matrices with CPOLY in the IMSL Library and PA16 in the Harwell library [760]; and companion matrices are the choice made by MATLAB [554]. Some underlying theory of backward error analysis for polynomials is presented in [228, 793]. What is going on here? Our presumption is that in principle, there is no reason why the matrix eigenvalue approach

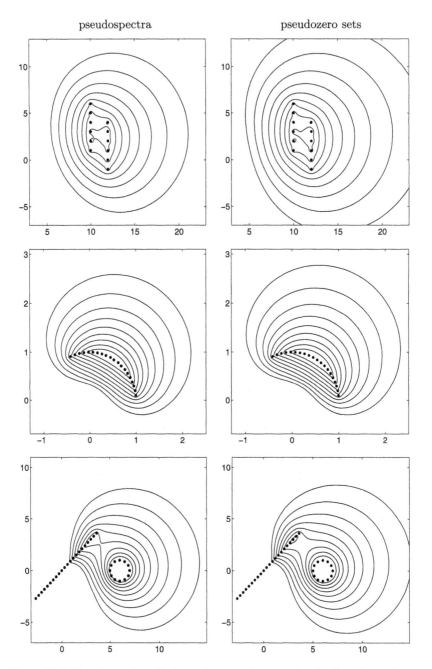

Figure 55.3: Pseudospectra of balanced companion matrices (left) and pseudozero scts for componentwise relative perturbations (right) for three polynomials p whose roots are plotted as dots. In each case $\varepsilon = 10^{-5}, 10^{-6}, \ldots, 10^{-13}$.

should be superior; after all, specialized zerofinding algorithms may take advantage of the mathematics of companion matrices without destroying their sparse structure as a general matrix eigenvalue solver will inevitably do. (The appendix of [760] shows how the Jenkins–Traub zerofinding algorithm [430] can be interpreted as a specialized sparsity-preserving Rayleigh quotient iteration on a companion matrix.) Rather, we suspect that despite widespread attention to polynomial zeros over the years, it is matrix eigenvalues that have benefitted most from the concentrated attention of numerical analysts obsessed with numerical stability. For forty years, no numerical problem has been studied more intensively by experts than matrix eigenvalues. The result has been computer programs exemplified by EISPACK and LAPACK that attain the very highest level in scientific software. For all its easy appeal, polynomial zerofinding is less important in applications and has been less carefully scrutinized by experts; as a result, it would seem that the standard software in the field is less robust.

56 · Markov chains and the cutoff phenomenon _____

The method of Markov chains reduces almost any discrete time dynamical process, linear or nonlinear, to the powers of a so-called *transition matrix*.[1] Thus it is no surprise that eigenvalues and eigenvectors have been a central tool in this field from the beginning. Yet many Markov chains exhibit significant transient phenomena, and in fact, it is sometimes the transient that is of greatest interest for applications. In such cases, the transition matrix is necessarily far from normal—or rather, since $\|\cdot\|_1$ rather than $\|\cdot\|_2$ is the right norm for these applications, the transition matrix necessarily fails to have a well-conditioned set of eigenvectors in the 1-norm.

A basic Markov chain describes a time-invariant, discrete time probabilistic process involving a finite set of states. This seemingly restrictive formulation has proved useful in innumerable applications. Among many other examples, Markov chains have been used to model magnetic phase transitions, mixing of fluids, population dynamics, satisfiability of boolean expressions, extinction of human languages, the theory of war, financial markets, queueing theory, and the 'small world' phenomenon. In this book we shall rather arbitrarily concentrate on two further applications: random walks (this section) and card shuffling (the next). We shall focus especially on the effect known as the 'cutoff phenomenon', which was made famous in the 1980s and 1990s by Diaconis and his coauthors [35, 206].

First, a warning! In the Markov chain literature, the state vector $\mathbf{u}$ is a row vector rather than a column vector, and a matrix $\mathbf{P}$ acts on it from the right rather than the left. In this and the next section we follow this convention, though it runs counter to our notation in the rest of the book. In these two sections, the 1-norm of a matrix is accordingly defined by a maximum absolute row sum, not a column sum:

$$\|\mathbf{P}\|_1 = \max_{\mathbf{u}\neq\mathbf{0}} \frac{\|\mathbf{u}\mathbf{P}\|_1}{\|\mathbf{u}\|_1} = \max_j \|\mathbf{P}(j,:)\|_1.$$

To define a Markov chain, we begin with a finite state space with N states. A *probability distribution* for this space is a row vector $\mathbf{u} \in \mathbb{R}^N$ satisfying

$$u_j \geq 0 \quad (1 \leq j \leq N), \qquad \|\mathbf{u}\|_1 = 1, \tag{56.1}$$

where u_j represents the probability that the system is in state j. At each step of the chain, the probabilities evolve according to multiplication by an $N \times N$ *transition matrix* $\mathbf{P}$:

$$\mathbf{u}^{(k+1)} = \mathbf{u}^{(k)}\mathbf{P}.$$

_____

[1] This section and the following one are adapted from [435].

The entry p_{ij} is the probability that if the chain is currently in state i, it moves to state j at the next step, and thus we have $0 \leq p_{ij} \leq 1$ for each i and j. Repeated steps of the chain are governed by powers of $\mathbf{P}$:

$$\mathbf{u}^{(k)} = \mathbf{u}^{(0)}\mathbf{P}^k.$$

Since probability is conserved, each row sum of $\mathbf{P}$ is equal to 1, and in particular, $\|\mathbf{P}\|_1 = 1$.

A *stationary probability distribution* is a row vector $\boldsymbol{\sigma} \in \mathbb{R}^N$ such that

$$\lim_{k \to \infty} \mathbf{u}\mathbf{P}^k = \boldsymbol{\sigma} \tag{56.2}$$

for any starting vector $\mathbf{u}$ satisfying (56.1). It is easily seen that $\boldsymbol{\sigma}$ is a stationary probability distribution if and only if $\mathbf{P}^\infty = \lim_{k \to \infty} \mathbf{P}^k$ exists and is the $N \times N$ matrix whose rows are all equal to $\boldsymbol{\sigma}$:

$$\mathbf{P}^\infty = \begin{pmatrix} \text{---} & \boldsymbol{\sigma} & \text{---} \\ \text{---} & \boldsymbol{\sigma} & \text{---} \\ & \vdots & \\ \text{---} & \boldsymbol{\sigma} & \text{---} \end{pmatrix}. \tag{56.3}$$

This limiting behavior is guaranteed to occur under the conditions that $\mathbf{P}$ is *irreducible* and *aperiodic,* conditions defined in standard references on Markov chains [273, 582, 746]. For information about Markov chains more closely tied to the present discussion, see [205] and [644].

If a stationary probability distribution $\boldsymbol{\sigma}$ exists, then it satisfies $\boldsymbol{\sigma}\mathbf{P} = \boldsymbol{\sigma}$ and is thus a left eigenvector of $\mathbf{P}$ for the eigenvalue 1. Since the row sums of $\mathbf{P}$ are 1, a corresponding right eigenvector is $N^{-1}(1,1,\ldots,1)^{\mathrm{T}}$. From (56.2) it follows that $\boldsymbol{\sigma}$ is the only normalized left eigenvector of $\mathbf{P}$ corresponding to the eigenvalue 1, and all other eigenvalues are smaller in absolute value. In particular, if λ_2 denotes the second largest eigenvalue in absolute value (not necessarily unique), then $|\lambda_2| < 1$. (The terms $\mathbf{u}\mathbf{P}^k$ in (56.2) can be thought of as iterates of the power method, which computes dominant eigenpairs; see §28.)

Here is an example with $N = 3$. Consider a random walk on the vertices of a triangle in which at each step, a particle moves with probability $1/2$ to each of the adjacent vertices. We have

$$\mathbf{P} = \begin{pmatrix} 0 & \frac{1}{2} & \frac{1}{2} \\ \frac{1}{2} & 0 & \frac{1}{2} \\ \frac{1}{2} & \frac{1}{2} & 0 \end{pmatrix}, \qquad \mathbf{P}^2 = \begin{pmatrix} \frac{1}{2} & \frac{1}{4} & \frac{1}{4} \\ \frac{1}{4} & \frac{1}{2} & \frac{1}{4} \\ \frac{1}{4} & \frac{1}{4} & \frac{1}{2} \end{pmatrix},$$

indicating, for example, that a particle at vertex 1 has probability $1/2$ of being at the same vertex again after two steps. In the limit we get

$$\mathbf{P}^\infty = \begin{pmatrix} \frac{1}{3} & \frac{1}{3} & \frac{1}{3} \\ \frac{1}{3} & \frac{1}{3} & \frac{1}{3} \\ \frac{1}{3} & \frac{1}{3} & \frac{1}{3} \end{pmatrix},$$

with each row equal to the vector $\boldsymbol{\sigma} = (\frac{1}{3}, \frac{1}{3}, \frac{1}{3})$, indicating that the probability is uniformly distributed on the vertices. Thus we see that for this matrix $\mathbf{P}$, the principal left and right eigenvectors are identical; this is expected, since the matrix is symmetric. (For nonsymmetric $\mathbf{P}$, left and right eigenvectors can differ, and hence $\boldsymbol{\sigma}$ need not be a uniform vector.)

The powers $\mathbf{P}^k$ satisfy $\|\mathbf{P}^k\|_1 = 1$ for all k, which does not tell us much. More interesting are the norms $\|\mathbf{P}^k - \mathbf{P}^\infty\|_1$. Let us define $\mathbf{A} \equiv \mathbf{P} - \mathbf{P}^\infty$. This matrix, which we shall call the *decay matrix,* represents the action of the Markov chain on the space spanned by the nondominant eigenvectors. (Here and throughout, it is not necessary for our matrices to be diagonalizable, but our discussion assumes diagonalizability for simplicity.) By induction it is readily shown that $\mathbf{A}^k = \mathbf{P}^k - \mathbf{P}^\infty$ for each $k \geq 1$. Our interest is thus in the behavior of the norms $\|\mathbf{A}^k\|_1$ as a function of k. For $k \geq 1$, $\|\mathbf{A}^k\|_1$ is equal to twice what probabilists call the *total variation* (*TV*) *norm* for this Markov chain.

For the example of a random walk on a triangle, we find

$$\mathbf{A} = \begin{pmatrix} -\frac{1}{3} & \frac{1}{6} & \frac{1}{6} \\ \frac{1}{6} & -\frac{1}{3} & \frac{1}{6} \\ \frac{1}{6} & \frac{1}{6} & -\frac{1}{3} \end{pmatrix}, \qquad \mathbf{A}^2 = \begin{pmatrix} \frac{1}{6} & -\frac{1}{12} & -\frac{1}{12} \\ -\frac{1}{12} & \frac{1}{6} & -\frac{1}{12} \\ -\frac{1}{12} & -\frac{1}{12} & \frac{1}{6} \end{pmatrix},$$

and in general, $\mathbf{A}^k = (-\frac{1}{2})^{k-1}\mathbf{A}$ for each $k \geq 1$, implying that $\mathbf{A}^k$ decreases steadily to 0 at the rate determined by the second eigenvalue $|\lambda_2| = 1/2$. Thus there are no transient effects for the random walk on a triangle, and the same is true for the random walk on an n-gon [206].

Let us turn to an example where transient effects are pronounced. A *random walk on the n-dimensional hypercube* is defined as follows (Figure 56.1). At each step, a particle located at a particular vertex either moves to one of the n adjacent vertices or remains where it is. Each event occurs with probability $1/(n + 1)$, and the steps are independent. On page 515, we shall see that this example is equivalent to another one, that of Ehrenfest urns.

The state space for this Markov chain is of dimension $N = 2^n$. The $N \times N$ transition matrix $\mathbf{P}$ is symmetric and sparse, with only $n+1$ nonzero entries in each row. To specify the matrix explicitly, one would have to pick an ordering of the vertices, but we need not do this, as our observations all pertain to the 1-norm or the 2-norm, which are ordering-independent.

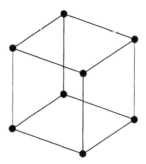

Figure 56.1: Schematic illustration of the hypercube of dimension n. At each step of the random walk, a particle at a vertex moves to one of the adjacent vertices or stays fixed, each with probability $1/(n+1)$.

It is easy to see in a rough way what happens with this example. Suppose the particle begins at a particular vertex, corresponding to an initial probability distribution $\mathbf{u}^{(0)}$ with the value 1 in one position and 0 elsewhere. At each step of the walk, the probability then diffuses around the hypercube. Obviously, $n-1$ steps are needed before it is even possible to get to the diagonally opposite vertex, and it is plausible that many more steps than this will be needed before a uniform distribution of probability on the vertices is approached.

In fact, $\frac{1}{4}n\log n$ steps are needed, as was proved by Diaconis, Graham, and Morrison [207]:

$$k_{\text{cutoff}} = \tfrac{1}{4}n\log n \qquad \text{(hypercube/Ehrenfest).} \qquad (56.4)$$

This result is illustrated in Figure 56.2, where $\|\mathbf{A}^k\|_1$ is plotted as a function of k for $n = 8, 64$, and 512. For the two larger values of n, the curve shows a plateau of norms almost exactly equal to 2 for a long range of early values of k. For example, $\|\mathbf{A}^{15}\|_1 \approx 1.999976$ for $n = 64$. Then, around the dashed line, the values fall off exponentially to 0. Diaconis et al. proved that for $k = \alpha n\log n$ (more precisely, take k to be the nearest integer to $\alpha n\log n$), $\|\mathbf{A}^k\|_1$ converges to 2 from below as $n \to \infty$ for each fixed $\alpha < 1/4$, and decays to 0 as $n \to \infty$ for any fixed $\alpha > 1/4$. In other words, as n increases, the curves of Figure 56.2 steepen to a step function.

For this example $\mathbf{A}$ is normal, implying $\|\mathbf{A}^k\|_2 = \|\mathbf{A}\|_2^k$ for all $k \geq 0$. Thus the cutoff behavior that is so pronounced in the 1-norm must be absent in the 2-norm. Figure 56.3 emphasizes this norm dependence by plotting $\|\mathbf{A}^k\|_1$ and $\|\mathbf{A}^k\|_2$ schematically on a log scale. Eventually, both curves are straight, but for the 1-norm, there is a long flat section before the curve turns downward. Readers used to assuming that all norms are equivalent for practical purposes should bear in mind that the 1- and 2-

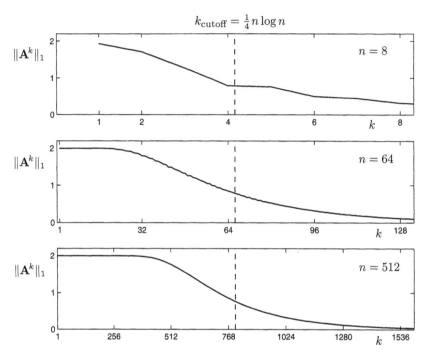

Figure 56.2: Illustration of the cutoff phenomenon for the random walk on the n-cube. These curves, including the chatter in the middle plot, are correct to plotting accuracy. The same curves also describe the problem of Ehrenfest urns. As $n \to \infty$, they steepen to a step at $k_{\text{cutoff}} = \frac{1}{4} n \log n$. Compare Figure 12.2.

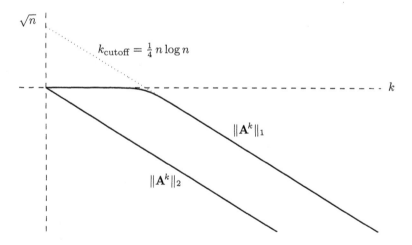

Figure 56.3: Schematic representation on a log scale of the cutoff phenomenon for the random walk on the n-cube. In the 2-norm, there is no cutoff, and the same will be true for any symmetric transition matrix.

norms of an $N \times N$ matrix may differ by as much as a factor of $\sqrt{N}$, and if $N = 2^{512}$, this is a lot of room for maneuver!

How were the data for Figure 56.2 computed, if N is so large? The answer is that we reduced the $N \times N$ matrix problem to an equivalent problem involving the matrix (56.6) of dimension $n+1$, to be explained in a moment.

Even without numerical computation, the eigenvalues and eigenvectors for the random walk on the n-cube can be determined by the methods of Fourier analysis on groups [207]. The eigenvalues of $\mathbf{P}$ are the evenly spaced real numbers

$$1 - \frac{2j}{n+1} \qquad (0 \leq j \leq n), \qquad (56.5)$$

with the jth distinct eigenvalue having multiplicity $\binom{n}{j}$. The eigenvalues of $\mathbf{A}$ are the same, except that the eigenvalue 1 corresponding to $j = 0$ is replaced by 0. Thus the asymptotic decay constant $|\lambda_2|$ in Figures 56.2 and 56.3 is $1 - 2/(n+1)$.

In the 2-norm, we have seen that the random walk on the hypercube has no transient effects. The left half of Figure 56.4 shows 2-norm pseudo-spectra for this problem with $n = 39$. Since the matrix is normal, $\sigma_\varepsilon(\mathbf{A})$ is equal to the set of points at distance $< \varepsilon$ from $\sigma(\mathbf{A})$. The right half of Figure 56.4, on the other hand, shows that the 1-norm pseudospectra bulge well away from the spectrum, an effect that grows more pronounced

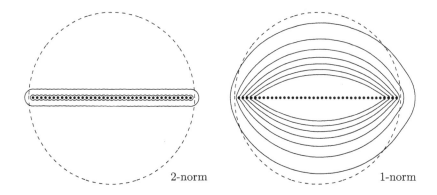

2-norm   1-norm

Figure 56.4: Pseudospectra $\sigma_\varepsilon(\mathbf{A})$ for the random walk on the n-cube with $n = 39$ ($\varepsilon = 10^{-1}, 10^{-1.5}, \ldots, 10^{-4}$); the dashed curve is the unit circle. The matrix $\mathbf{A}$ is normal, so in the 2-norm, on the left, the boundary of $\sigma_\varepsilon(\mathbf{A})$ is just the set of points at a distance ε from the spectrum. This image has little relevance to applications. In the 1-norm, on the right, the pseudospectra lie far from the eigenvalues; the 1-norm condition number of one matrix of eigenvectors is $\kappa_1(\mathbf{V}) = \|\mathbf{V}\|_1 \|\mathbf{V}^{-1}\|_1 \approx 10^6$, a figure that grows exponentially with n. This 1-norm image applies without change to the Ehrenfest urns problem.

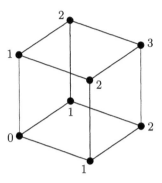

Figure 56.5: Compression of the n-cube, with state space of dimension 2^n, to a problem of dimension $n+1$. The numbers mark values of the new state variable, j, the distance along the graph from vertex 0.

as n increases. Note that the curves displayed correspond to values of ε as low as 10^{-4}, i.e., resolvent norms as great as 10^4.

The random walk on a hypercube, with state space of dimension 2^n, is equivalent to a different problem whose dimension is only $n + 1$. Suppose that instead of taking the state variable to be the vertex, we take it to be just the *distance* from a distinguished vertex called vertex 0, as indicated in Figure 56.5. Since all the vertices lie at a distance from vertex 0 in the range from 0 to n, the new state space has dimension $n + 1$.

The transition rule for the new chain is as follows. A vertex of the n-cube at a distance j from 0 has n neighbors, j of which are one step closer than it to 0 and $n - j$ of which are one step further away. Thus with probability $j/(n + 1)$, a particle at this vertex will move closer to 0 at the next step, with probability $(n - j)/(n + 1)$, it will move further away, and with probability $1/(n + 1)$, it will remain at the same distance. It follows that the Markov process on the n-cube induces a Markov process on the state variable j, and if we order the states from 0 to n, the transition matrix takes the form

$$
\mathbf{P} = \begin{pmatrix}
\dfrac{1}{n+1} & \dfrac{n}{n+1} & & & & \\[2mm]
\dfrac{1}{n+1} & \dfrac{1}{n+1} & \dfrac{n-1}{n+1} & & & \\[2mm]
& \dfrac{2}{n+1} & \dfrac{1}{n+1} & \dfrac{n-2}{n+1} & & \\[2mm]
& & \ddots & \ddots & \ddots & \\[2mm]
& & & \dfrac{n-1}{n+1} & \dfrac{1}{n+1} & \dfrac{1}{n+1} \\[2mm]
& & & & \dfrac{n}{n+1} & \dfrac{1}{n+1}
\end{pmatrix}. \tag{56.6}
$$

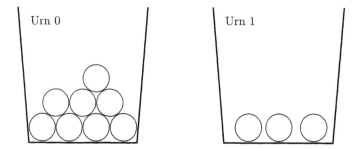

Figure 56.6: Ehrenfest urns containing n balls. At each step, a ball is selected at random and moved to the other urn. This corresponds to changing one coordinate of a particle at a vertex of the n-cube, thereby moving it to an adjacent vertex.

This is a tridiagonal, non-Hermitian matrix of dimension $n + 1$. Its eigenvalues are the same as in (56.5), but now, each only has multiplicity 1. The corresponding decay matrix $\mathbf{A}$ is dense, with the same eigenvalues except with 1 replaced by 0. Most importantly, by combining all vertices with the same j into a single state, we have conserved probability, and the 1-norms $\|\mathbf{A}^k\|_1$ and $\|(z - \mathbf{A})^{-1}\|_1$ are the same for this matrix as for the hypercube. (We used this fact to compute Figure 56.4.) The 2-norms are not conserved, but 2-norms are of little importance for Markov chains, and we shall not discuss them further. In the terminology of Part II of this book, (56.6) is a *twisted Toeplitz matrix*, or more precisely, an *asymptotically twisted Toeplitz matrix*, with symbol $f(x, \theta) = e^{-i\theta}(x/2\pi) + e^{i\theta}(1 - (x/2\pi))$; its 2-norm pseudospectra were presented in Figure 9.1.

Where in this discussion are the Ehrenfest urns? The answer comes from noting that our new Markov chain can be interpreted as follows (Figure 56.6). Consider n indistinguishable balls, each located in one of two urns, Urn 0 or Urn 1. At each step of a random process, a ball is selected at random and moved to the other urn (and with probability $1/(n + 1)$, the ball is not moved). The state variable is j, the number of balls in Urn 1, taking values from 0 to n. Paul and Tatiana Ehrenfest considered this problem in 1907 [230]. (The physical motivation is very interesting, but we shall not discuss it [439].) This problem is the same as before, with each ball representing one coordinate in the hypercube, and changing its urn corresponds to changing that coordinate from 0 to 1 or from 1 to 0.

We can now explain where the cutoff phenomenon comes from for the hypercube/Ehrenfest example. Figure 56.7 shows the probability distribution as a function of step number k, assuming there are $n = 100$ balls and they begin in Urn 0 at step 0.[2] At step 0, the distribution is a Kronecker delta function of height 1 at position $j = 0$. At step 1, most of

---

[2]Plots of this kind were first produced by Yohan Kim at Cornell University [435].

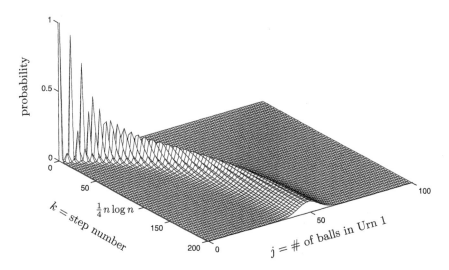

Figure 56.7: Evolution of the probability density function for the hyper-cube/Ehrenfest problem with $n = 100$. (The k axis here is undersampled for plotting purposes, so only every fourth step is visible.) The cutoff phenomenon results from the fact that the early and final states have exponentially small overlap. For this value of n, $k_{\text{cutoff}} = \frac{1}{4}n \log n \approx 115$. Compare Figure 12.1.

the probability, though not quite all, has moved to position $j = 1$. With further steps the distribution moves to the right and diffuses into approximately the shape of a Gaussian, and after 200 steps it is close to its final position centered at $j = 50$. The stationary distribution $\boldsymbol{\sigma}$ —the dominant left eigenvector of the matrix $\mathbf{P}$—can be derived from a simple probability argument: There are 2^n distinct ways to split n balls between the two urns, and $\binom{n}{j}$ distinct ways to put j balls in Urn 1. Thus, the entries in $\boldsymbol{\sigma}$ are given by a binomial distribution [207]:

$$\sigma_j = 2^{-n}\binom{n}{j}, \qquad 0 \le j \le n.$$

To explain the cutoff, the crucial point is that the location of the initial distribution along the j axis is different from that of the stationary distribution. Evidently this particular Markov chain can be interpreted as involving not just diffusion but also advection of probability. Consequently, until the moving pulse gets near the middle of the interval, it has exponentially small overlap with its final position, and the 1-norm of the difference between the current and the asymptotic state is exponentially close to the sum of the 1-norms of those two states:

$$\|\mathbf{u}^{(k)} - \mathbf{u}^{(\infty)}\|_1 \approx \|\mathbf{u}^{(k)}\|_1 + \|\mathbf{u}^{(\infty)}\|_1 = 2. \qquad (56.7)$$

The trajectory that the wave in Figure 56.7 approximates is given by the formula

$$x = \tfrac{1}{2}(1 - e^{-2t}). \tag{56.8}$$

One way to derive this formula is to show by induction that the mean of the probability distribution at step k—that is, the expected number of balls in Urn 1—is exactly

$$\mu_k = \frac{n}{2}\left(1 - \left(\frac{n-1}{n+1}\right)^k\right), \tag{56.9}$$

or equivalently,

$$\frac{1}{2} - \frac{\mu_k}{n} = \frac{1}{2}\left(\frac{n-1}{n+1}\right)^k,$$

which is consistent with the result $|\lambda_2| = 1 - 2/(n+1)$ of (56.5). With $x = \mu_k/n$ and $t = k/n$, (56.9) reduces to (56.8) in the limit $n \to \infty$.

The careful reader may be puzzled at this point. The arguments above, as well as Figure 56.7, suggest that the behavior of the hypercube/Ehrenfest problem scales in proportion to n. If there is a cutoff, it seems that it should occur at some step $k = \mathcal{O}(n)$. In fact, however, the formula (56.4) is $k_{\text{cutoff}} = \tfrac{1}{4}n \log n$. Where does the factor $\log n$ come from? The answer is that as $n \to \infty$, the width of the stationary probability distribution shrinks relative to n. For $\|\mathbf{u}^{(k)} - \mathbf{u}^{(\infty)}\|_1$ to be small, the location of $\mathbf{u}^{(k)}$ must be close to that of $\mathbf{u}^{(\infty)}$ not in an absolute sense, but relative to their widths. In other words, *the convergence criterion that defines the cutoff grows stricter as $n \to \infty$*. The cutoff condition implied by (56.8) is

$$e^{-2t} = \mathcal{O}(n^{-1/2}), \tag{56.10}$$

that is, $-2t = -\tfrac{1}{2}\log n$, hence $t \sim \tfrac{1}{4}\log n$, hence $k \sim \tfrac{1}{4}n \log n$.

We have shown that for the Ehrenfest urns problem, hence also for the random walk on a hypercube, the large-time asymptotic behavior is essentially a process of one Gaussian sliding along the x axis until it coincides with another. The Gaussian is the dominant left eigenvector of $\mathbf{P}$, and the error for large time, which is the second eigenvector of $\mathbf{P}$ or the dominant eigenvector of $\mathbf{A}$, must look like the difference of two nearby Gaussians, as suggested in Figure 56.8, with the shape of a tilted S. As more steps are taken, the two Gaussians align more closely and the amplitude of the S decays, but its width does not change.

This analysis makes clear the 'physical' significance of eigenvectors for this transient-dominated Markov chain. We have seen that the eigenvector whose decay governs the asymptotic behavior is an S curve located at the center of the interval. For small time, the probability distribution has nothing to do with this S curve. Nor does it have anything to do with the other eigenvectors, which are higher frequency oscillations (higher order

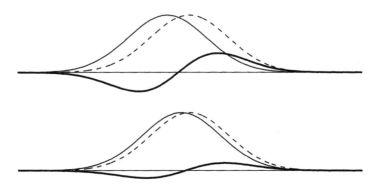

Figure 56.8: The difference of two nearby Gaussians is an S curve (heavy line), whose amplitude decays as the Gaussians slide on top of each other. This process describes the long-time convergence in the hypercube/Ehrenfest problem. The S curve is the dominant eigenvector of $\mathbf{A}$, which determines the long-time shape of the distribution, but it has nothing to do with the behavior for short time.

differences of Gaussians) that are also exponentially localized in the middle of the interval. In short, just as we found for advection-diffusion operators in §12, the eigenvectors of this problem bear no relationship to the behavior of the system for short times.

57 · Card shuffling

One example of the Markov chain cutoff phenomenon has become particularly famous: the riffle shuffle of a deck of cards. This problem was widely discussed in the popular press around the time of the publication of a beautiful paper by Bayer and Diaconis in 1992 [35, 458], which built upon earlier work by Aldous and others. Bayer and Diaconis proved that asymptotically as $n \to \infty$, it takes exactly $\frac{3}{2}\log_2 n$ riffle shuffles to randomize a deck of n cards [35]. At $1.4\log_2 n$ shuffles, they showed, the deck is nowhere near random (for large enough n).

Bayer and Diaconis analyzed a precisely defined Markov chain, proposed independently by Gilbert and Shannon (1955) and Reeds (1981), that has been widely accepted as a reasonable model of how humans actually shuffle. They describe the process as follows [35]:

> A deck of n cards is cut into two portions according to a binomial distribution; thus the chance that k cards are cut off is $\binom{n}{k}/2^n$ for $0 \le k \le n$. The two packets are then riffled together in such a way that cards drop from the left or right heaps with probability proportional to the number of cards in each heap.

These words define a Markov chain on the state space of permutations of the set $\{1, \ldots, n\}$, which has dimension $N = n!$. (For $n = 52$, $N \approx 8.1 \times 10^{67}$.) The associated transition and decay matrices $\mathbf{P}$ and $\mathbf{A}$ are of dimensions $n! \times n!$ and non-Hermitian. (We continue with the terminology and notation of the last section.) Bayer and Diaconis proved there is a cutoff at

$$k_{\text{cutoff}} = \tfrac{3}{2}\log_2 n \qquad \text{(riffle shuffle)} \qquad (57.1)$$

in the same sense as the cutoff at $\frac{1}{4}n\log n$ in the last section, namely, as measured via norms $\|\mathbf{A}^k\|_1$.

Just as with the hypercube/Ehrenfest problem of the last section, this Markov chain can be compressed to a problem of small dimension. There, we introduced as a state variable the distance j from a distinguished vertex. Here, our new state variable will be r, the *number of rising sequences*. Once more we take the definition from [35]:

> A rising sequence is a maximal subset of an arrangement of cards, consisting of successive face values displayed in order. Rising sequences do not intersect, so each arrangement of a deck of cards is uniquely the union of its rising sequences. For example, the arrangement A,5,2,3,6,7,4 consists of the two rising sequences A,2,3,4 and 5,6,7, interleaved together.

Suppose we start with a deck ordered from 1 to n, i.e., with one rising sequence. The first riffle shuffle will split that rising sequence into two (with an exceptionally small probability of keeping the deck in the same order). The second shuffle will most likely break two rising sequences into four, though with low probability it could yield only one, two, or three such sequences. While the early shuffles usually double the number of rising sequences, this trend cannot dominate indefinitely, as a deck with n rising sequences is just as far from random as a deck with only one. Bayer and Diaconis showed that the probability of having r rising sequences after k shuffles depends only on r; this probability is $\binom{2^k+n-r}{n}/2^{kn}$. Moreover, the riffle shuffle induces a Markov chain on the space of numbers of rising sequences, that is, numbers r in the range $1 \leq r \leq n$. The dimension of the compressed Markov chain is thus $N = n$. Bayer and Diaconis did not give a formula for the entries of the reduced $n \times n$ transition matrix $\mathbf{P}$, but the entries have been determined in unpublished work by Gessel (personal communication, August 1997; see also [35]) and again in [435]. The formula is

$$p_{ij} = 2^{-n}\binom{n+1}{2i-j}\frac{\alpha_j}{\alpha_i}, \tag{57.2}$$

where α_j denotes the number of permutations of $\{1, \ldots, n\}$ that have j rising sequences. These numbers α_j are known as *Eulerian numbers* and are given by the triangular recurrence

$$A_{r,k} = kA_{r-1,k} + (r-k+1)A_{r-1,k-1} \tag{57.3}$$

with $A_{1,1} = 1$, $A_{1,k} = 0$ for $k \neq 1$, and $\alpha_r = A_{n,r}$ [334, 457]. The stationary distribution for this Markov chain is

$$\boldsymbol{\sigma} = (\alpha_1, \alpha_2, \ldots, \alpha_n)/n!, \tag{57.4}$$

which gives us $\mathbf{P}^\infty$ by (56.3) and thence $\mathbf{A} = \mathbf{P} - \mathbf{P}^\infty$.

By assembling these formulas, we can compute the matrices $\mathbf{P}$ and $\mathbf{A}$ easily and investigate their properties. These matrices seem exceptionally interesting; they combine a nontrivial application with strong non-eigenvalue effects. For the sake of readers who may wish to explore these matrices themselves, we present in Figure 57.1 a brief MATLAB program for computing them. This program is adapted from [435], where further details are given and an analogy is presented between the compression from $n!$ to n and the coarse-grid approximations utilized in multigrid methods in numerical analysis.

The first thing we naturally choose to investigate are the powers $\|\mathbf{A}^k\|_1$. Figure 57.2 shows these norms for decks of size $n = 52, 208$, and 832. The cutoff phenomenon is pronounced, sharper than observed for the hypercube/Ehrenfest problem.

```
function [A,P] = riffle(n)

% riffle.m - compute the n x n decay and transition matrices
%            A and P for the riffle shuffle of a deck of n cards.
%            Gudbjorn Jonsson and Nick Trefethen, Cornell U., 1997.

% Logarithms of Eulerian numbers:
  a = zeros(1,n); anew = zeros(1,n);
  for j = 2:n
    anew(2:j-1) = log((2:j-1).*exp(a(2:j-1)-a(1:j-2)) ...
                                     +(j-1:-1:2))+a(1:j-2);
    a = anew;
  end

% Logarithms of binomial coefficients:
  b = zeros(1,n+2); bnew = zeros(1,n+2);
  for j = 2:n+2
    bnew(2:j-1) = log(exp(b(2:j-1)-b(1:j-2))+1)+b(1:j-2);
    b = bnew;
  end

% Transition matrix P:
  b = b-n*log(2);
  r = [b(1) -Inf*ones(1,n-1)];
  c = [b -Inf*ones(1,n-2)]';
  T = toeplitz(c,r);
  P = T(2:2:2*n,:);
  P = exp(P-a'*ones(1,n)+ones(n,1)*a);

% Stationary distribution and decay matrix A:
  v = eye(1,n); vnew = eye(1,n);
  for j = 2:n
    vnew(1) = v(1);
    vnew(2:j) = (2:j).*v(2:j)+(j-1:-1:1).*v(1:j-1);
    v = vnew/j;
  end
  Pinf = ones(n,1)*v;
  A = P - Pinf;
```

Figure 57.1: MATLAB program from [435] for building riffle shuffle matrices.

Next, we look at 1-norm pseudospectra of $\mathbf{A}$. Figure 57.3 shows that for $n = 52$, they once again bulge outside the unit circle near $z = 1$. Figure 57.4 shows closeups for various values of n.

The eigenvalues of $\mathbf{P}$ are known to be exactly 2^{-j} for $0 \leq j \leq n - 1$, and $\mathbf{A}$ has the same spectrum but with the eigenvalue 1 replaced by 0. The eigenvalues and eigenvectors, however, are highly ill-conditioned. Table 57.1 shows 1-norm eigenvalue condition numbers for various dimensions. The condition number of the jth eigenvalue 2^{-j} (see §52) is well approximated by the formula

$$\kappa_1(2^{-j}) \approx n^{3j/2}. \tag{57.5}$$

Figure 57.5 is an analogue of Figure 56.7. This figure reveals a striking similarity between the hypercube/Ehrenfest and riffle shuffle problems.

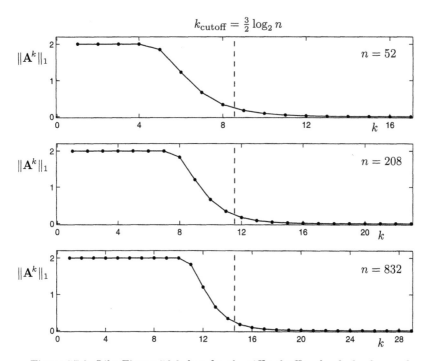

Figure 57.2: Like Figure 56.2, but for the riffle shuffle of a deck of n cards.

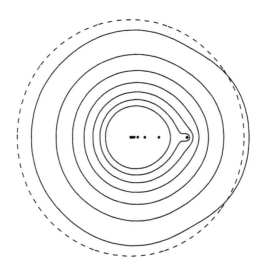

Figure 57.3: One-norm pseudospectra as in Figure 56.4, but for the riffle shuffle matrix with $n = 52$ ($\varepsilon = 10^{-1}, 10^{-1.5}, 10^{-2}, \ldots, 10^{-4}$). The condition number of one matrix of eigenvectors is $\kappa_1(\mathbf{V}) \equiv \|\mathbf{V}\|_1 \|\mathbf{V}^{-1}\|_1 \approx 10^{60}$.

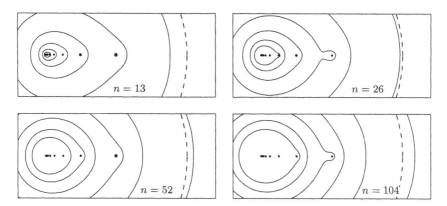

Figure 57.4: Closeups for $n = 13,\ 26,\ 52,\ 104$ and $\epsilon = 10^{-1}, 10^{-2}, \ldots, 10^{-6}$.

Table 57.1: One-norm eigenvalue condition numbers for the first three eigenvalues of $\mathbf{A}$ for decks of size $n = 13, 26, \ldots, 832$.

n	$\lambda = 1/2$	$\lambda = 1/4$	$\lambda = 1/8$
13	10.4	7.91×10^1	4.61×10^2
26	31.8	6.51×10^2	1.32×10^4
52	88.1	5.76×10^3	3.10×10^5
104	246.8	4.57×10^4	7.20×10^6
208	694.4	3.64×10^5	1.63×10^8
416	1959.	2.91×10^6	3.69×10^9
832	5534.	2.32×10^7	8.35×10^{10}

The major difference is that whereas for the hypercube/Ehrenfest problem the wave propagates smoothly from one value of j to the next, for the riffle shuffle the value of r approximately doubles at each of the early steps. This is why there is no factor $\mathcal{O}(n)$ in the formula for k_{cutoff}.

What does the trajectory of Figure 57.5 approach as $n \to \infty$? Here we get a surprise. The limit is not a smooth curve but a sharp step from $r/n \approx 0$ to $r/n \approx 1/2$ at $k \sim \log_2 n$. As with the hypercube/Ehrenfest problem, the first phase ($\log_2 n$ steps) is needed to move the pulse of probability roughly to the middle of the axis representing the number of rising sequences. The second phase ($\frac{1}{2}\log_2 n$ more steps) is then needed to align it closely with the target, whose width shrinks with n. The formula $k_{\text{cutoff}} = \frac{3}{2}\log_2 n$ can be derived by following the same pattern of argument used for the hypercube/Ehrenfest problem. First, the reason why $\sim \log_2 n$ steps are needed to achieve $r = \mathcal{O}(n)$ is that r approximately doubles at each step initially until it becomes of size $\mathcal{O}(n)$. Second, from this point on, with $|\lambda_2| = 1/2$ and the Gaussian pulse again of relative width $\mathcal{O}(n^{-1/2})$,

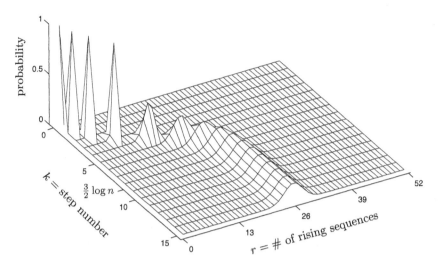

Figure 57.5: Like Figure 56.7, but for the riffle shuffle with $n = 52$. (There is no undersampling in this plot.) For this value of n, $k_{\text{cutoff}} = \frac{3}{2} \log_2 n \approx 8.55$.

the convergence is governed by the analogue of (56.10),

$$2^{-k} = \mathcal{O}(n^{-1/2}), \qquad (57.6)$$

that is, $-k = -\frac{1}{2} \log_2 n$. Thus we need $k = \frac{1}{2} \log_2 n$ additional steps for norm convergence, bringing the total to $\frac{3}{2} \log_2 n$.

Here is a summary of the observations made in this and the last section concerning Markov chains and the cutoff phenomenon:

- The hypercube/Ehrenfest and riffle shuffle cutoff phenomena share an elementary explanation: A probability wave must propagate from one place to another before convergence can occur. Eigenvalues and eigenvectors are irrelevant during this transient phase.

- These problems are appropriately analyzed with 1-, not 2-norms.

- Once the transient phase is past, the eigenvectors that govern the asymptotic convergence describe a process of alignment of Gaussians.

- For the hypercube/Ehrenfest problem, the probability wave reaches approximately the right position in $\mathcal{O}(n)$ steps, and $\sim \frac{1}{4} n \log n$ steps are needed for norm convergence.

- For the riffle shuffle problem, the probability wave reaches approximately the right position in $\sim \log_2 n$ steps, and $\sim \frac{3}{2} \log_2 n$ steps are needed for norm convergence.

- The transient effect has nothing to do with multiple or near-multiple eigenvalues. Indeed, for both problems in their compressed formulations, the eigenvalues are simple.

These are fascinating problems, in which all the interesting 'physics' seems to consist of transient behavior. Yet we note that other points of view can also be taken from which the physics would look rather different. In particular, one can analyze these randomization processes using techniques of information theory. For Ehrenfest urns, say, we begin with complete information that all the balls are in Urn 0; or for a deck of cards, we begin with the complete information that the cards are ordered from 1 to n. We then take steps of the Markov chain and analyze the rate of decay of this information to zero as randomization is approached. Such an analysis reveals no cutoff effect. Information is lost steadily from the very first step; each time you shuffle, the deck does indeed become more random! The absence of the cutoff phenomenon in this information theory analysis of Markov chains has been pointed out by Peter Doyle (personal communication, 1996) and is investigated in print in [718, 781].

We close with a question. Figures 57.3 and 57.4 show pseudospectra that contain points much further from the origin than the eigenvalues, implying that there must be some transient effects in these Markov chains. But this qualitative conclusion is a long way from the very precise actual situation, which is that $\|\mathbf{A}^k\|_1$ remains almost identically equal to 2 for small values of k. Theorems 16.4 and 16.5 do not get us very far. Can more precise information about $\|\mathbf{A}^k\|_1$ be inferred from these pseudospectra?

58 · Population ecology

Since the growth of a population over time is a dynamic process, it is no surprise that eigenvalues arise in the study of the stability and growth rates of communities. The models describing such processes typically involve nonnormal matrices, a fact with interesting biological implications. Discrete time single-species population models form one notable class of nonnormal examples [130, 455]. Here, though, we focus on multispecies systems described by continuous time differential equations.

Community models in population ecology predict the dynamic interaction of various competing and cooperating species in a restricted environment. The mathematical formulation addresses the question, Given initial population densities for all the species in this community, what will the populations be at some future time? The *Lotka–Volterra predator-prey equations* are the most celebrated model of this form. Suppose y_1 and y_2 denote the densities of two species, the first the prey (gazelles, say), and the second the predator (lions). Then these equations take the basic form

$$y_1' = y_1(1 - y_2), \qquad (58.1)$$
$$y_2' = y_2(y_1 - 1), \qquad (58.2)$$

where the primes denote derivatives with respect to time [568, Chap. 3]. Since population densities cannot be negative, $y_1, y_2 \geq 0$, this model implies that the lions prosper when the gazelle density exceeds 1, while the gazelles increase when the lion density drops below 1. The result is a cycle of growth and predation, a familiar early example in dynamical systems classes.

From the outset, one realizes that such models will at best provide a rough approximation of reality. This is an autonomous system of differential equations: time does not appear on the right-hand side. Thus, the model necessarily cannot account for temporal variation, such as seasonal changes. Furthermore, it only includes two species. What food source sustains the gazelles? What happens if this resource is depleted?

To allow for larger communities, potentially involving multiple levels of a food chain, one can generalize (58.1) to N different species, each with population density y_j described by the system of differential equations

$$y_j' = s_j\, y_j + \sum_{k=1}^{N} a_{jk}\, y_j y_k, \qquad (58.3)$$

where the s_j and a_{jk} are community-specific constants. The coefficient a_{jk} describes the amount that species j benefits or suffers from the presence of species k in the community.

This model can be analyzed by making the common, if not universally accepted, assumption that naturally occurring populations described by (58.3) will typically be in stable equilibrium.[1] Having made this assumption, one can follow the standard approach for analyzing stability of nonlinear systems, as sketched in Figure 33.3. At equilibrium, the population densities $\{y_j\}_{j=1}^N$ do not change with time, i.e., $y_j' = 0$ for $j = 1, \ldots, N$, and thus the equations (58.3) imply

$$-\mathbf{s} = \mathbf{A}\mathbf{y}, \tag{58.4}$$

where $\mathbf{y}$ is the vector of population densities. To evaluate stability, consider the Jacobian matrix $\mathbf{D}$ for the equations (58.3), built up from partial derivatives of the form

$$d_{jk} \equiv \frac{\partial y_j'}{\partial y_k} = \begin{cases} a_{jk}y_j, & j \neq k, \\ s_k + a_{kk}y_k + \sum_{\ell=1}^N a_{k\ell}y_\ell, & j = k. \end{cases}$$

For $\mathbf{y}$ at equilibrium, (58.3) and (58.4) imply that

$$\frac{\partial y_j'}{\partial y_k} = a_{jk}y_k$$

for all j, k, and thus the Jacobian matrix $\mathbf{D}$ is defined entrywise by $d_{jk} = a_{jk}y_k$. If an equilibrium point $\mathbf{y}$ is subjected to the small perturbation $\mathbf{p} = \mathbf{p}(0)$ at time $t = 0$, then this perturbation will develop according to

$$\mathbf{p}(t) = e^{t\mathbf{D}}\mathbf{p}(0). \tag{58.5}$$

The asymptotic rate at which generic perturbations grow or decay is governed by the spectral abscissa of the Jacobian, $\alpha(\mathbf{D}) = \sup_{\lambda \in \sigma(\mathbf{D})} \operatorname{Re} \lambda$. For the equilibrium point $\mathbf{y}$ to be stable, all perturbations must eventually decay, and thus $\alpha(\mathbf{D}) < 0$. Provided this is the case, $\alpha(\mathbf{D})$ is a measure of how rapidly the perturbed system returns the equilibrium system, and thus $-\alpha(\mathbf{D})$ is known as the *resilience* of the community (see [574] and references therein).[2] The more negative the spectrum of $\mathbf{D}$, the more quickly perturbations are damped out, and thus the more resilient the community is to small changes. From this perspective, robust communities are those that have their eigenvalues well to the left of the imaginary axis.

---

[1]This assumption arises from the premise that unstable communities would collapse under small perturbations to the environment, which inevitably occur, for example, from extreme weather or from interaction with species omitted from the model. Unstable communities are thus expected to be shortlived and rarely observed. For an empirical investigation of this assumption, see Yodzis [844].

[2]Note that for the two-species model (58.1)–(58.2), the Jacobian $\mathbf{D}$ at the stable point $\mathbf{y} = (1, 1)^{\mathrm{T}}$ has $\alpha(\mathbf{D}) = 0$, so the system is only neutrally stable.

As seen throughout this book, conclusions based on the eigenvalues of $\mathbf{D}$ alone can be misleading when $\mathbf{D}$ is nonnormal. In the population ecology context, Neubert and Caswell have discussed this point in detail [574], suggesting alternative measures of robustness to perturbations that incorporate nonnormal effects. A related notion of robust stability in a non-ecological context is considered in [115]. General techniques for analyzing linearized differential equations discussed in §§15 and 33 are appropriate.

Even when models of the type described here are applied to communities with a handful of species, the numerous constants in equation (58.3) may be difficult to estimate accurately. (For example, for the simplified four-species lake community model considered in [160, §4.3], it is even difficult to ascertain which entries in $\mathbf{A}$ should be nonzero.) Because of this uncertainty, many ecologists concede that (58.3) may not be useful for making quantitative predictions for specific communities, though perhaps general properties of communities can be deduced by understanding the qualitative behavior of (58.3) over different ensembles of parameters. Two landmark studies in this vein were conducted by Gardner and Ashby [301] and May [538], who constructed $\mathbf{D}$ to have negative elements on the main diagonal and random nonzero elements in some proportion of the off-diagonal elements. They were then interested in how various parameters, such as the proportion of nonzero entries (*connectance*), the system dimension, and the magnitude of these entries, affected the probability that such a system was stable.

A drawback of this approach is that such random distributions take no account of community structure; for example, they can potentially include unnatural cycles within a food web. As an alternative, Cohen et al. [160] have suggested the 'Lotka–Volterra cascade model', which imposes basic restrictions on the coefficients a_{jk} in (58.3). The entries in the system are ordered so that species j never preys upon species k when $j < k$; one says that the species are labeled according to ascending trophic levels. For any $j < k$, this model only allows four possibilities:

- $a_{jk} = 0$ and $a_{kj} > 0$, a 'donor-controlled' link;

- $a_{jk} < 0$ and $a_{kj} = 0$, a 'recipient-controlled' link;

- $a_{jk} < 0$ and $a_{kj} > 0$, the classic 'consumer–victim' link;

- $a_{jk} = 0$ and $a_{kj} = 0$, no dynamic interaction.

These requirements imply that all entries above the main diagonal are nonpositive, while all below are nonnegative. The entries of the main diagonal itself are always negative, $a_{jj} < 0$.

Let us consider a particular example from a food web for part of the river Thames shown in Figure 58.1, based on data from [158, p. 250]. The links in the web specify those pairs potentially corresponding to nonzero

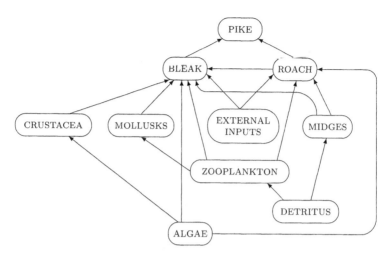

Figure 58.1: Food web for a section of the river Thames, adapted from a table in Cohen, Briand, and Newman [158]. The arrows indicate the direction of energy flow through the web.

entries in the matrix $\mathbf{A}$. Given these relationships, one must still determine the nature of each link, and indeed the magnitude of the nonzero coefficients. Rather than trying to apply ecological intuition to determine these parameters, we consider three extreme situations:

- all links are donor-controlled (lower triangular $\mathbf{A}$);

- all links are recipient-controlled (upper triangular $\mathbf{A}$);

- all links are consumer–victim.

Perhaps reality occurs somewhere between these extremes. In each case, we construct the matrix $\mathbf{A}$ to have the nonzero pattern suggested by Figure 58.1 and the sign pattern determined by the link type. Nonzero entries are assigned magnitudes randomly drawn from the uniform distribution on $[0,1]$. The entries of the stable point $\mathbf{y}$ are random numbers drawn uniformly from $[0,1]$. In Figure 58.2, we show the pseudospectra for these three realizations of $\mathbf{D}$ based on a random choice of the parameters that yields somewhat more nonnormality than is typical. The first two matrices are particularly far from normal, as reflected in the plot of $\|e^{t\mathbf{D}}\|$ in Figure 58.3. In the case of the donor-controlled links, perturbations can be amplified to more than 100 times their initial size. The consumer–victim model, though less resilient than the others, damps the perturbation out more quickly. (The oscillations seen in the consumer–victim curve could not occur if $\mathbf{D}$ were a normal matrix.) Far more extensive randomly generated experiments are detailed by Chen and Cohen [143].

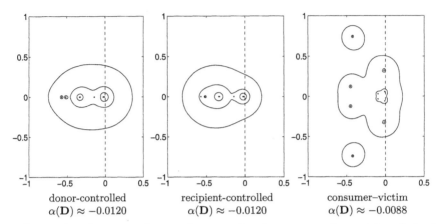

donor-controlled	recipient-controlled	consumer–victim
$\alpha(\mathbf{D}) \approx -0.0120$	$\alpha(\mathbf{D}) \approx -0.0120$	$\alpha(\mathbf{D}) \approx -0.0088$

Figure 58.2: Spectra and ε-pseudospectra for three community matrices derived from the Thames food web. In the left and middle plots, $\varepsilon = 10^{-1}, 10^{-2}, 10^{-3}$; on the right, $\varepsilon = 10^{-1}, 10^{-2}$. The dashed line in each plot is the imaginary axis.

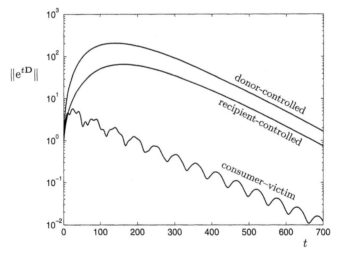

Figure 58.3: Transient growth, followed by asymptotic decay, for the Thames food web. Pseudospectra for the corresponding matrices are illustrated in Figure 58.2.

Here is a more extreme example of nonnormality in community matrix models. Consider an N-species community that is completely connected, reflecting interaction between all species, and again suppose the links are either all donor-controlled, all recipient-controlled, or all consumer–victim. Assume further that the nonzero entries of $\mathbf{D}$ all have magnitude 1, so that these three matrices take the general form

$$\mathbf{D_D} = \begin{pmatrix} -1 & 0 & 0 \\ 1 & -1 & 0 \\ 1 & 1 & -1 \end{pmatrix}, \quad \mathbf{D_R} = \begin{pmatrix} -1 & -1 & -1 \\ 0 & -1 & -1 \\ 0 & 0 & -1 \end{pmatrix}, \quad \mathbf{D_{CV}} = \begin{pmatrix} -1 & -1 & -1 \\ 1 & -1 & -1 \\ 1 & 1 & -1 \end{pmatrix}.$$

donor-controlled   recipient-controlled   consumer–victim

While this example is admittedly distant from nature's reality, it illustrates the potential for nonnormality contained in this class of models and highlights differences between the various types of links in the food web.

Any potential distinction between these types of links is hidden from standard eigenvalue analysis. It is clear that $\alpha(\mathbf{D_D}) = \alpha(\mathbf{D_R}) = -1$ since these two matrices are triangular. Note that $\mathbf{D_{CV}} + \mathbf{I}$ is skew-symmetric and thus has purely imaginary eigenvalues. It follows that all eigenvalues of $\mathbf{D_{CV}}$ have real part equal to -1, so we also have $\alpha(\mathbf{D_{CV}}) = -1$. All three systems thus appear considerably more resilient than the Thames examples studied earlier. From the perspective of eigenvalue analysis alone, that is all there is to say.

From the perspective of nonnormality, however, these three matrices could hardly be more different, as is illustrated by pseudospectral plots in Figure 58.4 for $N = 20$. First note that $\mathbf{D_{CV}}$ is a normal matrix since it is a shift of a skew-symmetric matrix. On the other hand, $\mathbf{D_D}$ and $\mathbf{D_R}$ are both nondiagonalizable. In both cases, we can look to the corner entries of the resolvent to understand the extent of the nonnormality, a common approach for analyzing triangular matrices; see §§36–38. (For symbol analysis of these Toeplitz matrices, see the discussion of (49.2) on p. 449.) For $\mathbf{D_D}$ we have

$$[(z - \mathbf{D_D})^{-1}]_{N1} = \frac{(z + 2)^{N-2}}{(z + 1)^N} \qquad (N \geq 2),$$

whose modulus goes to infinity with N whenever $\operatorname{Re} z \geq -3/2$. Thus for any $\varepsilon > 0$, in the large N limit, *any point in the right half-plane is an*

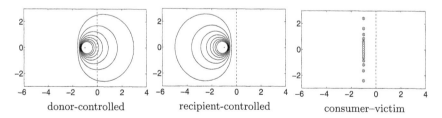

donor-controlled   recipient-controlled   consumer–victim

Figure 58.4: Spectra and ε-pseudospectra ($\varepsilon = 10^{-1}, \ldots, 10^{-8}$) for extreme community matrices. In all three examples, there are links between all $N = 20$ species in this fictitious web. On the left, the links are donor-controlled; in the middle, they are recipient-controlled; on the right, they are consumer–victim (some eigenvalues are off scale in this case). For all three examples, $\alpha(\mathbf{D}) = -1$.

ε-pseudoeigenvalue of $\mathbf{D}_D$. One can similarly compute that

$$[(z - \mathbf{D}_R)^{-1}]_{1N} = \frac{(-z)^{N-2}}{(-z-1)^N} \qquad (N \geq 2).$$

This case is closely related, but now this corner entry, and thus the resolvent norm, blows up whenever $\operatorname{Re} z \leq -1/2$. For points z with $\operatorname{Re} z > -1/2$, one can show that

$$\|(z - \mathbf{D}_R)^{-1}\| \leq \frac{1}{|z+1|}\left(1 + \frac{1}{|z+1| - |z|}\right)$$

independently of N, implying the loose bound

$$\max_{\operatorname{Re} z \geq 0} \|(z - \mathbf{D}_R)^{-1}\| \leq 3.$$

Thus, even when the dimension of the matrix is large, the resolvent norm on the imaginary axis is never big. The differing effects of nonnormality are evident from the plots of $\|e^{t\mathbf{D}}\|$ in Figure 58.5. There is large transient growth for $\mathbf{D}_D$, which becomes more acute for larger dimensions. For $\mathbf{D}_R$, the slope of the curve is approximately $-1/2$, only half as negative as one would expect based on eigenvalue analysis. (It must eventually reduce to -1 for this finite-dimensional example, but this is not observed on the scale shown here.) Finally, $\mathbf{D}_{CV}$ behaves as expected for a normal matrix, yielding a slope equal to $\alpha(\mathbf{D}_{CV}) = -1$ from the start.

We can summarize the effects of nonnormality in population ecology models as follows (see also [574]).

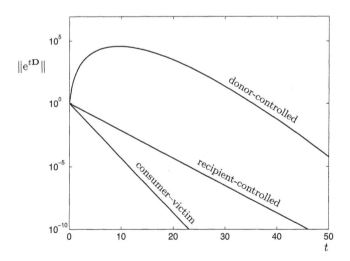

Figure 58.5: Matrix exponentials for the three extreme 20-species community matrices of Figure 58.4.

- Perturbations to stable equilibria can grow before they decay, leading to longer lasting effects than might be expected from the resilience alone.

- Such perturbations could potentially grow to an extent that makes the linearization inappropriate.

- Small errors in the community matrix, as would be expected in practice, can significantly change the stability properties.

- Judging community resilience based on eigenvalues alone can be misleading.

59 · The Papkovich–Fadle operator _____

The Papkovich–Fadle problem, studied since 1940 in both solid and fluid mechanics, concerns the biharmonic operator on a semi-infinite strip.[1] This is an example where the degree of nonnormality is mild—algebraic rather than exponential. Eigenfunction expansions turn out to be a reasonable tool for solving it, provided that one manipulates them in a way that does not depend on an implicit assumption of normality. Thus the Papkovich–Fadle problem is a case study in how to deal properly with expansions in nonorthogonal functions.

To set the stage, let us first consider an analogous but normal problem involving the Laplace operator. Let S be the semi-infinite strip $x > 0$, $-1 < y < 1$ (Figure 59.1). We seek a function $u(x, y)$ satisfying

$$\Delta u(x, y) = 0, \quad (x, y) \in S, \tag{59.1}$$

with boundary conditions

$$u(x, \pm 1) = 0 \tag{59.2}$$

and

$$u(0, y) = f(y). \tag{59.3}$$

Here Δ is the Laplacian $\Delta = \partial^2/\partial x^2 + \partial^2/\partial y^2$, and to make the problem well-posed we add the condition $\lim_{x \to \infty} u(x, y) = 0$. One can interpret $u(x, y)$ as the shape taken by an infinite membrane with height $f(y)$ at the left end of the strip.

A natural way to solve for $u(x, y)$ is by separation of variables. For simplicity we shall only consider boundary functions $f(y)$ that are even, which by symmetry guarantees solutions that are even with respect to y. (There is an analogous family of odd solutions, and a general solution for any f can be constructed as a superposition of an even and an odd component. Alternatively, the problem could have been posed on the half-strip $x > 0, 0 < y < 1$ with the Neumann boundary condition $u_y(x, 0) = 0$.) The solutions to (59.1)–(59.3) of the form $u(x, y) = X(x)Y(y)$ are the functions

$$\phi_k(x, y) = e^{-\frac{1}{2}k\pi x} \cos(\tfrac{1}{2}k\pi y), \qquad k = 1, 3, 5, \ldots, \tag{59.4}$$

and an arbitrary solution can be expanded as a series

$$u(x, y) = \sum_{k \in \mathbb{K}} a_k \phi_k(x, y), \tag{59.5}$$

_____

[1] This section originated in joint work with J. A. C. Weideman, summarized briefly in [773].

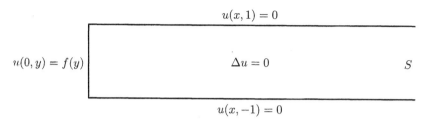

Figure 59.1: The Laplace evolution problem in a semi-infinite strip. The variable x plays the role of time, and f defines the initial data.

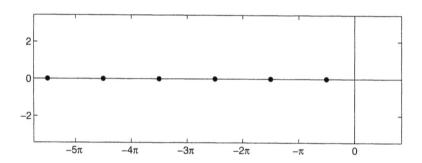

Figure 59.2: Spectrum of the Laplace evolution operator $\mathbf{L}$ for solutions even with respect to y. This operator is normal.

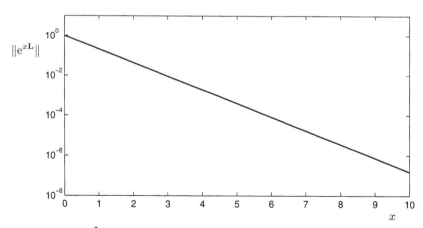

Figure 59.3: $\|e^{x\mathbf{L}}\|$ as a function of x. The slope of the line is determined by the spectral abscissa, $\alpha(\mathbf{L}) = -\pi/2$.

where $\mathbb{K}$ denotes the set of positive odd integers. Since the functions $\phi_k(0, y) = \cos(\frac{1}{2}k\pi y)$ are orthonormal in $L^2[-1, 1]$, the coefficients a_k are given simply by the Fourier integrals

$$a_k = \int_{-1}^{1} f(y) \cos(\tfrac{1}{2}k\pi y)\, dy. \qquad (59.6)$$

We can think of (59.1)–(59.3) as an evolution problem with respect to the 'time' variable x. If $u(x, \cdot)$ denotes a function of y at fixed x, i.e., $u(x, \cdot)(y) \equiv u(x, y)$, then the evolution problem can be written

$$\frac{d}{dx} u(x, \cdot) = \mathbf{L} u(x, \cdot), \qquad u(0, \cdot) = f, \qquad (59.7)$$

where $\mathbf{L}$ is a linear operator that we can define by means of (59.4)–(59.6). The following formula computes the Fourier coefficients of $u(x, \cdot)(y)$, inserts these in the series solution, and differentiates with respect to the space variable:

$$\mathbf{L} u(x, \cdot)(y) = \frac{d}{d\xi} \left(\sum_{k \in \mathbb{K}} e^{-\frac{1}{2}k\pi\xi} \cos(\tfrac{1}{2}k\pi y) \int_{-1}^{1} u(x, s) \cos(\tfrac{1}{2}k\pi s)\, ds \right) \Bigg|_{\xi=0}$$

$$= \sum_{k \in \mathbb{K}} -\tfrac{1}{2}k\pi \cos(\tfrac{1}{2}k\pi y) \int_{-1}^{1} u(x, s) \cos(\tfrac{1}{2}k\pi s)\, ds. \qquad (59.8)$$

We see readily that $\mathbf{L}$ is a self-adjoint operator on $L^2[-1, 1]$ whose spectrum consists of eigenvalues $-\frac{1}{2}k\pi$ with corresponding eigenfunctions $\cos(\frac{1}{2}k\pi y)$ (Figure 59.2). (If the factor k in (59.8) were changed to k^2, $\mathbf{L}$ would become an evolution operator for heat conduction.) The solution to (59.7) is $u(x, \cdot) = e^{x\mathbf{L}} f$, and since $\mathbf{L}$ is self-adjoint, the norms of the solution operators are determined by the eigenvalue with largest real part, $\|e^{x\mathbf{L}}\| = e^{-\frac{1}{2}\pi x}$, as shown in Figure 59.3.

To solve (59.1)–(59.3) on a computer, it would be natural to use the eigenfunction expansion. To make the problem finite, one could consider a finite set of eigenfunctions $\phi_1, \dots, \phi_N$ and a corresponding approximation

$$f(y) \approx \sum_{k \in \mathbb{K}} \widetilde{a}_k \cos(\tfrac{1}{2}k\pi y). \qquad (59.9)$$

Since the functions $\cos(\frac{1}{2}k\pi y)$ are orthogonal, the choice of coefficients $\widetilde{a}_k = a_k$ leads to an optimal approximation with respect to the 2-norm $\|\cdot\|$. In other words, simply truncating the infinite series $\{a_k\}$ is a reasonable procedure.

This is all there is to say about the Laplace evolution problem, except for the mention of one subtlety. Though one could hardly imagine a problem that seems better suited to the use of eigenfunction expansions, the fact is that these expansions are inefficient for some purposes. For example, suppose $f(y)$ is the function $f(y) = e - e^{y^2}$. Then the series coefficients decay at the rate $a_k = \mathcal{O}(k^{-3})$, and to achieve $\|f - f_N\| < 10^{-6}$, where f_N is the partial sum (59.9) with $\tilde{a}_k = a_k$, one needs $N \approx 600$. By contrast, in an expansion of $f(y)$ in Chebyshev polynomials the coefficients decrease geometrically and one achieves $\|f - f_N\| < 10^{-6}$ with $N = 13$. What causes such a dramatic difference? The answer is that the difficult part of the problem asymptotically is in the corners $x = 0$, $y = \pm 1$, where there is a singularity that can be represented only inefficiently by the eigenfunctions.[2] The observation that eigenfunction expansions may be inefficient was made famous by Orszag in the early 1970s [588] and is the foundation of the field of spectral methods for the numerical solution of PDEs (§§30, 43). For the present problem eigenfunctions are entirely suitable if the goal is the solution at some distance down the strip, but some other method is needed if one wants an accurate solution near the corners.

Now we turn to the Papkovich–Fadle problem. On the same semi-infinite strip S (Figure 59.4), we seek a function $u(x, y)$ satisfying

$$\Delta^2 u(x, y) = 0, \quad (x, y) \in S, \tag{59.10}$$

with boundary conditions

$$u(x, \pm 1) = u_y(x, \pm 1) = 0$$

and

$$u_y(0, y) = f(y), \quad u_x(0, y) = g(y). \tag{59.11}$$

Here Δ^2 is the biharmonic operator $\Delta^2 = \partial^4/\partial x^4 + 2\partial^4/\partial x^2 \partial y^2 + \partial^4/\partial y^4$, and for well-posedness we again add the condition $\lim_{x \to \infty} u(x, y) = 0$. Note that specifying $u_y(0, y)$ is equivalent to specifying $u(0, y)$; we assume that u is continuous on the boundary of S and thus that the compatibility condition $\int_{-1}^{1} f(y)\, dy = 0$ is satisfied.

In solid mechanics, $u(x, y)$ can be interpreted as the shape of a semi-infinite clamped plate. In fluid mechanics, it is the stream function of an incompressible fluid flow at zero Reynolds number (Stokes flow). The original papers in this field are those of Papkovich [600] and Fadle [254], and subsequent contributors include Smith [703], Benthem [53], Gaydon and Shepherd [303], Buchwald [112], Johnson and Little [434], Gupta [353], Bogy [66], Joseph and Sturges [437, 438], Gregory [345, 346, 347], and Spence [711, 712, 713]. (As discussed in several of these references, the

---

[2]Because of the periodicity of the cosine, the eigenfunctions are odd functions with respect to the points $y = 1$ and $y = -1$, whereas $f(y) = e - e^{y^2}$ does not have this property; in effect, we are trying to approximate a function with a discontinuous second derivative by cosines.

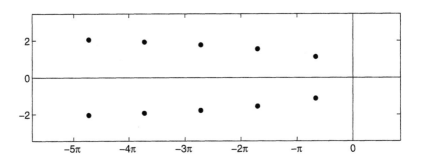

Figure 59.4: The Papkovich–Fadle evolution problem.

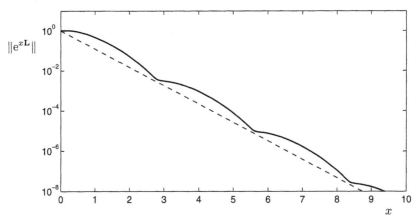

Figure 59.5: Spectrum of the Papkovich–Fadle operator $\mathbf{L}$ for solutions even with respect to y.

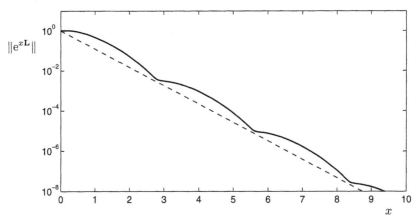

Figure 59.6: $\|e^{x\mathbf{L}}\|$ as a function of x. The dashed line corresponds to the spectral abscissa, $\alpha(\mathbf{L}) \approx -2.1062$.

above assumptions and boundary conditions are not the only physically reasonable ones. For example, many authors take data $u_{yy}(0, y)$ and $u_{xx}(0, y)$
instead of $u_y(0, y)$ and $u_x(0, y)$. These variations do not affect our main
points.)

For an interesting generalization of Papkovich–Fadle ideas to three-
dimensional flow in a cylindrical container with a rotating end wall, see [396].

Like the Laplace problem, the Papkovich–Fadle problem can be solved
by separation of variables. Restricting attention as before to solutions even
in y, we find that the solutions to (59.10) of the form $u(x, y) = X(x)Y(y)$
are the functions

$$\phi_k(x, y) = \mathrm{e}^{\lambda_k x} \phi_k(y), \qquad k = \pm 1, \pm 2, \ldots,$$

where

$$\phi_k(y) \;=\; \sin(\lambda_k)\cos(\lambda_k y) - y\cos(\lambda_k)\sin(\lambda_k y) \qquad (59.12)$$

and $\{\lambda_k\}$ are the roots in the open left half-plane of the equation $\lambda_k +$
$\sin(\lambda_k)\cos(\lambda_k) = 0$. The solutions to this equation come in complex conjugate pairs, which we denote by $\lambda_{\pm 1}, \lambda_{\pm 2}, \ldots$ in order of decreasing real
part, with $\mathrm{Im}\,\lambda_k > 0$ for $k > 0$ and $\mathrm{Im}\,\lambda_k < 0$ for $k < 0$. See Figure 59.5.

The eigenfunctions (59.12) are not orthogonal, and thus the Papkovich–
Fadle operator is nonnormal. But we must be precise about what we mean
by 'the Papkovich–Fadle operator', which we shall now denote by $\mathbf{L}$. In
keeping with the choice of first derivatives for the initial data in (59.11), it
is natural to define $\mathbf{L}$ via the equation

$$\frac{\mathrm{d}}{\mathrm{d}x}\begin{pmatrix} u_y(x, \cdot) \\ u_x(x, \cdot) \end{pmatrix} = \mathbf{L}\begin{pmatrix} u_y(x, \cdot) \\ u_x(x, \cdot) \end{pmatrix},$$

with solution

$$\begin{pmatrix} u_y(x, \cdot) \\ u_x(x, \cdot) \end{pmatrix} = \mathrm{e}^{x\mathbf{L}}\begin{pmatrix} f \\ g \end{pmatrix}.$$

Thus the data in the evolution problem are block 2-vectors and the solution
operator is a block 2×2 operator matrix whose form we do not explicitly
write out. (It would not do to let $(u, u_x)^{\mathrm{T}}$ be the evolution variable instead of $(u_y, u_x)^{\mathrm{T}}$, because u and u_x are dimensionally different and the
associated inner product would not make physical sense.) The equation
analogous to (59.5) is

$$\begin{pmatrix} u_y(x, y) \\ u_x(x, y) \end{pmatrix} = \sum_{|k|=1}^{\infty} a_k \begin{pmatrix} (\phi_k)_y \\ (\phi_k)_x \end{pmatrix}. \qquad (59.13)$$

Figure 59.6 plots $\|\mathrm{e}^{x\mathbf{L}}\|$ against x. The fact that the curve is not straight
reveals the nonnormality. Physically, the regular oscillations in this curve
are easy to interpret. If S is a semi-infinite channel of fluid, they correspond

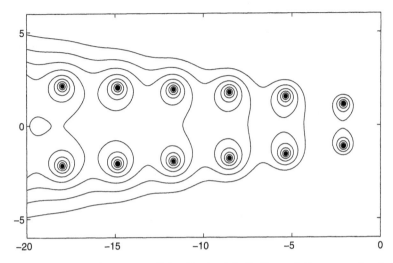

Figure 59.7: ε-pseudospectra of the Papkovich–Fadle evolution operator $\mathbf{L}$ for $\varepsilon = 10^{-0.6}, 10^{-0.8}, \ldots, 10^{-2.0}$.

to an infinite succession of counter-rotating 'Moffatt vortices'. (Since the amplitudes decay exponentially, only the first three or four of these can be observed in the laboratory.) If S is an infinite solid plate, they correspond to the back-and-forth pattern of bending one can induce in the plate (again with exponentially decaying amplitude) by clamping it at the end. For beautiful high-accuracy computations of these oscillations on finite square domains, see [64].

Figure 59.7 plots ε-pseudospectra of $\mathbf{L}$. The degree of nonnormality is mild, so the values of ε chosen for the plot are large: $\varepsilon = 10^{-0.6}$, $10^{-0.8}$, ..., $10^{-2.0}$. It is evident that the nonnormality becomes more pronounced as one moves into the left half-plane. This corresponds to the singularity in this problem at the corners of the strip, a phenomenon of small space scales that can only be resolved by the higher eigenfunctions. This singularity is well understood and is discussed, for example, in [29, §4.8].

We come now to the computational question, How can the Papkovich–Fadle eigenfunctions be used for computing solutions to (59.10)–(59.11)? (And how were Figures 59.6 and 59.7 produced?) Much of the literature cited above is devoted to this matter. One problem considered in a number of these papers is under what circumstances the series (59.13) converges. Another is the convergence of the analogous series in which the coefficients a_k, $|k| = 1, \ldots, N/2$, are replaced by approximations obtained by collocation in N points or by other analogous methods. Such a process can be interpreted as *truncation of an infinite system of equations;* in general it fails, and various acceleration and improvement devices have been proposed to remedy the situation [711, 713].

Alternatively, we can recognize from the start that if we are to use an expansion in N eigenfunctions, we must begin by solving an approximation problem [303]

$$
\begin{pmatrix} f(y) \\ g(y) \end{pmatrix} \approx \sum_{|k|=1}^{N/2} \widetilde{a}_k \begin{pmatrix} (\phi_k)_y \\ (\phi_k)_x \end{pmatrix}.
$$

This problem is the Papkovich–Fadle analogue of (59.9), but now, since the eigenfunctions are no longer orthogonal, there is no reason to choose $\widetilde{a}_k = a_k$. Instead the obvious procedure is to take coefficients $\widetilde{a}_k$ corresponding to a least-squares fit. Equivalently, we can speak of a projection of the data $(f(y), g(y))^{\mathrm{T}}$ onto the space spanned by the finite collection of eigenfunctions; such projections are the standard tool of spectral methods.

The mechanics of the projection process are a matter of straightforward numerical linear algebra. First we discretize the y variable by a number of points M—not equal to N as in most of the theoretical papers, but greater. (For example, we can take $M = 2N$. The linear algebra described here is equally valid if we leave the y variable continuous, i.e., $M = \infty$ [30].) Then, with $\mathbf{L}$ denoting the discrete analogue of $\mathbf{L}$, our aim is to project $\mathbf{L}$ onto the column space of an $M \times N$ matrix $\mathbf{V}$ whose columns are selected eigenvectors of $\mathbf{L}$, and hence satisfy $\mathbf{LV} = \mathbf{VD}$ for some $N \times N$ diagonal eigenvalue matrix $\mathbf{D}$. Let $\mathbf{V} = \mathbf{QR}$ be a QR (Gram–Schmidt) decomposition of $\mathbf{V}$, with $\mathbf{Q}$ of dimension $M \times N$ and $\mathbf{R}$ of dimension $N \times N$, so that we have $\mathbf{Q}^*\mathbf{V} = \mathbf{R}$ and $\mathbf{Q} = \mathbf{VR}^{-1}$ (cf. §40). Then the upper triangular matrix

$$
\mathbf{Q}^*\mathbf{LQ} = \mathbf{Q}^*\mathbf{LVR}^{-1} = \mathbf{Q}^*\mathbf{VDR}^{-1} = \mathbf{RDR}^{-1}
$$

is the representation of the desired projection with respect to the orthogonal basis of columns of $\mathbf{Q}$. Figures 59.6 and 59.7 were computed with matrices $\mathbf{RDR}^{-1}$ corresponding to parameters $N = 20$, $M = 40$.

These calculations are easily implemented in practice, for example, in MATLAB. The condition numbers of the finite bases of eigenfunctions are of modest size (e.g., $\kappa(\mathbf{V}) \approx 6.52$, 16.3, 44.5, 132 for $N = 4$, 8, 16, 32), so numerical stability is not much of a concern. If high accuracy very near the corners is required, spectral methods will be superior, but for most purposes, eigenfunction expansions are an excellent way to solve the Papkovich–Fadle problem.

60 · Lasers

Lasers, among the great technological discoveries of the twentieth century, are a fine illustration of the importance of eigenmodes of normal and near-normal systems. There are also some lasers, involving what are known as unstable resonators, that are mathematically far from normal. It was partly in connection with such problems that H. J. Landau of Bell Laboratories invented pseudospectra in the 1970s (see §6), and on the more physical side, the nonnormal aspects of this class of lasers have been emphasized since the 1980s by the well-known laser expert and textbook author A. E. Siegman.

A laser is an appealing system to the eigenvalue aficionado, for its operation depends on the interplay of two distinct eigenvalue problems. One of these, Hermitian, is governed by the Schrödinger operator that determines the energy states of the excited atoms or molecules in the lasing medium. If the laser emits light at a frequency corresponding to the gap between two states of a neon atom, for example, the frequency is mathematically the difference of two eigenvalues. The other eigenvalue problem involves the cavity in which light energy at this frequency accumulates, which is usually tuned to resonate very precisely so the output becomes a coherent beam. For most lasers this problem is close to normal, though it cannot be exactly normal since the mirror at one end of the cavity must reflect imperfectly to allow some light to leak out.

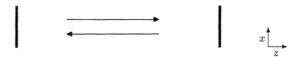

Figure 60.1: A laser cavity consists of a space between two mirrors, which may be flat or curved; one of the mirrors is often partially reflective to allow some light to escape. The cavity is filled with a solid, liquid, or gas populated by atoms or molecules in excited energy states.

Figure 60.1 suggests a typical laser cavity bounded by two mirrors separated by a distance L in the z direction, typically equal to thousands of wavelengths of light at the frequency of interest. For simplicity we take the configuration to be two-dimensional, with x as the transverse variable. (In the laboratory such a geometry could be realized by a wide cavity with mirrors in the shape of strips.) Let us imagine for a moment that the space between the mirrors is a vacuum. The governing equations are the free-space Maxwell equations, which reduce to the second-order wave equation for any of the vector components $E = E(x, z)$ of the electric field.[1] By confining our attention to the frequency $\omega = ck$ corresponding to a fixed wave number k, where c is the speed of light, we obtain the Helmholtz or reduced wave equation

$$E_{xx} + E_{zz} + k^2 u = 0,$$

where the double subscripts denote second partial derivatives with respect to the indicated variables. Since we are concerned with propagation approximately in the z direction, we now replace E by the new variable u defined by $E(x, z) = e^{-ikz} u(x, z)$. (As is customary with linear equations, we formulate the problem in complex variables; physical quantities are obtained at the end by taking real parts.) The differential equation becomes

$$u_{xx} + u_{zz} - 2iku_z = 0,$$

and if the wave propagation is approximately longitudinal, u will vary slowly with respect to z. For most lasers it is accordingly appropriate to make the *paraxial approximation*,[2] deleting the u_{zz} term to obtain

$$u_z = -\frac{i}{2k} u_{xx}. \tag{60.1}$$

We may regard this *paraxial equation* as an evolution equation with respect to the 'time' variable z for functions u defined on $\mathbb{R}$. In the usual L^2 inner product the operator $u \mapsto (-i/2k)u_{xx}$ is normal and purely dispersive; its spectrum is the nonnegative imaginary axis (or the nonpositive imaginary axis if $k < 0$). This operator has no eigenfunctions,[3] since the eigenfunctions 'ought' to be complex exponentials, which have infinite norm. Combining these not-quite eigenfunctions in a Fourier integral yields a *Huygens–Fresnel integral* representing the solution at position z in terms of the solution at $z = 0$:

$$u(x, z) = \sqrt{\frac{ik}{2\pi z}} \int_{-\infty}^{\infty} e^{-ik(x-s)^2/2z} u(s, 0) \, ds. \tag{60.2}$$

This integral operator mapping $u(x, 0)$ to $u(x, z)$ is unitary and hence normal and energy-conserving; its spectrum is the unit circle.[4]

---

[1] The classic text on the science of optics is the monograph by Born and Wolf [75]. Our treatment here is adapted from Chapter 16 et seq. of Siegman's book *Lasers* [685]. There are also many other good books on laser physics and applications.

[2] Equivalently, one can formulate the wave propagation problem in terms of Huygens's principle of spherical waves and then make the *Fresnel approximation* to these waves, again based on the assumption that the propagation is predominantly longitudinal.

[3] In lieu of eigenfunctions, it is often convenient to express solutions in terms of 'Gaussian beams', functions whose x dependence is given by Hermite polynomials times Gaussians and which evolve in z according to a simple formula.

[4] By deleting factors of i in (60.1) and (60.2), i.e., replacing dispersion by dissipation, we obtain the familiar representation of solutions of the heat equation in terms of convolution with a Gaussian kernel.

The most interesting physics—and the nonnormality—come when we introduce boundary conditions. Consider the simplest possible situation suggested by Figure 60.1, in which a two dimensional cavity of length L is open on the sides and bounded at the ends by two flat mirrors, which we take to have endpoints $x = \pm 1$. Imagine a packet of light at the left mirror as it begins to propagate to the right, with cross-sectional electric field given by $u_0 = u_0(x)$. The light propagates rightward to the second mirror, and there, the portion of it with $|x| \leq 1$ reflects, while the portion with $|x| > 1$ radiates to infinity and is lost. With repeated reflections this process of trimming energy with $|x| > 1$ takes place repeatedly, yielding a dynamical system that can be thought of as evolving in discrete time on the domain $-1 \leq x \leq 1$, with each time unit corresponding to the time L/c that it takes for light to propagate from one mirror to the other. In one such transit, the signal evolves according to the following truncation of (60.2):

$$u(x) = \mathbf{A}u_0(x) = \sqrt{\frac{\mathrm{i}F}{\pi}} \int_{-1}^{1} \mathrm{e}^{-\mathrm{i}F(x-s)^2} u_0(s)\,\mathrm{d}s, \qquad (60.3)$$

where the *Fresnel number* is defined by $F = k/2L$. Mathematically speaking, this is a compact integral operator on $L^2[-1,1]$, implying that it is bounded and has a countably infinite set of singular values decreasing to zero.

And so it is that the science of lasers is heavily concerned with the behavior of linear integral operators defined by kernels that are complex symmetric but non-Hermitian. This kind of analysis was introduced by Fox and Li in the early 1960s [284, 285, 690]. Whatever the geometry may be, the Fox and Li approach considers the operator $\mathbf{A}$ that describes the change of shape that takes place due to dispersion and diffraction at the edges as a wave of given frequency makes one transit across the cavity, or one round-trip. By taking powers $\mathbf{A}^k$, one models repeated reflections, and after many reflections, the signal will be dominated by the eigenmode whose eigenvalue has modulus $|\lambda|$ closest to 1. Of course, if $\mathbf{A}$ captured all of the physics of the system, the light would attenuate after repeated bounces and a laser cavity would be no more interesting than a pair of mirrors in a clothing store. In actuality, the space between the mirrors is filled with a medium that has been excited so that high energy levels are more highly populated than low ones. Consequently, by the process known as stimulated emission, the light as it passes through releases more light. Amplification takes place, bringing the total round trip amplification factor above 1. If all eigenmodes are equally amplified, the eigenmode of $\mathbf{A}$ with $|\lambda|$ closest to 1 will be the one that attains the highest amplification factor. A coherent oscillation quickly builds up to an amplitude where the excited atoms are sufficiently depleted to bring the amplification and decay

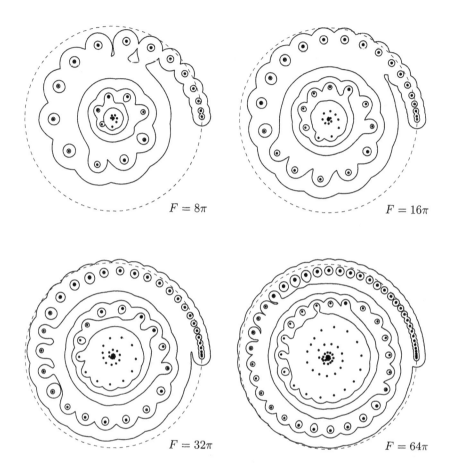

Figure 60.2: Eigenvalues and ε-pseudospectra of the operator (60.3) for $F = 8\pi, 16\pi, 32\pi, 64\pi$, with $\varepsilon = 10^{-1}, 10^{-1.5}, \ldots, 10^{-3}$ in each case; the dashed curve is the unit circle. The operator is nonnormal, but only mildly so; the dominant outer modes are nearly orthogonal.

into balance; and out comes an almost perfectly focused and coherent laser beam, locked into that dominant eigenmode of the resonant cavity.[5]

Figure 60.2, computed by a high-accuracy spectral discretization, shows

---

[5]More precisely, a laser mode is not just a bouncing wave packet but also extends through the z dimension as a standing wave. This means it must have a fixed phase at each point in space, which might seem to suggest that only eigenvalues λ of $\mathbf{A}$ with zero argument could be relevant. Fortunately, it is not u but E that must have a fixed phase, and the e^{-ikz} factor connecting the two oscillates so fast than any phase of λ will do; the value of k adjusts very slightly so that the e^{-ikz} factor compensates for any nonzero phase in λ.

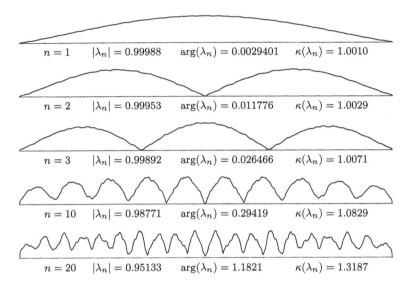

$n = 1$	$|\lambda_n| = 0.99988$	$\arg(\lambda_n) = 0.0029401$	$\kappa(\lambda_n) = 1.0010$
$n = 2$	$|\lambda_n| = 0.99953$	$\arg(\lambda_n) = 0.011776$	$\kappa(\lambda_n) = 1.0029$
$n = 3$	$|\lambda_n| = 0.99892$	$\arg(\lambda_n) = 0.026466$	$\kappa(\lambda_n) = 1.0071$
$n = 10$	$|\lambda_n| = 0.98771$	$\arg(\lambda_n) = 0.29419$	$\kappa(\lambda_n) = 1.0829$
$n = 20$	$|\lambda_n| = 0.95133$	$\arg(\lambda_n) = 1.1821$	$\kappa(\lambda_n) = 1.3187$

Figure 60.3: Eigenmodes 1, 2, 3, 10, 20 (absolute values) associated with the laser operator (60.3) with $F = 64\pi$. The horizontal axis is the interval $-1 \le x \le 1$, and the eigenvalues λ_n and their condition numbers are listed. The small-scale wiggles are not numerical artifacts but genuine. Laser engineers would generally plot the squares of these curves, since $|u(x)|^2$ represents power density.

the spectrum of the operator $\mathbf{A}$ of (60.3) for four values of F. One sees that the eigenvalues begin near 1 and wind in along elegant spirals toward the origin. Details of this behavior were worked out in the 1960s and 1970s by Vaĭnshteĭn, Hochstadt, and Cochran and Hinds, among others [157, 404, 787]. Some corresponding eigenmodes, approximately orthogonal but not exactly so, are shown in Figure 60.3. The dominant modes do not oscillate much in x, and by (60.1), this implies that they do not disperse rapidly, so that not much energy reaches the edges after one transit. Higher modes have more oscillations and faster dispersion, so that more energy is lost at the edges, making their values of $|\lambda|$ smaller. In Figure 60.2 we also see that as F increases, the eigenvalues hew closer to the unit circle and wind toward the origin more gradually. Again the explanation is in (60.1), which implies that as k increases, the dispersion diminishes so that again there is less loss at the edges.

Figure 60.2 shows pseudospectra as well as eigenvalues, and this is a set of plots of special importance in the history of our subject, for it was partly in connection with this operator and its unstable variant, which we shall introduce in a moment, that H. J. Landau invented pseudospectra of

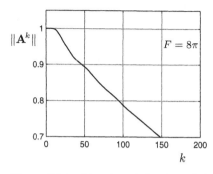
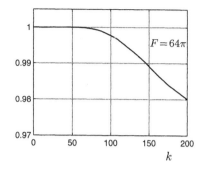

Figure 60.4: Transient behavior of the laser operator (60.3) for $F = 8\pi, 64\pi$. Some of the numbers are listed in Table 60.1.

Table 60.1: Norms of powers of the laser operator (60.3).

k	$\|\mathbf{A}^k\|$, $F = 8\pi$	$\|\mathbf{A}^k\|$, $F = 64\pi$
0	1	1
5	0.99999982	0.9999999999...
10	0.99914	0.9999999999...
20	0.973	0.9999999999...
40	0.911	0.99999970
80	0.827	0.99940
160	0.681	0.987
320	0.460	0.964
640	0.211	0.928

nonnormal operators in the mid-1970s [479, 480].[6] As described in §6, the only previous invention of this idea that we know of, in Varah's 1967 Ph.D. thesis, was motivated by purely numerical considerations. Thus Landau's analysis of (60.3) and related problems seems to have been the first occasion on which pseudospectra were recommended as a way of understanding the physical behavior of a particular operator. Now in fact, the pseudospectrally trained eye sees immediately from Figure 60.2 that in this problem, the nonnormality is quite mild. For the outer eigenvalues particularly, the ones that matter for laser physics, one cannot expect to encounter nonnormal effects of great magnitude.[7] Nevertheless, there is an interesting nonnormal effect here of small magnitude, and this is that as k increases,

---

[6]Images like this were first produced by Andrew Spratley (unpublished) at Oxford in 1998 and appeared in print for the first time in [774] and also on the cover of that *Acta Numerica* volume.

[7]Thus (60.3) is akin to the Gauss–Seidel iteration matrix for the classic discretization of the Laplace operator (§24): It is nonnormal, but the nonnormality is concentrated in the inner eigenvalues that do not matter for applications.

the value of $\|\mathbf{A}^k\|$ remains almost exactly equal to 1 for a while before eventually beginning its slow exponential descent (Figure 60.4 and Table 60.1). This behavior can be easily explained. A narrow beam aligned with the mirrors may bounce back and forth several times with exponentially little leakage at the edges, slowly widening due to dispersion. Only when it has widened to the spatial scale of the mirror width does the loss at the edges become significant. The parameter F determines the lower limit, caused by diffraction, on how narrow this initial beam can be and thus how long its eventual widening can be deferred.

Meanwhile, however, there is another class of laser problems for which the governing operators are more strongly nonnormal, as has been emphasized by A. E. Siegman in a succession of publications [459, 684, 686, 687, 688, 685]. These are the high-power lasers associated with what are known as 'unstable resonators', a topic Landau studied in one of his papers concerning lasers and pseudospectra in the 1970s [479].

Figure 60.5: In a practical laser, typically one or both mirrors is curved. A stable resonator is one in which rays tend to remain aligned, whereas an unstable resonator is one where they tend to diverge, giving magnification $M > 1$.

The distinction between stable and unstable resonators is purely geometric (Figure 60.5). Imagine that instead of the flat mirrors of Figure 60.1, we have curved mirrors at one or both ends of the cavity. Depending on the curvatures, an image as it passes through the system may remain of fixed size (apart from dispersion) or it may be magnified. The former case is called *stable* and the latter is *unstable*. For many purposes, stable laser cavities are preferred because they can support narrowly focused modes with sharp resolution and minimal losses. For high-power lasers, however, narrowly focused modes are disadvantageous because they do not extract energy from a large volume of the cavity. For this and other reasons, since the 1960s, some lasers have been based on unstable cavities [683].

For a wide class of unstable resonators, the Fox and Li style of analysis requires just a simple modification of (60.3): The Huygens–Fresnel operator

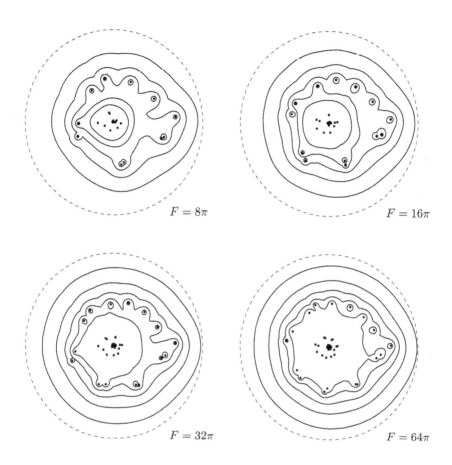

$F = 8\pi$ $F = 16\pi$

$F = 32\pi$ $F = 64\pi$

Figure 60.6: Repetition of Figure 60.2 for the unstable resonator (60.4) with $M = 2$. The eigenvalues are much smaller, but the operator is strongly nonnormal.

is now defined by

$$\mathbf{A}u_0(x) = \sqrt{\frac{\mathrm{i}F}{\pi}} \int_{-1}^{1} \mathrm{e}^{-\mathrm{i}FM(x/M-s)^2} u_0(s)\,\mathrm{d}s, \qquad (60.4)$$

where $M \geq 1$ is the magnification. This formula essentially goes back to the papers [98, 285] and is derived, for example, in [685, Chap. 22]. Figure 60.6 shows spectra and pseudospectra for this operator with $M = 2$ and $F = 8\pi, 16\pi, 32\pi, 64\pi$. We see that the eigenvalues are much smaller than in Figure 60.2, but the nonnormality is now pronounced. Figure 60.7 shows some eigenmodes, revealing that they wiggle more than those in Figure 60.3. The irregular structure of eigenmodes of unstable resonators has been recognized for many years, and indeed, it has been shown that

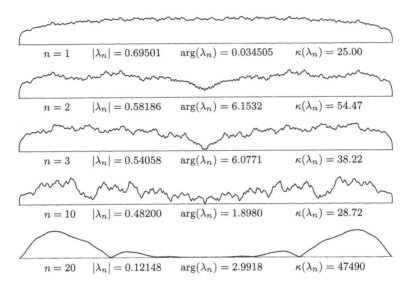

$n = 1$	$|\lambda_n| = 0.69501$	$\arg(\lambda_n) = 0.034505$	$\kappa(\lambda_n) = 25.00$
$n = 2$	$|\lambda_n| = 0.58186$	$\arg(\lambda_n) = 6.1532$	$\kappa(\lambda_n) = 54.47$
$n = 3$	$|\lambda_n| = 0.54058$	$\arg(\lambda_n) = 6.0771$	$\kappa(\lambda_n) = 38.22$
$n = 10$	$|\lambda_n| = 0.48200$	$\arg(\lambda_n) = 1.8980$	$\kappa(\lambda_n) = 28.72$
$n = 20$	$|\lambda_n| = 0.12148$	$\arg(\lambda_n) = 2.9918$	$\kappa(\lambda_n) = 47490$

Figure 60.7: Repetition of Figure 60.3 for the unstable resonator (60.4) with $F = 64\pi$ and $M = 2$. Now the modes are smooth approximations of fractals. Again, engineers would generally plot the squares of these curves, which would look still more irregular.

these modes are smoothed approximations of fractals and become true fractals in the zero-dispersion limit, $F \to \infty$ [446]. In fact, one can create a fractal pattern by essentially the same mechanism by pointing a video camera at a television screen [169].

Unlike the pseudospectra of Figure 60.2, those of Figure 60.6 suggest that the operator of (60.4) will exhibit significant transient behavior. Figure 60.8 and Table 60.2 confirm this prediction. As before, the initial norms $\|\mathbf{A}^k\|$ are exponentially close to 1, but now the cutoff is sharper and gets stronger rather than weaker as $F \to \infty$. A difference is that the transient lasts for fewer steps since now a narrow beam must double in width with each passage through the cavity rather than just widening algebraically. This figure is very similar to our portrayal of the 'cutoff phenomenon' in the riffle shuffling of decks of cards (Figure 57.2); the associated pseudospectra of Figure 60.6 are also much like their riffle shuffle counterparts (Figure 57.3). In both cases the transient effect corresponds to a wave whose coordinate doubles at each step before eventually coming near its final position (a beam of light for lasers, a wave of probability of rising sequence numbers for shuffling), and it is this doubling that makes the length of the transient grow logarithmically with the problem parameter (Fresnel number F for lasers, number of cards n for shuffling). Other systems in this book in which the norms of an evolution operator stay very

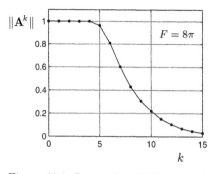
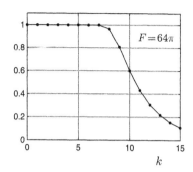

Figure 60.8: Repetition of Figure 60.4 for the unstable resonator operator (60.4) with $M = 2$.

Table 60.2: Norms of powers of the laser operator (60.4) with $M = 2$.

k	$\|\mathbf{A}^k\|$, $F = 8\pi$	$\|\mathbf{A}^k\|$, $F = 64\pi$
0	1	1
2	0.9999999999...	0.9999999999...
4	0.99947	0.9999999999...
6	0.809	0.999999933
8	0.430	0.965
10	0.219	0.601

close to 1 for a time before eventually decreasing include weakly dissipative convection-diffusion operators (§12) and Gauss–Seidel sweeps in the 'upwind' direction (§25). In all of these problems, the initial behavior of the system is controlled by advection but the asymptote is controlled by diffusion or leakage out of a boundary.

These nonnormal effects have a big impact on the physics of lasers based on unstable resonators. One's first thought in looking at Figure 60.8 might be that because of the transient, these lasers would have difficulty settling into dominant modes, and indeed, this prediction was made by Landau in his papers. However, this is not observed in practice. The speed of light is so great that any initial effects of this kind vanish almost instantly.

The actual physical consequence of nonnormality in unstable laser cavities is more subtle, having to do with the *noise* that is observed when they operate. The idea is as follows. Ideally, one would like a laser to emit a perfectly coherent beam. However, even if the cavity is a perfect resonator, this will not be possible because some photons will be emitted spontaneously by the lasing medium with the correct frequency but with a random phase, causing the phase of the overall signal to drift slightly. As a consequence, one never observes a frequency sharper than a certain minimal linewidth given by a formula devised by two of the Nobel Prize–winning laser pioneers, the *Schawlow–Townes formula* [665].

Now in a laser based on an unstable resonator, just as in the stable case, photons will be emitted spontaneously at the correct frequency but with random phases. The effect of such a photon, however, may now be much enhanced if it is emitted in a form that excites transient behavior as in Figure 60.8, for then it may linger a long time before decaying, stimulating emission of additional photons that further distort the phase.[8] Speaking for simplicity (and omitting rigor) as if A is a matrix diagonalizable in the form $A = V\Lambda V^{-1}$, with the dominant eigenmode being the first column v_1 of V, suppose the spontaneously emitted photon corresponds to a vector x with $\|x\| = 1$. Then $c = V^{-1}x$ is the vector of coefficients of x in a (nonorthogonal) expansion in the columns of V, i.e., the eigenvectors of A. In particular, the coefficient c_1 associated with v_1 is w_1^*x, where w_1^* is the first row of V^{-1}, and we can maximize this coefficient by taking $x = w_1/\|w_1\|$, which gives

$$|c_1| = \|w_1\|. \tag{60.5}$$

This is a familiar phenomenon in nonnormal dynamics: One gets the greatest response by choosing inputs based not on the eigenvectors of the system itself, but on the eigenvectors of the adjoint.

From mathematics like this comes a fundamental result for laser physics: In an unstable resonator, the Schawlow–Townes linewidth must be increased by a factor known as the *Petermann excess noise factor K*, which in the notation above takes the form

$$K = \|w_1\|^2. \tag{60.6}$$

This conclusion has been established both theoretically [57, 577, 608, 686, 687] and experimentally [145, 798]. In practice, K may be as large as hundreds or thousands. We recognize $\|w_1\|$ as the *condition number* $\kappa(\lambda_1)$ of the eigenvalue λ_1 of A (§52). Thus an equivalent statement is that the excess noise factor is the square of the condition number:

$$K = \kappa(\lambda_1)^2. \tag{60.7}$$

In the system of Figure 60.3, for example, the Petermann excess noise factor will be $K \approx (25.00)^2 = 625$. Perhaps (60.7) suggests a new interpretation of the notion of excess noise, for a quantity appears in the original Schawlow–Townes formula that is interpreted as the square of the bandwidth of the oscillation frequency in the laser cavity. Multiplying that quantity by K could be interpreted as incorporating the algebraic sensitivity of the eigenvalues directly into the Schawlow–Townes formula.

---

[8]Related mathematics of stochastically forced nonnormal systems has been investigated in [25, 262, 263].

In this section we have discussed nonnormality introduced in optical systems by boundary conditions, but just as in other parts of this book, such as §§8, 11, and 22, it is equally possible to find such effects caused by variable coefficients rather than boundaries. Petermann's original discussion of the excess noise factor was in a context of this kind (a gain-guided duct), and Siegman emphasizes that other variable coefficient optical systems also have analogous nonnormal effects, such as those associated with the notion of 'optical twist' [689].

References

Numbers at the end of each entry indicate pages on which that reference is cited.

[1] L. V. Ahlfors. *Complex Analysis.* McGraw-Hill, New York, third edition, 1979. (*180, 181*)

[2] M. Ahues, A. Largillier, and B. V. Limaye. *Spectral Computations for Bounded Operators.* Chapman and Hall/CRC, Boca Raton, FL, 2001. (*294*)

[3] C. M. Aiken, A. M. Moore, and J. H. Middleton. The nonnormality of coastal ocean flows around obstacles, and their response to stochastic forcing. *J. Phys. Oceanography*, 32:2955–2974, 2002. (*227*)

[4] C. M. Aiken, A. M. Moore, and J. H. Middleton. Non-normal perturbation growth in idealised island and headland wakes. *Dyn. Atmos. Oceans*, 37:171–195, 2003. (*227*)

[5] R. Alam and S. Bora. Effect of linear perturbation on spectra of matrices. *Linear Algebra Appl.*, 368:329–342, 2003. (*465*)

[6] E. Anderson, Z. Bai, C. Bischof, S. Blackford, J. Demmel, J. Dongarra, J. Du Croz, A. Greenbaum, S. Hammarling, A. McKenney, and D. Sorensen. *LAPACK User's Guide.* SIAM, Philadelphia, third edition, 1999. (*371, 503*)

[7] N. Anderson, E. B. Saff, and R. S. Varga. An extension of the Eneström–Kakeya theorem and its sharpness. *SIAM J. Math. Anal.*, 12:10–22, 1981. (*260*)

[8] P. W. Anderson. Absence of diffusion in certain random lattices. *Phys. Rev.*, 109:1492–1505, 1958. (*339*)

[9] P. Andersson, M. Berggren, and D. S. Henningson. Optimal disturbances and bypass transition in boundary layers. *Phys. Fluids*, 11:134–150, 1999. (*225*)

[10] A. Antkowiak and P. Brancher. Transient energy growth for the Lamb–Oseen vortex. *Phys. Fluids*, 16:L1–L4, 2004. (*228*)

[11] V. I. Arnol'd. Modes and quasimodes. *Funct. Anal. Appl.*, 6:94–101, 1972. (*41, 102*)

[12] W. E. Arnoldi. The principle of minimized iterations in the solution of the matrix eigenvalue problem. *Quart. Appl. Math.*, 9:17–29, 1951. (*263*)

[13] U. M. Ascher, R. M. M. Mattheij, and R. D. Russell. *Numerical Solution of Boundary Value Problems for Ordinary Differential Equations.* SIAM, Philadelphia, 1995. (*98, 113*)

[14] U. M. Ascher and L. R. Petzold. *Computer Methods for Ordinary Differential Equations and Differential-Algebraic Equations.* SIAM, Philadelphia, 1998. (*423*)

[15] S. F. Ashby. CHEBYCODE: A FORTRAN implementation of Manteuffel's adaptive Chebyshev algorithm. Technical Report 1203, Department of Computer Science, University of Illinois at Urbana-Champaign, 1985. (*255, 256*)

[16] A. Aslanyan and E. B. Davies. Spectral instability for some Schrödinger operators. *Numer. Math.*, 85:525–552, 2000. (*98*)

[17] O. Axelsson. *Iterative Solution Methods.* Cambridge University Press, Cambridge, 1994. (*244*)

[18] J. S. Baggett. Pseudospectra of an operator of Hille and Phillips. Technical Report 94-15, Interdisciplinary Project Center for Supercomputing, ETH Zürich, 1994. (*189, 190*)

[19] J. S. Baggett, T. A. Driscoll, and L. N. Trefethen. A mostly linear model of transition to turbulence. *Phys. Fluids*, 7:833–838, 1995. (*213*)

[20] J. S. Baggett and L. N. Trefethen. Low-dimensional models of subcritical transition to turbulence. *Phys. Fluids*, 9:1043–1053, 1997. (*208, 213*)

[21] J. Baglama, D. Calvetti, G. H. Golub, and L. Reichel. Adaptively preconditioned GMRES algorithms. *SIAM J. Sci. Comput.*, 20:243–269, 1998. (*255*)

[22] Z. Bai, D. Day, J. Demmel, and J. Dongarra. A test matrix collection for non-Hermitian eigenvalue problems (release 1.0). Manuscript, October 1996. (*270*)

[23] Z. Bai, J. Demmel, J. Dongarra, A. Ruhe, and H. van der Vorst, editors. *Templates for the Solution of Algebraic Eigenvalue Problems: A Practical Guide*. SIAM, Philadelphia, 2000. (*264, 372, 381*)

[24] Z. Bai, J. Demmel, and A. McKenney. On computing condition numbers for the nonsymmetric eigenproblem. *ACM Trans. Math. Software*, 19:202–223, 1993. (*479*)

[25] B. Bamieh and M. Dahleh. Energy amplification in channel flows with stochastic excitation. *Phys. Fluids*, 13:3258–3269, 2001. (*552*)

[26] E. Barbier. Note sur le problème de l'aiguille et le jeu du joint couvert. *J. Math. Pures Appl.*, 5:273–286, 1860. (*181, 182*)

[27] D. Barkley and L. S. Tuckerman. Stability analysis of perturbed plane Couette flow. *Phys. Fluids*, 11:1187–1195, 1999. (*228*)

[28] J. Barkmeijer, T. Iversen, and T. N. Palmer. Forcing singular vectors and other sensitive model structures. *Quart. J. Royal Met. Soc.*, 129:2401–2423, 2003. (*227*)

[29] G. K. Batchelor. *An Introduction to Fluid Dynamics*. Cambridge University Press, Cambridge, 1967. (*540*)

[30] Z. Battles and L. N. Trefethen. An extension of MATLAB to continuous functions and operators. *SIAM J. Sci. Comput.*, 25:1743–1770, 2004. (*541*)

[31] F. L. Bauer. On the field of values subordinate to a norm. *Numer. Math.*, 4:103–113, 1962. (*172, 174*)

[32] F. L. Bauer and C. T. Fike. Norms and exclusion theorems. *Numer. Math.*, 2:137–141, 1960. (*20, 475*)

[33] H. Baumgärtel. *Analytic Perturbation Theory for Matrices and Operators*. Birkhäuser, Basel, 1985. (*473, 477*)

[34] C. A. Bavely and G. W. Stewart. An algorithm for computing reducing subspaces by block diagonalization. *SIAM J. Numer. Anal.*, 16:359–367, 1979. (*468*)

[35] D. Bayer and P. Diaconis. Trailing the dovetail shuffle to its lair. *Ann. Appl. Prob.*, 2:294–313, 1992. (*508, 519, 520*)

[36] R. Beals and C. Fefferman. On local solvability of linear partial differential equations. *Ann. Math.*, 97:482–498, 1973. (*126, 129*)

[37] R. M. Beam and R. F. Warming. The asymptotic spectra of banded Toeplitz and quasi-Toeplitz matrices. *SIAM J. Sci. Comput.*, 14:971–1006, 1993. (*57, 79, 465*)

[38] C. Beattie. Harmonic Ritz and Lehmann bounds. *Elect. Trans. Numer. Anal.*, 7:18–39, 1998. (*246*)

[39] C. Beattie, M. Embree, and J. Rossi. Convergence of restarted Krylov subspaces to invariant subspaces. *SIAM J. Matrix Anal. Appl.*, 25:1074–1109, 2004. (*274, 277*)

[40] C. A. Beattie, M. Embree, and D. C. Sorensen. Convergence of polynomial restart Krylov methods for eigenvalue computations. *SIAM Review.* To appear. (*274, 277*)

[41] C. Beattie and I. C. F. Ipsen. Inclusion regions for matrix eigenvalues. *Linear Algebra Appl.*, 358:281–291, 2003. (*435*)

[42] C. Bekas and E. Gallopoulos. Cobra: Parallel path following for computing the matrix pseudospectrum. *Parallel Comp.*, 27:1879–1896, 2001. (*396*)

[43] C. Bekas and E. Gallopoulos. Parallel computation of pseudospectra by fast descent. *Parallel Comp.*, 28:223–242, 2002. (*396*)

[44] C. Bekas, E. Gallopoulos, and V. Simoncini. Pseudospectra computation of large matrices. Manuscript, 2004. (*396*)

[45] R. Bellman. Limit theorems for non-commutative operations. I. *Duke Math. J.*, 21:491–500, 1954. (*355*)

[46] A. Ben-Artzi and I. Gohberg. Dichotomy of systems and invertibility of linear ordinary differential operators. *Oper. Theory: Adv. Appl.*, 56:90–119, 1992. (*98, 113*)

[47] C. M. Bender and S. Boettcher. Real spectra in non-Hermitian Hamiltonians having $\mathcal{PT}$ symmetry. *Phys. Rev. Lett.*, 80:5243–5246, 1998. (*98, 108*)

[48] C. M. Bender, S. Boettcher, and P. N. Meisinger. $\mathcal{PT}$-symmetric quantum mechanics. *J. Math. Phys.*, 40:2201–2229, 1999. (*98, 108*)

[49] E. S. Benilov. Explosive instability in a linear system with neutrally stable eigenmodes. Part 2. Multi-dimensional disturbances. *J. Fluid Mech.*, 501:105–124, 2004. (*124, 125, 406*)

[50] E. S. Benilov, S. B. G. O'Brien, and I. A. Sazonov. A new type of instability: Explosive disturbances in a liquid film inside a rotating horizontal cylinder. *J. Fluid Mech.*, 497:201–224, 2003. (*124, 125, 406*)

[51] P. Benner and H. Faßbender. An implicitly restarted symplectic Lanczos method for the Hamiltonian eigenvalue problem. *Linear Algebra Appl.*, 263:75–111, 1997. (*399*)

[52] P. Benner, V. Mehrmann, and H. Xu. A numerically stable, structure preserving method for computing the eigenvalues of real Hamiltonian or symplectic pencils. *Numer. Math.*, 78:329–358, 1998. (*399, 491*)

[53] J. P. Benthem. A Laplace transform method for the solution of semi-infinite and finite strip problems in stress analysis. *Quart. J. Mech. Appl. Math.*, 16:413–429, 1963. (*537*)

[54] C. A. Berger. A strange dilation theorem (Abstract 625–152). *Notices Amer. Math. Soc.*, 12:590, 1965. (*168*)

[55] E. Bergstrand. Determination of the velocity of light. In S. Flügge, editor, *Handbuch der Physik*, volume 24, pages 1–43. Springer-Verlag, Berlin, 1956. (*495*)

[56] C. Bernardi and Y. Maday. *Approximations Spectrales de Problèmes aux Limites Elliptiques*. Springer-Verlag, Berlin, 1992. (*408*)

[57] M. V. Berry. Mode degeneracies and the Petermann excess-noise factor for unstable lasers. *J. Mod. Optics*, 50:63–81, 2003. (*552*)

[58] M. V. Berry and J. P. Keating. The Riemann zeros and eigenvalue asymptotics. *SIAM Review*, 41:236–266, 1999. (*333*)

[59] A. L. Bertozzi and M. P. Brenner. Linear stability and transient growth in driven contact lines. *Phys. Fluids*, 9:530–539, 1997. (*228*)

[60] T. Betcke and L. N. Trefethen. Computations of eigenvalue avoidance in planar domains. *PAMM*, 4:634–635, 2004. (*377*)

[61] T. R. Bewley. Flow control: New challenges for a new Renaissance. *Prog. Aero. Sci.*, 37:21–58, 2001. (*227*)

[62] R. T. Birge. The general physical constants. *Rep. Progr. Phys.*, 8:90–134, 1941. (*495, 496*)

[63] Å. Björck. *Numerical Methods for Least Squares Problems*. SIAM, Philadelphia, 1996. (*448*)

[64] P. E. Bjørstad and B. P. Tjøstheim. High precision solutions of two fourth order eigenvalue problems. *Computing*, 63:97–107, 1999. (*540*)

[65] L. Boberg and U. Brosa. Onset of turbulence in a pipe. *Z. Naturforschung*, 43a:697–726, 1988. (*202, 207, 208, 497*)

[66] D. B. Bogy. Solution of the plane end problem for a semi-infinite elastic strip. *Zeit. Ang. Math. Phys.*, 26:749–769, 1975. (*537*)

[67] H. F. Bohnenblust and S. Karlin. Geometrical properties of the unit sphere of Banach algebras. *Ann. Math.*, 62:217–229, 1955. (*173*)

[68] R. Boisvert, R. Pozo, K. Remington, B. Miller, and R. Lipman. Matrix market, 2004. Web site: http://math.nist.gov/MatrixMarket. (*270, 392*)

[69] D. Boley. Computing rank-deficiency of rectangular matrix pencils. *Sys. Control Lett.*, 9:207–214, 1987. (*435*)

[70] D. Boley. Estimating the sensitivity of the algebraic structure of pencils with simple eigenvalue estimates. *SIAM J. Matrix Anal. Appl.*, 11:632–643, 1990. (*430*)

[71] F. F. Bonsall, B. E. Cain, and H. Schneider. The numerical range of a continuous mapping of a normed space. *Aeq. Math.*, 2:86–93, 1968. (*173*)

[72] F. F. Bonsall and J. Duncan. *Numerical Ranges of Operators on Normed Spaces and of Elements of Normed Algebras*, volume 2 of *London Math. Soc. Lecture Note Series*. Cambridge University Press, Cambridge, 1971. (*166, 174*)

[73] F. F. Bonsall and J. Duncan. *Numerical Ranges II*, volume 10 of *London Math. Soc. Lecture Note Series*. Cambridge University Press, Cambridge, 1973. (*166*)

[74] D. Borba, K. S. Riedel, W. Kerner, G. T. A. Huysmans, M. Ottaviani, and P. J. Schmid. The pseudospectrum of the resistive magnetohydrodynamics operator: Resolving the Resistive Alfvén Paradox. *Phys. Plasmas*, 1:3151–3160, 1994. (*12, 221, 228*)

[75] M. Born and E. Wolf. *Principles of Optics*. Cambridge University Press, Cambridge, seventh edition, 1999. First edition published in 1959. (*495, 543*)

[76] D. Borthwick and A. Uribe. On the pseudospectra of Berezin–Toeplitz operators. *Methods Appl. Anal.*, 10:31–65, 2003. (*64, 67, 82*)

[77] A. Bottaro and P. Luchini. Görtler vertices: Are they amenable to local eigenvalue analysis? *Europ. J. Mech. B*, 18:47–65, 1999. (*226*)

[78] A. Böttcher. Pseudospectra and singular values of large convolution operators. *J. Int. Eqs. Appl.*, 6:267–301, 1994. (*30, 58, 87*)

[79] A. Böttcher. Infinite matrices and projection methods. In P. Lancaster, editor, *Lectures on Operator Theory and Its Applications*, volume 3 of *Fields Institute Monographs*. American Mathematical Society, Providence, RI, 1996. (*33, 58*)

[80] A. Böttcher. Transient behavior of powers and exponentials of large Toeplitz matrices. *Elect. Trans. Numer. Anal.*, 18:1–41, 2004. (*26, 145, 243, 404*)

[81] A. Böttcher, M. Embree, and V. I. Sokolov. On large Toeplitz band matrices with an uncertain block. *Linear Algebra Appl.*, 366:87–97, 2003. (*465*)

[82] A. Böttcher, M. Embree, and V. I. Sokolov. The spectra of large Toeplitz band matrices with a randomly perturbed entry. *Math. Comp.*, 72:1329–1348, 2003. (*79, 465*)

[83] A. Böttcher, M. Embree, and L. N. Trefethen. Piecewise continuous Toeplitz matrices and operators: Slow approach to infinity. *SIAM J. Matrix Anal. Appl.*, 24:484–489, 2002. (*56*)

[84] A. Böttcher and S. Grudsky. Toeplitz band matrices with exponentially growing condition numbers. *Elect. J. Lin. Alg.*, 5:104–125, 1999. (*56, 58*)

[85] A. Böttcher and S. M. Grudsky. *Toeplitz Matrices, Asymptotic Linear Algebra, and Functional Analysis*. Birkhäuser, Basel, 2000. (*50*)

[86] A. Böttcher and S. M. Grudsky. Can spectral value sets of Toeplitz band matrices jump? *Linear Algebra Appl.*, 351–352:99–116, 2002. (*33*)

[87] A. Böttcher and S. Grudsky. Toeplitz matrices with slowly growing pseudospectra. In S. Samko, A. Lebre, and A. F. dos Santos, editors, *Factorization, Singular Operators, and Related Problems*, pages 43–54. Kluwer, Dordrecht, 2003. (*56*)

[88] A. Böttcher and S. M. Grudsky. Asymptotically good pseudomodes for Toeplitz matrices and Wiener–Hopf operators. In I. Gohberg, A. F. dos Santos, F.-O. Speck, F. S. Teixeira, and W. Wendland, editors, *Operator Theoretical Methods and Applications to Mathematical Physics: The Erhard Meister Memorial Volume*. Birkhäuser, Basel, 2004. (*56*)

[89] A. Böttcher and S. M. Grudsky. *Spectral Properties of Banded Toeplitz Matrices*. SIAM, Philadelphia. To appear. (*50*)

[90] A. Böttcher, S. M. Grudsky, and B. Silbermann. Norms of inverses, spectra, and pseudospectra of large truncated Wiener–Hopf operators and Toeplitz matrices. *New York J. Math.*, 3:1–31, 1997. (*33*)

[91] A. Böttcher and B. Silbermann. *Invertibility and Asymptotics of Toeplitz Matrices*. Akademie-Verlag, Berlin, 1983. (*50*)

[92] A. Böttcher and B. Silbermann. *Analysis of Toeplitz Operators*. Springer-Verlag, Berlin, 1990. (*50*)

[93] A. Böttcher and B. Silbermann. *Introduction to Large Truncated Toeplitz Matrices*. Springer-Verlag, New York, 1999. (*12, 30, 50, 53, 55, 58*)

[94] P. Bougerol and J. Lacroix. *Products of Random Matrices with Applications to Schrödinger Operators*. Birkhäuser, Boston, 1985. (*355*)

[95] L. S. Boulton. Non-self-adjoint harmonic oscillator, compact semigroups and pseudospectra. *J. Oper. Theory*, 47:413–429, 2002. (*98*)

[96] N. Bourbaki. *Eléments d'Histoire des Mathématiques*. Hermann, Paris, 1969. (*6*)

[97] G. Boutry, M. Elad, G. H. Golub, and P. Milanfar. The generalized eigenvalue problem for non-square pencils using a minimal perturbation approach. *SIAM J. Matrix Anal. Appl.* To appear. (*430, 436*)

[98] G. D. Boyd and H. Kogelnik. General confocal resonator theory. *Bell Sys. Tech. J.*, 41:1347–1369, 1962. (*549*)

[99] J. P. Boyd. *Chebyshev and Fourier Spectral Methods*. Dover, Mineola, NY, second edition, 2001. (*289, 405*)

[100] S. Boyd and V. Balakrishnan. A regularity result for the singular values of a transfer matrix and a quadratically convergent algorithm for computing its L_∞-norm. *Sys. Control Lett.*, 15:1–7, 1990. (*397*)

[101] S. Boyd and L. Vandenberghe. *Convex Optimization.* Cambridge University Press, Cambridge, 2004. (*280*)

[102] T. Braconnier, F. Chatelin, and J. C. Dunyach. Highly nonnormal eigenproblems in the aeronautical industry. *Japan J. Indust. Appl. Math.*, 12:123–136, 1995. (*228*)

[103] T. Braconnier and N. J. Higham. Computing the field of values and pseudospectra using the Lanczos method with continuation. *BIT*, 36:422–440, 1996. (*392*)

[104] T. Braconnier, R. A. McCoy, and V. Toumazou. Using the field of values for pseudospectra generation. Technical Report TR/PA/97/28, CERFACS, Toulouse, September 1997. (*377*)

[105] K. Braman, R. Byers, and R. Mathias. The multishift QR algorithm. Part I: Maintaining well-focused shifts and level 3 performance. *SIAM J. Matrix Anal. Appl.*, 23:929–947, 2002. (*486*)

[106] K. Braman, R. Byers, and R. Mathias. The multishift QR algorithm. Part II: Aggressive early deflations. *SIAM J. Matrix Anal. Appl.*, 23:948–973, 2002. (*486*)

[107] S. Brandt and H. D. Dahmen. *The Picture Book of Quantum Mechanics.* Springer-Verlag, New York, third edition, 2001. (*493*)

[108] P. Brenner and V. Thomée. On rational approximation of semigroups. *SIAM J. Numer. Anal.*, 16:683–694, 1979. (*310*)

[109] E. Brézin and A. Zee. Non-hermitean delocalization: Multiple scattering and bounds. *Nuc. Phys. B*, 509 [FS]:599–614, 1998. (*339, 340, 345*)

[110] P. W. Brouwer, P. G. Silvestrov, and C. W. J. Beenakker. Theory of directed localization in one dimension. *Phys. Rev. B*, 56:4333–4335, 1997. (*339, 340*)

[111] M. Brühl. A curve tracing algorithm for computing the pseudospectrum. *BIT*, 36:441–454, 1996. (*394, 395*)

[112] V. T. Buchwald. Eigenfunctions of plane elastostatics. I. The strip. *Proc. Roy. Soc. London A*, 277:385–400, 1964. (*537*)

[113] G. Buffon. Essai d'arithmétique morale. *Supplément á l'Histoire Naturelle*, 4:46–123, 1777. (*181*)

[114] R. Buizza, J. Tribbia, F. Molteni, and T. N. Palmer. Computation of optimal unstable structures for a numerical weather prediction model. *Tellus*, 45A:388–407, 1993. (*227*)

[115] J. V. Burke, A. S. Lewis, and M. L. Overton. Two numerical methods for optimizing matrix stability. *Linear Algebra Appl.*, 351–352:117–145, 2002. (*528*)

[116] J. V. Burke, A. S. Lewis, and M. L. Overton. A nonsmooth, nonconvex optimization approach to robust stabilization by static output feedback and low-order controllers. Proceedings of ROCOND 2003, Milan, June 2003. (*155*)

[117] J. V. Burke, A. S. Lewis, and M. L. Overton. Robust stability and a criss-cross algorithm for pseudospectra. *IMA J. Numer. Anal.*, 23:359–375, 2003. (*394, 397, 399, 400, 401*)

[118] J. V. Burke, A. S. Lewis, and M. L. Overton. Pseudospectral components and the distance to uncontrollability. *SIAM J. Matrix Anal. Appl.*, 26:350–361, 2004. (*430, 433, 435*)

[119] J. V. Burke and M. L. Overton. Stable perturbations of nonsymmetric matrices. *Linear Algebra Appl.*, 171:249–273, 1992. (*451*)

[120] K. Burrage and J. Erhel. On the performance of various adaptive preconditioned GMRES strategies. *Numer. Lin. Alg. Applics.*, 5:101–121, 1998. (*255*)

[121] K. M. Butler and B. F. Farrell. Three-dimensional optimal perturbations in viscous shear flow. *Phys. Fluids A*, 4:1637–1650, 1992. (*202, 225*)

[122] P. L. Butzer and H. Berens. *Semi-Groups of Operators and Approximation.* Springer-Verlag, Berlin, 1967. (*188*)

[123] R. Byers. A bisection method for measuring the distance of a stable matrix to the unstable matrices. *SIAM J. Sci. Stat. Comput.*, 9:875–881, 1988. (*397, 399, 403, 451*)

[124] R. Byers. Detecting nearly uncontrollable pairs. In *Signal Processing, Scattering and Operator Theory, and Numerical Methods (Amsterdam, 1989)*. Birkhäuser, Boston, 1990. (*433, 435*)

[125] A. Calderón, F. Spitzer, and H. Widom. Inversion of Toeplitz matrices. *Illinois J. Math.*, 3:490–498, 1959. (*53*)

[126] D. Calvetti, L. Reichel, and D. C. Sorensen. An implicitly restarted Lanczos method for large symmetric eigenvalue problems. *Elect. Trans. Numer. Anal.*, 2:1–21, 1994. (*255, 392*)

[127] C. Canuto, M. Y. Hussaini, A. Quarteroni, and T. A. Zang. *Spectral Methods in Fluid Dynamics.* Springer-Verlag, New York, 1988. (*289, 298, 408*)

[128] Z.-H. Cao. A note on the convergence behavior of GMRES. *Appl. Num. Math.*, 25:13–20, 1997. (*246*)

[129] J. F. Carpraux, J. Erhel, and M. Sadkane. Spectral portrait for non-Hermitian large sparse matrices. *Computing*, 53:301–310, 1994. (*392*)

[130] H. Caswell. *Matrix Population Models: Construction, Analysis, and Interpretation.* Sinauer, Sunderland, MA, second edition, 2001. (*3, 463, 503, 526*)

[131] A. Cauchy. Note sur divers théorèmes relatifs à la rectification des courbes, et à la quadrature des surfaces. *Comptes Rendus*, 13:1060–1065, 1841. (*181*)

[132] G. D. Chagelishvili, A. D. Rogava, and D. Tsiklauri. Compressible hydrodynamic shear flows with anisotropic thermal pressure: Non-modal study in waves and instabilities. *Phys. Plasmas*, 4:1182–1195, 1997. (*228*)

[133] F. Chaitin-Chatelin and V. Frayssé. *Lectures on Finite Precision Computations.* SIAM, Philadelphia, 1996. (*12, 45, 446, 465, 487, 491*)

[134] F. Chaitin-Chatelin and A. Harrabi. About definitions of pseudospectra of closed operators in Banach spaces. Technical Report TR/PA/98/08, CERFACS, August 1998. (*13, 30, 33*)

[135] S. Chandrasekhar. *Hydrodynamic and Hydromagnetic Stability.* Oxford University Press, Oxford, 1961. (*3, 195*)

[136] C.-L. Chang and M. R. Malik. Oblique-mode breakdown and secondary instability in supersonic boundary layers. *J. Fluid Mech.*, 273:323–360, 1994. (*226*)

[137] P. Chang, R. Saravanan, T. DelSole, and F. Wang. Predictability of linear coupled systems. Part I: Theoretical analyses. *J. Climate*, 17:1474–1486, 2004. (*227*)

[138] S. J. Chapman. Subcritical transition in channel flows. *J. Fluid Mech.*, 451:35–97, 2002. (*113, 213, 214*)

[139] F. Chatelin. *Spectral Approximation of Linear Operators.* Academic Press, New York, 1983. (*139, 294*)

[140] F. Chatelin. Resolution approchée d'équations sur ordinateur. Technical report, Laboratoire de Satistique Théorique et Appliquée, Université Pierre et Marie Curie, Paris, 1989. (*45*)

[141] F. Chatelin. La passion des nombres. *Science et Avenir*, 502:18, December 1998. (*45*)

[142] F. Chatelin and T. Braconnier. About the qualitative computation of Jordan forms. *Z. Angew. Math. Mech.*, 74:105–113, 1994. (*45, 479*)

[143] X. Chen and J. E. Cohen. Transient dynamics and food-web complexity in the Lotka-Volterra cascade model. *Proc. Roy. Soc. London B*, 268:869–877, 2001. (*529*)

[144] E. W. Cheney. *Introduction to Approximation Theory.* McGraw-Hill, New York, 1966. (*279*)

[145] Y.-J. Cheng, C. G. Fanning, and A. E. Siegman. Experimental observation of a large excess quantum noise factor in the linewidth of a laser oscillator having nonorthogonal modes. *Phys. Rev. Lett.*, 77:627–630, 1996. (*552*)

[146] S. Childress and A. D. Gilbert. *Stretch, Twist, Fold: The Fast Dynamo.* Springer-Verlag, Berlin, 1995. (*227*)

[147] R. C. Y. Chin and T. A. Manteuffel. An analysis of block successive overrelaxation for a class of matrices with complex spectra. *SIAM J. Numer. Anal.*, 25:564–585, 1988. (*238*)

[148] G. E. Cho and C. D. Meyer. Markov chain sensitivity measured by mean first passage times. *Linear Algebra Appl.*, 316:21–28, 2000. (*464*)

[149] G. E. Cho and C. D. Meyer. Comparison of perturbation bounds for the stationary distribution of a Markov chain. *Linear Algebra Appl.*, 335:137–150, 2001. (*464*)

[150] J. M. Chomaz, P. Huerre, and L. G. Redekopp. Models of hydrodynamic resonances in separated shear flows. In *Preliminary Proceedings of the Sixth Symposium on Turbulent Shear Flow, Toulouse, France.* 1987. (*123*)

[151] A. K. Chopra. *Dynamics of Structures: Theory and Applications to Earthquake Engineering.* Prentice-Hall, Upper Saddle River, NJ, second edition, 2001. (*3*)

[152] G. M. Cicuta, M. Contedini, and L. Molinari. Enumeration of simple random walks and tridiagonal matrices. *J. Phys. A*, 35:1125–1146, 2002. (*357*)

[153] C. W. Clenshaw and A. R. Curtis. A method for numerical integration on an automatic computer. *Numer. Math.*, 2:197–205, 1960. (*412*)

[154] R. W. Clough and J. Penzien. *Dynamics of Structures.* McGraw-Hill, New York, second edition, 1993. (*3*)

[155] L. A. Coburn. The C^*-algebra generated by an isometry. *Bull. Amer. Math. Soc.*, 73:722–726, 1967. (*53*)

[156] L. A. Coburn. The C^*-algebra generated by an isometry. II. *Trans. Amer. Math. Soc.*, 137:211–217, 1969. (*53*)

[157] J. A. Cochran and E. W. Hinds. Eigensystems associated with the complex-symmetric kernels of laser theory. *SIAM J. Appl. Math.*, 26:776–786, 1974. (*546*)

[158] J. E. Cohen, F. Briand, and C. M. Newman. *Community Food Webs: Data and Theory.* Springer-Verlag, Berlin, 1990. (*528, 529*)

[159] J. E. Cohen, H. Kesten, and C. M. Newman, editors. *Random Matrices and Their Applications*, volume 50 of *Contemporary Mathematics.* American Mathematical Society, Providence, RI, 1986. (*334*)

[160] J. E. Cohen, T. Łuczak, C. M. Newman, and Z.-M. Zhou. Stochastic structure and nonlinear dynamics of food webs: Qualitative stability in a Lotka–Volterra cascade model. *Proc. Roy. Soc. Lond. B*, 240:607–627, 1990. (*528*)

[161] J. B. Conway. *A Course in Functional Analysis.* Springer-Verlag, New York, 1985. (*139*)

[162] P. Corbett and A. Bottaro. Optimal control of nonmodal disturbances in boundary layers. *Theor. Comp. Fluid Dynamics*, 15:65–81, 2001. (*227*)

[163] P. Corbett and A. Bottaro. Optimal linear growth in swept boundary layers. *J. Fluid Mech.*, 435:1–23, 2001. (*225*)

[164] C. Cossu and J. M. Chomaz. Global measures of local convective instabilities. *Phys. Rev. Lett.*, 78:4387 4390, 1997. (*123, 124, 228*)

[165] C. Cossu, J.-M. Chomaz, P. Huerre, and M. Costa. Maximum spatial growth of Görtler vertices. *Flow Turb. Combust.*, 65:369–392, 2000. (*226*)

[166] M. Couette. Études sur le frottement des liquides. *Ann. Chim. Phys.*, 21:433–510, 1890. (*198*)

[167] R. Courant, K. Friedrichs, and H. Lewy. Über die partielle Differenzengleichungen der mathematichen Physik. *Math. Annal.*, 100:32–74, 1928. (Translated as 'On the partial difference equations of mathematical physics', *IBM J. Res. Devel.*, 11:215–234, 1967.) (*296*)

[168] R. Courant and D. Hilbert. *Methods of Mathematical Physics*, volume 1. Wiley-Interscience, New York, 1953. Translated from 1937 German second edition. (*6*)

[169] J. Courtial and M. J. Padgett. Monitor-outside-a-monitor effect and self-similar fractal structure in the eigenmodes of unstable optical resonators. *Phys. Rev. Lett.*, 85:5320–5323, 2000. (*550*)

[170] J. D. Crawford. Pseudo-spectra of the linearized Vlasov equation (abstract). *Bull. Amer. Phys. Soc.*, 41:1606–1607, 1996. (*228*)

[171] M. Crouzeix, B. Philippe, and M. Sadkane. The Davidson method. *SIAM J. Sci. Comput.*, 15:62–76, 1994. (*264*)

[172] C. F. Curtiss and J. O. Hirschfelder. Integration of stiff equations. *Proc. Nat. Acad. Sci.*, 38:235–243, 1952. (*314*)

[173] G. Dahlquist. Convergence and stability in the numerical integration of ordinary differential equations. *Math. Scand.*, 4:33–53, 1956. (*314*)

[174] G. Dahlquist. A special stability problem for linear multistep methods. *BIT*, 3, 1963. (*314*)

[175] O. Dauchot and F. Daviaud. Finite amplitude perturbation and spots growth mechanism in plane Couette flow. *Phys. Fluids*, 7:335–343, 1995. (*198*)

[176] E. R. Davidson. The iterative calculation of a few of the lowest eigenvalues and corresponding eigenvectors of large real symmetric matrices. *J. Comp. Phys.*, 17:87–94, 1975. (*264*)

[177] E. B. Davies. *One-Parameter Semigroups*. Academic Press, London, 1980. (*148*)

[178] E. B. Davies. *Spectral Theory and Differential Operators*. Cambridge University Press, Cambridge, 1995. (*3*)

[179] E. B. Davies. Pseudo-spectra, the harmonic oscillator and complex resonances. *Proc. Roy. Soc. London A*, 455:585–599, 1999. (*13, 34, 98, 103, 105*)

[180] E. B. Davies. Semi-classical states for non-self-adjoint Schrödinger operators. *Comm. Math. Phys.*, 200:25–41, 1999. (*34, 98, 103, 105*)

[181] E. B. Davies. Pseudospectra of differential operators. *J. Oper. Theory*, 43:243–262, 2000. (*87, 95, 96*)

[182] E. B. Davies. Spectral properties of random non-self-adjoint matrices and operators. *Proc. Roy. Soc. London A*, 457:191–206, 2001. (*339, 347*)

[183] E. B. Davies. Non-self-adjoint differential operators. *Bull. London Math. Soc.*, 34:513–532, 2002. (*98*)

[184] E. B. Davies. Spectral theory. Manuscript, 2003. (*472*)

[185] E. B. Davies. Computing the decay of a simple reversible sub-Markov semigroup. *London Math. Soc. J. Comp. Math.*, 7:1–20, 2004. (*95, 117, 120, 122*)

[186] E. B. Davies. Semi-classical analysis and pseudospectra. Preprint arXiv.math.SP/ 0402172, 2004. (*93*)

[187] E. B. Davies. Semigroup growth bounds. *J. Oper. Theory*. To appear. (*140, 151*)

[188] E. B. Davies and A. B. J. Kuijlaars. Spectral asymptotics of the non-self-adjoint harmonic oscillator. *J. London. Math. Soc.*, (2) 70:420–426, 2004. (*98, 106*)

[189] P. J. Davis and P. Rabinowitz. *Methods of Numerical Integration*. Academic Press, New York, second edition, 1984. (*412*)

[190] E. J. Davison, editor. *Benchmark problems for control system design*. Report of the IFAC Technical Committee. Pergamon, Oxford, 1990. (*155*)

[191] L. de Luca, M. Costa, and C. Caramiello. Energy growth of initial perturbations in two-dimensional gravitational jets. *Phys. Fluids*, 14:289–299, 2002. (*228*)

[192] W. R. Dean. Fluid motion in a curved channel. *Proc. Roy. Soc. London A*, 121:402–420, 1928. (*225*)

[193] K. Dekker and J. G. Verwer. *Stability of Runge–Kutta Methods for Stiff Nonlinear Differential Equations*. North-Holland, Amsterdam, 1984. (*167, 320*)

[194] E. Delabaere and D. T. Trinh. Spectral analysis of the complex cubic oscillator. *J. Phys. A*, 33:8771–8796, 2000. (*108*)

[195] J. W. Demmel. *A Numerical Analyst's Jordan Canonical Form*. Ph.D. thesis, University of California, Berkeley, 1983. (*44, 468*)

[196] J. W. Demmel. A counterexample for two conjectures about stability. *IEEE Trans. Auto. Control*, AC-32:340–343, 1987. (*44, 45, 452, 461*)

[197] J. W. Demmel. On condition numbers and the distance to the nearest ill-posed problem. *Numer. Math.*, 51:251–289, 1987. (*448*)

[198] J. Demmel. Nearest defective matrices and the geometry of ill-conditioning. In M. G. Cox and S. Hammarling, editors, *Reliable Numerical Computation*, pages 35–55. Oxford University Press, Oxford, 1990. (*44*)

[199] J. W. Demmel. *Applied Numerical Linear Algebra*. SIAM, Philadelphia, 1997. (*475, 486*)

[200] J. Demmel and B. Kågström. The generalized Schur decomposition of an arbitrary pencil $A - \lambda B$: Robust software with error bounds and applications. Part I: Theory and algorithms. *ACM Trans. Math. Software*, 19:160–174, 1993. (*429*)

[201] N. Dencker. The resolution of the Nirenberg–Treves conjecture. *Ann. Math.* To appear. (*129*)

[202] N. Dencker, J. Sjöstrand, and M. Zworski. Pseudospectra of semiclassical (pseudo-) differential operators. *Comm. Pure Appl. Math.*, 57:384–415, 2004. (*98, 103, 104, 114, 126*)

[203] J. Descloux. Bounds for the spectral norm of functions of matrices. *Numer. Math.*, 5:185–190, 1963. (*147, 445*)

[204] G. Di Lena and D. Trigiante. On the spectrum of families of matrices with applications to stability problems. In A. Bellen, C. W. Gear, and E. Russo, editors, *Numerical Methods for Ordinary Differential Equations*, volume 1386 of *Lecture Notes in Mathematics*, pages 36–53. Springer-Verlag, Berlin, 1989. (*310*)

[205] P. Diaconis. *Group Representations in Probability and Statistics*. IMS, Hayward, CA, 1988. (*509*)

[206] P. Diaconis. The cutoff phenomenon in finite Markov chains. *Proc. Nat. Acad. Sci.*, 93:1659–1664, 1996. (*508, 510*)

[207] P. Diaconis, R. L. Graham, and J. A. Morrison. Asymptotic analysis of a random walk on a hypercube with many dimensions. *Random Struct. Alg.*, 1:294–313, 1990. (*511, 513, 516*)

[208] J. Dieudonné. *History of Functional Analysis*. North-Holland, Amsterdam, 1981. (*6*)

[209] M. Dimassi and J. Sjöstrand. *Spectral Asymptotics in the Semi-Classical Limit*, volume 268 of *London Math. Soc. Lecture Note Series*. Cambridge University Press, Cambridge, 1999. (*99*)

[210] R. C. DiPrima and G. J. Habetler. A completeness theorem for non-selfadjoint eigenvalue problems in hydrodynamic stability. *Arch. Rat. Mech. Anal.*, 34:218–227, 1969. (*217*)

[211] G. Domokos and P. Holmes. On nonlinear boundary-value problems: Ghosts, parasites and discretizations. *Proc. Roy. Soc. London A*, 459:1535–1561, 2003. (*98, 99, 113*)

[212] P. Dorey, C. Dunning, and R. Tateo. Spectral equivalences, Bethe ansatz equations, and reality properties in $\mathcal{PT}$-symmetric quantum mechanics. *J. Phys. A*, 34:5679–5704, 2001. (*108*)

[213] R. G. Douglas. *Banach Algebra Techniques in Operator Theory*. Springer-Verlag, New York, second edition, 1998. (*50*)

[214] P. G. Drazin and W. H. Reid. *Hydrodynamic Stability*. Cambridge University Press, Cambridge, 1981. (*167, 195, 197, 215, 219, 225*)

[215] T. A. Driscoll, K.-C. Toh, and L. N. Trefethen. From potential theory to matrix iterations in six steps. *SIAM Review*, 40:547–578, 1998. (*250, 259*)

[216] T. A. Driscoll and L. N. Trefethen. Pseudospectra for the wave equation with an absorbing boundary. *J. Comp. Appl. Math.*, 69:125–142, 1996. (*141, 155*)

[217] M. Dubiner. Asymptotic analysis of spectral methods. *J. Sci. Comp.*, 2:3–31, 1987. (*298, 301*)

[218] I. S. Duff, A. M. Erisman, and J. K. Reid. *Direct Methods for Sparse Matrices*. Oxford University Press, Oxford, 1986. (*391*)

[219] J. J. Duistermaat and J. Sjöstrand. A global construction for pseudo-differential operators with non-involutive characteristics. *Invent. Math.*, 20:209–225, 1973. (*98, 103*)

[220] G. E. Dullerud and F. Paganini. *A Course in Robust Control Theory: A Convex Approach*. Springer-Verlag, New York, 2000. (*451*)

[221] N. Dunford and J. T. Schwartz. *Linear Operators, Part I: General Theory*. Wiley-Interscience, New York, 1958. (*27, 139, 186*)

[222] H. Dym and H. P. McKean. *Fourier Series and Integrals*. Academic Press, New York, 1972. (*155*)

[223] P. J. Eberlein. Solution to the complex eigenproblem by a norm reducing Jacobi type method. *Numer. Math.*, 14:232–245, 1970. (*442*)

[224] P. J. Eberlein. A note on the matrices denoted B_n. *SIAM J. Appl. Math.*, 20:87–92, 1971. (*487, 488*)

[225] A. Edelman. Eigenvalues and condition numbers of random matrices. *SIAM J. Matrix Anal. Appl.*, 9:543–560, 1988. (*338*)

[226] A. Edelman. *Eigenvalues and Condition Numbers of Random Matrices*. Ph.D. thesis, MIT, Cambridge, MA, 1989. (*333, 334*)

[227] A. Edelman, E. Kostlan, and M. Shub. How many eigenvalues of a random matrix
       are real? *J. Amer. Math. Soc.*, 7:247–267, 1994. (*333*)

[228] A. Edelman and H. Murakami. Polynomial roots from companion matrix eigen-
       values. *Math. Comp.*, 64:763–776, 1995. (*505*)

[229] D. E. Edmunds and W. D. Evans. *Spectral Theory and Differential Operators*.
       Oxford University Press, Oxford, 1987. (*90*)

[230] P. Ehrenfest and T. Ehrenfest. Über zwei bekannte Einwände gegen das Boltz-
       mannsche H-Theorem. *Phys. Zeit.*, 8:311–314, 1907. (*515*)

[231] M. Eiermann. On semiiterative methods generated by Faber polynomials. *Numer.
       Math.*, 56:139–156, 1989. (*255*)

[232] M. Eiermann. Fields of values and iterative methods. *Linear Algebra Appl.*,
       180:167–197, 1993. (*168*)

[233] M. Eiermann, O. G. Ernst, and O. Schneider. Analysis of acceleration strategies
       for restarted minimal residual methods. *J. Comp. Appl. Math.*, 123:261–292, 2000.
       (*255*)

[234] M. Eiermann and W. Niethammer. On the construction of semiiterative methods.
       SIAM J. Numer. Anal., 20:1153–1160, 1983. (*231*)

[235] M. Eiermann, W. Niethammer, and R. S. Varga. A study of semiiterative methods
       for nonsymmetric systems of linear equations. *Numer. Math.*, 47:505–533, 1985.
       (*255*)

[236] S. C. Eisenstat, H. C. Elman, and M. H. Schultz. Variational iterative methods for
       nonsymmetric systems of linear equations. *SIAM J. Numer. Anal.*, 20:345–357,
       1983. (*248*)

[237] R. Eising. Between controllable and uncontrollable. *Sys. Control Lett.*, 4:263–264,
       1984. (*435*)

[238] T. Ellingsen and E. Palm. Stability of linear flow. *Phys. Fluids*, 18:487–488, 1975.
       (*201, 224*)

[239] H. C. Elman and M. P. Chernesky. Ordering effects on relaxation methods applied
       to the discrete one-dimensional convection-diffusion equation. *SIAM J. Numer.
       Anal.*, 30:1268–1290, 1993. (*238*)

[240] H. C. Elman, Y. Saad, and P. E. Saylor. A hybrid Chebyshev Krylov subspace
       algorithm for solving nonsymmetric systems of linear equations. *SIAM J. Sci.
       Stat. Comput.*, 7:840–855, 1986. (*256*)

[241] H. C. Elman and R. L. Streit. Polynomial iteration for nonsymmetric indefinite
       linear systems. In J. P. Hennart, editor, *Numerical Analysis*, volume 1230 of
       Lecture Notes in Mathematics, pages 103–117. Springer-Verlag, Berlin, 1986. (*255,
       256*)

[242] L. Elsner and K. D. Ikramov. Normal matrices: An update. *Linear Algebra Appl.*,
       285:291–303, 1998. (*18, 442*)

[243] L. Elsner and M. H. C. Paardekooper. On measures of nonnormality of matrices.
       Linear Algebra Appl., 92:107–123, 1987. (*446*)

[244] M. Embree. *Convergence of Krylov Subspace Methods for Non-Normal Matrices*.
       D.Phil. thesis, Oxford University, 1999. (*251, 253*)

[245] M. Embree. The tortoise and the hare restart GMRES. *SIAM Review*, 45:259–266,
       2003. (*245*)

[246] M. Embree and L. N. Trefethen. Growth and decay of random Fibonacci sequences.
       Proc. Roy. Soc. London A, 455:2471–2485, 1999. (*356, 357, 390*)

[247] M. Embree and L. N. Trefethen. Generalizing eigenvalue theorems to pseudospectra theorems. *SIAM J. Sci. Comput.*, 23:583–590, 2001. (*260*)

[248] K.-J. Engel and R. Nagel. *One-Parameter Semigroups for Linear Evolution Equations.* Springer-Verlag, New York, 2000. (*148, 149, 155, 186, 187, 191*)

[249] O. G. Ernst. Residual-minimizing Krylov subspace methods for stabilized discretizations of convection-diffusion equations. *SIAM J. Matrix Anal. Appl.*, 21:1079–1101, 2000. (*168, 251*)

[250] G. Faber. Über Tschebyscheffsche Polynome. *J. Reine Angew. Math.*, 150:79–106, 1920. (*278*)

[251] V. Faber, A. Greenbaum, and D. E. Marshall. The polynomial numerical hulls of Jordan blocks and related matrices. *Linear Algebra Appl.*, 374:231–246, 2003. (*284*)

[252] V. Faber, W. Joubert, E. Knill, and T. A. Manteuffel. Minimal residual method stronger than polynomial preconditioning. *SIAM J. Matrix Anal. Appl.*, 17:707–729, 1996. (*247*)

[253] D. K. Faddeev and V. N. Faddeeva. *Computational Methods of Linear Algebra.* Freeman, San Francisco, 1963. Translated by Robert C. Williams. (*255*)

[254] J. Fadle. Die Selbstspannungs-Eigenwertfunktionen der quadratische Scheibe. *Ing.-Arch.*, 11:125–149, 1941. (*537*)

[255] N. A. Farkova. The use of Faber polynomials to solve systems of linear equations. *USSR Comput. Math. Math. Phys.*, 28(6):22–32, 1988. Translated from the Russian: *Z. Vycisl. Mat. Mat. Fiz.*, 28(11):1634–1648, 1988. (*255*)

[256] B. Farrell. Modal and non-modal baroclinic waves. *J. Atmos. Sci.*, 41:668–673, 1984. (*227*)

[257] B. Farrell. Optimal excitation of neutral Rossby waves. *J. Atmos. Sci.*, 45:163–172, 1988. (*227*)

[258] B. F. Farrell. Optimal excitation of perturbations in viscous shear flow. *Phys. Fluids*, 31:2093–2102, 1988. (*218*)

[259] B. F. Farrell. Optimal excitation of baroclinic waves. *J. Atmos. Sci.*, 46:1193–1206, 1989. (*227*)

[260] B. F. Farrell. Small error dynamics and the predictability of atmospheric flows. *J. Atmos. Sci.*, 47:2409–2416, 1990. (*227*)

[261] B. F. Farrell and P. J. Ioannou. Stochastic dynamics of baroclinic waves. *J. Atmos. Sci.*, 50:4044–4057, 1993. (*12*)

[262] B. F. Farrell and P. J. Ioannou. Stochastic forcing of the linearized Navier–Stokes equations. *Phys. Fluids A*, 5:2600–2609, 1993. (*228, 552*)

[263] B. F. Farrell and P. J. Ioannou. Variance maintained by stochastic forcing of non-normal dynamical systems associated with linearly stable shear flows. *Phys. Rev. Lett.*, 72:1188–1191, 1994. (*552*)

[264] B. F. Farrell and P. J. Ioannou. Turbulence suppression by active control. *Phys. Fluids*, 8:1257–1268, 1996. (*227*)

[265] B. F. Farrell and P. J. Ioannou. Optimal excitation of magnetic fields. *Astrophys. J.*, 522:1079–1087, 1999. (*227*)

[266] B. F. Farrell and P. J. Ioannou. Perturbation growth and structure in time dependent flows. *J. Atmos. Sci.*, 56:3622–3639, 1999. (*227*)

[267] B. F. Farrell and A. M. Moore. An adjoint method for obtaining the most rapidly growing perturbation to oceanic flows. *J. Phys. Oceanography*, 22:338–349, 1992. (*227*)

[268] P. A. Farrell. Flow conforming iterative methods for convection dominated flows. In C. Brezinski, editor, *IMACS Annals on Computing and Applied Mathematics*, volume 1.2, pages 681–686. Baltzer, Basel, 1989. (*238*)

[269] P. A. Farrell, A. Hegarty, J. J. H. Miller, E. O'Riordan, and G. I. Shishkin. *Robust Computational Techniques for Boundary Layers*. Chapman and Hall/CRC, Boca Raton, FL, 2000. (*122*)

[270] S. Fedotov, I. Bashkirtseva, and L. Ryashko. Stochastic analysis of a non-normal dynamical system mimicking a laminar-to-turbulent subcritical transition. *Phys. Rev. E*, 66:066310–1–0663106, 2002. (*207*)

[271] J. Feinberg and A. Zee. Non-Hermitian localization and delocalization. *Phys. Rev. E*, 59:6433–6443, 1999. (*339, 345, 357*)

[272] J. Feinberg and A. Zee. Spectral curves of non-hermitian hamiltonians. *Nuc. Phys. B*, 552 [FS]:599–623, 1999. (*339, 345*)

[273] W. Feller. *An Introduction to Probability Theory and Its Applications*, volume 1. Wiley, New York, third edition, 1968. (*509*)

[274] B. Fischer. *Polynomial Based Iteration Methods for Symmetric Linear Systems*. Wiley-Teubner, Chichester, England, 1996. (*255*)

[275] B. Fischer and J. Modersitzki. An algorithm for complex linear approximation based on semi-infinite programming. *Numer. Alg.*, 5:287–297, 1993. (*280*)

[276] B. Fischer, A. Ramage, D. J. Silvester, and A. J. Wathen. On parameter choice and iterative convergence for stabilised discretisations of advection–diffusion problems. *Comp. Methods Appl. Mech. Eng.*, 179:179–195, 1999. (*251, 470, 471*)

[277] B. Fischer and L. Reichel. A stable Richardson iteration method for complex linear systems. *Numer. Math.*, 54:225–242, 1988. (*255*)

[278] J. P. Flaherty, C. E. Seyler, and L. N. Trefethen. Large-amplitude transient growth in the linear evolution of equatorial spread *F* with a sheared zonal flow. *J. Geophys. Res.*, 104A:6843–6857, 1999. (*228*)

[279] N. H. Fletcher and T. D. Rossing. *The Physics of Musical Instruments*. Springer-Verlag, New York, second edition, 1998. (*3*)

[280] J. M. Floryan. On the Görtler instability of boundary layers. *Prog. Aero. Sci.*, 28:235–271, 1991. (*225*)

[281] A. S. Fokas and B. Pelloni. Two-point boundary value problems for linear evolution equations. *Math. Proc. Cambridge Phil. Soc.*, 131:521–543, 2001. (*97*)

[282] B. Fornberg. *A Practical Guide to Pseudospectral Methods*. Cambridge University Press, Cambridge, 1996. (*289, 296, 405*)

[283] R. C. Foster. Structure and energetics of optimal Ekman layer perturbations. *J. Fluid Mech.*, 333:97–123, 1997. (*227*)

[284] A. G. Fox and T. Li. Resonant modes in a maser interferometer. *Bell Sys. Tech. J.*, 40:453–488, 1961. (*544*)

[285] A. G. Fox and T. Li. Modes in a maser interferometer with curved and tilted mirrors. *Proc. IEEE*, 51:80–89, 1963. (*544, 549*)

[286] J. G. F. Francis. The QR transformation: A unitary analogue to the LR transformation—Part 1. *Computer J.*, 4:265–271, 1961. (*486*)

[287] J. G. F. Francis. The QR transformation—Part 2. *Computer J.*, 4:332–345, 1962. (*486*)

[288] W. L. Frank. Computing eigenvalues of complex matrices by determinant evaluation and by methods of Danilewski and Wielandt. *J. Soc. Ind. Appl. Math.*, pages 378–392, 1958. (*487*)

[289] S. P. Frankel. Convergence rates of iterative treatments of partial differential equations. *Math. Tables Aids Comput.*, 4:65–75, 1950. (*236*)

[290] V. Fraysseé, M. Gueury, F. Nicoud, and V. Toumazou. Spectral portraits for matrix pencils. Technical Report TR/PA/96/19, CERFACS, Toulouse, August 1996. (*424*)

[291] V. Fraysseé and V. Toumazou. A note on the normwise perturbation theory for the regular generalized eigenproblem. *Numer. Lin. Alg. Applics.*, 5:1–10, 1998. (*429*)

[292] R. W. Freund. Quasi-kernel polynomials and their use in non-Hermitian matrix iterations. *J. Comp. Appl. Math.*, 43:135–158, 1992. (*246*)

[293] I. Frigaard and C. Nouar. On three-dimensional linear stability of Poiseuille flow of Bingham fluids. *Phys. Fluids*, 15:2843–2851, 2003. (*226*)

[294] D. Funaro. *Polynomial Approximation of Differential Equations.* Springer-Verlag, Berlin, 1992. (*408*)

[295] H. Furstenberg and H. Kesten. Products of random matrices. *Ann. Math. Stat.*, pages 457–469, 1960. (*355*)

[296] R. Gabriel. Matrizen mit maximaler Diagonale bei unitärer Similarität. *J. Reine Angew. Math.*, 307/308:31–52, 1979. (*442*)

[297] E. Gallestey. Computing spectral value sets using the subharmonicity of the norm of rational matrices. *BIT*, 38:22–33, 1998. (*393*)

[298] E. Gallestey, D. Hinrichsen, and A. J. Pritchard. Spectral value sets of closed linear operators. *Proc. Roy. Soc. London A*, 456:1397–1418, 2000. (*30*)

[299] F. R. Gantmacher. *The Theory of Matrices*, volume 2. Chelsea, New York, 1959. (*428*)

[300] P. R. Garabedian. An unsolvable equation. *Proc. Amer. Math. Soc.*, 25:207–208, 1970. (*126, 131*)

[301] M. R. Gardner and W. R. Ashby. Connectance of large dynamic (cybernetic) systems: Critical values for stability. *Nature*, 228:784, 1970. (*528*)

[302] R. J. Gathmann, M. Si-Ameur, and F. Mathey. Numerical simulations of three-dimensional natural transition in the compressible confined shear layer. *Phys. Fluids A*, 5:2946–2968, 1993. (*226*)

[303] F. A. Gaydon and W. M. Shepherd. Generalized plane stress in a semi-infinite strip under arbitrary end-load. *Proc. Roy. Soc. London A*, 281:184–206, 1964. (*537, 541*)

[304] C. W. Gear. *Numerical Initial Value Problems in Ordinary Differential Equations.* Prentice-Hall, Englewood Cliffs, NJ, 1971. (*314*)

[305] L. Gearhart. Spectral theory for contraction semigroups on Hilbert space. *Trans. Amer. Math. Soc.*, 236:385–394, 1978. (*187*)

[306] T. Gebhardt and S. Grossmann. Chaos transition despite linear stability. *Phys. Rev. E*, 50:3705–3711, 1994. (*213*)

[307] S. Geman. A limit theorem for the norm of random matrices. *Ann. Probability*, 8:252–261, 1980. (*336*)

[308] S. Geman. The spectral radius of large random matrices. *Ann. Probability*, 14:1318–1328, 1986. (*336*)

[309] M. I. Gil'. *Norm Estimations for Operator-Valued Functions and Applications.* Dekker, New York, 1995. (*147*)

[310] J. Ginibre. Statistical ensembles of complex, quaternion, and real matrices. *J. Math. Phys.*, 6:440–449, 1965. (*334*)

[311] V. L. Girko. Circular law. *Theory Prob. Appl.*, 29:694–706, 1985. (*334*)

[312] V. L. Girko. Elliptic law. *Theory Prob. Appl.*, 30:677–690, 1986. (*337*)

[313] V. L. Girko. *Theory of Random Determinants*. Kluwer, Dordrecht, 1990. (*334, 337*)

[314] B. W. Glickfeld. On an inequality of Banach algebra geometry and semi-inner space theory. *Illinois J. Math.*, 14:76–81, 1970. (*173*)

[315] S. K. Godunov. *Modern Aspects of Linear Algebra*. Scientific Books, Novosibirsk, 1997. English translation published by the American Mathematical Society, Providence, RI, 1999. (*43, 489*)

[316] S. K. Godunov, A. G. Antonov, O. P. Kirilyuk, and V. I. Kostin. *Guaranteed Accuracy of the Solution to Systems of Linear Equations in Euclidean Spaces*. Nauka, Moscow, 1988. In Russian. English translation of second, revised edition published as *Guaranteed Accuracy in Numerical Linear Algebra* by Kluwer, Dordrecht, 1993. (*43*)

[317] S. K. Godunov, O. P. Kirilyuk, and V. I. Kostin. Spectral portraits of matrices. Technical Report Preprint 3, Inst. of Math., Sib. Branch of USSR Acad. Sci., 1990. In Russian. (*43, 147, 487*)

[318] S. K. Godunov and V. S. Ryabenkii. *Theory of Difference Schemes: An Introduction*. North-Holland, Amsterdam, 1964. Russian original published by Gosudarstv. Izdat. Fiz.-Mat. Lit., Moscow, 1962. (*42*)

[319] J. P. Goedbloed and P. H. Sakanaka. New approach to magnetohydrodynamic stability. I: A practical stability concept. *Phys. Fluids*, 17:908–918, 1974. (*228*)

[320] S. Goedecker. Remark on algorithms to find roots of polynomials. *SIAM J. Sci. Comput.*, 15:1059–1063, 1994. (*505*)

[321] I. C. Gohberg. On the application of the theory of normed rings to singular integral equations. *Usp. Matem. Nauk*, 7:149–156, 1952. (*53*)

[322] I. C. Gohberg and I. A. Fel'dman. *Convolution Equations and Projection Methods for Their Solution*. American Mathematical Society, Providence, RI, 1974. Translated by F. M. Goldware from the 1971 Russian edition. (*50*)

[323] I. C. Gohberg and M. G. Krein. *Introduction to the Theory of Linear Nonselfadjoint Operators in Hilbert Space*. American Mathematical Society, Providence, RI, 1969. (*27, 472*)

[324] I. Y. Goldsheid and B. A. Khoruzhenko. Distribution of eigenvalues in non-Hermitian Anderson models. *Phys. Rev. Lett.*, 80:2897–2900, 1998. (*339, 340, 344, 346*)

[325] I. Y. Goldsheid and B. A. Khoruzhenko. Eigenvalue curves of asymmetric tridiagonal matrices. *Elect. J. Prob.*, 5:1–28, 2000. (*339, 340, 344*)

[326] G. H. Golub. *The Use of Chebyshev Matrix Polynomials in the Iterative Solution of Linear Equations Compared to the Method of Successive Overrelaxation*. Ph.D. thesis, University of Illinois, 1959. (*232*)

[327] G. H. Golub and C. F. Van Loan. *Matrix Computations*. Johns Hopkins University Press, Baltimore, third edition, 1996. (*336, 365, 372, 384, 426, 430, 431, 444, 447, 448, 450, 486, 487*)

[328] G. H. Golub and J. H. Wilkinson. Ill-conditioned eigensystems and the computation of the Jordan canonical form. *SIAM Review*, 18:578–619, 1976. (*468, 479, 487, 489*)

[329] H. Görtler. Über eine dreidimensionale Instabilität laminarer Grenzschichten an konkaven Wänden. *Nachr. Ges. Wiss. Göttingen, N. F.*, 2:1–26, 1940. (*225*)

[330] D. Gottlieb, M. Y. Hussaini, and S. A. Orszag. Introduction: Theory and applications of spectral methods. In R. G. Voigt, D. Gottlieb, and M. Y. Hussaini, editors, *Spectral Methods for Partial Differential Equations*, pages 1–54. SIAM, Philadelphia, 1984. (*410*)

[331] D. O. Gough, J. W. Leibacher, P. H. Scherrer, and J. Toomre. Perspectives in helioseismology. *Science*, 272:1281–1283, 1996. (*3*)

[332] J.-M. Gracia and I. de Hoyos. Nearest pair with more nonconstant invariant factors and pseudospectrum. *Linear Algebra Appl.*, 298:143–158, 1999. (*435*)

[333] W. B. Gragg and L. Reichel. On the application of orthogonal polynomials to the iterative solution of linear systems of equations with indefinite or non-Hermitian matrices. *Linear Algebra Appl.*, 88/89:349–371, 1987. (*255*)

[334] R. L. Graham, D. E. Knuth, and O. Patashnik. *Concrete Mathematics*. Addison-Wesley, Reading, MA, second edition, 1994. (*520*)

[335] L. Grammont and A. Largillier. On ε-spectra and stability radii. *J. Comp. Appl. Math.*, 147:453–469, 2002. (*385*)

[336] I. Grants and G. Gerbeth. Experimental study of non-normal nonlinear transition to turbulence in a rotating magnetic field driven flow. *Phys. Fluids*, 15:2803–2809, 2003. (*228*)

[337] J. F. Grcar. Operator coefficient methods for linear equations. Technical Report SAND89-8691, Sandia National Laboratories, November 1989. (*58*)

[338] A. Greenbaum. *Iterative Methods for Solving Linear Systems*. SIAM, Philadelphia, 1997. (*3, 244, 254*)

[339] A. Greenbaum. Using the Cauchy integral formula and the partial fractions decomposition of the resolvent to estimate $\|f(A)\|$. Manuscript, August 2000. (*253*)

[340] A. Greenbaum. Generalizations of the field of values useful in the study of polynomial functions of a matrix. *Linear Algebra Appl.*, 347:233–249, 2002. (*284, 285*)

[341] A. Greenbaum. Card shuffling and the polynomial numerical hull of degree k. *SIAM J. Sci. Comput.*, 25:408–416, 2003. (*283, 284, 285, 286*)

[342] A. Greenbaum, V. Pták, and Z. Strakoš. Any nonincreasing convergence curve is possible for GMRES. *SIAM J. Matrix Anal. Appl.*, 17:465–469, 1996. (*12*)

[343] A. Greenbaum and L. N. Trefethen. Do the pseudospectra of a matrix determine its behavior? Technical Report TR 93-1371, Computer Science Department, Cornell University, Ithaca, NY, August 1993. (*437, 440*)

[344] A. Greenbaum and L. N. Trefethen. GMRES/CR and Arnoldi/Lanczos as matrix approximation problems. *SIAM J. Sci. Comput.*, 15:359–368, 1994. (*248, 269, 278, 279*)

[345] R. D. Gregory. Green's functions, bi-linear forms, and completeness of the eigenfunctions for the elastostatic strip and wedge. *J. Elasticity*, 9:283–309, 1979. (*537*)

[346] R. D. Gregory. The semi-infinite strip $x \geq 0$, $-1 \leq y \leq 1$; completeness of the Papkovich–Fadle eigenfunctions when $\phi_{xx}(0, y)$, $\phi_{yy}(0, y)$ are prescribed. *J. Elasticity*, 10:57–80, 1980. (*537*)

[347] R. D. Gregory. The traction boundary value problem for the elastostatic semi-infinite strip; existence of solution, and completeness of the Papkovich–Fadle eigenfunctions. *J. Elasticity*, 10:295–327, 1980. (*537*)

[348] G. Greiner and R. Nagel. One-parameter semigroups on Banach spaces: Spectral theory. In R. Nagel, editor, *One-Parameter Semigroups of Positive Operators*, volume 1184 of *Lecture Notes in Mathematics*, pages 60–97. Spinger-Verlag, Berlin, 1986. (*187*)

[349] L. A. Gribov and W. J. Orville-Thomas. *Theory and Methods of Calculation of Molecular Spectra.* Wiley, Chichester, England, 1988. (*3*)

[350] G. R. Grimmett and D. R. Stirzaker. *Probability and Random Processes.* Oxford University Press, Oxford, second edition, 1992. (*336*)

[351] R. Grone, C. R. Johnson, E. M. Sa, and H. Wolkowicz. Normal matrices. *Linear Algebra Appl.*, 87:213–225, 1987. (*18, 442*)

[352] M. Gu. New methods for estimating the distance to uncontrollability. *SIAM J. Matrix Anal. Appl.*, 21:989–1003, 2000. (*433, 435*)

[353] S. C. Gupta. A note on the semi-infinite strip subjected to a concentrated load. *J. Appl. Mech.*, 41:1119–1120, 1974. (*537*)

[354] K. E. Gustafson and D. K. M. Rao. *Numerical Range: The Field of Values of Linear Operators and Matrices.* Springer-Verlag, New York, 1997. (*166, 172*)

[355] B. Gustafsson. The convergence rate for difference approximations to mixed initial boundary value problems. *Math. Comp.*, 29:396–406, 1975. (*329*)

[356] B. Gustafsson, H.-O. Kreiss, and J. Oliger. *Time Dependent Problems and Difference Methods.* Wiley, New York, 1995. (*322*)

[357] B. Gustafsson, H.-O. Kreiss, and A. Sundström. Stability theory of difference approximations for mixed initial boundary value problems. II. *Math. Comp.*, 26:649–686, 1972. (*322, 328, 329*)

[358] K. Gustafsson. Control theoretic techniques for stepsize selection in explicit Runge–Kutta methods. *ACM Trans. Math. Software*, 17:533–554, 1991. (*320*)

[359] K. Gustafsson, M. Lundh, and G. Söderlind. A *PI* stepsize control for the numerical solution of ordinary differential equations. *BIT*, 28:270–287, 1988. (*320*)

[360] L. H. Gustavsson. Energy growth of three-dimensional disturbances in plan Poiseuille flow. *J. Fluid Mech.*, 224:241–260, 1991. (*202*)

[361] M. H. Gutknecht. An iterative method for solving linear equations based on minimum norm Pick-Nevanlinna interpolation. In C. K. Bhui, L. L. Schumaker, and J. D. Ward, editors, *Approximation Theory V*, pages 371–374. Academic Press, Orlando, 1986. (*255*)

[362] M. H. Gutknecht and D. Loher. Preconditioning by similarity transformations: Another valid option? Abstract, GAMM Workshop on Numerical Linear Algebra, 2001. (*253*)

[363] J. Hadamard. *Lectures on Cauchy's Problem in Linear Partial Differential Equations.* Yale University Press, New Haven, CT, 1923. (*131*)

[364] E. Hairer, S. P. Nørsett, and G. Wanner. *Solving Ordinary Differential Equations I: Nonstiff Problems.* Springer-Verlag, Berlin, second edition, 1993. (*168, 315*)

[365] E. Hairer and G. Wanner. *Solving Ordinary Differential Equations II: Stiff and Differential-Algebraic Problems.* Springer-Verlag, Berlin, 1991. (*167, 303, 314, 315, 320, 423*)

[366] G. Hall and D. J. Higham. Analysis of stepsize selection schemes for Runge–Kutta codes. *IMA J. Numer. Anal.*, 8:305–310, 1988. (*320*)

[367] P. R. Halmos. *A Hilbert Space Problem Book.* Springer-Verlag, New York, second edition, 1982. (*50, 166, 168*)

[368] S. Hammarling and J. H. Wilkinson. The practical behaviour of linear iterative methods with particular reference to S.O.R. Technical Report NPL Report NAC 69, National Physical Laboratory, Teddington, England, 1976. (*232*)

[369] H. Han, V. P. Il'in, R. B. Kellogg, and W. Yuan. Analysis of flow directed iterations. *J. Comp. Math.*, 10:57–76, 1992. (*238, 243*)

[370] C. R. Handy. Generating converging bounds to the (complex) discrete states of the $P^2 + iX^3 + i\alpha X$ Hamiltonian. *J. Phys. A*, 34:5065–5081, 2001. (*108*)

[371] A. Hanifi, P. J. Schmid, and D. S. Henningson. Transient growth in compressible boundary layer flow. *Phys. Fluids*, 8:826–837, 1996. (*226*)

[372] P. C. Hansen. *Rank-Deficient and Discrete Ill-Posed Problems: Numerical Aspects of Linear Inversion.* SIAM, Philadelphia, 1998. (*448*)

[373] A. Harrabi. On the approximation of pseudospectra of nonnormal operators by discretization. Part I: The first derivative operator. Technical Report TR/PA/98/37, CERFACS, Toulouse, 1998. (*408*)

[374] A. Harrabi. On the approximation of pseudospectra of nonnormal operators by discretization. Part II: The convection-diffusion operator. Technical Report TR/PA/98/38, CERFACS, Toulouse, 1998. (*120, 408*)

[375] A. Harrabi. Pseudospectre d'une suite d'opérateurs bornés. *RAIRO Math. Model. Num. Anal.*, 32:671–680, 1998. (*30, 33, 294, 408*)

[376] C. M. Harris and A. G. Piersol, editors. *Harris' Shock and Vibration Handbook.* McGraw-Hill, New York, fifth edition, 2002. (*3*)

[377] N. Hatano and D. R. Nelson. Localization transitions in non-Hermitian quantum mechanics. *Phys. Rev. Lett.*, 77:570–573, 1996. (*12, 339, 348*)

[378] N. Hatano and D. R. Nelson. Vortex pinning and non-Hermitian quantum mechanics. *Phys. Rev. B*, 56:8651–8673, 1997. (*339, 348*)

[379] N. Hatano, T. Watanabe, and J. Yamasaki. Localization, resonance and non-Hermitian quantum mechanics. *Physica A*, 314:170–176, 2002. (*339, 341*)

[380] P. Henrici. Bounds for iterates, inverses, spectral variation and fields of values of non-normal matrices. *Numer. Math.*, 4:24–40, 1962. (*147, 444, 445*)

[381] P. Henrici. *Discrete Variable Methods in Ordinary Differential Equations.* Wiley, New York, 1962. (*314*)

[382] I. Herbst. The spectrum of Hilbert space semigroups. *J. Operator Theory*, 10:87–94, 1983. (*187*)

[383] V. Heuveline and M. Sadkane. Arnoldi–Faber method for large non Hermitian eigenvalue problems. *Elect. Trans. Numer. Anal.*, 5:62–76, 1997. (*255*)

[384] T. Hey and P. Walters. *The New Quantum Universe.* Cambridge University Press, Cambridge, 2003. (*7*)

[385] D. J. Higham. Mean-square and asymptotic stability of numerical methods for stochastic and differential equations. *SIAM J. Numer. Anal.*, 38:753–769, 2000. (*345*)

[386] D. J. Higham and B. Owren. Nonnormality effects in a discretized nonlinear reaction-convection-diffusion equation. *J. Comp. Phys.*, 124:309–323, 1996. (*210*)

[387] D. J. Higham and L. N. Trefethen. Stiffness of ODEs. *BIT*, 33:285–303, 1993. (*319*)

[388] N. J. Higham. Matrix nearness problems and applications. In M. J. C. Gover and S. Barnett, editors, *Applications of Matrix Theory*, pages 1–27. Oxford University Press, Oxford, 1989. (*447, 451*)

[389] N. J. Higham. Algorithm 694: A collection of test matrices in MATLAB. *ACM Trans. Math. Software*, 17:289–305, 1991. (*172*)

[390] N. J. Higham. The Test Matrix Toolbox for MATLAB (Version 3.0). Numerical Analysis Report 276, Manchester Centre for Computational Mathematics, September 1995. (*371*)

[391] N. J. Higham. *Accuracy and Stability of Numerical Algorithms*. SIAM, Philadelphia, second edition, 2002. (*165, 447, 486, 487*)

[392] N. J. Higham and P. A. Knight. Matrix powers in finite precision arithmetic. *SIAM J. Matrix Anal. Appl.*, 16:343–358, 1995. (*165, 486*)

[393] N. J. Higham and F. Tisseur. A block algorithm for matrix 1-norm estimation, with an application to 1-norm pseudospectra. *SIAM J. Matrix Anal. Appl.*, 21:1185–1201, 2000. (*380*)

[394] N. J. Higham and F. Tisseur. More on pseudospectra for polynomial eigenvalue problems and applications in control theory. *Linear Algebra Appl.*, 351–352:435–453, 2002. (*428, 429, 430, 435, 436*)

[395] E. Hille and R. S. Phillips. *Functional Analysis and Semi-Groups*. American Mathematical Society, Providence, RI, revised edition, 1957. First edition by Hille, 1948. (*27, 29, 35, 148, 158, 175, 185, 188*)

[396] C. P. Hills. Eddies induced in cylindrical containers by a rotating end wall. *Phys. Fluids*, 13:2279–2286, 2001. (*539*)

[397] D. Hinrichsen and B. Kelb. Spectral value sets: A graphical tool for robustness analysis. *Sys. Control Lett.*, 21:127–136, 1993. (*44, 455, 465*)

[398] D. Hinrichsen and A. J. Pritchard. Stability radii of linear systems. *Systems Control Lett.*, 7:1–10, 1986. (*451, 455*)

[399] D. Hinrichsen and A. J. Pritchard. Real and complex stability radii: A survey. In D. Hinrichsen and B. Mårtensson, editors, *Control of Uncertain Systems*. Birkhäuser, Boston, 1990. (*451, 455, 465*)

[400] D. Hinrichsen and A. J. Pritchard. On spectral variations under bounded real matrix perturbations. *Numer. Math.*, 60:509–524, 1992. (*44, 459*)

[401] D. Hinrichsen and A. J. Pritchard. Stability of uncertain systems. In U. Helmke, R. Mennicken, and J. Saurer, editors, *Systems and Networks: Mathematical Theory and Applications*, volume 1, pages 159–182. Academie-Verlag, Berlin, 1994. (*12*)

[402] I. I. Hirschman, Jr. The spectra of certain Toeplitz matrices. *Illinois J. Math.*, 11:145–159, 1967. (*53*)

[403] D. Ho, F. Chatelin, and M. Bennani. Arnoldi–Tchebychev procedure for nonsymmetric matrices. *RAIRO Math. Model. Numer. Anal*, 24:53–65, 1990. (*255*)

[404] H. Hochstadt. On the eigenvalues of a class of integral equations arising in laser theory. *SIAM Review*, 8:62–65, 1966. (*546*)

[405] M. E. Hochstenbach. A Jacobi–Davidson type SVD method. *SIAM J. Sci. Comput.*, 23:606–628, 2001. (*392*)

[406] B. Hof, A. Juel, and T. Mullin. Scaling of the turbulence transition threshold in a pipe. *Phys. Rev. Lett.*, 91:244502-1–244502-4, 2003. (*214*)

[407] M. Högberg and D. S. Henningson. Linear optimal control applied to instabilities in spatially developing boundary layers. *J. Fluid Mech.*, 470:151–179, 2002. (*227*)

[408] Y. P. Hong and C.-T. Pan. Rank-revealing QR factorizations and the singular value decomposition. *Math. Comp.*, 58:213–232, 1992. (*448*)

[409] L. Hörmander. Differential equations without solutions. *Math. Ann.*, 140:169–173, 1960. (*98, 103, 126*)

[410] L. Hörmander. Differential operators of principal type. *Math. Ann.*, 140:124–146, 1960. (*98, 103*)

[411] L. Hörmander. *The Analysis of Linear Partial Differential Operators. III: Pseudodifferential Operators.* Springer-Verlag, Berlin, 1985. (*90*)

[412] L. Hörmander. *The Analysis of Linear Partial Differential Operators. IV: Fourier Integral Operators.* Springer-Verlag, Berlin, 1985. (*126*)

[413] L. Hörmander. On the solvability of pseudodifferential equations. In M. Morimoto and J. Kawai, editors, *Structure of Solutions of Differential Equations*, pages 183–213. World Scientific, Singapore, 1996. (*126, 129*)

[414] R. A. Horn and C. R. Johnson. *Matrix Analysis.* Cambridge University Press, Cambridge, 1985. (*3, 12, 16, 439, 444, 468, 501*)

[415] R. A. Horn and C. R. Johnson. *Topics in Matrix Analysis.* Cambridge University Press, Cambridge, 1991. (*17, 139, 166, 437*)

[416] A. S. Householder. *The Theory of Matrices in Numerical Analysis.* Blaisdell, New York, 1964. (*444*)

[417] J. S. Howland. On a theorem of Gearhart. *Integral Equations Operator Theory*, 7:138–142, 1984. (*187*)

[418] V. E. Howle and L. N. Trefethen. Eigenvalues and musical instruments. *J. Comp. Appl. Math.*, 135:23–40, 2001. (*5*)

[419] H. Hristova, S. Roch, P. J. Schmid, and L. S. Tuckerman. Transient growth in Taylor–Couette flow. *Phys. Fluids*, 14:3475–3484, 2002. (*225*)

[420] F. Huang. Characteristic conditions for exponential stability of linear dynamical systems in Hilbert spaces. *Ann. Diff. Eq.*, 1:45–53, 1985. (*187*)

[421] P. Huerre and P. A. Monkewitz. Local and global instabilities in spatially developing flows. *Ann. Rev. Fluid Mech.*, 22:473–537, 1990. (*123*)

[422] M. Huhtanen. Ideal GMRES can be bounded from below by three factors. Technical Report A412, Helsinki University of Technology Institute of Mathematics, 1999. (*248*)

[423] L. S. Hultgren and L. H. Gustavsson. Algebraic growth of disturbances in a laminar boundary layer. *Phys. Fluids*, 24:1000–1004, 1981. (*225*)

[424] W. Hundsdorfer and J. G. Verwer. *Numerical Solution of Time-Dependent Advection-Diffusion-Reaction Equations.* Springer-Verlag, Berlin, 2003. (*302*)

[425] K. J. in 't Hout and M. N. Spijker. Analysis of error growth, via stability regions in numerical initial value problems. *BIT*, 43:363–385, 2003. (*306, 310, 311*)

[426] A. Iserles. *A First Course in the Numerical Analysis of Differential Equations.* Cambridge University Press, Cambridge, 1996. (*295*)

[427] R. G. Jacobs and P. A. Durbin. Simulations of bypass transition. *J. Fluid Mech.*, 428:185–212, 2001. (*225*)

[428] R. A. Janik, M. A. Nowak, G. Papp, and I. Zahaed. Localization transitions from free random variables. *Acta Phys. Polon. B*, 30:45–58, 1999. (*339*)

[429] H. Jeffreys and B. Jeffreys. *Methods of Mathematical Physics.* Cambridge University Press, Cambridge, third edition, 2000. (*495*)

[430] M. A. Jenkins and J. F. Traub. A three-stage variable-shift iteration for polynomial zeros and its relation to generalized Rayleigh iteration. *Numer. Math.*, 14:252–263, 1970. (*507*)

[431] Z. Jia. Refined iterative algorithms based on Arnoldi's process for large unsymmetric eigenproblems. *Linear Algebra Appl.*, 259:1–23, 1997. (*435*)

[432] Z. Jia and G. W. Stewart. An analysis of the Rayleigh–Ritz method for approximating eigenspaces. *Math. Comp.*, 70:637–647, 2001. (*267, 272*)

[433] C. R. Johnson. Numerical determination of the field of values of a general complex matrix. *SIAM J. Numer. Anal.*, 15:595–602, 1978. (*172*)

[434] M. W. Johnson, Jr. and R. W. Little. The semi-infinite elastic strip. *Quart. Appl. Math.*, 22:335–344, 1965. (*537*)

[435] G. F. Jónsson and L. N. Trefethen. A numerical analyst looks at the "cutoff phenomenon" in card shuffling and other Markov chains. In D. F. Griffiths, D. J. Higham, and G. A. Watson, editors, *Numerical Analysis 1997*, pages 150–178. Addison Wesley Longman, Harlow, Essex, UK, 1998. (*12, 508, 515, 520, 521*)

[436] D. D. Joseph. *Stability of Fluid Motions.* Springer-Verlag, Berlin, 1976. Two volumes. (*167, 195, 321*)

[437] D. D. Joseph. The convergence of biorthogonal series for biharmonic and Stokes flow edge problems. Part I. *SIAM J. Appl. Math.*, 33:337–347, 1977. (*537*)

[438] D. D. Joseph and L. Sturges. The convergence of biorthogonal series for biharmonic and Stokes flow edge problems. Part II. *SIAM J. Appl. Math.*, 34:7–26, 1978. (*537*)

[439] M. Kac. Random walk and the theory of Brownian motion. *Amer. Math. Monthly*, 54:369–391, 1947. (*515*)

[440] B. Kågström and A. Ruhe. An algorithm for numerical computation of the Jordan normal form of a complex matrix. *ACM Trans. Math. Software*, 6:398–419, 1980. (*468*)

[441] W. Kahan. Numerical linear algebra. *Canadian Math. Bull.*, 9:757–801, 1966. (*447, 450*)

[442] W. Kahan. Conserving confluence curbs ill-condition. Technical Report 6, Department of Computer Science, University of California, Berkeley, 1972. (*448, 479*)

[443] T. Kailath. *Linear Systems.* Prentice-Hall, Englewood Cliffs, NJ, 1980. (*3, 435*)

[444] E. Kalnay. *Atmospheric Modeling, Data Assimilation and Predictability.* Cambridge University Press, Cambridge, 2003. (*227*)

[445] V. Karlin and G. Makhviladze. Computational analysis of the steady states of the Sivashinsky model of hydrodynamic flame instability. *Combust. Theory Modell.*, 7:87–108, 2003. (*228*)

[446] G. P. Karman, G. S. McDonald, G. H. C. New, and J. P. Woerdman. Fractal modes in unstable resonators. *Nature*, 402:138, 1999. (*550*)

[447] M. Karow. *Geometry of Spectral Value Sets.* Ph.D. thesis, Universität Bremen, 2003. (*459*)

[448] T. Kato. *Perturbation Theory for Linear Operators.* Springer-Verlag, Berlin, corrected second edition, 1980. (*13, 27, 35, 41, 117, 139, 439, 473, 477, 479, 481*)

[449] J. B. Keller. Semiclassical mechanics. *SIAM Review*, 27:485–504, 1985. (*99, 102*)

[450] J. B. Keller and S. I. Rubinow. Asymptotic solution of eigenvalue problems. *Ann. Physics*, 9:24–75, 1960. (*102*)

[451] Lord Kelvin. Stability of fluid motion—rectilinear motion of viscous fluid between two parallel plates. *Phil. Mag.*, 24:188–196, 1887. (*207*)

[452] W. Kerner. Large-scale complex eigenvalue problems. *J. Comp. Phys.*, 85:1–85, 1989. (*228*)

[453] W. Kerner, K. Lerbinger, and K. Riedel. Resistive Alfvén spectrum of tokamaklike configurations in straight cylindrical geometry. *Phys. Fluids*, 29:2975–2987, 1986. (*228*)

[454] I. M. Khabaza. An iterative least-square method for solving large sparse matrices. *Computer J.*, 6:202–206, 1963. (*261*)

[455] S. Kirkland. An eigenvalue region for Leslie matrices. *SIAM J. Matrix Anal. Appl.*, 13:507–529, 1992. (*526*)

[456] L. Knizhnerman. Error bounds for the Arnoldi method: A set of extreme eigenpairs. *Linear Algebra Appl.*, 296:191–211, 1999. (*272*)

[457] D. E. Knuth. *The Art of Computer Programming*, volume 3: *Sorting and Searching*. Addison-Wesley, Reading, MA, second edition, 1998. (*520*)

[458] G. Kolata. In shuffling cards, 7 is winning number. *New York Times*, 1990. Sec. C, p. 1, January 9. (*519*)

[459] A. Kostenbauder, Y. Sun, and A. E. Siegman. Eigenmode expansions using biorthogonal functions: Complex-valued Hermite–Gaussians. *J. Opt. Soc. Amer. A*, 14:1780–1790, 1997. (*548*)

[460] V. I. Kostin. On definition of matrices' spectra. In M. Durand and F. E. Dabaghi, editors, *High Performance Computing II*. North-Holland, Amsterdam, 1991. (*43, 394*)

[461] V. I. Kostin and S. I. Razzakov. On convergence of the power orthogonal method of spectrum computing. *Trans. Inst. Math. Sib. Branch Acad. Sci.*, 6:55–84, 1985. In Russian. (*43*)

[462] I. Koutis and E. Gallopoulos. Exclusion regions and fast estimation of pseudospectra. Manuscript, 2002. (*393, 394*)

[463] J. F. B. M. Kraaijevanger, H. W. J. Lenferink, and M. N. Spijker. Stepsize restrictions for stability in the numerical solution of ordinary and partial differential equations. *J. Comp. Appl. Math.*, 20:67–81, 1987. (*310*)

[464] M. G. Krein. Integral equations on a half-line whose kernel depends on the difference of its arguments. *Amer. Math. Soc. Trans. Ser. 2*, 2:163–288, 1962. (*53*)

[465] H.-O. Kreiss. Über die Stabilitätsdefinition für Differenzengleichungen die partielle Differentialgleichungen approximieren. *BIT*, 2:153–181, 1962. (*158, 176, 298, 301*)

[466] H.-O. Kreiss. Difference approximations for the initial-boundary value problem for hyperbolic differential equations. In D. Greenspan, editor, *Numerical Solutions of Nonlinear Differential Equations*, pages 141–166. Wiley, New York, 1966. (*328*)

[467] H.-O. Kreiss. Stability theory for difference approximations of mixed initial boundary value problems. I. *Math. Comp.*, 22:703–714, 1968. (*328, 329*)

[468] H.-O. Kreiss. Difference approximations for initial boundary-value problems. *Proc. Roy. Soc. London A*, 323:255–261, 1971. (*328*)

[469] H.-O. Kreiss and J. Lorenz. Resolvent estimates and quantification of nonlinear stability. *Acta Math. Sinica, English Series*, 16:1–20, 2000. (*210*)

[470] H. O. Kreiss and L. Wu. On the stability definition of difference approximations for the initial boundary value problem. *Appl. Numer. Math.*, 12:213–227, 1993. (*310*)

[471] V. N. Kublanovskaya. On some algorithms for the solution of the complete eigenvalue problem. *USSR Comp. Math. Math. Phys.*, 1(3):637–657, 1962. Translated from the Russian: *Z. Vycisl. Mat. Mat. Fiz.*, 1(4):550–570, 1961. (*486*)

[472] H. Lamb. *Hydrodynamics*. Cambridge University Press, Cambridge, sixth edition, 1932. First edition published 1879. (*495*)

[473] J. D. Lambert. *Numerical Methods for Ordinary Differential Systems: The Initial Value Problem*. Wiley, Chichester, England, 1991. (*303*)

[474] P. Lancaster and P. Psarrakos. On the pseudospectra of matrix polynomials. Numerical Analysis Report 445, Manchester Centre for Computational Mathematics, February 2004. (*429*)

[475] P. Lancaster and M. Tismenetsky. *Theory of Matrices*. Academic Press, Orlando, FL, second edition, 1985. (*468*)

[476] C. Lanczos. Solution of systems of linear equations by minimized iterations. *J. Res. Nat. Bur. Standards*, 49:33–53, 1952. (*278*)

[477] M. T. Landahl. A note on an algebraic instability of inviscid parallel shear flows. *J. Fluid Mech.*, 98:243–251, 1980. (*224*)

[478] H. J. Landau. On Szegő's eigenvalue distribution theorem and non-Hermitian kernels. *J. d'Analyse Math.*, 28:335–357, 1975. (*42, 58*)

[479] H. J. Landau. Loss in unstable resonators. *J. Opt. Soc. Amer.*, 66:525–529, 1976. (*42, 371, 547, 548*)

[480] H. J. Landau. The notion of approximate eigenvalues applied to an integral equation of laser theory. *Quart. Appl. Math.*, 35:165–172, 1977. (*42, 371, 547*)

[481] H. Langer and C. Tretter. Spectral properties of the Orr–Sommerfeld operator. *Proc. Roy. Soc. Edinburgh A*, 127:1245–1261, 1997. (*217*)

[482] L. László. An attainable lower bound for the best normal approximation. *SIAM J. Matrix Anal. Appl.*, 15:1035–1043, 1994. (*445*)

[483] A. Laub and K. Meyer. Canonical forms for symplectic and Hamiltonian matrices. *Celestial Mech.*, 9:213–238, 1974. (*491*)

[484] E. Lauga and T. R. Bewley. The decay of stabilizability with Reynolds number in a linear model of spatially developing flows. *Proc. Roy. Soc. London A*, 459:2077–2095, 2003. (*227*)

[485] P.-F. Lavallée. *Nouvelles Approches de Calcul du ε-Spectre de Matrices et de Faisceaux de Matrices*. Ph.D. thesis, Université de Rennes 1, 1997. (*424*)

[486] P.-F. Lavallée, A. Malyshev, and M. Sadkane. Spectral portrait of matrices by block diagonalization. In L. Vulkov, J. Waśniewski, and P. Yalamov, editors, *Numerical Analysis and Its Applications*, volume 1196 of *Lecture Notes in Computer Science*, pages 266–273. Springer-Verlag, Berlin, 1997. (*470*)

[487] P. D. Lax. *Linear Algebra*. Wiley, New York, 1997. (*377*)

[488] P. D. Lax and R. D. Richtmyer. Survey of the stability of linear finite difference equations. *Comm. Pure Appl. Math.*, 9:267–293, 1956. (*297*)

[489] B. Le Bailly and J. P. Thiran. Computing complex polynomial Chebyshev approximants on the unit circle by the real Remez algorithm. *SIAM J. Numer. Anal.*, 36:1858–1877, 1999. (*280*)

[490] S. L. Lee. Nonnormality and the effectiveness of left- and right-preconditioning. Slides from the IMACS International Symposium on Iterative Methods in Linear Algebra, 1995. (*253*)

[491] S. L. Lee. A practical upper bound for departure from normality. *SIAM J. Matrix Anal. Appl.*, 16:462–468, 1995. (*445*)

[492] S. L. Lee. Best available bounds for departure from normality. *SIAM J. Matrix Anal. Appl.*, 17:984–991, 1996. (*445*)

[493] N. J. Lehmann. Zur Verwendung optimaler Eigenwerteingrenzungen bei der Lösung symmetrischer Matrizenaufgaben. *Numer. Math.*, 8:42–55, 1966. (*435*)

[494] R. B. Lehoucq, D. C. Sorensen, and C. Yang. *ARPACK Users' Guide: Solution of Large-Scale Eigenvalue Problems with Implicitly Restarted Arnoldi Methods*. SIAM, Philadelphia, 1998. (*246, 263, 270, 389, 392, 435*)

[495] H. W. J. Lenferink and M. N. Spijker. On the use of stability regions in the numerical analysis of initial value problems. In K. Strehmel, editor, *Numerical Treatment of Differential Equations*. Teubner, Leipzig, 1988. (*310*)

[496] H. W. J. Lenferink and M. N. Spijker. On the use of stability regions in the numerical analysis of initial value problems. *Math. Comp.*, 57:221–237, 1991. (*310*)

[497] N. Lerner. Solving pseudo-differential equations. In L. Tatsien, editor, *Proceedings of the International Congress of Mathematicians 2002*, volume 3. World Scientific, Singapore, 2002. (*126, 129*)

[498] R. J. LeVeque and L. N. Trefethen. Advanced problems, #6462. *Amer. Math. Monthly*, 91:371, 1984. (*180*)

[499] R. J. LeVeque and L. N. Trefethen. On the resolvent condition in the Kreiss matrix theorem. *BIT*, 24:584–591, 1984. (*177, 179, 180, 181, 184*)

[500] H. Lewy. An example of a smooth linear partial differential equation without solution. *Ann. Math.*, 66:155–158, 1957. (*126, 131*)

[501] X. Li. An adaptive method for solving nonsymmetric linear systems involving applications of SCPACK. *J. Comp. Appl. Math.*, 44:351–370, 1992. (*255, 256*)

[502] V. B. Lidskii. Perturbation theory of non-conjugate operators. *U.S.S.R. Comput. Math. Math. Phys.*, 6(1):73–85, 1966. Translated from the Russian: *Z. Vycisl. Mat. i Mat. Fiz.*, 6(1):52–60, 1966. (*477*)

[503] J. Liesen. Computable convergence bounds for GMRES. *SIAM J. Matrix Anal. Appl.*, 21:882–903, 2000. (*247*)

[504] J. Liesen and Z. Strakoš. GMRES convergence analysis for a convection-diffusion model problem. *SIAM J. Sci. Comput.* To appear. (*251, 471*)

[505] J. Lighthill. *Waves in Fluids*. Cambridge University Press, Cambridge, 1978. (*492*)

[506] M. J. Lighthill. Group velocity. *J. Inst. Maths. Appl.*, 1:1–28, 1965. (*495*)

[507] J. Lim and J. Kim. A singular value analysis of boundary layer control, 2004. (*227*)

[508] C. C. Lin. *The Theory of Hydrodynamic Stability*. Cambridge University Press, Cambridge, 1955. (*195*)

[509] W. Liu and J. A. Domaradzki. Direct numerical simulation of transition to turbulence in Görtler flow. *J. Fluid Mech.*, 246:267–299, 1993. (*226*)

[510] X. Liu, G. Strang, and S. Ott. Localized eigenvectors from widely spaced matrix modifications. *SIAM J. Discrete Math.*, 16:479–498, 2003. (*465*)

[511] P. W. Livermore and A. Jackson. On magnetic energy instability in spherical stationary flows. Manuscript, 2003. (*227, 228*)

[512] J. Locker. *Spectral Theory of Non-Self-Adjoint Two-Point Differential Operators*. American Mathematical Society, Providence, RI, 2000. (*90*)

[513] G. Lohmann. Pseudospectrum of equatorial waves. Manuscript, 2000. (*227*)

[514] G. Lohmann, R. Gerdes, and D. Chen. Stability of the thermohaline circulation in a simple coupled model. *Tellus*, 48A:465–476, 1996. (*227*)

[515] G. Lohmann and J. Schneider. Dynamics and predictability of Stommel's box model: A phase-space perspective with implications for decadal climate variability. *Tellus*, 51A:326–336, 1999. (*227*)

[516] G. G. Lorentz. *Approximation of Functions*. Chelsea, New York, 1986. (*279*)

[517] E. N. Lorenz. A study of the predictability of a 28-variable atmospheric model. *Tellus*, 17:321–333, 1965. (*227*)

[518] C. Lubich and O. Nevanlinna. On resolvent conditions and stability estimates. *BIT*, 31:293–313, 1991. (*310*)

[519] P. Luchini. Reynolds-number-independent instability of the boundary layer over a flat surface: Optimal perturbations. *J. Fluid Mech.*, 404:289–309, 2000. (*225*)

[520] S. H. Lui. Computation of pseudospectra by continuation. *SIAM J. Sci. Comput.*, 18:565–573, 1997. (*373, 434*)

[521] S.-H. Lui. A pseudospectral mapping theorem. *Math. Comp.*, 72:1841–1854, 2003. (*21*)

[522] G. Lumer. Semi-inner-product spaces. *Trans. Amer. Math. Soc.*, 100:29–43, 1961. (*172, 174*)

[523] G. Lumer and R. S. Phillips. Dissipative operators in a Banach space. *Pacific J. Math.*, 11:679–698, 1961. (*175*)

[524] T. A. Manteuffel. The Tchebychev iteration for nonsymmetric linear systems. *Numer. Math.*, 28:307–327, 1977. (*255, 256*)

[525] T. A. Manteuffel. Adaptive procedure for estimating parameters for the nonsymmetric Tchebychev iteration. *Numer. Math.*, 31:183–208, 1978. (*256*)

[526] T. A. Manteuffel and J. S. Otto. On the roots of the orthogonal polynomials and residual polynomials associated with a conjugate gradient algorithm. *Numer. Lin. Alg. Applics.*, 1:449–475, 1994. (*246*)

[527] T. A. Manteuffel and S. V. Parter. Preconditioning and boundary conditions. *SIAM J. Numer. Anal.*, 27:656–694, 1990. (*253*)

[528] T. A. Manteuffel and G. Starke. On hybrid iterative methods for nonsymmetric systems of linear equations. *Numer. Math.*, 73:489–506, 1996. (*168, 256*)

[529] V. A. Marčenko and L. A. Pastur. Distribution of eigenvalues for some sets of random matrices. *Math. USSR Sbornik*, 1:457–483, 1967. (*334*)

[530] O. A. Marques and V. Toumazou. Spectral portrait computation by a Lanczos method (augmented matrix version). Technical Report TR/PA/95/05, CERFACS, 1995. (*392*)

[531] O. A. Marques and V. Toumazou. Spectral portrait computation by a Lanczos method (normal equation version). Technical Report TR/PA/95/02, CERFACS, 1995. (*392*)

[532] R. S. Martin, G. Peters, and J. H. Wilkinson. The QR algorithm for real Hessenberg matrices. *Numer. Math.*, 14:219–231, 1970. (*487*)

[533] R. S. Martin and J. H. Wilkinson. Similarity reduction of a general matrix to Hessenberg form. *Numer. Math.*, 12:349–368, 1968. (*487*)

[534] A. Martinez. *An Introduction to Semiclassical and Microlocal Analysis*. Springer-Verlag, New York, 2002. (*99*)

[535] J. C. Mason and D. C. Handscomb. *Chebyshev Polynomials*. Chapman and Hall/CRC, Boca Raton, FL, 2003. (*278*)

[536] A. M. Mathai. *Jacobians of Matrix Transformations and Functions of a Matrix Argument*. World Scientific, Singapore, 1997. (*367*)

[537] M. Matsubara and P. H. Alfredsson. Disturbance growth in boundary layers subjected to free-stream turbulence. *J. Fluid Mech.*, 430:149–168, 2001. (*225*)

[538] R. M. May. Will a large complex system be stable? *Nature*, 238:413–414, 1972. (*528*)

[539] E. W. Mayer and E. Reshotko. Evidence for transient disturbance growth in a 1961 pipe-flow experiment. *Phys. Fluids*, 9:242–244, 1997. (*224*)

[540] J. M. McNamee. A bibliography on roots of polynomials. *J. Comp. Appl. Math.*, 47:391–394, 1993. Online access to updated bibliography is available at http://www.elsevier.nl/homepage/sac/cam/mcnamee/. (*505*)

[541] K. Meerbergen and R. Morgan. Inexact methods (section 11.2). In *Templates for the Solution of Algebraic Eigenvalue Problems*, pages 339–352. SIAM, Philadelphia, 2000. (*392*)

[542] K. Meerbergen and D. Roose. Matrix transformations for computing rightmost eigenvalues of large sparse non-symmetric eigenvalue problems. *IMA J. Numer. Anal.*, 16:297–346, 1996. (*426*)

[543] B. Mehlig and J. T. Chalker. Statistical properties of eigenvectors in non-Hermitian Gaussian random matrix ensembles. *J. Math. Phys.*, 41:3233–3256, 2000. (*337*)

[544] M. L. Mehta. *Random Matrices*. Academic Press, San Diego, second edition, 1991. (*333, 334*)

[545] E. Mengi and M. Overton. Algorithms for the computation of the pseudospectral radius and the numerical radius of a matrix. *IMA J. Numer. Anal.* To appear. (*172, 394, 397, 402, 403*)

[546] R. Mennicken and M. Möller. *Non-self-adjoint Boundary Eigenvalue Problems*. Elsevier, Amsterdam, 2003. (*90*)

[547] A. Meseguer. Energy transient growth in the Taylor–Couette problem. *Phys. Fluids*, 14:1655–1660, 2002. (*225*)

[548] C. D. Meyer. *Matrix Analysis and Applied Linear Algebra*. SIAM, Philadelphia, 2000. (*468*)

[549] D. Mezher. A graphical tool for driving the parallel computation of pseudospectra. In *Proceedings of the 15th International Conference on Supercomputing*, pages 270–276. ACM Digital Library, 2001. Software available at http://pauillac.inria.fr/cdrom/www/ppat/eng.htm. (*396*)

[550] D. Mezher and B. Philippe. PAT—A reliable path-following algorithm. *Numer. Alg.*, 29:131–152, 2002. (*396*)

[551] G. A. Mezincescu. Some properties of eigenvalues and eigenfunctions of the cubic oscillator with imaginary coupling constant. *J. Phys. A*, 33:4911–4916, 2000. (*108*)

[552] J. J. H. Miller, E. O'Riordan, and G. I. Shishkin. *Numerical Methods for Singular Perturbation Problems: Error Estimates in the Maximum Norm for Linear Problems in One and Two Dimensions*. World Scientific, Singapore, 1996. (*122*)

[553] S. Mizohata. Solutions nulles et solutions non analytiques. *J. Math. Kyoto Univ.*, 1:271–302, 1962. (*126*)

[554] C. Moler. ROOTS—Of polynomials, that is. *MathWorks Newsletter*, 5(1):8–9, 1991. (*505*)

[555] C. Moler and C. Van Loan. Nineteen dubious ways to compute the exponential of a matrix, twenty-five years later. *SIAM Review*, 45:3–49, 2003. (*138*)

[556] F. Molteni, R. Buizza, T. N. Palmer, and T. Petroliagis. The ECMWF ensemble prediction system: Methodology and validation. *Quart. J. Roy. Meteor. Soc.*, 122:73–119, 1996. (*227*)

[557] A. Moore and R. Kleeman. The nonnormal nature of El Niño and intraseasonal variability. *J. Climate*, 12:2965–2982, 1999. (*227*)

[558] A. M. Moore et al. A comprehensive ocean prediction and analysis system based on the tangent linear and adjoint of a regional ocean model. *Ocean Model.*, 7:227–258, 2004. (*227*)

[559] A. M. Moore, C. L. Perez, and J. Zavala-Garay. A non-normal view of the wind-driven ocean circulation. *J. Phys. Oceanogr.*, 32:2681–2705, 2002. (*227*)

[560] R. B. Morgan. Implicitly restarted GMRES and Arnoldi methods for nonsymmetric systems of equations. *SIAM J. Matrix Anal. Appl.*, 21:1112–1135, 2000. (*255*)

[561] J. Moro, J. V. Burke, and M. L. Overton. On the Lidskii–Vishik–Lyusternik perturbation theory for eigenvalues of matrices with arbitrary Jordan structure. *SIAM J. Matrix Anal. Appl.*, 18:793–817, 1997. (*478*)

[562] P. M. Morse and H. Feshbach. *Methods of Theoretical Physics.* McGraw Hill, New York, 1953. Two volumes. (*473*)

[563] P. M. Morse and K. U. Ingard. *Theoretical Acoustics.* McGraw-Hill, New York, 1968. (*3*)

[564] K. W. Morton. Stability of finite-difference approximations to a diffusion-convection equation. *Int. J. Num. Meth. Engrg.*, 15:677–683, 1980. (*122, 311*)

[565] K. W. Morton. *Numerical Solution of Convection-Diffusion Problems.* Chapman and Hall, London, 1996. (*122*)

[566] R. G. Mosier. Root neighborhoods of a polynomial. *Math. Comp.*, 47:265–273, 1986. (*504*)

[567] R. J. Muirhead. *Aspects of Multivariate Statistical Theory.* Wiley, New York, 1982. (*333, 334*)

[568] J. D. Murray. *Mathematical Biology I: An Introduction.* Springer-Verlag, Berlin, third edition, 2002. (*526*)

[569] N. M. Nachtigal, S. C. Reddy, and L. N. Trefethen. How fast are nonsymmetric matrix iterations? *SIAM J. Matrix Anal. Appl.*, 13:778–795, 1992. (*45, 245*)

[570] N. M. Nachtigal, L. Reichel, and L. N. Trefethen. A hybrid GMRES algorithm for nonsymmetric linear systems. *SIAM J. Matrix Anal. Appl.*, 13:796–825, 1992. (*45, 58, 254, 256, 261*)

[571] M. Nagata. Three-dimensional finite-amplitude solutions in plane Couette flow: Bifurcation from infinity. *J. Fluid Mech.*, 217:519–527, 1990. (*225*)

[572] D. R. Nelson and N. M. Shnerb. Non-Hermitian localization and population biology. *Phys. Rev. E*, 58:1383–1403, 1998. (*339, 348*)

[573] Y. Nesterov and A. Nemirovskii. *Interior Point Polynomial Methods in Convex Programming.* SIAM, Philadelphia, 1994. (*280*)

[574] M. G. Neubert and H. Caswell. Alternatives to resilience for measuring the responses of ecological systems to perturbations. *Ecology*, 78:653–665, 1997. (*12, 527, 528, 532*)

[575] O. Nevanlinna. *Convergence of Iterations for Linear Equations.* Birkhäuser, Basel, 1993. (*30, 138, 168, 248, 284*)

[576] O. Nevanlinna. Hessenberg matrices in Krylov subspaces and the computation of the spectrum. *Numer. Func. Anal. Optim.*, 16:443–473, 1995. (*284*)

[577] G. H. C. New. The origin of excess noise. *J. Mod. Optics*, 52:799–810, 1995. (*552*)

[578] L. Nirenberg and F. Treves. Solvability of a first order linear partial differential equation. *Comm. Pure Appl. Math.*, 16:331–351, 1963. (*126*)

[579] L. Nirenberg and F. Treves. On local solvability of linear partial differential equations. I. Necessary conditions. *Comm. Pure Appl. Math.*, 23:1–38, 1970. (*126, 129*)

[580] L. Nirenberg and F. Treves. On local solvability of linear partial differential equations. II. Sufficient conditions. *Comm. Pure Appl. Math.*, 23:1–38, 1970. Correction: ibid., 24:279–288, 1971. (*126, 129*)

[581] N. Nirschl and H. Schneider. The Bauer field of values of a matrix. *Numer. Math.*, pages 335–365, 1964. (*173*)

[582] J. R. Norris. *Markov Chains.* Cambridge University Press, Cambridge, 1997. (*3, 509*)

[583] G. G. O'Brien, M. A. Hyman, and S. Kaplan. A study of the numerical solution of partial differential equations. *J. Math. & Phys.*, 29:223–251, 1951. (*296*)

[584] I. Olkin and A. R. Sampson. Jacobians of matrix transformations and induced functional equations. *Linear Algebra Appl.*, 5:257–276, 1972. (*367*)

[585] P. J. Olsson and D. S. Henningson. Optimal disturbance growth in watertable flow. *Stud. Appl. Math.*, 94:183–210, 1995. (*228*)

[586] G. Opfer and G. Schober. Richardson's iteration for nonsymmetric matrices. *Linear Algebra Appl.*, 58:343–361, 1984. (*255*)

[587] W. M. F. Orr. The stability or instability of the steady motions of a perfect liquid and of a viscous liquid. Part I: A perfect liquid. Part II: A viscous liquid. *Proc. Roy. Irish Acad. A*, 27:9–138, 1907. (*198, 216*)

[588] S. A. Orszag. Accurate solution of the Orr–Sommerfeld stability equation. *J. Fluid Mech.*, 50:689–703, 1971. (*10, 203, 217, 537*)

[589] S. Osher. Stability of difference approximations of dissipative type for mixed initial-boundary value problems. I. *Math. Comp.*, 23:335–340, 1969. (*328*)

[590] S. Osher. Systems of difference equations with general homogeneous boundary conditions. *Trans. Amer. Math. Soc.*, 137:177–201, 1969. (*328*)

[591] A. M. Ostrowski. *Solution of Equations and Systems of Equations.* Academic Press, New York, second edition, 1966. (*504*)

[592] M. L. Overton. *Numerical Computing with IEEE Floating Point Arithmetic.* SIAM, Philadelphia, 2001. (*487, 503*)

[593] C. Paige and C. Van Loan. A Schur decomposition for Hamiltonian matrices. *Linear Algebra Appl.*, 41:11–32, 1981. (*491*)

[594] C. C. Paige. Properties of numerical algorithms related to computing controllability. *IEEE Trans. Auto. Control*, 26:130–138, 1981. (*435*)

[595] C. C. Paige, B. N. Parlett, and H. A. van der Vorst. Approximate solutions and eigenvalue bounds from Krylov subspaces. *Numer. Lin. Alg. Applics.*, 2:115–133, 1995. (*246*)

[596] C. C. Paige and M. A. Saunders. Solution of sparse indefinite systems of linear equations. *SIAM J. Numer. Anal.*, 12:617–629, 1975. (*255*)

[597] K. J. Palmer. Exponential dichotomies and transversal homoclinic points. *J. Diff. Eq.*, 55:225–256, 1984. (*98, 113*)

[598] T. N. Palmer. Extended-range atmospheric prediction and the Lorenz model. *Bull. Amer. Met. Soc.*, 74:49–65, 1993. (*227*)

[599] T. N. Palmer, R. Gelaro, J. Barkmeijer, and R. Buizza. Singular vectors, metrics, and adaptive observations. *J. Atmos. Sci.*, 55:633–653, 1998. (*227*)

[600] P. F. Papkovich. Über eine Form der Lösung des byharmonischen Problems für das Rechteck. *C. R. (Dokl.) Acad. Sci. USSR*, 27:334–338, 1970. (*537*)

[601] B. N. Parlett. *The Symmetric Eigenvalue Problem*. Prentice-Hall, Englewood Cliffs, NJ, 1980. (*268, 278*)

[602] B. N. Parlett and C. Reinsch. Balancing a matrix for calculation of eigenvalues and eigenvectors. *Numer. Math.*, 13:293–304, 1969. (*503*)

[603] S. V. Parter. Stability, convergence, and pseudo-stability of finite-difference equations for an over-determined problem. *Numer. Math.*, 4:277–292, 1962. (*298*)

[604] D. Pathria and L. N. Trefethen. Eigenvalues and pseudo-eigenvalues for convection-diffusion problems. II. Upwind-downwind effects. Unpublished manuscript, 1993. (*238, 242*)

[605] V. Paulsen. *Completely Bounded Maps and Operator Algebras*. Cambridge University Press, Cambridge, 2002. (*441*)

[606] A. Pazy. *Semigroups of Linear Operators and Applications to Partial Differential Equations*. Springer-Verlag, New York, 1983. (*27, 35, 138, 148, 149, 175*)

[607] C. Pearcy. An elementary proof of the power inequality for the numerical radius. *Michigan Math. J.*, 13:284–291, 1966. (*168*)

[608] K. Petermann. Calculated spontaneous emission factor for double-heterostructure injection lasers with gain-induced waveguiding. *IEEE J. Quant. Elect.*, QE-15:566–570, 1979. (*552*)

[609] G. Peters and J. H. Wilkinson. Practical problems arising in the solution of polynomial equations. *J. Inst. Math. Appl.*, 8:16–35, 1971. (*503*)

[610] L. Petzold. Differential/algebraic equations are not ODE's. *SIAM J. Sci. Stat. Comput.*, 3:367–384, 1982. (*423*)

[611] S. Poedts and A. D. Rogava. Does spiral galaxy IC 342 exhibit shear induced wave transformations!? *Astron. Astrophys.*, 385:32–38, 2002. (*228*)

[612] K. Pravda-Starov. A general result about the pseudo-spectrum of Schrödinger operators. *Proc. Roy. Soc. London A*, 460:471–477, 2004. (*98*)

[613] A. J. Pritchard and S. Townley. Robustness of linear systems. *J. Diff. Eq.*, 77:254–286, 1989. (*451*)

[614] A. Prothero and A. Robinson. On the stability and accuracy of one-step methods for solving stiff systems of ordinary differential equations. *Math. Comp.*, 28:145–162, 1974. (*320*)

[615] J. Prüss. On the spectrum of C_0-semigroups. *Trans. Amer. Math. Soc.*, 284:847–857, 1984. (*187*)

[616] L. Qiu, B. Bernhardsson, A. Rantzer, E. J. Davison, P. M. Young, and J. C. Doyle. A formula for computation of the real stability radius. *Automatica*, 31:879–890, 1995. (*455, 459*)

[617] Lord Rayleigh. On progressive waves. *Proc. Lond. Math. Soc.*, 9:21–26, 1877. (*494*)

[618] Lord Rayleigh. *The Theory of Sound*. Macmillian, London, 1877, 1878. 2 volumes. (*6, 494*)

[619] Lord Rayleigh. On the instability of jets. *Proc. London Math. Soc.*, 10:4–13, 1879. Reprinted in *Scientific Papers*, volume 1, Cambridge University Press, Cambridge, 1899. (*195*)

[620] S. C. Reddy. *Pseudospectra of Operators and Discretization Matrices and an Application to Stability of the Method of Lines*. Ph.D. thesis, MIT, Cambridge, MA, 1991. (*46*)

[621] S. C. Reddy. Pseudospectra of Wiener–Hopf integral operators and constant-coefficient differential operators. *J. Int. Eqs. Appl.*, 5:369–403, 1993. (*87, 90, 92, 96*)

[622] S. C. Reddy and D. S. Henningson. Energy growth in viscous channel flows. *J. Fluid Mech.*, 252:209–238, 1993. (*167, 202*)

[623] S. C. Reddy, P. J. Schmid, J. S. Baggett, and D. S. Henningson. On stability of streamwise streaks and transition thresholds in plane channel flows. *J. Fluid Mech.*, 365:269–303, 1998. (*214*)

[624] S. C. Reddy, P. J. Schmid, and D. S. Henningson. Pseudospectra of the Orr–Sommerfeld operator. *SIAM J. Appl. Math.*, 53:15–47, 1993. (*107, 169, 215, 216, 219, 221, 382, 426*)

[625] S. C. Reddy and L. N. Trefethen. Lax-stability of fully discrete spectral methods via stability regions and pseudo-eigenvalues. *Comp. Methods Appl. Mech. Eng.*, 80:147–164, 1990. (*45, 289, 291, 306, 310*)

[626] S. C. Reddy and L. N. Trefethen. Stability of the method of lines. *Numer. Math.*, 62:235–267, 1992. (*45, 306, 307, 308, 310*)

[627] S. C. Reddy and L. N. Trefethen. Pseudospectra of the convection-diffusion operator. *SIAM J. Appl. Math.*, 54:1634–1649, 1994. (*95, 120, 121*)

[628] P. Redparth. The pseudospectral properties of non-self-adjoint Schrödinger operators in the semi-classical limit. Preprint arXiv.math.SP/0007172, July 2000. (*107, 222*)

[629] M. Reed and B. Simon. *Methods of Modern Mathematical Physics IV: Analysis of Operators.* Academic Press, New York, 1978. (*473*)

[630] M. Reed and B. Simon. *Methods of Modern Mathematical Physics I: Functional Analysis.* Academic Press, San Diego, revised and enlarged edition, 1980. (*3, 27*)

[631] L. Reichel. The application of Leja points to Richardson iteration and polynomial preconditioning. *Linear Algebra Appl.*, 154–156:389–414, 1991. (*257*)

[632] L. Reichel and L. N. Trefethen. Eigenvalues and pseudo-eigenvalues of Toeplitz matrices. *Linear Algebra Appl.*, 162–164:153–185, 1992. (*45, 56, 58, 62, 79, 165, 250, 257, 467*)

[633] E. Reshotko. Boundary layer stability and transition. *Ann. Rev. Fluid Mech.*, 8:311–349, 1976. (*225*)

[634] E. Reshotko. Transient growth: A factor in bypass transition. *Phys. Fluids*, 13:1067–1075, 2001. (*225, 226*)

[635] E. Reshotko and A. Tumin. The blunt-body paradox—a case for transient growth. In H. F. Fasel and W. S. Saric, editors, *Laminar-Turbulent Transition*, pages 403–408. Springer-Verlag, New York, 2000. (*226*)

[636] O. Reynolds. An experimental investigation of the circumstances which determine whether the motion of water shall be direct or sinuous, and of the law of resistance in parallel channels. *Phil. Trans. Roy. Soc.*, 174:935–982, 1883. (*205*)

[637] L. F. Richardson. The approximate arithmetical solution by finite differences of physical problems involving differential equations, with an application to the stresses in a masonry dam. *Phil. Trans. Royal Soc. London*, 210:307–357, 1911. (*296*)

[638] L. F. Richardson. *Weather Prediction by Numerical Process.* Cambridge University Press, London, 1922. Reprinted by Dover, New York, 1965. (*296*)

[639] R. D. Richtmyer and K. W. Morton. *Difference Methods for Initial-Value Problems.* Wiley-Interscience, New York, second edition, 1967. (*3, 176, 177, 297, 298, 307*)

[640] K. S. Riedel. Generalized epsilon-pseudospectra. *SIAM J. Numer. Anal.*, 31:1219–1225, 1994. (*378, 424, 425, 426*)

[641] F. Riesz and B. Sz.-Nagy. *Functional Analysis*. Frederick Ungar, New York, 1955. Translation of the second French edition. (*139, 439, 441*)

[642] V. A. Romanov. Stability of plane-parallel Couette flow. *Funkcional Anal. i Proložen*, 7:62–73, 1973. Translated in *Functional Anal. Appl.* 7:137–146, 1973. (*199*)

[643] H.-G. Roos, M. Stynes, and L. Tobiska. *Numerical Methods for Singularly Perturbed Differential Equations*. Springer-Verlag, Berlin, 1996. (*122*)

[644] J. S. Rosenthal. Convergence rates for Markov chains. *SIAM Review*, 37:387–405, 1995. (*509*)

[645] A. Ruhe. Closest normal matrix finally found! *BIT*, 27:585–598, 1987. (*442, 443*)

[646] A. Ruhe. The rational Krylov algorithm for large nonsymmetric eigenvalues — mapping the resolvent norms (pseudospectrum). Unpublished manuscript, March 1995. (*390, 424*)

[647] M. A. Rutgers, X.-l. Wu, and W. I. Goldburg. The onset of two-dimensional grid generated turbulence in flowing soap films. *Phys. Fluids*, 8:S7, 1996. (*196*)

[648] Y. Saad. Variations on Arnoldi's method for computing eigenelements of large unsymmetric matrices. *Linear Algebra Appl.*, 34:269–295, 1980. (*263, 272, 277, 388*)

[649] Y. Saad. Projection methods for solving large sparse eigenvalue problems. In B. Kågström and A. Ruhe, editors, *Matrix Pencils*, volume 973 of *Lecture Notes in Mathematics*, pages 121–144. Springer-Verlag, Berlin, 1983. (*278*)

[650] Y. Saad. Chebyshev acceleration techniques for solving nonsymmetric eigenvalue problems. *Math. Comp.*, 42:567–588, 1984. (*255*)

[651] Y. Saad. Least squares polynomials in the complex plane and their use for solving nonsymmetric linear systems. *SIAM J. Numer. Anal.*, 24:155–169, 1987. (*255, 256*)

[652] Y. Saad. Krylov subspace methods on supercomputers. *SIAM J. Sci. Stat. Comput.*, 10:1200–1232, 1989. (*255*)

[653] Y. Saad. *Numerical Methods for Large Eigenvalue Problems*. Manchester University Press, Manchester, 1992. (*264, 267, 381*)

[654] Y. Saad. *Iterative Methods for Sparse Linear Systems*. SIAM, Philadelphia, second edition, 2003. (*244, 254, 255*)

[655] Y. Saad and M. H. Schultz. GMRES: A generalized minimal residual algorithm for solving nonsymmetric linear systems. *SIAM J. Sci. Stat. Comput.*, 7:856–869, 1986. (*244, 248*)

[656] E. B. Saff and V. Totik. *Logarithmic Potentials with External Fields*. Springer-Verlag, Berlin, 1997. (*278*)

[657] N. D. Sandham, N. A. Adams, and L. Kleiser. Direct simulation of breakdown to turbulence following oblique instability waves in a supersonic boundary layer. *Appl. Sci. Res.*, 54:223–224, 1995. (*226*)

[658] L. A. Santaló. *Integral Geometry and Geometric Probability*. Addison-Wesley, Reading, MA, 1976. (*183*)

[659] J. M. Sanz-Serna and J. G. Verwer. Stability and convergence at the pde/stiff ode interface. *Appl. Numer. Math.*, 5:117–132, 1989. (*310*)

[660] W. S. Saric, H. L. Reed, and E. J. Kerschen. Boundary-layer receptivity to freestream disturbances. *Ann. Rev. Fluid Mech.*, 34:291–319, 2002. (*202, 225*)

[661] W. S. Saric, H. L. Reed, and E. B. White. Stability and transition of three-dimensional boundary layers. *Ann. Rev. Fluid Mech.*, 35:413–440, 2003. (*224*)

[662] M. Sato, T. Kawai, and M. Kashiwara. Microfunctions and pseudo-differential equations. In H. Komatsu, editor, *Hyperfunctions and Pseudo-Differential Equations*, volume 287 of *Lecture Notes in Mathematics*, pages 265–529. Springer-Verlag, Berlin, 1973. (*103*)

[663] P. E. Saylor. An adaptive algorithm for Richardson's iteration. In D. R. Kincaid and L. J. Hayes, editors, *Iterative Methods for Large Linear Systems*, pages 215–233. Academic Press, Boston, 1990. (*256*)

[664] P. E. Saylor and D. C. Smolarski. Implementation of an adaptive algorithm for Richardson's method. *Linear Algebra Appl.*, 154–156:615–646, 1991. (*255, 256*)

[665] A. L. Schawlow and C. H. Townes. Infrared and optimal masers. *Phys. Rev.*, 112:1940–1949, 1958. (*552*)

[666] L. I. Schiff. *Quantum Mechanics*. McGraw-Hill, New York, third edition, 1968. (*3*)

[667] P. J. Schmid. Energy amplification of steady disturbances in growing boundary layers. In U. Frisch, editor, *Advances in Turbulence VII*. Kluwer, Dordrecht, 1998. (*225*)

[668] P. J. Schmid. Linear stability theory and bypass transition in shear flows. *Phys. Plasmas*, 7:1788–1794, 2000. (*225*)

[669] P. J. Schmid and D. S. Henningson. *Stability and Transition in Shear Flows*. Springer-Verlag, New York, 2001. (*3, 195, 197, 213, 215, 217, 219, 497*)

[670] P. J. Schmid, D. S. Henningson, M. R. Khorrami, and M. R. Malik. A study of eigenvalue sensitivity for hydrodynamic stability operators. *Theor. Comp. Fluid Dynamics*, 4:227–240, 1993. (*226, 228*)

[671] P. J. Schmid and H. K. Kytömaa. Transient and asymptotic stability of granular shear flow. *J. Fluid Mech.*, 264:255–275, 1994. (*228*)

[672] P. Schmidt and F. Spitzer. The Toeplitz matrices of an arbitrary Laurent polynomial. *Math. Scand.*, 8:15–38, 1960. (*53, 55, 57*)

[673] L. F. Shampine and C. W. Gear. A user's view of solving stiff ordinary differential equations. *SIAM Review*, 21:1–17, 1979. (*315*)

[674] H. Shapiro. A survey of canonical forms and invariants for unitary similarity. *Linear Algebra Appl.*, 147:101–167, 1991. (*469*)

[675] K. V. Sharp and R. J. Adrian. Transition from laminar to turbulent flow in liquid filled microtubes. *Exp. Fluids*, 36:741–747, 2004. (*228*)

[676] A. A. Shkalikov. The limit behavior of the spectrum for large parameter values in a model problem. *Math. Notes*, 62:796–799, 1997. (*107*)

[677] A. A. Shkalikov. How to define the Orr–Sommerfeld operator? *Moscow Univ. Math. Bull.*, 53(4):36–43, 1998. (*107, 216*)

[678] A. A. Shkalikov. Spectral portraits and the resolvent growth of a model problem associated with the Orr–Sommerfeld equation. Preprint arXiv.math.FA/0306342, June 2003. (*222*)

[679] A. A. Shkalikov and C. Tretter. Kamke problems. Properties of the eigenfunctions. *Math. Nachr.*, 170:251–275, 1994. (*217*)

[680] A. A. Shkalikov and C. Tretter. Spectral analysis for linear pencils $N - \lambda P$ of ordinary differential operators. *Math. Nachr.*, 179:275–305, 1996. (*217*)

[681] N. M. Shnerb and D. R. Nelson. Winding numbers, complex currents, and non-Hermitian localization. *Phys. Rev. Lett.*, 80:5172–5175, 1998. (*339, 340, 348*)

[682] M. Shub. *Global Stability of Dynamical Systems*. Springer-Verlag, New York, 1987. (*67*)

[683] A. E. Siegman. Unstable optical resonators for laser applications. *Proc. IEEE*, 53:277–287, 1965. (*548*)

[684] A. E. Siegman. Orthogonality properties of optical resonator eigenmodes. *Opt. Commun.*, 31:369–373, 1979. (*548*)

[685] A. E. Siegman. *Lasers*. University Science Books, Mill Valley, CA, 1986. (*543, 548, 549*)

[686] A. E. Siegman. Excess spontaneous emission in non-Hermitian optical systems. I. Laser amplifiers. *Phys. Rev. A*, 39:1253–1263, 1989. (*548, 552*)

[687] A. E. Siegman. Excess spontaneous emission in non-Hermitian optical systems. II. Laser oscillators. *Phys. Rev. A*, 39:1264–1268, 1989. (*548, 552*)

[688] A. E. Siegman. Lasers without photons—or should it be lasers with too many photons? *Appl. Phys. B*, 60:247–257, 1995. (*12, 548*)

[689] A. E. Siegman. Laser beams and resonators: Beyond the 1960s. *IEEE J. Select. Topics Quant. Elect.*, 6:1389–1399, 2000. (*553*)

[690] A. E. Siegman. Laser beams and resonators: The 1960s. *IEEE J. Select. Topics Quant. Elect.*, 6:1380–1388, 2000. (*544*)

[691] J. W. Silverstein. The smallest eigenvalue of a large dimensional Wishart matrix. *Ann. Probability*, 85:1364–1368, 1985. (*335, 365*)

[692] V. Simoncini. Remarks on non-linear spectral perturbation. *BIT*, 39:350–365, 1999. (*465, 479*)

[693] V. Simoncini and E. Gallopoulos. Convergence properties of block GMRES and matrix polynomials. *Linear Algebra Appl.*, 247:97–119, 1996. (*246*)

[694] V. Simoncini and E. Gallopoulos. Transfer functions and resolvent norm approximation of large matrices. *Elect. Trans. Numer. Anal.*, 7:190–201, 1998. (*246*)

[695] A. M. Sinclair. *Continuous Semigroups in Banach Algebras*, volume 63 of *London Math. Soc. Lecture Note Series*. Cambridge University Press, Cambridge, 1982. (*155*)

[696] F. H. Slaymaker and W. F. Meeker. Measurements of the tonal characteristics of carillon bells. *J. Acoust. Soc. Amer.*, 26:515–522, 1954. (*5*)

[697] G. L. G. Sleijpen and H. A. van der Vorst. A Jacobi–Davidson iteration method for linear eigenvalue problems. *SIAM J. Matrix Anal. Appl.*, 17:401–425, 1996. (*264, 392*)

[698] SLICOT control and systems library. http://www.win.tue.nl/wgs/slicot.html. (*155*)

[699] S. Smale. On the efficiency of algorithms of analysis. *Bull. (N.S.) Amer. Math. Soc.*, 13:87–121, 1985. (*338*)

[700] B. T. Smith, J. M. Boyle, J. J. Dongarra, B. S. Garbow, Y. Ikebe, V. C. Klema, and C. B. Moler. *Matrix Eigensystem Routines—EISPACK Guide*, volume 6 of *Lecture Notes in Computer Science*. Springer-Verlag, Berlin, second edition, 1976. (*503*)

[701] G. D. Smith. *Numerical Solution of Partial Differential Equations: Finite Difference Methods*. Oxford University Press, Oxford, third edition, 1985. (*238*)

[702] J. C. Smith. Solutions to advanced problems: An inequality for rational functions. *Amer. Math. Monthly*, 92:740–741, 1985. (*180*)

[703] R. C. T. Smith. The bending of a semi-infinite strip. *Austral. J. Sci. Res.*, 5:227–237, 1952. (*537*)

[704] T. R. Smith, J. Moehlis, and P. Holmes. Low-dimensional models for turbulent plane Couette flow in a minimal flow unit. *J. Fluid Mech.* To appear. (*213*)

[705] D. C. Smolarski. Optimum semi-iterative method for the solution of any linear algebraic system with a square matrix. Technical Report 1077, Department of Computer Science, University of Illinois at Urbana-Champaign, 1981. (*256*)

[706] D. C. Smolarski and P. E. Saylor. An optimum iterative method for solving any linear system with a square matrix. *BIT*, 28:163–178, 1988. (*256*)

[707] A. Sommerfeld. Ein Beitrag zur hydrodynamischen Erklaerung der turbulenten Fluessigkeitsbewegungen. In *Proceedings of the 4th International Congress of Mathematicians, Rome 1908*, pages 116–124. Accademia dei Lincei, Rome, 1909. (*198, 216*)

[708] H. J. Sommers, A. Crisanti, H. Sompolinsky, and Y. Stein. Spectrum of large random asymmetric matrices. *Phys. Rev. Lett.*, 60:1895–1898, 1998. (*337*)

[709] D. C. Sorensen. Implicit application of polynomial filters in a k-step Arnoldi method. *SIAM J. Matrix Anal. Appl.*, 13:357–385, 1992. (*263, 277, 389, 435*)

[710] D. C. Sorensen. Numerical methods for large eigenvalue problems. *Acta Numerica*, 11:519–584, 2002. (*435*)

[711] D. A. Spence. Mixed boundary value problems for the elastic strip: The eigenfunction expansion. Technical Report 1863, Math. Res. Ctr., University of Wisconsin, 1978. (*537, 541*)

[712] D. A. Spence. A note on the eigenfunction expansion for the elastic strip. *SIAM J. Appl. Math.*, 42:155–173, 1982. (*537*)

[713] D. A. Spence. A class of biharmonic end-strip problems arising in elasticity and Stokes flow. *IMA J. Appl. Math.*, 30:107–139, 1983. (*537, 541*)

[714] M. N. Spijker. Stepsize restrictions for stability of one-step methods in the numerical solution of initial value problems. *Math. Comp.*, 45:377–392, 1985. (*310*)

[715] M. N. Spijker. On a conjecture by LeVeque and Trefethen related to the Kreiss matrix theorem. *BIT*, 31:551–555, 1991. (*177, 180*)

[716] M. N. Spijker and F. A. J. Straetmans. Error growth analysis via stability regions for discretizations of initial value problems. *Appl. Numer. Math.*, 37:442–464, 1997. (*310, 311*)

[717] H. B. Squire. On the stability for three-dimensional disturbances of viscous fluid flow between parallel walls. *Proc. Roy. Soc. London A*, 142:621–628, 1933. (*198, 219*)

[718] D. Stark, A. Ganesh, and N. O'Connell. Information loss in riffle shuffling. *Combin. Probab. Comput.*, 11:79–95, 2002. (*525*)

[719] G. Starke. Fields of values and the ADI method for non-normal matrices. *Linear Algebra Appl.*, 180:199–218, 1993. (*168*)

[720] G. Starke and R. S. Varga. A hybrid Arnoldi-Faber iterative method for non-symmetric systems of linear equations. *Numer. Math.*, 63:213–240, 1993. (*255, 256*)

[721] L. A. Steen. Highlights in the history of spectral theory. *Amer. Math. Monthly*, 80:359–381, 1973. (*6*)

[722] J. I. Steinfeld, J. S. Francisco, and W. L. Hase. *Chemical Kinetics and Dynamics*. Pearson Educ. POD, 2nd edition, 1998. (*3*)

[723] G. W. Stewart. Error bounds for approximate invariant subspaces of closed linear operators. *SIAM J. Numer. Anal.*, 8:796–808, 1971. (*480*)

[724] G. W. Stewart. Error and perturbation bounds for subspaces associated with certain eigenvalue problems. *SIAM Review*, 15:727–764, 1973. (*480, 491*)

[725] G. W. Stewart. Perturbation theory for rectangular matrix pencils. *Linear Algebra Appl.*, 208/209:297–301, 1994. (*429, 430*)

[726] G. W. Stewart. *Matrix Algorithms, Volume I: Basic Decompositions*. SIAM, Philadelphia, 1998. (*448*)

[727] G. W. Stewart. *Matrix Algorithms, Volume II: Eigensystems*. SIAM, Philadelphia, 2001. (*267, 272, 490*)

[728] G. W. Stewart. On the powers of a matrix with perturbations. *Numer. Math.*, 96:363–376, 2003. (*165, 486*)

[729] G. W. Stewart and J.-g. Sun. *Matrix Perturbation Theory*. Academic Press, San Diego, 1990. (*20, 264, 386, 428, 429, 431, 447, 491*)

[730] J. J. Stoker. *Water Waves*. Wiley-Interscience, New York, 1957. (*495*)

[731] S. D. Stoller, W. Happer, and F. J. Dyson. Transverse spin relaxation in inhomogeneous magnetic fields. *Phys. Rev. A*, 44:7459–7477, 1991. (*107, 222*)

[732] M. H. Stone. *Linear Transformations in Hilbert Space*. Americal Mathematical Society, New York, 1932. (*169*)

[733] G. Strang. Wiener–Hopf difference equations. *J. Math. Mech.*, 13:85–96, 1964. (*328*)

[734] G. Strang. Implicit difference methods for initial-boundary value problems. *J. Math. Anal. Appl.*, 16:188–198, 1966. (*328*)

[735] R. Sureshkumar, M. D. Smith, R. C. Armstrong, and R. A. Brown. Linear stability and nonlinear dynamics of viscoelastic flows by time-dependent numerical simulations. *Non-Newtonian Fluid Mech.*, 82:57–104, 1999. (*226*)

[736] J. D. Swearingen and R. F. Blackwelder. The growth and breakdown of streamwise vortices in the presence of a wall. *J. Fluid Mech.*, 182:255–290, 1987. (*225*)

[737] E. Tadmor. The equivalence of L_2-stability, the resolvent condition, and strict H-stability. *Linear Algebra Appl.*, 41:151–159, 1981. (*177*)

[738] R. Tagg. The Couette–Taylor problem. *Nonlin. Sci. Today*, 4:1–25, 1994. (*198*)

[739] A. Takayama. *Mathematical Economics*. Cambridge University Press, Cambridge, second edition, 1985. (*3*)

[740] H. Tal-Ezer. A pseudospectral Legendre method for hyperbolic equations with an improved stability condition. *J. Comp. Phys.*, 67:145–172, 1986. (*298, 301*)

[741] H. Tal-Ezer. Spectral methods in time for hyperbolic equations. *SIAM J. Numer. Anal.*, 23:11–26, 1986. (*15, 292*)

[742] H. Tal-Ezer. Polynomial approximation of functions of matrices and applications. *J. Sci. Comput.*, 4:25–60, 1989. (*255*)

[743] P. T. P. Tang. A fast algorithm for linear complex Chebyshev approximation. *Math. Comp.*, 51:721–739, 1988. (*280*)

[744] G. I. Taylor. Experiments with rotating fluids. *Proc. Camb. Phil. Soc.*, 20:326–329, 1921. (*198, 225*)

[745] G. I. Taylor. Stability of a viscous liquid contained between two rotating cylinders. *Phil. Trans. Roy. Soc. A*, 223:289–343, 1923. (*198, 225*)

[746] H. M. Taylor and S. Karlin. *An Introduction to Stochastic Modeling*. Academic Press, San Diego, revised edition, 1984. (*509*)

[747] M. Teytel. How rare are multiple eigenvalues? *Comm. Pure Appl. Math.*, 52:917–934, 1999. (*377*)

[748] V. Thomée. Stability theory for partial difference operators. *SIAM Review*, 11:152–195, 1969. (*298*)

[749] C. J. Thompson. Initial conditions for optimal growth in a coupled ocean-atmosphere model of ENSO. *J. Atmos. Sci.*, 55:537–557, 1998. (*227*)

[750] G. L. Thompson and R. L. Weil. The roots of matrix pencils $(Ay = \lambda By)$: Existence, calculations, and relations to game theory. *Linear Algebra Appl.*, 5:207–226, 1972. (*430, 436*)

[751] D. J. Thouless. Electrons in disordered systems and the theory of localization. *Phys. Reports*, 13:93–142, 1974. (*339, 357*)

[752] P. Tilli. Locally Toeplitz sequences: Spectral properties and applications. *Linear Algebra Appl.*, 278:91–120, 1998. (*74*)

[753] N. Tillmark and P. H. Alfredsson. Experiments on transition in plane Couette flow. *J. Fluid Mech.*, 235:89–102, 1992. (*198*)

[754] F. Tisseur and N. J. Higham. Structured pseudospectra for polynomial eigenvalue problems, with applications. *SIAM J. Matrix Anal. Appl.*, 23:187–208, 2001. (*424, 429*)

[755] M. J. Todd. Semidefinite optimization. *Acta Numerica*, 10:515–560, 2001. (*280*)

[756] O. Toeplitz. Zur Theorie der quadratischen und bilinearen Formen von unendlichvielen Veränderlichen. *Math. Ann.*, 70:351–376, 1911. (*53*)

[757] K.-C. Toh. *Matrix Approximation Problems and Nonsymmetric Iterative Methods.* Ph.D. thesis, Cornell University, Ithaca, NY, 1996. (*247, 278*)

[758] K.-C. Toh. GMRES vs. ideal GMRES. *SIAM J. Matrix Anal. Appl.*, 18:30–36, 1997. (*247*)

[759] K.-C. Toh, M. J. Todd, and R. H. Tütüncü. SDPT3—a Matlab software package for semidefinite programming. *Opt. Meth. Softw.*, 11:545–581, 1999. Software available at http://www.math.nus.edu.sg/~mattohkc/sdpt3.html. (*250, 280*)

[760] K.-C. Toh and L. N. Trefethen. Pseudozeros of polynomials and pseudospectra of companion matrices. *Numer. Math.*, 68:403–425, 1994. (*505, 507*)

[761] K.-C. Toh and L. N. Trefethen. Calculation of pseudospectra by the Arnoldi iteration. *SIAM J. Sci. Comput.*, 17:1–15, 1996. (*269, 285, 386, 387, 430, 434*)

[762] K.-C. Toh and L. N. Trefethen. The Chebyshev polynomials of a matrix. *SIAM J. Matrix Anal. Appl.*, 20:400–419, 1998. (*278, 280*)

[763] K.-C. Toh and L. N. Trefethen. The Kreiss matrix theorem on a general complex domain. *SIAM J. Matrix Anal. Appl.*, 21:145–165, 1999. (*183, 311*)

[764] V. Toumazou. *Portraits Spectraux de Matrices: Un Outil d'Analyse de la Stabilité.* Ph.D. thesis, Université Henri Poincaré, Nancy, 1996. (*424*)

[765] A. E. Trefethen, L. N. Trefethen, and P. J. Schmid. Spectra and pseudospectra for pipe Poiseuille flow. *Comp. Methods Appl. Mech. Eng.*, 1926:413–420, 1999. (*205*)

[766] L. N. Trefethen. Wave packet pseudomodes of variable coefficient differential operators. *Proc. Roy. Soc. London A.* To appear. (*98, 103, 104, 112*)

[767] L. N. Trefethen. Group velocity in finite difference schemes. *SIAM Review*, 24:113–136, 1982. (*497*)

[768] L. N. Trefethen. Group velocity interpretation of the stability theory of Gustafsson, Kreiss, and Sundström. *J. Comp. Phys.*, 49:199–217, 1983. (*322, 325, 326*)

[769] L. N. Trefethen. Instability of difference models for hyperbolic initial boundary value problems. *Comm. Pure Appl. Math.*, 37:329–367, 1984. (*322, 325*)

[770] L. N. Trefethen. Lax-stability vs. eigenvalue stability of spectral methods. In K. W. Morton and M. J. Baines, editors, *Numerical Methods for Fluid Dynamics III*, pages 237–253. Clarendon Press, Oxford, 1988. (*289, 291, 298*)

[771] L. N. Trefethen. Approximation theory and numerical linear algebra. In J. C. Mason and M. G. Cox, editors, *Algorithms for Approximation II*. Chapman and Hall, London, 1990. (*45, 249*)

[772] L. N. Trefethen. Pseudospectra of matrices. In D. F. Griffiths and G. A. Watson, editors, *Numerical Analysis 1991*, pages 234–266. Longman Scientific and Technical, Harlow, Essex, UK, 1992. (*46, 58, 71, 236, 371, 381, 451, 452*)

[773] L. N. Trefethen. Pseudospectra of linear operators. *SIAM Review*, 39:383–406, 1997. (*30, 46, 141, 534*)

[774] L. N. Trefethen. Computation of pseudospectra. *Acta Numerica*, 8:247–295, 1999. (*108, 371, 375, 382, 384, 405, 547*)

[775] L. N. Trefethen. *Spectral Methods in MATLAB*. SIAM, Philadelphia, 2000. (*10, 93, 289, 290, 298, 405, 408, 410, 411, 412*)

[776] L. N. Trefethen and D. Bau, III. *Numerical Linear Algebra*. SIAM, Philadelphia, 1997. (*3, 12, 17, 336, 365, 367, 372, 430, 486*)

[777] L. N. Trefethen and S. J. Chapman. Wave packet pseudomodes of twisted Toeplitz matrices. *Comm. Pure Appl. Math.*, 57:1233–1264, 2004. (*62, 67, 74, 75, 77, 79, 103*)

[778] L. N. Trefethen, M. Contedini, and M. Embree. Spectra, pseudospectra, and localization for random bidiagonal matrices. *Comm. Pure Appl. Math.*, 54:595–623, 2001. (*339, 341, 344, 346, 347, 348, 350*)

[779] L. N. Trefethen and R. S. Schreiber. Average-case stability of Gaussian elimination. *SIAM J. Matrix Anal. Appl.*, 11:335–360, 1990. (*367*)

[780] L. N. Trefethen, A. E. Trefethen, S. C. Reddy, and T. A. Driscoll. Hydrodynamic stability without eigenvalues. *Science*, 261:578–584, 1993. (*12, 121, 135, 199, 200, 202, 204, 207, 213, 214, 378, 438, 497*)

[781] L. N. Trefethen and L. M. Trefethen. How many shuffles to randomize a deck of cards? *Proc. Roy. Soc. London A*, 456:2561–2568, 2000. (*525*)

[782] L. N. Trefethen and M. R. Trummer. An instability phenomenon in spectral methods. *SIAM J. Numer. Anal.*, 24:1008–1023, 1987. (*15, 45, 289, 293, 298, 301, 305*)

[783] F. Treves. On the existence and regularity of solutions of linear partial differential equations. In *Partial Differential Equations*, volume 23 of *Proceedings of Symposia in Pure Mathematics*. American Mathematics Society, Providence, RI, 1973. (*126*)

[784] A. Tumin and E. Reshotko. Spatial theory of optimal disturbances in boundary layers. *Phys. Fluids*, 13:2097–2104, 2001. (*225*)

[785] A. M. Turing. Rounding-off errors in matrix processes. *Quart. J. Mech. Appl. Math.*, 1:287–308, 1948. (*486*)

[786] J. L. Ullman. A problem of Schmidt and Spitzer. *Bull. Amer. Math. Soc.*, 73:883–885, 1967. (*53*)

[787] L. A. Vaĭnshteĭn. Open resonators for lasers. *Soviet Physics JETP*, 17:709–719, 1963. (*546*)

[788] A. van der Sluis. Condition numbers and equilibration of matrices. *Numer. Math.*, 14:14–23, 1969. (*19, 444*)

[789] A. van der Sluis. Perturbations of eigenvalues of non-normal matrices. *Comm. ACM*, 18:30–36, 1975. Corrigendum: *Comm. ACM* 18: 180, 1975. (*445*)

[790] H. A. van der Vorst. Computational methods for large eigenvalue problems. In P. G. Ciarlet and J. L. Lions, editors, *Handbook of Numerical Analysis*, volume 8. Elsevier, Amsterdam, 2002. (*264*)

[791] H. A. van der Vorst. *Iterative Krylov Methods for Large Linear Systems*. Cambridge University Press, Cambridge, 2003. (*244, 254*)

[792] H. A. van der Vorst and C. Vuik. The superlinear convergence of GMRES. *J. Comp. Appl. Math.*, 48:327–341, 1993. (*252*)

[793] P. Van Dooren and P. Dewilde. The eigenstructure of an arbitrary polynomial matrix: Computational aspects. *Linear Algebra Appl.*, 50:545–579, 1983. (*505*)

[794] J. L. M. van Dorsselaer. Pseudospectra for matrix pencils and stability of equilibria. *BIT*, 37:833–845, 1997. (*424*)

[795] J. L. M. van Dorsselaer. Several concepts to investigate strongly nonnormal eigenvalue problems. *SIAM J. Sci. Comput.*, 24:1031–1053, 2003. (*221, 424*)

[796] J. L. M. van Dorsselaer, J. F. B. M. Kraaijevanger, and M. N. Spijker. Linear stability analysis in the numerical solution of initial value problems. *Acta Numerica*, 2:199–237, 1993. (*12, 16, 30, 167*)

[797] M. Van Dyke. *An Album of Fluid Motion*. Parabolic Press, Stanford, CA, 1982. (*196*)

[798] M. A. van Eijkelenborg, A. M. Lindberg, M. S. Thijssen, and J. P. Woerdman. Resonance of quantum noise in an unstable cavity laser. *Phys. Rev. Lett.*, 77:4314–4317, 1996. (*552*)

[799] C. Van Loan. How near is a stable matrix to an unstable matrix? In R. A. Brualdi, D. H. Carlson, B. N. Datta, C. R. Johnson, and R. J. Plemmons, editors, *Contemporary Mathematics*, volume 47, pages 465–478. American Mathematical Society, Providence, RI, 1985. (*452*)

[800] C. F. Van Loan. Generalizing the singular value decomposition. *SIAM J. Numer. Anal.*, 13:76–83, 1976. (*426*)

[801] J. M. Varah. The computation of bounds for the invariant subspaces of a general matrix operator. Technical Report CS 66, Computer Science Department, Stanford University, 1967. (*41*)

[802] J. M. Varah. On the separation of two matrices. *SIAM J. Numer. Anal.*, 16:216–222, 1979. (*41, 44, 46, 480*)

[803] J. M. Varah. A generalization of the Frank matrix. *SIAM J. Sci. Stat. Comput.*, 7:835–839, 1986. (*487, 488*)

[804] R. S. Varga. *Matrix Iterative Analysis*. Prentice-Hall, Englewood Cliffs, NJ, 1962. Second edition published by Springer-Verlag, New York, 2000. (*231, 232, 241*)

[805] J. G. Verwer and J. M. Sanz-Serna. Convergence of method of lines approximations to partial differential equations. *Computing*, 33:297–313, 1984. (*310*)

[806] K. Veselić. Exponential decay of semigroups in Hilbert space. *Semigroup Forum*, 55:325–331, 1997. (*147*)

[807] K. Veselić. Estimating the operator exponential. *Linear Algebra Appl.*, 280:241–244, 1998. (*147*)

[808] K. Veselić. Bounds for exponentially stable semigroups. *Linear Algebra Appl.*, 358:309–333, 2003. (*147*)

[809] D. Viswanath. Random Fibonacci sequences and the number 1.13198824.... *Math. Comp.*, 69:1131–1155, 1999. (*354*)

[810] D. Viswanath and L. N. Trefethen. Condition numbers of random triangular matrices. *SIAM J. Matrix Anal. Appl.*, 19:564–581, 1998. (*359*)

[811] M. I. Višik and L. A. Lyusternik. Regular degeneration and boundary layer for linear differential equations with small parameter. *Uspehi Mat. Nauk (N.S.)*, 12:3–122, 1957. English translation in Amer. Math. Soc. Trans. (Ser. 2) 20 (1962) 239–364. (*41*)

[812] J. von Neumann and E. Wigner. Über das Verhalten von Eigenwerten bei Adiabatischen Prozessen. *Z. Phys. A*, 30:467–470, 1929. (*377*)

[813] C. B. Vreugdenhil and B. Koren. *Numerical Methods for Advection-Diffusion Problems*. Vieweg, Braunschweig, 1993. (*122*)

[814] F. Waleffe. Transition in shear flows. Nonlinear normality versus non-normal linearity. *Phys. Fluids*, 7:3060–3066, 1995. (*214*)

[815] J. L. Walsh. *Interpolation and Approximation by Rational Functions in the Complex Domain*. American Mathematical Society, Providence, RI, fifth edition, 1969. (*278*)

[816] E. Wegert and L. N. Trefethen. From the Buffon needle problem to the Kreiss matrix theorem. *Amer. Math. Monthly*, 101:132–139, 1994. (*176, 177, 183*)

[817] J. A. C. Weideman and S. C. Reddy. A MATLAB differentiation matrix suite. *ACM Trans. Math. Software*, 26:465–519, 2000. (*405*)

[818] E. B. White. Transient growth of stationary disturbances in a flat plate boundary layer. *Phys. Fluids*, 14:4429–4439, 2002. (*225*)

[819] E. B. White and E. Reshotko. Roughness-induced transient growth in a flat-plate boundary layer. AIAA Paper 2002-0138, 2002. (*228*)

[820] G. B. Whitham. *Linear and Nonlinear Waves*. Wiley-Interscience, New York, 1974. (*492*)

[821] H. Widom. Toeplitz matrices. In I. I. Hirschman, Jr., editor, *Studies in Real and Complex Analysis*, pages 179–209. Mathematical Association of America, Washington, D.C., 1965. (*50*)

[822] H. Widom. Extremal polynomials associated with a system of curves in the complex plane. *Adv. Math.*, 3:127–232, 1969. (*278*)

[823] H. Wielandt. Inclusion theorems for eigenvalues. *National Bureau of Standards Applied Mathematics Series*, 29:75–78, 1953. (*443*)

[824] E. P. Wigner. Random matrices in physics. *SIAM Review*, 9:1–23, 1967. (*334*)

[825] J. H. Wilkinson. Error analysis of direct methods of matrix inversion. *J. Assoc. Comput. Mach.*, 8:281–330, 1961. (*367*)

[826] J. H. Wilkinson. *Rounding Errors in Algebraic Processes*. Prentice-Hall, Englewood Cliffs, NJ, 1963. (*486, 503*)

[827] J. H. Wilkinson. *The Algebraic Eigenvalue Problem*. Oxford University Press, Oxford, 1965. (*20, 44, 71, 458, 469, 474, 486, 487, 490, 501*)

[828] J. H. Wilkinson. Modern error analysis. *SIAM Review*, 13:548–568, 1971. (*486*)

[829] J. H. Wilkinson. The perfidious polynomial. In G. H. Golub, editor, *Studies in Numerical Analysis*, pages 1–28. Mathematical Association of America, Washington, D.C., 1984. (*503*)

[830] J. H. Wilkinson. Sensitivity of eigenvalues II. *Utilitas Math.*, 30:243–286, 1986. (*16, 43*)

[831] J. H. Wilkinson and C. Reinsch. *Handbook for Automatic Computation*, volume II: Linear Algebra. Springer-Verlag, Berlin, 1971. (*486, 487*)

[832] J. P. Williams. Spectra of products and numerical ranges. *J. Math. Anal. Appl.*, 17:214–220, 1967. (*173*)

[833] H. L. Willke, Jr. Stability in time-symmetric flows. *J. Math. & Phys.*, 46:151–163, 1967. (*228*)

[834] A. Wintner. Zur theorie der beschränkten Bilinearformen. *Math. Z.*, 30:228–282, 1929. (*53*)

[835] M. P. H. Wolff. On the approximation of operators and the convergence of the spectra of the approximants. *Oper. Theory: Adv. Appl.*, 103:279–283, 1998. (*30*)

[836] M. P. H. Wolff. Discrete approximation of unbounded operators and approximation of their spectra. *J. Approx. Theory*, 113:229–244, 2001. (*294, 408*)

[837] T. G. Wright. *Algorithms and Software for Pseudospectra*. D.Phil. thesis, Oxford University, 2002. (*140, 144, 371, 380, 390, 416, 417, 418, 419, 420, 430*)

[838] T. G. Wright. EigTool, 2002. Software available at http://www.comlab.ox.ac.uk/pseudospectra/eigtool. (*21, 140, 144, 160, 371, 375, 390, 397, 416, 434*)

[839] T. G. Wright and L. N. Trefethen. Computing Lyapunov constants for random recurrences with smooth coefficients. *J. Comp. Appl. Math.*, 132:331–340, 2001. (*357*)

[840] T. G. Wright and L. N. Trefethen. Large-scale computation of pseudospectra using ARPACK and eigs. *SIAM J. Sci. Comput.*, 23:591–605, 2001. (*390*)

[841] T. G. Wright and L. N. Trefethen. Pseudospectra of rectangular matrices. *IMA J. Numer. Anal.*, 22:501–519, 2002. (*430, 432, 434, 435, 436*)

[842] J. Yamasaki and N. Hatano. A new algorithm of analyzing the metal–insulator transition of the Anderson model. *Comput. Phys. Comm.*, 147:263–266, 2002. (*341, 392*)

[843] S. Yoden and M. Nomura. Finite-time Lyapunov stability analysis and its application to atmospheric predictability. *J. Atmos. Sci.*, 50:1531–1543, 1993. (*227*)

[844] P. Yodzis. The stability of real ecosystems. *Nature*, 289:674–676, 1981. (*527*)

[845] K. Yosida. On the differentiability and the representation of one-parameter semigroups of linear operators. *J. Math. Soc. Japan*, 1:15–21, 1948. (*175*)

[846] D. M. Young, Jr. *Iterative Methods for Solving Partial Differential Equations of Elliptic Type*. Ph.D. thesis, Harvard University, Cambridge, MA, 1950. (*236, 241*)

[847] J. Zabczyk. A note on C_0-semigroups. *Bull. Acad. Polon. Sci.*, 23:895–898, 1975. (*185*)

[848] K. Zhou with J. C. Doyle. *Essentials of Robust Control*. Prentice-Hall, Upper Saddle River, NJ, 1998. (*451*)

[849] D. W. Zingg. Aspects of linear stability analysis for higher-order finite-difference methods. AIAA Paper 97-1939, 1997. (*329*)

[850] M. Zworski. A remark on a paper of E. B. Davies. *Proc. Amer. Math. Soc.*, 129:2955–2957, 2001. (*126*)

[851] M. Zworski. Numerical linear algebra and solvability of partial differential equations. *Comm. Math. Phys.*, 229:293–307, 2002. (*98, 126*)

Index